T0255241

End Times Bible Handbook

An Invitation from the God of Reason!

"Come now, and let us reason together, saith the LORD: though your sins be as scarlet, they shall be as white as snow; though they be red like crimson, they shall be as wool". (Isaiah 1:18)

God Declared the Past Before it happened.

CHAD TROWELL

authorHOUSE®

AuthorHouse™
1663 Liberty Drive
Bloomington, IN 47403
www.authorhouse.com
Phone: 833-262-8899

Published by AuthorHouse 08/29/2020

ISBN: 978-1-5049-2942-4 (sc)
ISBN: 978-1-5049-2943-1 (e)

Library of Congress Control Number: 2015913305

Print information available on the last page.

Introduction

From the pre-human history of the earth to this present day, the Bible is God's confirmable record of what God has said and what God has done. Just as God's prophecies about the past have been correct, so will prophecies about near future events.

The next event on God's prophetic calendar is the Rapture of the Church. The Rapture is God's evacuation of all **born-again believers** before the seven years of Tribulation begins. Whatever year it happens, the Rapture of the Church will occur at the end of Rosh Hashanah, *"at the last trump."* Rosh Hashanah is the Hebrew New Year for counting years and **Epochs**. The Rapture will mark the end of the Church Age and the beginning of "The Day of The Lord".

I believe that the Temple construction will begin at the Rapture and continue for the following three days. At the same time the Antichrist and the False Prophet will begin a series of planned, fake miracles intended to deceive the world into believing a lie that involves the Rapture.

The nine days following the Rapture are a part of the "**Ten Days of Awe**", a Jewish observance that begins with Rosh Hashanah and ends with Yom Kippur. I believe that those last nine days of "The Ten Days of Awe" will begin with the construction of the **Third Temple** while at the same time, the Antichrist and the False Prophet will begin a series of fake miracles, signs and wonders cleverly planned to explain the Rapture and gain control with a lie. The Antichrist and the False Prophet will use fake miracles to deceive and persuade the world to follow the Antichrist. The Ten Days of Awe will climax on Yom Kippur.

Speaking of the Antichrist, the Bible says that ... *"a crown was given unto him"* (Revelation 6:2). I believe that the "crown" spoken of here is the control of the United States and the United Nations. I believe that will happen on Yom Kippur, nine days after the Rapture. That will be the day that the Antichrist will be revealed as the twelfth Imam and promise to enforce the Oslo Accords for seven years. The United Nations will say "*peace and safety*" but after sundown will be "*sudden destruction*" (1Thessalonians 5:3) as the Four Horsemen begin their dreadful ride.

Also known as the 70th "*Week*" of Daniel, the Tribulation will begin with "***sudden destruction***" just after sundown ending Yom Kippur. I believe that the borders will be closed and travel prohibited. The following seven years will be the worst time in human history, future included. One fourth of Humanity will die during the first three and a half years from Jihad, starvation and disease. By today's population that would be nearly 2 **b**illion people. The last three and a half years will be much worse!

The Church is commanded to "*watch*" for the signs of this day as it approaches. We are told what to watch for and we are offered a special reward if we do. There is much misinformation about the Rapture, the Day of the Lord, as well as Creation, Salvation and many other things in the Bible.

The End Times Bible Handbook documents the misinformation by presenting evidence from the Bible, reconciled with the historical record and science. Before you finish the first chapter, you will learn exactly how you can "***know***" for sure that you "**have**" the "***Free Gift***" of **Eternal Life**; and are ***born again*** by Just Faith in Jesus Christ's sacrifice on the cross of Calvary, and nothing else; and are saved forever and will be taken at the Rapture when it occurs.

www.endtimesbiblehandbook.org

Chad Trowell is available for...Sunday Schools;
Your Group Bible Study and...

The Creation and Bible Prophecy Conference

An invitation from God, to come and reason with him.

Come now, and let us reason together, saith the LORD: though your sins be as scarlet, they shall be as white as snow; though they be red like crimson, they shall be as wool. (Isaiah 1:18)

Topics Include...

God simple plan of salvation explained.
Prophecy and science in the Bible
Questions answered from the Bible

To have a custom fit program for your organization...
Email us at...

www.endtimesbiblehandbook.org

The Creation and Bible Prophecy Conference

About the Author

Chad was born again in the summer of 1967. He was persuaded to go to a summer Bible Camp by promises of surfing, skiing, scuba diving and girls. No mention was made about 5 hours of Bible study daily. The summer camp was sponsored by Florida Bible College.

For the first time, Chad heard the other side of the story concerning Bible accuracy, evolution, science in the Bible, Bible Prophecy and God's simple plan of salvation.

There was excitement in the air about something that just happened in Jerusalem. In an auditorium with multiple T.V.'s Chad saw Moshe Dayan walk into Jerusalem and take control of the Temple site for the first time in 2,000 years. It was as though someone just stepped on the moon. There was another discussion about Bible Prophecy and Chad became a believer!

Upon believing Chad was born again; sealed with the Holy Spirit; passed from death unto to life and he became a child of God with a citizenship in Heaven!

Within two weeks of believing, God relocated Chad all the way across town to 4 blocks away from the Tampa Youth Ranch, the home of Dr. Hank Lindstrom.

(continued in back of book)

www.endtimesbiblehandbook.org

Table of Contents

1

An Invitation From The God Of Reason...

1.1 "Come now, and let us reason together"...

Our Creator has invited everyone, to come to him through ***reason***. Our God is a God of love, compassion and **reason**. He is in fact, **The God of Reason!** He Created Reason! Reason is all about knowing Him!

*"Come now, and **let us reason together, saith the LORD**: though your sins be as scarlet, they shall be as white as snow; though they be red like crimson, they shall be as wool"*. (Isaiah 1:18)

Sir Isaac Newton once wrote..."He who thinks half-heartedly will not believe in God; but he who really thinks has to believe in God." **(Newton)**

Newton is recognized as the one who reasoned out the laws of the Universe. We call it physics. He invented Calculus and Authored **"Principia"**, his book of physics and scripture.

"Principia" was called the most influential Book of science, ever written by the Royal Society of England. It sounds reasonable to assume that this man knows what he is talking about. After all, Newtonian Physics has taken us all over our solar system and beyond.

Newtonian Physics has allowed us to detect, see and understand things beyond our human senses. Just as the atom is invisible to the senses, so is God. Newton thought it through and decided that there was extraordinary evidence and abundant proof that there is a Creator, like the Bible describes.

"In the absence of any other proof, the thumb alone would convince me of God's existence." **(Isaac Newton)**

(Romans 1:20) *"For the* ***invisible things*** *of him from the creation of the world are clearly seen, being understood by the things that are made, even his eternal power and Godhead; so that* ***they are without excuse:"***

God has provided us with physical evidence that we can see to confirm what we cannot see. Newton gave us the tools of math and physics to further expand our access and understanding of God's evidence (Hebrews 11: 1) through science. (Isaiah 48:3-6)

God Created mankind so that He could have fellowship with us, but **God has an enemy**. Satan is a Created being called a "cherub" who was Created perfect, by God. His original name was "*Lucifer*". Lucifer was placed on this earth and given the title of "*the god of this world*" (2Corinthians 4:4) long before man was Created. (ETBH Chapters 13-17)

Lucifer, a cherub, ruled over a race of Angels who were also Created. They looked like men and have been mistaken for being men(Hebrews 13:2), but they are not human. They may be unrecognized among us today. (ETBH Chapter 11)

We call them "Angels" but the original meaning of the word "angel means messenger, representative or ambassador. They don't have wings or halos. **There are no baby or lady angels**. They built their cities here when the earth was first Created and still young(Jeramiah 4:23-26), long before man was Created. (2Peter 3:6)

Lucifer was not satisfied with being "*the god of this world*", he wanted to be god of the universe. (Isaiah 14: 12-15) Lucifer led one third of the earth's population in an attack against God. God defeated them, cast them back to the earth and changed Lucifer's name to satan. The angels who rebelled against God are now called demons. (ETBH Ch. 1.16) Lucifer's war left copious geological evidence throughout the world.

God destroyed everything on the planet's surface and condemned satan and his followers to hell in the not so distant future. **According to N.A.S.A.**'s interpretation of the geological record, an event matching the Bible's description of Lucifer's war took place about sixty six million years ago. The ruins of those ancient angelic cities are probably at the bottoms of the oceans and are already being discovered… (ETBH Ch. 12)

(Revelation 20: 13) *"And **the sea gave up the dead** which were in it; and death and hell delivered up the dead which were in them: and they were judged every man according to their works"*

Dr. Lindstrom believed that the dead from the sea were probably from ancient, angelic cities whose ruins may still be beneath the oceans. Many scientists, Hank included, believe that what was once the bottoms of the ocean is now dry land. (ETBH Chapters 12–14)

God placed man on the earth to have fellowship with us. Satan is still the *"god of this world"*, he hates God and he hates the human race. The only way that he has to get back at God is by harming the human race, so he did. (ETBH Chapters 15–17) In the Garden of Eden satan corrupted the human race. Now, the whole human race is born into sin. (Romans 3: 23)

No sin can enter God's presence. **Our sin separates us from God**. (Isaiah59:2) Sin is anything that is less perfect than God. Just one little sin makes us guilty of all! (James 2:10) People can look good on the outside and be the most evil of all. (Matthew 16:15) God doesn't look on outward appearances, God looks on the heart. (Luke 16:15)

Sin in the human race blocks communication and fellowship with our Creator. **Where there is sin there is death**. (ETBH chapters 7-19)

(Romans 6:23) *"For the wages of **sin is death**, but the **Gift** of God is Eternal Life through Jesus Christ our Lord"*.

We must be as righteous as God to be in His presence. There is nothing that we can do to achieve that. (Titus 3:5) God loves us and has provided a way for the human race to be reunited with him at no cost to ourselves. (ETBH Ch. 2)

God has already done everything that needs to be done and offers his righteousness as a **free Gift** to all those who will **believe God's record about His Son.**

(1John 5:10, 11) "*he that believeth not God hath made him a liar; because he believeth not the* ***record*** *that God gave of his Son. "And this is the record, that God hath given to us eternal life, and this life is in his Son*"

God has provided a way that eliminates the sin problem, no matter how much or how great the sin may be! Eternal Life is "***the Gift of God"***(Ephesians 2:8,9). That means that it is free! It cannot be earned and cannot be lost, it must be received as a gift through faith in what Jesus Christ the LORD did for us on the Cross at Calvary and that alone!

After being beaten and hung on the Cross to die a slow, painful and humiliating **death**, Jesus said… "***it is finished***"(John 19:30) The Greek word "*teleō*" is an accounting term that means "paid in full".

Our Creator took on a human body and sacrificed himself for the sins of the whole world. When Jesus said…"*it is finished*", He was referring to the fact that He just finished paying for every bodies sins on the Cross. "***for the wages of sin is death***"(Romans 6:23)

If you believe that and put your faith in what Jesus did for you on the Cross of Calvary, then you are saved. If you have not already done so, you need to talk to God in the quietness of your own thoughts, He will hear you. Tell him that you believe and trust in Him to save you and give you eternal life.

Just start talking to him. When you do, God gives to you a new birth; ; a new nature; You are passed from death unto life; God becomes your father forever after; God gives you a citizenship in Heaven as a free gift the very moment that you understand and tell God that you believe…

(Romans 6:23) *"For the wages of sin is **death**; but **the gift of God is eternal life** through Jesus Christ our Lord"*

The **death** of Jesus on the Cross paid for our sins. The resurrection of Jesus Christ on the third day proves that God can raise the dead.

Eternal Life is a Gift; it is free; it has nothing to do with your behavior; past, present or future!

(Ephesians 2:8,9)*"For by grace are ye saved through **faith**; and that not of yourselves: it is **the gift of God**, Not of works, lest any man should boast".*

Eternal Life is free and it begins the moment that you **believe "*the record that God gave of his Son*"** (1John 5:10) and it can never end, because Eternal Life is Eternal!

(1John 5:10)*"He that **believeth** on the Son of God **hath** the witness in himself: he that **believeth not** God **hath made him a liar**; because he **believeth not** the **record** that God gave of his Son".*

This passage just said that if you believe the record that God gave of His Son then you have the "witness" living in side you at the moment that you believe. The "witness" is another Biblical name for the Holy Spirit. If you don't believe God's record, then you are calling God a liar. (vs. 10)

(1John 5:11) *"And this is the **record**, that God hath **given** to us eternal life, and this life is in his Son".*

(1John 5:12) *"He that hath the Son hath life; and he that hath not the Son of God hath not life"*

According to this passage, you receive Jesus and Eternal Life the moment that you believe God's Record.

(1John 5:13) *"These things have I written unto you that **believe** on the name of the Son of God; **that ye may know that ye have eternal life**, and that ye may **believe** on the name of the Son of God".*

You can know that you have Eternal Life the moment that you believe God's record about His only begotten Son. The name Jesus indicates that He is the one spoken of by the prophets. You **have** the Son of God and you **have** Eternal Life and you **have** the "witness" living inside you **even if you stop believing**. That is how you become "born again", it's when you believe and forever after regardless of anything. There is no way to become "unborn", it is irreversible.

These passages state that you can "***know"*** that you "***have***", "***eternal Life***" the moment that you **believe God's record about his Son**. (1John 5:10) Eternal Life is not something that you get sometime in the future. Eternal Life is the Gift of God and you receive it as a free, paid for in full, gift of God the instant that you **believe the record** that God gave of His Son Jesus Christ the LORD. (1John 5:10)

Jesus said that you receive Eternal Life **the instant that you believe** God's record about his Son…

(John 6: 47)"*Verily, verily, I say unto you, He that* ***believeth*** *on me* ***hath*** *everlasting life*".

The Old King James, old English word "***hath***" is the same as the modern day word "**has**". They are both present tense verbs that mean to possess right now.

Only Believe on Jesus! That is all that you can do. If you are not sure about where you are going to spend eternity, then you need to settle it now! Just talk to God in the quietness of your own heart, in your thoughts…

Tell God that you believe that Jesus is God in the flesh; who came to the earth to pay for all human sin, past, present and future on the cross of Calvary 2000 years ago. Tell God that you believe His record about how He would take on a human body and save mankind. He did that when He shed His blood on the cross of Calvary 2000 years ago and rose from the dead three days later!

Tell God that you **believe** that he died on the cross to pay for all sins, past, present and future; and that you are trusting in what he did for you on the Cross of Calvary 2,000 years ago and that **you are trusting in that <u>alone</u>** to save you **and <u>nothing</u> else**. Just like the thief on the cross.

Admit that you have nothing to offer God in addition to what Jesus Christ did on the cross for our sins. God says that When you do that, you are saved **forever!** It cannot be undone! If you haven't already done that before, then you better do it right now!

You are *"born again"* the moment that you believe. You become a child of God, you have a new relationship with God. God becomes your father and you become his child by your saving faith. If you as a new born-again believer will seek to know God's will for your life, He will reveal it to you through the Bible. God will intervene in your life like a father. God rewards his children for their obedience and disciplines them when they need it.

As your father God commands you to grow in knowledge and understanding of His Word and to fellowship with other believers. God also commands us to tell others and study to answer their questions. God pays a wage for obedience and disciplines those who need it. As a believer, God hears your prayers and forgives the sins that you confess. If you do not confess your sins to God privately and regularly then you should expect God's discipline.

Over the millenniums, many have accepted God's invitation to come and reason with him. Among those who have accepted God's invitation … *"Come now, and **let us <u>reason</u> together saith the LORD**"* (Isaiah 1:18)… are some of the greatest scientific geniuses, who have made major contributions to science. **Galileo Galilei**(1564-1642)**, Blaise Pascal**(1623-1662) and **Sir Isaac Newton**(1642-1727) just to mention a few.

I don't know if this is more than just a coincidence, but Sir Isaac Newton was born the same year that Galileo Galilei died. (1642) Galileo Galilei died on January 8th and Newton was born on December 25th. Newton was a Christmas baby.

Galileo once said…

…"I do not feel obliged to believe that the same God who has endowed us with sense, **reason**, and intellect has intended us to forgo their use." **(Galileo)**

The LORD *Yĕhovah* (Strong's #H3068) is **The God of <u>Reason</u>!** He gave us the ability to reason and to understand. Those are channels that God Created and Uses to Communicate with us...

However, the "infallible Pope" in Galileo's day disagreed with him and gave him two choices; either recant or be tortured to death. Galileo recanted and spent the rest of his life under house arrest.

In the beginning God also Created other worlds… (Hebrews 1: 2 ; 11:3). (See ETBH 8.5) Contrary to the Catholic Church, the earth is not the center of the Universe and God never said that it was.

The Bible has been misrepresented by false teachers claiming to be God's representatives. God calls them wolves in sheep's clothing.

(Matthew 7: 15) *"Beware of false prophets, which come to you in sheep's clothing, but inwardly they are ravening wolves"*

Contrary to popular false teachings, the Bible does not teach that God Created everything in 6 days (ETBH Ch. 7, 8). Angels don't have wings, no lady angels; no baby angels; no Bible reference of Angels ever singing anything, (ETBH chapter 10) and you should not forbid to marry, just to mention a few. (1Timothy 4:1-3)

God said that the earth is round (Isaiah 40:22); God hangs the earth on "nothing" (Job26:7) and it is God that expands the Universe!

That was six days of **<u>restoration</u>** not Creation. Furthermore God never calls the earth young. He always calls the earth old. The Bible does **not** teach that the earth is 6,000 years old, but it does teach that humans have only been on the earth for 6,000 years and are not the first occupants of this planet. (ETBH chapters 7-18) According to the Bible, the earth is old and has a pre-human history...

(2Peter 3:5) *"For this they willingly are ignorant of, that by the word of God the heavens **were of old**, and the **earth standing out of the water and in the water**"*

(2Peter 3:6)*"whereby **the world that then was**, being **overflowed with water, perished**"*

(2Peter 3:7) *"but the heavens and the earth, which are **now**, by the same word are kept in store, reserved unto fire against the day of judgment and perdition of ungodly men".*

Scientists believe that our planet is about 4.6 billion years old and the Universe is about 13.8 billion years old. According to the Bible that may be right.

Ussher made some mistakes in his chronology of the Bible which has misled more than a few and caused some confusion about what the Bible actually does say about the Creation and the age of the earth. The Bible does not say that the earth is 6,000 years old. We are not the first ones here. There was a pre-human civilization here first; there has been two global floods and Jesus is about to return for the Church.

(**Blaise Pascal**) "Truth is so obscure in these times, and falsehood so established, that, unless we love the truth, we cannot know it"

1.2 Sir Isaac Newton Believed In The Literal Interpretation Of The Bible.

"A man may imagine things that are false, but he can only understand things that are true,". **(Newton)**

Newton made predictions about the future of science; the future of Israel and the future of mankind based on his literal interpretation of the Bible...

"About "***<u>the time of the end</u>***", a body of men will be raised up who will turn their attention to the Prophecies, and insist upon their **literal interpretation**, in the midst of much clamor and **opposition.**" **(Newton)**

Just as Newton predicted, the few of us who still believe in **the literal interpretation of the Bible** will face scorn and opposition from every direction. Newton said… **"in the midst of <u>much</u> clamor and <u>opposition</u>".** This book embraces the literal interpretation of the Bible like Newton and many other outstanding scientists, Archeologists, Historians and Scholars.

At a time when nobody believed that Israel could or would ever go back to their land, Newton predicted that they would. Newton's prediction was based entirely on his literal interpretation of the Bible and it happened just like the Bible described and on the right day too! (ETBH chapters 28 and 29.)

Newton wrote a commentary on the book of Daniel at a time when theologians were largely avoiding such a task. The reason for avoiding this task is found in (Daniel 12:4).

"*But thou, O Daniel,* ***shut up the words, and seal the book, even to <u>the time of the end</u>****: many shall run to and fro, and knowledge shall be increased*" (Daniel 12:4).

The book of Daniel is very interesting because it was intended for the Generation of the "***the time of the end***"! We have been living in that time since May 14, 1948. (ETBH chapter 29) That passage is talking about us. Notice that verse 4 says…

"***Many shall run to and fro, and knowledge shall be increased***" (Daniel 12:4)

Newton made predictions about the future of science based on his literal interpretation of the Bible. This passage inspired some of that.

That passage is talking about a time of rapid travel and a technology explosion like we have going on today! That is the time when the book of Daniel will be understood more fully.

One prediction Newton made based on the Bible, was that human technology would become very advanced in the "*last days*" compared to his time. At the time when such a thing was considered impossible, Newton said that men would learn how to travel at speeds beyond a mile per minute (60mph) and travel will be increased greatly. (Daniel 12:4)

Newton's arch rival and critic was the famous French atheist philosopher Voltaire. Voltaire criticized Newton and condescendingly asserted that at such high speeds the air would tear the flesh right off the bones.

Voltaire is also credited with saying that by the next turn of the century, Christianity will become extinct and "the Bible would be a book found only in museums". However by the turn of the century Voltaire was extinct and his home became the headquarters for the Geneva Bible society. Voltaire the atheist publically scorned Newton and was a chronic irritation to one of the greatest, undisputed scientific Geniuses of human history.

"*The fool hath said in his heart, There is no God. They are corrupt, they have done abominable works, there is none that doeth good*" (Psalms 14:1)

Newton's faith in the word of God was based on the Observation, Fact Gathering and Sound Reasoning of the greatest contributor to human science to ever live. He lived on this earth about 84 years and he wrote more about Gods word than he did about math and science combined

Newton believed in the literal Creator of the Bible and wanted to know more about him and his Creation. Newton's search to understand the laws of "Nature" or Physics stemmed from his search to know the Creator of those laws.

Newton wanted to know about his Creator and he even prayed to his Creator for that knowledge. Newton once said…

…"All my discoveries have been made in answer to prayer". **(Newton)**

1.3 The Bible Claims Over 3,000 Times To Be The Word Of God!

No other book claims to be the word of God. The Koran claims to be the word of Mohamed, <u>**not**</u> God, big difference! This consistent claim, that is unique to the Bible, is substantiated by revelations that could only come from the real God described in the Bible!

God says that he offers prophecy(Isaiah 48: 3-5) and reveals "*hidden things*" (Isaiah 48: 2)to confirm that the Bible is authentically the word of God as it claims over 3000 times to be.

We all know what prophecy is but "*hidden things*" may sound a little mysterious. Actually, this is talking about science that was to be discovered sometime in the future. That time in the future has arrived. Not only are prophecies being fulfilled daily but we are discovering scientific statements about Geography, Geology, Astronomy, Biology, Physiology and Genetics in the Bible that were there way before human science had such knowledge… Yet it is there!

The Bible contains amazing revelations about science and future technologies that would be discovered much later. Some of the physical laws were scoffed at because of man's lack of understanding at the time. Man is playing catch up to the real science in the Bible. The only problem is that men just seem to forget that **God said it first!**

Three thousand times is a lot of repetition! Either the Bible is the word of God as it claims to be, or it's just one very big and complex lie! A lie propagated over a very long period of time, involving all the known world up till now. The possibility of men cooking up the Bible on their own is less than plausible because of its history and it's contents.

You can open your Bible to almost any place and find the claim that God is doing the talking and the prophet is only doing the writing…

*"But I certify you, brethren that the gospel which was preached of me is **<u>not after man</u>**"*. (Galatians 1:11)

*"For **I neither received it of man,** neither was I taught it, but by **the revelation of <u>Jesus Christ</u>**"*. (Galatians 1:12)

The Bible consistently claims to be… "*<u>The Word of God</u>*"!

God wants all men to have saving faith which God defines as **believing**… …***"the <u>record</u> that God gave of his Son"*** (1John 5:10)

*"He that **<u>believeth</u>** on the Son of God hath the witness in himself: he that **believeth <u>not</u>** God <u>hath made him a liar</u>; **because he <u>believeth not</u> the <u>record</u>** that God gave of his Son"*. (1John 5:10)

According to that passage, those who have **<u>not</u> believed** God's record about His Son are calling the Holy Spirit a liar! According to the Bible, the Holy Spirit is dealing with every person on an individual basis. The Holy Spirit provides truth that will lead to a saving faith in Jesus Christ to all those who choose to respond to the truth that is provided. If a person responds well and wants more truth, God will get it to them.

*"**And this is the record**, that <u>God hath **given to us**</u> eternal life, and this life is in his Son"* (1John 5:11)

When we believe like the thief on the cross, that we have nothing to offer to God in exchange for his forgiveness; but we are willing to **believe** that Jesus is God in the Flesh; who came to the earth and offered himself as the only sacrifice for **all** of our sins… that Jesus suffered, bled and died for us; was buried for us; and that he rose again on the third day, *"according to the Scriptures"*, to prove that he is alive for ever more and that He freely gives eternal life to all those who **believe** that God did all of that for them. If you are trusting in that **<u>alone</u>** for your Salvation, then you are Saved Forever more! It is that simple. It is a Gift; it is eternal and cannot be undone.

When we believe God's record, we receive the Holy Spirit instantly! The moment that we believe God's Record, the Holy Spirit seals us until the

Day of Redemption. It happens before you get a chance to tell anyone because it is Just Faith alone!

"He that hath the Son hath life; and he that hath not the Son of God hath not life". (1John 5:12)

If you believe God's record, then you are born again; you receive the Holy Spirit and you have Jesus! You received him by Faith. Just Faith!

God says that you can know that you have eternal life right now! It is FREE and can never be taken away, no matter what! It is irreversible and permeant! If you haven't believed already, you could be born again before you finish reading the next verse if you **believe** what you read. Then you can **know** that you **have** eternal life!

(1John 5: 13*) "These things have I written unto you that **believe** on the name of the Son of God; that ye may **know** that ye **have eternal** life, and that ye may believe on the name of the Son of God".*

When you **believe** on Jesus, you are born again, Sealed with the Holy Spirit; passed from death unto life; your name is called out in Heaven; people in Heaven that you may or may not know yet will hear all about it; there will be cheering; shouts of joy; people, angels and the Heavenly Hosts will all be interested in knowing all about you; you will be big news in Heaven! I suppose that there is some sort of mass communication up there that is much better than the internet.

Just Faith, It is a Gift! It is Free!

1.4 Remarkable Claims Require Remarkable Evidence!

The ancient manuscripts that make up the Bible are comprised of very old copies of copies that have been around for a very long time. Their origins in Antiquity and global distribution are widely accepted facts! Without reputable disputation is the fact that the HOLY Scriptures have been around the Globe many times over the thousands of years since they began to be written.

The Bible was written by 35-40 writers over a period of about 1,600 years. All of the writers claimed to be writing the Words as the Holy Spirit gave them. That is a pretty big claim to be made and it is unique to the Bible! Remarkable claims require Remarkable evidence and that is exactly what God offers!

Alexander the Great conquered all of the known world of his time however he bypassed Jerusalem on his way to Egypt where he established his city and its library and put his name on both. **Alexander wanted the ancient Hebrew Scriptures translated into his language**, so before he left Egypt, he gave the order. His library translated the entire Old Testament from Hebrew to Greek and used 72 scholars to do the job which began soon after Alexander left Egypt. At the time, the Hebrew Scriptures were already considered to be ancient. The translation is called **The Septuagint**.

I think that it is important to note that Alexander the Great did not have the Quran translated because, Contrary to Muslim claims, it did not yet exist.

Starting out with 72 scholars dedicated to the translation of just the Old Testament alone was extraordinary! At the height of the library's activity it employed almost 100 scholars. The Greek translation of the Torah, which is the first five books of the Bible were completed by the middle of the third century B.C. The remainder was completed early second century B.C. which contained ancient prophecies about many things that have been fulfilled since. Over twenty five percent of the Bible is prophecy or history written in advance!

God claims to have existed before he Created the Universe and he says that he will continue to exist after our Universe wears out and he is finished with it.

"And, Thou, Lord, in the beginning hast laid the foundation of the earth; and the heavens are the works of thine hands": **(Hebrews 1:10)**

"They shall perish; but thou remainest; and they all shall wax old as doth a garment"; **(Hebrews 1:11)**

"And as a vesture shalt thou fold them up, and they shall be changed: but thou art the same, and thy years shall not fail". **(Hebrews 1:12)**

This passage is about 2700 years old and it states the now accepted **Law of Entropy**. It states that the Universe is slowly running down. It had a beginning and it will have an end. Modern "Science" did not accept this until recently. Once again, God said it first.

God describes the Universe as "finite", however, God describes himself as "infinite". God says that he existed **before** he Created time and space. He also says that he will still exist **after** the Universe is gone.

God claims to be outside of our Universe; outside of time and space. God claims to know the end even from the beginning…

*"**Declaring the end from the beginning, and from ancient times <u>the things that are not yet done</u>**, saying, My counsel shall stand, and I will do all my pleasure":* (Isaiah 46:10)

…"the things that are not yet done"… is a reference to prophesied future events that we should be aware of and watching for.

God has recorded the earth's history from the beginning of Creation through this present time and into the future. The Biblical record is confirmed by Archeology, Geology, Astronomy and History just to name a few. (ETBH chapters 7-17)

God has described human history in advance. Recorded events occurred hundreds and even thousands of years after they were written down by God's prophets and they happened at their appointed times. (ETBH

Chapters 27-29) All of this is done to confirm God's word. The Bible has accurately foretold the future of Israel!

The fact that God has a proven track record of accurately foretelling the future is confirmation that what God says about his relationship to time must be true! Now that is remarkable!

God describes the Universe that He Created quite well for all of the sceptics and scorners. (ETBH chapters 7-18) It seems reasonable to expect that he would know all about his own Creation, its **past, present and future**. God has revealed secrets of his Creation that only he could know. In areas like biology, geology, genetics and astronomy, God has revealed himself for anyone who is looking.

God has always had the policy of confirming his word with proof or evidence to satisfy even the most ardent of skeptics.

(Isaiah 48:3-6)…

(3) *"I have declared the former things from the beginning and they went forth out of my mouth, and I showed them; I did them suddenly, and they came to pass."*

(4) *"Because I knew that thou art obstinate, and thy neck is an iron sinew, and thy brow brass;"*

(5) *"I have even from the beginning declared it to thee; before it came to pass I showed it thee: lest thou shouldest say, Mine idol hath done them, and my graven image, and my molten image, hath commanded them"*.

God has an established record of accurately foretelling the future. God has declared in writing events which were still future when it was written. Those records still remain as evidence that God said it first. But few people ever hear about it.

Nobody believed that Israel could ever return to the Promised Land in "*the last days*" like the Bible said. So the Catholic Church decided that all of those promises that God made to Abraham, Isaac and Jacob were just going to waist. So the "infallible Pope" decided to claim all of that Promised Land for the Catholic Church.

The Vatican decided to liberate Jerusalem from the Muslims and make a land grab for themselves. But the land that they were grabbing was God's land and he has already given it away to the descendants of Abraham, Isaac and Jacob!

Even though nobody seemed to believe Israel could ever return to their land, they did! And not only that, but they did it on the very day that God said they would. Some of these prophecies were written from 3400 years ago and some from about 2600 years ago. The most recent prophecies were written 2000 years ago! (ETBH chapters 26-29)

God said that he would remove his people off of their land for a period of time for past sins. That time was completed on May 13, 1948. The very next day, the nation of Israel was recognized by the United Nations General Assembly, May 14, 1948! That kind of accuracy is what characterizes Bible prophecy. A whole nation was born overnight as described in scripture! (ETBH chapters 26-29)

(Isaiah66: 8) *"Who hath heard such a thing? who hath seen such things? **Shall the earth be made to bring forth <u>in one day</u>? or shall a nation be born at once?** for as soon as Zion travailed, she brought forth her children"*.

This is exactly what happened, right down to the very day.

God has also revealed end times technologies and events which were considered to be impossible until recently. Until now, such predictions were a source of mockery. A two hundred million man army crossing over a dried up Euphrates river from China into the middle East was unthinkable when it was written. The population size and technologies required to meet Biblical descriptions of Armageddon simply did not exist until now! But the Bible said it would be like this in the End Times. The Bible says that the "last days" began about 70 Hebrew years ago when Israel was reborn.

The Bible has recorded science thousands of years before men had such knowledge. The Bible has accurately described the technology that we are experiencing every day. Such Biblical prophecies have been the source of much scorn over the centuries.

(6) *"Thou hast heard, see all this; and will not ye declare it? I have showed thee* ***new things from this time****, even* ***hidden things****, and thou didst not know them."* (Isaiah 48:3-6)

The End Time Military technologies described in the Bible are here today and yet the scorners continue to scorn. (ETBH chapter 31)

Whenever man's science disagreed with the Bible, eventually science catches up and recognizes that the Bible had the story right all along, but the world hides their mistakes.

God wants to have a relationship with every person who was ever born into the world. But sin cannot exist in God's presence and must be dealt with so he sends prophets, and he sends Angels and then finally God himself took on a human body and came to the earth as his own Son!

God announced in writing and documented his Devine appointments in advance! (ETBH chapters 24, 26 and 37) God said what he was going to do, when he was going to do it and why. Just so that nobody would miss his important dates God gave plenty of details so the world would recognize His only begotten Son.

(Acts 4:12) *"Neither is there salvation in any other: for* ***there is none other name*** *under heaven given among men, whereby we must be saved"*

(John 14: 6) *"Jesus saith unto him,* ***I am the way****, the truth, and the life:* ***no man cometh unto the Father, but by me****"*.

God recorded what would happen centuries and even millenniums before the events were scheduled. The point of all this is to persuade His people to **believe** his word. He wanted them to believe that the whole world needed a Savior and that **Jesus is the only way!** But God's people continued to ignore the God of Abraham, Isaac, and Jacob and pursued false gods instead. They did not listen then, and they are not listening now.

Jesus was the necessary payment that must be made for the sins of all mankind, but look at who had to do it! It was **his own people**, and **that was prophesied too!**

"And one shall say unto him, ***What are these wounds in thine hands? Then he shall answer, Those with which I was wounded in the house of my friends"***. (Isaiah 13:6)

This was written centuries before it happened!

God said that he would come to the earth; be born of a virgin; in the town of ***Bethlehem Ephratah*** ...

"But thou, Bethlehem Ephratah, though thou be little among the thousands of Judah, yet out of thee shall he come forth unto me that is to be ruler in Israel; whose goings forth have been from of old, ***from everlasting****"*. (Mica 5:2)

God knew in advance what he would have to endure to gain our salvation. God foretold every detail centuries in advance. There were two Bethlehem's, but the Bible names the right one, *Bethlehem Ephratah.*

The point of all of this is to provide a lost world with knowledge about Eternal Life so that they can believe and be saved. The Bible says...

"*For the* ***wages of sin is death****; but* ***the gift of God*** *is* ***eternal life*** *through Jesus Christ our Lord*". (Romans 6: 23)

Notice that it says that Eternal Life is the **Gift** of God. That means that God has done all of the work and is offering it to you for **free**! It is a free Gift. That is where God does the giving and the believer does the receiving. God never reverses that order! Eternal Life is a Gift, **not** a reward!

"*For by grace are ye saved through* ***faith****, and that* ***not of yourselves****, it is the* ***gift*** *of God, not of works lest any man should boast*" (Ephesians 2:8 and 9)

Eternal Life is a Gift, you receive it when you believe it. Jesus paid it all. It is free! You don't have to promise or give anything or give up anything to be saved. You don't have to give your heart to Jesus; you don't have to give your life to Jesus or give anything else to be saved.

God does the giving because it is the Gift of God. God gave his son so that anyone who **believes** in him will **receive** eternal life as a free **gift.** Just Faith! Eternal Life comes from God doing the giving and the believer doing the receiving. God never reverses that order! Eternal Life is a Gift, **not** a reward!

Eternal Life can't be earned and can't be lost, it is the Gift of God and he will never take it back! It is free. It is Eternal! You get it the moment you tell God that you believe on Jesus and are trusting him to save you! You don't have a chance of telling someone that you are saved before you are actually saved because you are saved the instant that you believe and trust God to save you.

Confession to men is not necessary to be saved. It has nothing to do with it. There are degrees of reward in heaven for those who want to please their heavenly father but eternal life is not a reward, it is a gift. God pays a wage to believers who choose to serve their heavenly Father and rewards them for service, but that is a separate deal. You don't have to serve God to be saved, but God rewards you if you do.

1.5 Sir Isaac Newton is The Real Science Guy....

Sir Isaac Newton (1643-1727) published **Philosophae Naturalis Principia Mathematica** In 1687, which is considered to be the most influential book of science ever written. Also known as "Principia" for short, Newton's book was based on the Bible. Newton believed in the literal interpretation of the Bible!

Sir Isaac Newton was a well-respected English mathematician who invented Calculus and discovered the laws of motion; gravity and Physics. He was the first to express the laws of physics in mathematical terms.

Newton discovered the Science of Physics and Astrophysics that took us to space and sent space probes beyond our solar system. Newton was also an; Alchemist; Astronomer; Inventor and Natural Philosopher; but most importantly, Sir Isaac Newton was a born again believer!

In a 2005 survey of scientists in Britain's Royal Society, members were asked who had the greatest influence on science. Newton was regarded to be first and much more influential than Albert Einstein.

In his ground breaking, historical achievement titled **Philosophiae Naturalis Principia Mathematica** (1687), also known as "**Principia**" for short, Newton described his laws of motion. Furthermore, Newton outlined the laws of Creation and a little bit about its Creator. Newton viewed the world in a way that is natural to surroundings, history, and reality. His book, Principia was declared to be the most influential science book in history by the Royal Society of England.

A story about American engineers' dealing with problems during the space race reflects the respect they had for Newton. Scientists were faced with the problem of launching an enormous amount of fuel and equipment to the moon and bringing some of it back. When Scientists were considering ways to reduce weight, and were stumped on this problem for quite some time, one engineer remembered a famous quote from Newton and suggested a multistage rocket. The quote was…

"*If I have seen further, it is by standing on the shoulders of Giants.*" (Sir Isaac Newton)

Besides being a world renowned scientific Genius, Newton was a believer in Jesus Christ and a Bible scholar as well. Newton wrote more about the Bible than he did about math and science combined.

"*Thus saith the LORD, Let not the wise man glory in his wisdom, neither let the mighty man glory in his might, let not the rich man glory in his riches:*" (Jeremiah 9:23)

"But let him that glorieth glory in this, that he understandeth and knoweth me, that I am the LORD which exercise lovingkindness, judgment, and righteousness, in the earth: for in these things I delight, saith the LORD." (Jeremiah 9:24)

The Bible inspired Newton's "Scientific Method" approach to "Natural Law". "Natural Law" and reason came from Newton's search of the Bible for answers about the Creator and his Creation.

"*The heart of the righteous studieth to answer*"... (Proverbs 15: 28)

Newton recognized that the physical world was governed by invisible rules of order and he knew who made the rules. Rules can't be seen directly, so Newton sought to understand those rules by indirect means.

"For by him were all things created, that are in heaven, and that are in earth, ***visible and invisible****, whether they be thrones, or dominions, or principalities, or powers: all things were created by him, and for him":* (Colossians 1:16)

Newton gave us The Three Laws Of Motion: The first law is the Law of Inertia.

The Second Law is Acceleration. Newton looked at (t) time, (d) distance and (m) mass and put them into an equation. Force = Mass x Acceleration, whereas Force is measured by a basic unit of force called a "newton". A "newton" is defined as ...1newton = Mass (1 Kilogram) x Distance (1 Meter) x Time^2 (Seconds) N=M x D x T^2 (F=M A) 2nd law

Newton's third Law states that for every action there is an equal and opposite reaction.

Newton understood physics and turned it into an organized science with mathematical expressions of physical laws. Now we can understand those laws and use them to build machines and all sorts of things. Newton gave us Calculus; the three laws of motion and much more.

The understanding of physical forces which cannot be seen directly but can be demonstrated indirectly through experimentation is what modern technology is built on. Newton taught the world to do that and his inspiration came from the Bible!

Newton discovered the science that has sent space probes beyond the far reaches of our solar system!

Modern day technology is built on Newton's ability to indirectly detect and understand invisible forces as they act! It has been said that Newton had that keen ability, more so than any other man in the field of science to this day!

Sir Isaac Newton was asked how he discovered the law of gravity. He replied, **"By thinking about it all the time."**

Newton was a real Genius, one of the greatest scientific minds of human history and he believed that the Bible is the word of God!

Sir Isaac Newton wrote more about the Bible than he did about math and science combined. Newton invented calculus; discovered gravity; discovered the laws of Physics and he made predictions based entirely on his literal interpretation of God's Word. Some of those predictions have already come to pass and right on schedule.

(Jeremiah 9:23)*"Thus saith the LORD, Let not the wise man glory in his wisdom, neither let the mighty man glory in his might, let not the rich man glory in his riches:*

But let him that glorieth glory in this, that he understandeth and knoweth me, that I am the LORD which exercise lovingkindness, judgment, and righteousness, in the earth: for in these things I delight, saith the LORD." (Jeremiah 9:24)

1.6 Newton Prayed To The God Of Reason!

Newton once said…

"All my discoveries have been made in answer to prayer". (Newton)

Newton may not have known God by the name "God of Reason", but he knew for sure that we have a reasonable God and Newton spoke with him often. Newton said that he prayed to God for answers to his questions.

"Truth is the offspring of silence and meditation." **(Newton)**

"If I have ever made any valuable discoveries, it has been due more to patient attention, than to any other talent" **(Newton)**

"No great discovery was ever made without a bold guess." **(Newton)**

It is interesting to note that **"*longsuffering*" is one of the fruits of the Spirit.**

"But the fruit of the Spirit is love, joy, peace, ***longsuffering****, Gentleness, goodness,* ***faith****",* (Galatians 5:22, 23)

The word "*longsuffering*" is translated from the Greek word "***makrothymia***", which is found in the Strong's Greek Dictionary under the number **"G3115".** The first definition for **"*Makrothymia*"** is …"patience, endurance, constancy, steadfastness, perseverance", and that is a fruit of the Spirit represented by the word translated as "longsuffering"

Observation and reason followed by experimentation and measuring are the basis for the scientific method that Sir Isaac Newton employed. Newton's ability to understand things that cannot be seen directly and devise experiments to confirm his reasoning is part of what Newton contributed to the science of physics. Newton prayed for God to reveal truth to him.

"Come now, and ***let us <u>reason</u> together, <u>saith the LORD</u>"*** **(Isaiah 1:18)**

Sir Isaac Newton was asked how he discovered the law of gravity. He replied, "By thinking about it all the time." (Newton)

(Romans 1:20) *"For the invisible things of him from the creation of the world are clearly seen, being understood by the things that are made, even his eternal power and Godhead; so that they are without excuse":*

"If others would think as hard as I did, then they would get similar results". (Newton)

Newton was a thinker with a proven track record in physics. It would be wise to consider what he said about God and the Bible.

Another time Newton said…

"Gravity explains the motions of the planets, but it cannot explain who sets the planets in motion." **(Newton)**

The Bible says…

"For by him were all things created, that are in heaven, and that are in earth, visible and invisible"… **(Colossians 1:16)**

"This most beautiful system of the sun, planets and comets, could only proceed from the counsel and dominion of an intelligent and powerful Being." **(Newton; the Principia: Mathematical Principles of Natural Philosophy, July 5, 1687)** This is considered by secularists to be the most influential book of science ever written.

Jesus said that He is the light of the world. Newton purchased his first prism in the year (1666 AD). That was the year that Newton began his search to understand light, literally. Some people call it "Newton's Year of Light".

1.7 God's Creation Confirms His Word.

"In the absence of any other proof, the thumb alone would convince me of God's existence." **(Isaac Newton)**

The Bible contains God's record about the Creation, fall and Salvation of humanity! It is written in terms that we can understand.

Newton's work in the field of science has enabled people to develop modern technology and to explore and live in space. We should consider what he said about God as well.

Newton believed in the literal interpretation of the Bible. His predictions were based on applying scientific method to scripture just like he did with the laws of Physics. He wrote more about the Bible than he did about math and science combined. He also wrote about Bible Prophecy. He maintained that the Bible has Accurately foretold the future from ancient times, before it happened. Some of those prophecies are still future, some in the near future. One of Newton's books is …

The Prophecies of Daniel and the Apocalypse by Sir Isaac Newton, This book deals with a comparison of the Book of Daniel to the Book of Revelation. (ETBH chapter 42).

Newton argues in favor of the existence of God based on the scientific and prophetic accuracy of the Bible and common sense.

"For the wrath of God is revealed from heaven against all ungodliness and unrighteousness of men, ***who hold the truth in unrighteousness****;" (Romans 1:18)*

"Because that which may be known of God is manifest in them; for God hath shewed it unto them". (Romans 1:19)

"For the ***invisible things of him from the creation of the world are clearly seen****, being understood by the things that are made, even his eternal power and Godhead; so that* ***they are without excuse****:"* (Romans 1:20)

"Because that, when they knew God, they glorified him not as God, neither were thankful; but ***became vain in their imaginations****, and* ***their foolish heart was darkened.****"* (Romans 1:21)

"Professing themselves to be wise, they became fools", (Romans 1:22)

"And changed the glory of the incorruptible God into an image made like to corruptible man, and to birds, and four-footed beasts, and creeping things" (Romans 1:23)

Our world requires a Creator! To deny that is to deny reality and choose fantasy because it seems to feel good now. There is no scientific basis for a belief in biological evolution.

Sir Isaac Newton once said… "A man may imagine things that are false, but he can only understand things that are true, for if the things be false, the apprehension of them is not understanding". **(Sir Isaac Newton)**

All evidence confirms the Biblical record of Creation and the Geological Ages that have transpired sense the beginning. However, there has been plenty of hoaxes which have contradicted Scripter.

There are many "scientific" hoaxes in the area of biological evolution. The first "Ape-man" or "missing link" was called the Piltdown Man. It fooled the scientific community for about a quarter of a century before being proven to be a hoax. During the time that "science" accepted the fossils as real, the perpetrators made money in the lecture circuits and publishing books. Others got the same idea during that time. Charles Dawson seemed to be the leader. I sure would like to know if he was a Mason. I say that because the Free Mason's cooked up Darwin's ideas and financed his work. (See chapter 47.4-6)

Sir Isaac Newton once said…

"How came the bodies of animals to be contrived with so much art, and for what ends were their several parts? Was the eye contrived without skill in Optics, and the ear without knowledge of sounds? ...and these things being rightly dispatch's, does it not appear from phenomena that there is a Being incorporeal, living, intelligent?" **(Newton)**

The living word authenticated the authority of his word with his Creation record found only in the Bible. Jesus endorsed it and he claims to be God Almighty in flesh, the one who did the Creating!

"*Because that which may be known of God* ***is manifest <u>in</u> them;*** *for God hath shewed it unto them*". (Romans 1:19)

God uses his Creation to proclaim his existence to every man that comes into this world. Order cannot exist without an organizer! To suggest otherwise is pure Lunacy.

Every time that you see a solar or lunar eclipse, you are seeing a sign from God, letting you know that He is there. The probability of our sun, our moon and our planet all being the right size, distance, proportions and orbits to be synchronized to produce solar and lunar eclipses like they do

1.8 God Created Us For His Fellowship.

The whole idea of fellowship is when one or more persons are working together for a common goal. Our Creator's goals should be our goals too, but they usually are not. Someday soon, that will change.

You can't have fellowship with God if you are rebelling against him! The greater the sin, the greater the feeling of guilt is. The feeling of guilt tends to drive people away from God. But God forgives all!

How can you have fellowship with someone that you are opposing? How can you have fellowship with someone that you are rebelling against? You can't have fellowship, but you can have salvation anytime that you call upon the name of the Lord Jesus Christ.

It almost seems like there is a universal understanding that God dislikes somethings more than others. Universally, we all understand that some things are just really wrong and perverse. The greater the perversion, the greater the feeling of guilt is. Our feelings can easily be redirected by an outside intelligent force, but the Word of God is forever and so are those who hear.

The feelings of guilt or rebellion creates barriers between people who are lost and the Truth that will set them free and give them Eternal Life. The feeling of guilt or rebellion creates a barrier between God and his Creation!

Satan is God's Enemy he is also called "the father of lies". He blinds the minds of unbelievers with lies and creates negative feelings toward the truth and that becomes a barrier. When we believe those lies we sin. God is always ready to forgive all sin and give a new birth by faith and give Eternal Life Freely to anyone who believes! Although we will always continue to sin, God deals with sin in the life of his children just like a perfect father.

1.9 Who am I? How Did I Get Here? What Is My Purpose?

Knowing God as our Creator makes a lot of sense. When we know our Creator, we know our purpose. Until then, you are lost and spiritually dead!

... "***Thus saith the LORD***, *Let not the wise man glory in his wisdom, neither let the mighty man glory in his might, let not the rich man glory in his riches*": (Jeremiah 9:23)

"*But let him that glorieth glory in this, that he* ***understandeth*** *and* ***knoweth me****, that* ***I am the LORD*** *(Yĕhovah) which exercise* ***lovingkindness, judgment****, and* ***righteousness****, in the earth: for in these things I delight,* ***saith the LORD***". (Jeremiah 9:24)

Notice Who is doing the talking here. This is God speaking!

The LORD (*Yĕhovah*) is **The God Of Reason** and he loves us and wants us to know him!

The Three most basic questions for every man, woman and child are ..."**Who am I? What is my purpose? How did I get here?**"... The first two questions are dependent on the third..."How did I get here? "Am I the result of some meaningless accident? Did matter become organized without any outside intelligent direction or intervention at any point along the way? Is that even possible?" ... If not, then God must have Created me along with everything else for His purpose!

So **the big question is ..."<u>Was I Created or was I the result of a meaningless accident</u>? ...God says that the answer is <u>obvious</u>**...

"Because that which may be known of God is manifest in them; for God hath shewed it unto them" (Romans 1:19)

"For the invisible things of him from the creation of the world ***are clearly seen****, being understood by the things that are made, even his eternal power and Godhead; so that* ***<u>they are without excuse</u>****:"* (Romans 1:20)

"The meek shall eat and be satisfied: they shall praise the LORD that seek him: ***your heart shall live for ever****"*. (Psalm 22:26)

1.10 Biblical Faith Is Not Blind, It Is based on facts and sound reason.

Satan gives a false impression about what God has said by what false Prophets teach. They always say and do the opposite of what God says. Contrary to popular belief, Biblical faith is <u>not</u> blind, and it is not based on feelings either; <u>however</u> it <u>is</u> based on <u>fact</u> and sound reasoning. Because Biblical Faith Is based on reality and reason, it actually becomes an eye opener. Nothing blind about it.

Our God is also the God of Reason. He appeals to us through reason.

"Where the senses fail us, reason must step in." **(Galileo Galilei)**

Hebrews chapter eleven is called the "faith chapter" and in the very first verse of the chapter, Biblical faith is defined… the verse reads like this…

*"Now faith is the **substance** of things hoped for, the **evidence** of things not seen."* (Hebrews 11:1)

The Greek word translated as "***substance***" is "Hypostasis", (Strong's # G5287);

Some of the Strong's definitions for that word are; "that which has actual existence; that which has foundation, is firm; a substance, real being; the substantial quality, nature, of a person or thing"

…and according to the note (L) for that passage, in the margin of the Old Scofield reference Bible on page 1301, is better translated as **"*substantiating.*"** The meaning here has to do with gathering, verifying and considering the facts to determine what is real.

The Greek word translated as "***evidence***" is "*elegchos*" and is found in the Strong's Greek Dictionary under #G1650 which came From G1651 and is defined as; "**proof**, conviction:—**evidence**, reproof".

The word is used in the KJV in the following manner; **a proof, that by which a thing is proved or tested**".

Once again on that same page 1301 in the Scofield Reference Bible is a note about this word "***evidence***" in the margin under (M)

The word is better translated as **"proof"** or proving your belief or confirming the facts.

So this passage could better be read …" *faith is the substantiating of things hoped, the proving with **evidence the things not seen.***" (Hebrews 11:1) This passage is emphasizing that our faith is all about making sure that what we "know" is actually "so". This is where Newton got his inspiration from.

This passage is talking about verifying the facts so you can have confidence in what you believe. You must have **<u>a good reason</u>** to believe something.

This verse defines Biblical Faith as substantiating with evidence or proof the claims and promises of the Bible. We are confirming the unseen with what can be seen. This is done in science all the time. We do it daily. As you can see, **Biblical Faith is <u>not</u> blind, but it is based on evidence and sound reasoning**!

"come let us <u>reason</u> together saith the LORD" (Isaiah 1:18).

This passage is talking about substantiating Biblical claims by gathering evidence for proof of the "fact". The spirit world cannot be seen directly, but there is plenty of indirect evidence.

The Bible says…

…*"**the heart of <u>the righteous studieth to answer</u>**:"* (Proverbs 15:28)

God confirms his word by telling us things that only God could know. God says in (Isaiah 48:3-6) that he has already foretold things that would happen in the future. Some of it has already happened, right on time. Israel became a nation the day after the first curse ended. God described it long before it actually happened. The events that God said **would** happen, **did happen**, when God said it would! God says…in (Isaiah 48:3-6)

…*"**I have declared the former things from the beginning**; and they went forth out of my mouth, and **I shewed them**; I did them suddenly, and **they came to pass**"*... (Isaiah 48:3)

The Bible is God's record of what he has declared and when He declared it**.** The Bible says…

…*"I have <u>declared</u> the former things <u>from the beginning</u>"*; (Isaiah 48:3)

In other words, God is saying… look at all of the things that I said would happen and they did happen just like I said it would. That is verifying proof that the Bible is the authentic Word of God! God has foretold the future and He did what He said He would do.

*"...and they went forth out of my mouth, and **I shewed them**; I did them suddenly, **and they came to pass**"...* (Isaiah 48:3)

God is saying that he has foretold things from the beginning; things that we can look back at and see; things that happened just like God said they would. (ETBH chapters 23-27) God has recorded details about the history of Israel, the Messiah and the world up to our present time and beyond. Only the Bible consistently and accurately tells the future.

The Bible is God's record of what He said would happen; when it would happen and also when He said it. Isaiah was written about 2600 years ago. Some books are much older. **The Bible has pre-told the history of Israel, the Messiah and more!**

1.11 God Calls People "Brass Headed And Stiff Necked"

Long before men could have known about it, God reveals his knowledge about his Creation, the history of the earth and the future.

God demonstrates that he is God by giving us a detailed, written description of events that happened long ago, before man had the knowledge or was even Created.

God describes events which occurred long before man was Created and long before men could discover them. (Isaiah 48:3-6)

*"I have declared the former things from the beginning; and they went forth out of my mouth, and I shewed them; I did them suddenly, and **they came to pass**"* (Isaiah 48:3)

God says that He told us in advance, in writing a number of events which came to past just like He said. Events such as the history of Israel and the promised Messiah. Jesus fulfilled over 300 prophecies about his birth; life; death; burial and resurrection.

"Because I knew that thou art obstinate, and thy neck is an iron sinew, and thy brow brass" (Isaiah 48: 4)

"I have even from the beginning declared it to thee; ***before*** ***it came to pass I shewed it thee****: lest thou shouldest say…"* (Isaiah 48: 5)

God says that people are Generally hard headed by nature. God says here that he reveals science and hidden knowledge from ancient times, long before man could have known about it. God leaves no room for excuses. Only God could be the Author of the Bible.

"Thou hast heard, see all this; and ***will not ye declare it? I have shewed thee new things from this time****, even* ***hidden things****, and* ***thou didst not know them****"*. (Isaiah 48: 6)

Many are too proud to admit that God said it first. Many will not even look at the evidence because they don't want to believe. They are afraid to come to God because the Bible has been so misrepresented to them. They have heard so many lies from Satan's minions about the Bible and what it says that they become angry and frustrated. Often people figure that they are going to hell anyway, so why worry about it. **Satan spreads lies about the Bible and about God to invoke a negative emotional response to stop men from seeking and finding the truth**. Atheists have given up the search for the truth because of their feelings. Atheists are not considering the same data that a born again believer is considering.

Satan sets up "straw men" arguments to misrepresent God and the teachings of the Bible. Straw men arguments are wrong and easily defeated, but the Bible is misrepresented as saying something that it did not say. That is one of the ways Satan attacks the authenticity of God's word. This happens when men don't reverently consider what God has actually said. God said not to add to or take away from His word!

Many today have bad feelings about the Bible, based on bad information or a bad experience with a religious group! Consequently many believe Satan's lies and are blinded from the truth. They often become frustrated, discouraged and quit looking. The Truth is that ***"the Gift of God is Eternal Life"!*** Eternal Life is **Free! Jesus paid for it all on the Cross**! It is a Gift! It is that Simple! **Just Faith!**

If you have questions, ask God and read your Bible! If you pray for it and search for it, you will find the Truth and you are reading some of it right now!

Many are afraid that they may be too evil; or they don't want to become a "hypocrite" by making promises that they can't keep. You don't make promises to become born again or become saved. God is the one who makes the promise to those who believe his promise. God makes the promise and God keeps His promise. Eternal Life is the Gift of God, Jesus paid the price in full on the Cross of Calvary! When you believe it, you receive it. It is that simple.

This book was written so that people would **believe the Gospel**, receive Eternal Life and become a Child of God! Once you are saved, you are always saved! You can't become unborn, once you are born again, it is irreversible! A new relationship begins when your Creator becomes your spiritual father at the moment that you believe! **God deals with the sins of the believer like a father!**

God provides us with evidence that His word is true and His promises are sure so that we will trust in his promises.

"Because I knew that thou art obstinate, and thy neck is an iron sinew, and thy brow brass;" (Isaiah 48:4)…

God has declared that men are obstinate, and hard headed. Men often tend to reject truth and accept lies for emotional reasons, but God appeals to us through **reason**. God continues on to say…

"I have even from the beginning declared it to thee; ***before*** *it came to pass I shewed it thee: lest thou shouldest say, Mine idol hath done them, and my graven image, and my molten image, hath commanded them."* (Isaiah 48:5)…

God is trying to reason with unbelieving men. God says that he has declared the future from the beginning. No religion can do that. Not Catholicism; Pentecostalism; 20th Century liberal Protestantism; Hinduism; Islam or any other religion has or can tell the future.

Remember that God said that He was already there **before** the beginning, before he Created the Universe. (John 1:1) From the beginning, God has revealed the future and scientific knowledge that man could not have known about.

God offers proof, lots of it! God has a proven track record to consider! (ETBH chapters 26-29) *God goes on to say…*

…"Thou hast heard, see all this; and will not ye declare it? I have shewed thee new things from this time, even hidden things, ***and thou didst not know them****".* (Isaiah 48:6)

God is saying again that He has provided a verifiable record of what He said and when He said it. The Bible has accurately recorded events **before** they happened; described the time we are living in and what is about to happen next! God refers to men's excitement about seeing and hearing new things but not admitting that **God said it first**.

This passage mentions… ***"new things from this time, even hidden things****"*… that is a reference to science or knowledge that God recorded in ancient times **before man had such knowledge**. God goes on to make the point… ***"and thou didst not know them****"*! In other words…

God said it first! This book is written to demonstrate that point repeatedly so that the reader will **believe** and **receive** the **free gift of eternal life**! And then maybe take the next step and go on to earn eternal rewards. Either the Bible is what it claims to be or it isn't! There is no room for middle ground.

Either it is all true or it is all false. The evidence of the Bible being true is overwhelming! You just need to take the time and look! Keep reading.

"Jesus saith unto him, ***I am the way, the truth, and the life****:* ***no man cometh unto the Father, but by me"***. (John 14:6)

God wants us to know the **only** way to Eternal Life is Jesus Christ. Jesus is God in human flesh who offered himself for our sins and gives us Eternal Life as a Free Gift, if we trust in what He did for us on the cross and we believe what he said. Jesus said…

"Verily, verily, I say unto you, He that ***believeth*** *on me* ***hath everlasting*** *life"* (John 6:47)

It is that simple! Just Faith!

1.12 God Has Always Confirmed His Word.

There are many examples in the past of God offering proof to confirm his word. One example is the miracles that Moses did before Israel and the Pharaoh of Egypt.

When God sent Moses to speak to Israel, in Exodus chapter 4:1 Moses brought up the issue of unbelief. The verse reads like this…

"[1]*And Moses answered and said, But, behold, they will not believe me, nor hearken unto my voice: for they will say, The LORD hath not appeared unto thee*". (Exodus 4:1)

God explains to Moses in the following verses 2-9 how he would enable Moses to confirm his message with miracles and signs. Moses performed the miracles before Israel, and in (Exodus 4:31), it says that the people believed because of the miracles.

When it came time to speak to pharaoh, God gave Moses even greater power. In fact, God hardened the heart of Pharaoh just so that he would have the opportunity to show even greater miracles.

"And I will harden Pharaoh's heart, and multiply my signs and my wonders in the land of Egypt. But Pharaoh shall not hearken unto you, that I may lay my hand upon Egypt, and bring forth mine armies, and my people the children of Israel, out of the land of Egypt by great judgments. And the Egyptians shall know that I am the LORD," (Exodus 7:3-5)

Nevertheless in Exodus12:38, the Bible says that when the Israelites left Egypt, a *"mixed multitude"* of non-Jewish believers left with them. This "*mixed multitude*" was convinced, by God, confirming, his word by these miracles, and also by accurately prophesying when and where they would occur.

God said that his miracles would be great enough to reach the whole world and it has! Unfortunately the Pharaoh never believed. You know the story. The Pharaoh let Israel go but then changed his mind and gathered up his army in anger and sought to kill Israel. Today you can see the remains of the Egyptian army's chariots at the bottom of the Red Sea. It is believed that the remains of the Pharaoh's chariot has been discovered and is on display in a museum. This can be seen on the internet.

Many times in the past, God used miracles, signs, and Prophecy to confirm his word. Another example of God using miracles, signs and prophesy to persuade men to believe, was when God sent out the apostles to preach the Gospel of the New Testament…

…*"and they went forth, and preached every where, the Lord working with them, and **confirming the word with signs** following. Amen."* (Mark 16:20)

I don't know of an example of God ever asking anyone to believe something without good reason. Remember what Galileo said…

…"I do not feel obliged to believe that the same God who has endowed us with sense, **reason**, and intellect has intended us to forgo their use." **(Galileo)**

The Church Age began at the feast of Pentecost, which was fifty days after Christ was crucified. God confirmed the message of the New Testament with miraculous gifts and signs. These gifts were prophesied by Joel the prophet and were a sampling of future events, but <u>not</u> the fulfillment.

These signs were necessary for the acceptance of the New Testament. But after the New Testament was completed,… the miracles and signs became unnecessary and therefore ceased.

"Charity never faileth: but whether there be prophecies, they shall fail; whether there be tongues, ***they shall <u>cease</u>****; whether there be knowledge, it shall vanish away."* (1Corinthians 13:8)

The New Testament was completed by around 96AD. That's probably why the early church fathers of the first century <u>never</u> mentioned these miraculous gifts in their writings. At that point in time, the miraculous gifts were already a thing of the past. The completed and accepted Bible rendered these miraculous gifts unnecessary and they ceased just like the Bible said they would.

Since those miraculous gifts ceased, God has continued his policy of confirming his word through other means. God offers proof that what he says is true. He did in the past, and he still does today with prophecy being fulfilled and science in the Bible revealed.

1.13 Biblical Creation Makes Sense!

God Created a built in repair system in every living thing. The rate of defense and repair, in part, determines how long we live. The Aging process also had to be Created, otherwise we would not age. We don't Age because we simply "wear out".

We "age" because **<u>we were Genetically pre-programed to age and die.</u>** Our free radical defenses and our immune defenses all slowly shut down at a Genetically controlled rate. We were designed to age in a way that makes us realize that we have a limited amount of time in this world to think about and answer the big question, "Is There a God to save me?!"

If we come down with a dreaded disease that kills us slowly, then we will have that warning time to think about that question and seek God. If we truly want to know our maker, we should just tell him so. He will answer that prayer in a way that can be understood. (Keep reading)

We were designed for a purpose. God's Purpose! Our primary function is to know God, but so many people have been blinded and deceived by the god of this world, by so many lies about so many things it is difficult to lead people to a saving belief in Jesus! The human race has been deceived by the god of this world who is also the father of lies...

*"Ye are of your father **the devil**, and the lusts of your father ye will do. **He was a murderer** from the beginning, and abode not in the truth, because there is no truth in him. When he speaketh a lie, he speaketh of his own: for **he is a liar, and <u>the father of it</u>**"...* (John 8:44)

Lucifer was the name of the Covering Cherub who had his name changed to "Satan" after he led a rebellion against God. This passage also says that the devil was a murderer from the beginning. So we must ask ourselves the question, "beginning of what"?

The context is talking about this world, so it is talking about the beginning of **<u>this</u>** world, the former world was destroyed. When this world began, Satan already had blood on his hands from the former world. (See chapters 6-20)

"How art thou fallen from heaven, O Lucifer, son of the morning! how art thou cut down to the ground, which didst weaken the nations"! (Isaiah 14:12)

To this day, he is still called... ***"the god of this world"*** ; The Prince of the Power of the Air" and the ***"Father of Lies"***, but his former name of

"Lucifer" was dropped by God. His new name is "Satan" which means Adversary of God. He was here first. We are new comers on <u>his</u> planet. (ETBH chapters 6 - 20)

He persuaded a third of this planets former occupants to participate in a rebellion against God. This "Angelic", pre-human race was destroyed by "Lucifer's War". (ETBH chapter 17)

"Wherein in time past ye walked according to the course of this world, according to ***the prince of the power of the air****, the spirit that now worketh in the children of disobedience":* (Ephesians 2:2)

"In whom ***the god of this world*** *hath* ***blinded the minds*** *of them which* ***believe not****, lest the light of the glorious gospel of Christ, who is the image of God, should shine unto them".* (2 Corinthians 4:4)

Satan "blinds" the minds of unbelievers with lies. Remember that Satan is the "father of lies".

We are living in Satan's domain! His number one goal is the destruction of the human race. That is the primary, motivation of Satan and his fallen demons. They are doomed, they know it. They also know that the only thing that they can do to hurt God is to hurt us. Satan does not want for the human race to understand that...

...*"the* ***gift*** *of God is eternal life through Jesus Christ our Lord".* (Romans 6:23)

Satan and his demons want every human being to be destroyed in Hell along with them, so they have cooked up a bunch of lies to keep people from finding out that the Bible is true and Eternal Life is free! It has already been paid for in full! All we can do is believe it and receive it! It is free! It is a Gift! Just Faith! Can't be lost! That is Good News!

One of Satan's big lies is Evolution. The belief that life could just happen without a Creator is a New Age Fantasy, a Religion based on lies, lots of them! Biological Evolution is not a scientific theory, yet many teachers, theologians and some scientists call it a fact. From the beginning, Evolutionists have used their credentials to commit one fraud after another to promote their New Age Doctrine. For more than a century now, evolutionists have propagated their lie with more lies.

The fact is that Darwinian Evolution is the brain child of Erasmus Darwin, a 33rd degree Mason and grandfather of Charles Darwin. Charles Darwin refined the idea of "Natural Selection", but for the most part he was promoting his grandfather's work without giving him proper credit.

Things tend toward randomness, except when it interferes with a particular New Age belief that feels good now… God appeals to our reason but Satan plays our feelings like a musical instrument that he has mastered.

Pantheistic, Pagan religions are masquerading as science and anything else they can infiltrate. A deluge of intellectual filth is being forced down the throats of people all over the world. Evil is being promoted everywhere, it seems. Evil drives people from God! The Bible talks about a time when men would call evil to be good and good will be called evil…

"Woe unto them that call evil good, and good evil; that put darkness for light, and light for darkness; that put bitter for sweet, and sweet for bitter" (Isaiah 5:20)

The time is here! Today they openly… *"call evil good, and good evil; that put darkness for light, and light for darkness;"*! (Isaiah 5:20)

The most important thing to know about at any time is, **who Created us and Why?** We know it. It is intuitive. **Is there a God? That is the big question that we must all answer before we die! <u>Is there a God?!</u>**

…*"**the heart of <u>the righteous studieth to answer</u>**: but the mouth of the wicked poureth out evil things".* (Proverbs 15:28)

Creation made sense to Newton.

1.14 Creation Or Chaos? Consider The Implications.

Consider the implications of random chaos becoming highly organized by chance, without an organizer. The implications of random chance being the "organizer" of everything means that we are all just the result of random accidents.

Therefore, we have no purpose under such a scenario. There is no good or bad, no right or wrong. That is Darwinian, Pantheism and is the basis of anarchy, revolution, terror and Genocide throughout its history.

National tragedies occurred where such beliefs were embraced. Consider the French Revolution and the "Reign Of Terror"; consider the Stalin Genocide; consider Hitler and the Nazi terror; consider Cambodia and the killing fields; Consider Bosnia. There is a common thread of evil that runs through these acts of Genocide and that common thread is Darwinian Evolution. (see chapter 1.24 and 35.3-.5)

Now let's consider the implications of a Creator.

Newton looked at the Creation and the Creator and wrote about both. He developed mathematical expressions for the laws of physics and he invented calculus. His science is the foundation that modern technology is built upon.

Newton's Christian view of the world took us to the moon and has sent probes all over our solar system. Christian schools, hospitals and colleges were spread all over the world. They taught that freedom and liberty were gifts from God. The American Revolution was based on Biblical values such as freedom and liberty...

"Now the Lord is that Spirit: and ***where the Spirit of the Lord is, there is liberty****"*. (2Corinthians 3:17)

Where the Spirit of the Lord is in control, there is liberty, freedom, justice, and prosperity for all. The secret of our success in the United States of

America is the foundation of Jesus Christ and the freedoms and other benefits that come with it.

Mohamed Ali was a very white, Muslim slave trader who captured and sold slaves for a living, just like the ones who would sell slaves to America just a few centuries later. Muslim slave traders sold slaves to Europeans who brought them into America.

Churches who taught against slavery and Christians who openly opposed slavery were threatened and some were even killed for talking against it. Cassius Marcellus Clay the emancipationist from Kentucky had many attempts on his life. But eventually God's people prevailed and slavery was abolished in the U.S. That was a Christian achievement.

Pantheism and Darwinism are followed by anarchy, terror, Genocide and a New World Order of tyranny, and it is always called a "revolution".

On the other hand Newton found that Natural Laws confirms the Bible record. Newton declared that the Creation is proof of a Supreme Creator.

Newton once said..."Blind metaphysical necessity, which is certainly the same always and everywhere, could produce no variety of things. All that diversity of natural things which we find suited to different times and places could arise from nothing but the ideas and will of a Being, necessarily existing." **Isaac Newton**... The **Principia**: Mathematical Principles of Natural Philosophy published in 1687, 1713 and 1726

A sense of right and wrong is confirmed with natural evidence. Freedom and equality are natural products of the "Golden Rule" "*And as ye would that men should do to you, do ye also to them likewise*" (Luke 6:31)

The Natural Law of Order and Disorder tells us that things naturally tend to wear out. They grow old. Things tend toward randomness or decreased order. Only intelligent intervention can create order. Newton once said... "Whence arises all that **order and beauty** we see in the world?" **(Isaac Newton, Opticks)**

Without intelligent intervention, things tend toward disorder, naturally. That is a universal, scientific concept that anyone can understand from simple observation of life. Everything is winding down. That is what the Bible says and what science has recently acknowledged, but God said it first.

At a point in time, someone had to wind everything up and organize it! This is common sense! A source of intelligence and power beyond anything humans could comprehend is required to Create and maintain our universe! That kind of God is described in the Bible and he says that we must all be born again! Jesus said…

*"Except a man be **born again**, he cannot see the kingdom of God."*. (John 3:3)

*"Nicodemus saith unto him, How can a man be **born** when he is old"* (John 3:4)

"That which is born of the flesh is flesh; and that which is born of the Spirit is spirit" (John 3:6)

"The wind bloweth where it listeth, and thou hearest the sound thereof, but canst not tell whence it cometh, and whither it goeth: so is every one that is born of the Spirit" (John 3:8)

"And as Moses lifted up the serpent in the wilderness, even so must the Son of man be lifted up :"(John 3:14)

*"That whosoever **believeth** in him should not perish, but have eternal life"* (John 3:15)

Just Faith!

Sir Isaac Newton believed in the God of the Bible. He was born again and he wrote more about that than he did anything else. Which brings us to consider the consequences of there being a God. **<u>God</u>** determines what is right and wrong, good or bad, **<u>and he wrote it all in the Bible!</u>** God is perfect, but we aren't! Eternal Life is free and begins the moment that you believe the record that God gave of his son. Eternal Life cannot end and God cannot lie! Trust him now if you haven't already!

1.15 We Have A Limited Perspective, God Does Not!

God said…

"*For my thoughts are not your thoughts, neither are your ways my ways, saith the LORD For as the heavens are higher than the earth, so are my ways higher than your ways, and my thoughts than your thoughts*". **(Isaiah 55: 8,9)**

Newton once said…

"As a blind man has no idea of colors, so we have no idea how the all-wise God perceives and understands all things" **(Newton)**.

Our Creator is omniscient (all-knowing), Omnipresent, Omnipotent, Eternal and a whole lot more. This passage indicates that God's knowledge is far beyond ours.

Some things that God has said may seem foolish to an unbeliever because they fail to consider that there are things about this Universe that we just don't understand yet. Man's science has not caught up with God's science! But the Bible is a short cut to learning about our earth; our sun; our universe and all of God's Creation. Anyone who pretends that he knows more than God does is blinded by pride. God has revealed himself through his Creation. At least that is one of the ways.

Anyone who can Create all of this is all powerful; all knowing; timeless; righteous; and that is the God of the Bible! Only a delusional fool would criticize such a God for doing what he decided to do. He must have a reason that you don't understand yet! Do ya think? Don't get mad and give up and quit seeking and trying to understand the truth because of your feelings. Satan is playing the feelings of all of humanity just like a musical instrument that he has mastered.

The Power and Creative Intelligence required to organize and sustain the four dimensions that we are aware of is beyond human comprehension! Yet the Bible indicates that there is more than our human senses can discern. The Bible said that centuries before Louis Pasteur was born. That information has been in the Bible for about 2,000 years. It has been scorned and made fun of, but I think that is about to change.

Today Physicists are researching the idea of other dimensions. That is currently the focal point of CERN, the world's biggest and most expensive Science Experiment. Research is currently being done at the CERN Facility in Europe to better understand things that the Bible discussed 3500 years ago. Things that critiques like Voltaire publicly scorned!

Other dimensions is the only way to explain the things that the Bible talks about. They are learning a lot at CERN right now. Unfortunately there seems to be an occult connection with CERN. There are ancient demonic statues, ritualistic pagan dancing and emblems surrounding this place. What does all of that have to do with science? (ETBH chapters 7 and 8)

"For by him were all things created, that are in heaven, and that are in earth, ***visible and invisible****, whether they be thrones, or dominions, or principalities, or powers: all things were created by him, and for him":* (Colossians 1:16)

God even views time differently…

"But, beloved, be not ignorant of this one thing, that one day is with the Lord as a thousand years, and a thousand years as one day". (2Peter 3:8)

1.16 God Has An Enemy.

Satan is the "*god of this world*", that is the title that God gave to him. (2Cor 4:4) He still holds that title today. Jesus will take away that title at the Battle of Armageddon.(Revelation 11:15) If the Rapture were to happen this year, that battle would be about 7 years away.

Satan lived on this earth and ruled over all of the Angels from the Beginning of Creation until he led a rebellion against God. At the time he was known as Lucifer, but God changed his name to Satan. Satan means "adversary of God" (ETBH Chapters 8-17)

. Satan took one third of the angels with him to attack God. His home planet, the earth, was destroyed. After being in darkness, flooded and frozen for an unspecified amount of time, God restored the earth and repopulated it with new species of air breathing animals and man.

Satan has other names and titles. Satan is also called the god of this world; the father of lies; the prince of the power of the air; the enemy and the destroyer, just to mention a few. Just consider Satan as the enemy of God, (Yeshua or Jesus) and man.

Satan is the enemy of God and man. Satan is the opposite of God. He is the original Anarchist. What God calls good, satan calls evil. What God calls evil, satan calls good. He opposes what God calls good. Satan masquerades under other names. Satan is the father of lies and deceit.

Satan and his demons are condemned to hell and their time is almost up and they know it. But their corrupt and perverse hearts want vengeance against God because God planted the human race on this earth six thousand years ago. The earth used to be their home before the dinosaurs were killed. According to NASA that happened due to an asteroid strike about sixty five million years ago in the Yucatán Peninsula.

For a more in-depth discussion of this subject... (ETBH chapters 11-17.)

As an attack against God, satan introduced sin into the once perfect human race because he does not want God to have us. Satan wants to destroy the whole human race with sin and he uses our feelings achieve his goals.

1.17 Everyone Has Sinned, Nobody Is As Righteous As God!

Everyone has broken at least one of the commandments.

"For there is not a just man upon earth, that doeth good, and sinneth not". (Ecclesiastes7:20)

"*For **all** have sinned, and come short of the glory of God*"; (Romans 3:23)

No sin can enter Heaven, God's Righteousness is required!

"And be found in him, not having mine own righteousness, which is of the law, but that which is through the faith of Christ, the righteousness which is of God by faith:" (Philippians 3:9)

The Law Reveals the Flaw. Nobody is as perfect as God, but that is how perfect you must be. The law shows that we are all guilty and need a Savior. Jesus is the Savior and the law directs us to Him.

"Wherefore the law was our schoolmaster to bring us unto Christ, that we might be justified by faith". (Galatians 3:24)

The law is one of God's witnesses against all sinners except those who are "born again" by faith.

Our righteousness is the very best that we have to offer God, and he says that is not good enough!

"*But we are all as an unclean thing, and **all our righteousnesses are as filthy rags**; and we all do fade as a leaf; and our iniquities, like the wind, have taken us away*". (Isaiah 64:6)

You need God's Righteousness, and he only offers it one way! He offers it as a Gift. **It is the Gift of God!** It is Free! It can't be earned!

"*For by **grace are ye saved** through **faith**; and that **not of yourselves**: it is **the gift of God**, **not of works**, lest any man should boast*": (Ephesians 2:8, 9)

An old hymn said it well… "Jesus paid it all! All to him I owe, sin has left a crimson stain but he has washed it white as snow".

All you have to do, all you can do is ***believe*** it and ***receive*** it as a free gift*!!!* ***Believe*** that Jesus did what the Prophets said he was going to do. The Prophets said it thousands of years **before it happened**! The very year, day and hour of Christ's crucifixion was announced by God long before it happened! Only ***Believe*** God's **record about his son** and be saved! Only Believe!

You need God's righteousness and it only comes as a Gift by faith, just faith.

Everyone needs God's righteousness, nothing else is good enough!

"*For he hath made him to be sin for us, who knew no sin; that we might be made the **righteousness of God** in him*". (2Corinthians 5:21)

Hell is called the "*second death*" and that is where the world is headed without Jesus. You get Jesus the moment that you believe and receive the Gift of God.

"For the wages of sin is death; but the gift of God is eternal life through Jesus Christ our Lord". (Romans 6: 23)

Jesus has paid for all of the sins of the human race. God offers his righteousness as a **free gift** and forgiveness for everything that any human has done, **except for unbelief**! You must **believe the Gospel** to be forgiven! If you don't, you won't! When you believe the Gospel, you are obeying the Gospel.

The term "***Obey not** the gospel*" is a phrase that means to "***believe not***" the Gospel. The Gospel says **Believe** and be saved. **Believing** the Gospel is the same as obeying the Gospel. Believing is the only way that anyone can get saved! There is no other way! **Just Faith Alone** is the **<u>only</u>** way.

*"**The Gospel of Christ** … is **The Power of God** unto **Salvation to Every One That <u>Believeth</u>"…*** (Romans 1:16)

*"In flaming fire taking vengeance on them that know not God, and that **obey not <u>the gospel</u> of our Lord Jesus Chris**"* (2 Thessalonians 1:8)

*"For the time is come that judgment must begin at the house of God: and if it first begin at us, what shall the end be of them that **obey <u>not</u> the gospel of God**?* (1 Peter 4:17)

The term "***obey <u>not</u> the gospel of God***" is talking about unbelievers, those who **"*<u>believe not</u>*"**.

*"He that **believeth** on him is not condemned: but he that **believeth not** is condemned already, **because he hath not <u>believed</u>** in the name of the only begotten Son of God"*. (John 3:18)

Belief or unbelief are the only two options. If you **believe the Gospel** then you are obeying the gospel and are **not condemned!** If you don't believe the Gospel, then you are calling God a liar (1John 5:10)and are already condemned! Please don't die like that, Hell is real!

1.18 Heaven Is A Perfect Place. No Sin Can Enter Heaven

Our Creator is perfect! That is his standard!

*"Therefore hath the LORD watched upon the evil, and brought it upon us: for the LORD our **God is righteous in <u>all</u> his works** which he doeth: for **we <u>obeyed not</u> his voice**"*. (Daniel 9:14)

We have a righteous God! He calls our disobedience "sin". God Loves us but He Hates our sin because our sin separates us from Him! God cannot hear our prayers when we let sin get in the way. God says…

…*"**But your iniquities have separated between you and your God,** and your sins have hid his face from you, <u>that he will not hear</u>"*. (Isaiah 59:2)

Sin cannot exist in God's presence.

"For thou art not a God that hath pleasure in wickedness: neither shall evil dwell with thee."(Psalms 5: 4)

Heaven is a perfect place, no sin can enter in.

*"Nevertheless we, according to his promise, look for new heavens and a new earth, **wherein dwelleth righteousness**"*. (2Peter 3:13)

No sin can enter Heaven!

*"<u>And there shall in no wise enter into it any thing that defileth, neither whatsoever worketh abomination, **or maketh a lie**</u>: but **they which are written in the Lamb's book of life**"*. (Revelation 21:27)

Not even a **<u>lie</u>** can enter Heaven! It sounds like we need to make sure that our names never leave ***"the Lamb's book of life"***! Everyone's name is written in the Lambs Book of Life until they die because Jesus paid for all sin. If anyone dies without Jesus, their name is stricken from the book and they are lost forever!

This passage said that Heaven is a Perfect Place! Sin cannot exist in God's presence! No sin can enter Heaven, not ever, because where there is sin, there is death!

*"For **the wages of sin is death**"*. (Romans 6:23)

"*He that **overcometh**, the same shall be clothed in white raiment; and **<u>I will not blot out his name out of the book of life</u>**, but I will confess his name before my Father, and before his angels*". (Revelation 3:5)

God says that if you "***overcome***" your name will **<u>not</u>** be blotted out of the Lambs book of Life. So you need to become an "**overcomer**" now!

"*Who is he that **overcometh** the world, but **he that <u>believeth</u>** that Jesus is the Son of God?"* (1John 5:5)

God says that if you **<u>believe</u> his record** of his Son, then you have "**overcome the world**". Only believe, that is all. **Just Faith!** Nothing more.

1.19 Our Sin Separates Us From God

God loves the sinner but he hates our sin because sin separates us from him.

"*For thou art not a God that hath pleasure" in wickedness: **neither shall evil dwell with thee***". (Psalms 5:4)

In the Bible, "death" implies separation. There are two kinds of "death".

The First Death Is Physical. That is the separation of the body from the spirit.

The second death is spiritual. That is the **separation** of the spirit from our Creator in a place called the Lake of Fire!

"*And death and hell were cast into the lake of fire. This is the **second death**.*" (Revelation 20:14)

If you are only **<u>born once</u>,** then you will **<u>die twice</u>**! If you are **<u>born twice</u>,** then you will only **<u>die once</u>**! (Those who are raptured will **not see death**.)

Sin cannot exist in the presence of our Creator, because **where there is sin there is death**. Sin must be dealt with or it will pollute the universe!

There is no avoiding this conflict between good and evil! It has to do with the nature of God and the nature of the Universe that he Created!

Remember, sin means… "to be less perfect than God"…"***all have sinned***"… (Romans 3:23)

"***The wages of sin is death*** *but* ***the <u>gift</u> of God is <u>Eternal Life</u> through "<u>Jesus Christ our Lord</u>***". (Romans 6:23)

"Wages" is what is due the imperfect man. Notice that the Bible says… "***wages of <u>sin</u>***" **(singular)**. That means that **just one little sin condemns <u>anyone</u>!**

<u>Absolute perfection is required!</u> This is not talking about a contest between you and your neighbor or anything like that, we are talking about God's requirement of absolute perfection! God compares you to himself! Manmade religion can't help!

Satan Introduced sin into the world (ETBH chapters 5, and14, 15 and 16) and God is not allowing it to infect the rest of the universe.

1.20 Jesus Is Our Creator And Savior In Human Form!

According to the Prophets that accurately declared the past back when it was all, still future, according to them, Jesus is God in the flesh! Just think about it. If you Created a world of people and assorted other intelligent, spiritual beings, wouldn't it make sense to become one of them so that you could communicate with them on a more personal level? That is what the Bible is telling us!

"For unto us a child is born, unto us a son is given: and the government shall be upon his shoulder: and his name shall be called Wonderful, Counsellor, ***<u>The mighty God, The everlasting Father</u>****, The Prince of Peace"* (Isaiah 9:6)

According to (Isaiah 9:6), Jesus is God, who took on the form of a man.

According to the Bible, it was Jesus who spoke to Moses from the burning bush on Mount Sinai.

"*And Moses said unto God, Behold, when I come unto the children of Israel, and shall say unto them, The God of your fathers hath sent me unto you; and they shall say to me, What is his name? what shall I say unto them*"? (Exodus 3:13)

"*And God said unto Moses,* ***I AM THAT I AM****: and he said, Thus shalt thou say unto the children of Israel,* ***I AM hath sent me unto you****".*

(Exodus 3:14)

When the Pharisees came looking for Jesus in the garden at Gethsemane, Jesus identified himself as **"I Am"**. This indicates that it was Jesus who spoke to Moses from the burning bush.

(John 18:4) "*Jesus therefore, knowing all things that should come upon him, went forth, and said unto them, Whom seek ye?*"

(John 18:5) "*They answered him, Jesus of Nazareth. Jesus saith unto them,* ***I am*** (he). *And Judas also, which betrayed him, stood with them*".

As soon then as he had said unto them, **I am** (he), they went backward, and fell to the ground (John 18:6)

The word (he) following "I am" spoken by Jesus is not in the original manuscripts. It was added later by the translators. The Pharisees recognized "**I Am**" as the identifying name for the God of Abraham, Isaac and Jacob that spoke to Moses out of the burning bush. That is why they all fell to the ground. The word (he) was not spoken.

Later on the Apostle Thomas spoke to Jesus and called Him "*my LORD and my God*". (John 20: 27-29)

Jesus did not bear witness of himself. (John 5:31) God bear witness of Jesus through the prophets in minute detail.

The prophets said that the Messiah would God in flesh and the sin bearer.

"All we like sheep have gone astray; we have turned every one to his own way; and the LORD hath laid on him the iniquity of us all." (Isaiah 53: 6)

According to (Isaiah 53: 6) God said that He would provide a sin bearer for **everyone**.

According to (Daniel 9:26) the Messiah will be killed for someone else's sins, but not his own. In the same chapter, God reveals the day that this sin bearer would present himself to Israel. (Daniel 9:25)

"After threescore and two weeks shall Messiah be cut off, ***but not for himself****"* (Daniel 9:26)

God took on human flesh in the name of Jesus Christ, which is the Greek and English translation for the Hebrew name Yeshua which is found in the old testament about 135 times.

"he that <u>believeth</u> on me <u>hath</u> Everlasting Life"(John 6:47)

The only requirement that God gives for us to receive Eternal Life is to believe what God has said about Jesus, his only begotten Son.

"In the beginning ***<u>was</u>*** *the Word, and the Word was with God, and the Word was God".* (John 1:1)

<u>Before</u> the universe was Created, the Word **already <u>was</u>** and his name is Jesus or in the Hebrew, it is Yeshua.

"And the Word was made flesh, and dwelt among us, and we beheld his glory, the glory as of the only begotten of the Father, full of grace and truth". (John 1:14)

"He was in the world, and the world was made by him, and the world knew him not" (John 1: 10)

Jesus Christ is Yeshua the messiah, Almighty God the Creator in flesh who sacrificed himself for all the sins of all mankind because he loves us and wants to have fellowship with his Creation.

Jesus is God in human form! Jesus made it clear that He came down from heaven.

"And he said unto them, Ye are from beneath; I am from above: ye are of this world; I am not of this world".(John 8:23)

Jesus is **not an Angel**! Jesus is ***"much higher"*** than Angels are!

*"Being **made so much better than the angels**, as HE hath by inheritance obtained a **more excellent name** than they"*. (Hebrews 1:4)

It says right here that Jesus is *"made **so much better** than the angels"*

The words ***"so much"*** implies a **big** difference!

Strong's G4181 *polymerōs*

Strong's Definitions; "τοσοῦτος tosoûtos, tos-oo'-tos; from τόσος tósos (so much; apparently from G3588 and G3739) and G3778 (including its variations); so vast as this, i.e. such (in quantity, amount, number of space):—as large, so great (long, many, much), these many."

Not only are they **not the same**, but there is a **<u>big</u> difference** and furthermore the difference is "***better"*. "Much Better"!**

"Better" Strong's G2909 - *keratin*

"κρείττων kreíttōn, krite'-tohn; comparative of a derivative of G2904; stronger, i.e. (figuratively) better, i.e. nobler:—best, better."

Not only is He ***"Much better"***, but HE is… **"*SO Much Better*!"**

Furthermore, His Name is ...***"more excellent" than any Angel's name.*** ***"Excellent"*** means ...

Strong's Definitions # G1313 – *diaphoros, dee-af'-or-os;* from G1308; *varying; also surpassing:—differing, divers, more excellent.*

There are groups that say that Jesus is not God. They are wrong! They say that Jesus was an Angel. That is a lie! Jesus is God! The Watch Tower Society (A.K.A. "The Jehovah's Witnesses") and the Mormons (A.K.A. "The Church of the Latter Days Saints") are two of these groups. These groups are not what their names are claiming! They have mislabeled themselves!

If you don't Believe that Jesus Christ is LORD, then you are lost!

The term "LORD" in all capitals, can only mean **"GOD ALMIGHTY, CREATOR OF HEAVEN AND EARTH!"**

Jesus is no ordinary person! The Prophets, The Angels, His Mother, His family, His Disciples and Jesus Himself all declared that Jesus is God in human form and that is part of God's record about his Son!

God became a man, he became his own Son! Nobody else can do that! He paid for our sin and rose again on the third day to prove that he really is who he says that he is! That is the Gospel Record. If you don't believe God's record, then you need to **"Change your Mind"**! If you don't believe the Gospel Record of God's Son then you are in danger of...eternal separation from God in a place called Hell or the Lake of Fire! It is time to **"Change your Mind"** and believe the Truth!

Unbelief is a violation against God. Anyone who does not want to know the Truth will be left to the father of lies. This is a time of deception. Don't reject Truth about God's Son, ever! If you reject Truth when God sends it to you, he may not resend it. But Satan can use the opportunity to send a bunch of lies!

"And they said, ***Believe*** ***on the Lord Jesus Christ, and thou shalt be saved,"*** (Acts 16:31)

DO IT NOW! It is Free! DON'T WAIT!

God the Creator offered himself for our sins

"*For he hath made him to be sin for us, who knew no sin; that we might be made* ***the righteousness of God*** *in him*". (2 Corinthians 5:21).

Only God's righteousness is good enough! Our righteousness is "*as filthy rags*" (Isaiah 64:6)

Jesus was tortured, crucified to death, buried and rose again after the third day to pay for the sins of the world. Jesus/Yeshua has already paid for all of our sins. All he requires is that we acknowledge who he is and what he did for us and trust in him **alone**. Only Believe! Just faith! That is it!

"*He that* ***believeth*** *on the Son of God hath the witness in himself: he that* ***believeth not*** *God hath made him a liar; because he believeth not* ***the record*** *that God gave of his Son.*" (1John 5:10)

Don't call God a liar! **Believe his record now!** Trust him and you will be born again and sealed by the Holy Spirit forever! It happens the moment that you believe. God's Gospel record is the list of facts found in the Bible about Jesus that we just mentioned.

"***Whosoever believeth*** *that Jesus is the Christ is born of God*" (1John 5:1)

"Verily, verily, I say unto you, He that ***believeth*** *on me* ***HATH*** *everlasting life"*. (John 6:47)

Everlasting life is free. You can **Know** that you have it the moment that you **believe** and it never ends!

"*These things have I written unto you that* ***believe*** *on the name of the Son of God; that ye may* ***<u>KNOW</u>*** *that ye* ***<u>HAVE</u> ETERNAL*** *life, and that ye may* ***believe*** *on the name of the Son of God"*. (1John 5:13)

You get the new birth the instant that you believe God's record of his only Son Jesus. The new birth happens before you get a chance to tell anyone and it is irreversible. Confession before men is not necessary for salvation, but confession to God is.

Confession is necessary to earn rewards after you have believed and are already born again, but salvation is not a reward, it is a gift and that is the only way that you can receive it!

1.21 If You Have A Question, Ask God, Newton Did.

"All my discoveries have been made in answer to prayer". (Newton)

If you really want to know the **truth** about God, why not take the first logical step in the most important quest of everyone's life? You don't have to be a Genius to figure out this one. Any child could figure out how to solve the question, **"is there a God"?** EASY! You just need to **<u>ask him</u>!** Yes, that is right, just ask God. "Are you there"? If you humble yourself before him he will hear you.

Ask God, if he is really there? Does he really care as the Bible says? God desires to hear from you more than you can understand right now. **He is listening and eagerly** waiting to hear from you! When you ask God questions about who he is and how you fit in, he makes a promise to get you the Truth! It may be a little piece at a time or it may be a lot at once. The answer may come now or later. Either way, you need to be paying attention when it comes! That is why this book was written.

God promises to get to you the Truth that you need, in a way that you can understand! Understanding is important!

"*When any one heareth the word of the kingdom, and* ***understandeth it*** ***not****,* *then cometh the wicked one, and catcheth away that which was sown in his heart. This is he which received seed by the way side*" (Matthew 13:19)

"*But he that received seed into the good ground is he that heareth the word, and* ***understandeth it****; which also beareth fruit, and bringeth forth, some an hundredfold, some sixty, some thirty*". (Matthew 13:23)

Once you start asking God to send you the Truth and you are diligently paying attention to the Truth that He gives to you, God will give you more.

Don't get sidetracked! Jesus said that he is the Truth!!

1.22 Anyone Can Make An Appointment With God, Anytime!

Anywhere in the world, at any time in history, anyone can make an Appointment with God, even a child!

As a child, I was humbled and in awe of the Creation. With the help of my mother and father, I realized early that it took a lot to Create our Universe and all of the life that we can see. The Magnitude of the Organizing, Creative and Sustaining Power and the Intelligence required to Create and direct all of everything that we know, demands an awesome God! The Creation of God's Universe was far beyond anything that Man could even begin to understand! **I was left in awe of God!** That sense of awe is what I felt as I tried to understand my Universe. How could anyone not be awestruck by the Creation? I am not the only one Awestruck by the Creation of Yeshua! Here are some quotes from **(Sir Isaac Newton)**

"This most beautiful system of the sun, planets and comets, could only proceed from the counsel and dominion of an intelligent and powerful Being." **(Sir Isaac Newton)**

"Gravity explains the motions of the planets, but it cannot explain who sets the planets in motion." **(Sir Isaac Newton)**

Throughout the Bible, are God's accounts of how he has responded to people searching for the truth.

They may not know it but they are really searching for Jesus....

***"Come let us Reason together saith the LORD"** though your **sins be as scarlet,** **they shall be as white as snow**; though they be red like crimson, they shall be as wool".* (Isaiah 1:18)

"Sin" means "to be less perfect than God". The Greek word used in the New Testament which is associated with "**Strong's Dictionary #G264 –** is **..."*hamartanō*"...**and it literally means... "to miss the mark of God's perfection".

"*For **all have sinned,** and come short of the glory of God*"; (Romans 3:23)

The Bible says that our Creator is perfect! His home is perfect! No sin can exist in his presence because where there is sin there is death! The Bible says that everyone has sinned and is less perfect than God.

Sin has a penalty!

God is warning us of imminent danger and freely offers himself as the only solution! Eternal Life is **The Gift of God**. That means that it is **free! God does the giving** and the **believer does the receiving!** You can't give anything to God or Jesus to be saved!

Salvation is where God does the giving and the Believer does the receiving. God never reverses that order, but Satan does! **You don't give anything to be saved!** You do not give your life; not your heart; not your promises; not your money,! You have nothing to give to God in exchange for perfection! God says that your best deeds are as filthy rags to him! So God is offering everyone a free gift that he has already paid for, in full, about 2,000 years ago! Eternal Life is the Gift of God, it is free; it can't be earned. Receive it as a gift from God right now by faith in Jesus Christ if you haven't already!

"But as many as ***received him****, to them* ***gave he*** *power to become the sons of God, even* ***to them that believe*** *on his name"* (John 1:12)

Notice in verse (12) who is doing the giving and who is doing the receiving. Once again, when it comes to salvation, it is always God who does the giving and the believer who does the receiving. **God never reverses this order**, but Satan does!

If someone tells you to **give** anything to be saved, that man is a liar because God says that Eternal Life is a Gift and It is free! God doesn't ask anyone to do anything to be saved or be born again… **except believe on Jesus Christ as your Savior**. Just Faith! That is all!

You Can Have An Appointment With God Right Now!

When you ask God to send you Truth about who he is and what your purpose is, then **you have an appointment with God!** God will answer your Spirit-led prayers and questions with loving care! He promises to. God will find a believer somewhere in the World in the right place, at the right time with the right answers and in a way that you can understand! That is the whole point of writing this book that you are holding. Someone's prayer for you is answered, maybe yours. After reading this book, go to... **Endtimesbiblehandbook.org** and get your questions answered. Share your questions and concerns with us.

Make sure that you are paying attention when God sends you Truth! (That is **The First Principle of Truth!**) God says that… *"Ye have not, because ye ask not"*… (James 4:2)… so ask him.

When God answers your prayers and reveals Truth, you need to give it respect, write it down! God uses his people to spread Truth. Truth is not cheap, but God gives it away **freely**. God is displeased when people disrespect the Truth! When God sends you Truth, you should cherish it; hold on to it and honor it! If you do, then God will give you more.

If you don't respect the Truth that you have, God will stop sending it! If you reject and disrespect God's Truth, he will allow Satan to send you lies!

Satan's lies will take away the Truth that you had. Satan will neutralize or destroy you with lies! Hold on to the Truth that you have! Pray and search for more! Pray for God's protection without ceasing!

1.23 Think Of The Lake Of Fire As A Toxic Waste Dump For Sin.

"And death and hell were cast into the lake of fire. This is the ***second death****"*. (Revelation 20:14)

"Then shall he say also unto them on the left hand, Depart from me, ye cursed, into everlasting fire, ***prepared for the devil and his angels****"*. (Matthew 25:41)

Hell was not Created for people, it was Created for Satan and his angels.

"For if God spared not the angels that sinned, but cast them down to hell, and delivered them into chains of darkness, to be reserved unto judgment"; (2Peter 2:4)

Just like a toxic waste dump contains pollution, the Lake of Fire will contain sin. Soon God will clean up this part of his universe and the Lake of Fire is where sin will be kept for eternity! The question is, will you be in there?

"The wages of sin is death *but* ***the gift of God is eternal life*** *through Jesus Christ our Lord!"* (Romans 6:23)

Notice that Eternal Life is a **gift**. That means that God is doing the giving and the **believer** is doing the **receiving**. In other words, **eternal life is free** to all those that **believe** that Jesus is God in flesh and Jesus paid it all, there is nothing else left to do but believe!

Hell was Created for the Devil and his angels and is currently empty.

The Bible says that angels will be judged by Believers (humans) to determine the degree of punishment in hell for eternity.

The lost (humans) will be judged by Jesus to determine their degree of punishment in hell for eternity.

*"Of **how much sorer punishment**, suppose ye, shall he be thought worthy, who hath trodden under foot the Son of God, and hath counted the blood of the covenant, wherewith he was sanctified, an unholy thing, and **hath done despite unto the Spirit of grace**?"* (Hebrews 10:29)

God is going to hold every unbeliever accountable for the amount of truth that they reject! Also, God will hold unbelievers accountable for the way they treated God's Children on this earth. God is also going to hold unbelievers accountable for the way they treated Israel.

There are degrees of punishment in hell and there are degrees of reward in HeavenJ

2

"The Gift Of God Is Eternal Life"...!

2.1 Eternal Life Is A Free Gift To Those Who Believe God's Record!

"The Gift of God" is a Biblical term used exclusively for "**Eternal Life**".

*"For the wages of sin is death; but **the gift of God is eternal life** through Jesus Christ our Lord".* (Romans 6:23)

The **"Gift of God"** is, by definition, where **God does the giving** and the **Believer does the receiving** and Eternal Life is the **Gift**. **It is that simple!**

*"But as many as **received** him, **to them gave he** power to become the sons of God, even **to them that believe** on his **name**:"* (John 1:12)

Notice that the only requirement to become a son of God is to believe on Jesus name. Notice also **who is doing the giving** and **who is doing the receiving**! God never changes that order! If anyone tells you that you must give anything to God to be saved, then it is no longer a Gift! Those who seek salvation are not asked to give anything to be saved! Nothing, nothing at all! They are asked to **believe** and **receive** the **free gift of eternal life**, through Jesus Christ the Lord.

Grace is unmerited favor, a gift. By definition you can't earn grace because it is no longer grace if you have to do anything to get it. The same is true about a gift. By definition a gift is no longer a gift if you have to do anything for it.

Sin entered into the whole human race through **one man named Adam.** The payment for all of the sins of the human race from Adam on into eternity was made on the Cross of Calvary **by one Man, the God/man Jesus Christ.**

"But not as the offence, so also is the ***free gift****. For if through the offence of* ***one****,*(Adam) ***many*** *be dead much more the* ***grace*** *of God, and the* ***gift by grace****, which is by* ***one man****, Jesus Christ, hath abounded* ***unto many****".* (Romans 5:15)

"And not as it was by ***one*** *that sinned, so is* ***the gift****: for the judgment was by* ***one to condemnation****, but the* ***free gift*** *is of* ***many*** *offences unto justification".* (Romans 5:16)

"Therefore as by the ***offence of one****, judgment came upon* ***all men*** *to condemnation; even so by the righteousness of* ***one*** *the* ***free gift*** *came upon* ***all men*** *unto justification of life".* (Romans 5:18)

Eternal Life is a free gift that God gives freely to all those who **believe** God's record about his Son. God's record about His Son begins in Genesis. It tells us in advance that God would take on a Human body. His Name would be Yeshua or in English that would be Jesus. About three hundred and thirty three prophecies give us the details of Jesus birth, life and crucifixion on the cross of Calvary.

The day that Jesus would preset himself to Israel as the "Lamb Of God" was told in advance. The prophet Daniel said that the Temple would be destroyed after Messiah was killed for others. That has all happened 2,000 years ago. Jesus said on the cross, "it is finished". That means that He finished paying for all of the sins of the human race; past, present and future! Eternal Life is a gift. Jesus paid for it all on the Cross of Calvary 2,000 years ago. The Gift of God it Eternal Live and it is free to all of those who believe! That is what God's record about His only Son says. We are justified by faith and that is all. That is why it is called "Just Faith"!

You have Eternal Life the moment that you **believe** God's record about His only begotten Son; you can know that you have it, it is eternal; you can't lose it and it can't be taken away. It is a gift and God is not an Indian giver.

If you are trying to be saved then you are not trusting to be saved. Eternal Life is a Gift offered by God; it is free; it can't be earned and it cannot be lost! It can only be received by understanding and believing **God's record of his Son**. It is a simple record that says …Jesus Paid For all human sins! (John 3:16) sums it up…

"For God so loved the world, that ***he gave*** *his only begotten Son, that whosoever believeth in him should not perish, but have everlasting life"*(John 3:16)

Notice again that God is doing the giving

A record is a list of facts. In this case we are talking about God's list of facts about His Son. It is that simple, God gave his Son for us. If we believe, we have Eternal life, a new birth and a new father! That is the Record.

"He that ***believeth*** *on the Son of God hath the witness in himself: he that* ***believeth not*** *God hath made him a liar;* ***because he believeth not the <u>record</u> that God gave of his Son****"* (1John5:10)

If you don't believe God's record of his Son then you are calling God a liar. That is Blasphemy and it is what sends unbelievers to hell!

*"****And <u>this is the record</u>,*** *that* ***God hath <u>given</u> to us eternal life,*** *and* ***this <u>life is in his Son</u>****"* (1John5:11)

That is the record. It is a list of facts about Jesus. Everyone needs to understand and believe certain facts about Jesus.

"He that hath the Son hath life; and he that hath not the Son of God hath not life". (1John5:12)

If you don't have certain facts about Jesus right, then you don't have the Jesus of the Bible and are not saved! God's record is the one that you want to go by. There are many false records and a false Jesus for everyone. If you **believe God's record** then *"you may **know** that you **have** __**eternal**__ life"!*

*"These things have I written unto you that **believe** on the name of the Son of God; **that ye may __know__ that ye __have__ eternal life**, and that ye may believe on the name of the Son of God"*. (1John5:13)

You can **Know** that you **Have**, **Eternal Life when** you **believe** the record that God gave of his son. That is what God's record says.

That is the only way that you can know that you have Eternal Life, it has to be a Gift that you receive at some point in time.

Some preachers talk about different kinds of belief. The Bible makes no such distinction when talking about salvation. In the Greek language, there are two kinds of belief. One is emotional in nature and the other is an intellectual belief. I believe that God accepts them both.

God's record about Jesus goes back to the beginning of Creation! (John 1:1) God knew what he was going to have to do to save mankind from the beginning! Yet He Created us to have fellowship with Him. Then He Loved us all the way to the Cross! I believe that He would do it again.

Before any of us were born, God Loved us all! He already had a plan to save the lost, even before Adam was Created! He gave us free will to make our own choices. He foreknew what would happen and he did it anyway. He gave himself as a sacrifice for our sins.

2.2 "Jesus said... You Must Be Born Again"!

*We need a new birth to inherit everlasting life. Jesus said that we get it when we **believe** on him...*

*"Jesus answered and said unto him, Verily, verily, I say unto thee, **Except a man be born again, he cannot see the kingdom of God**". (John 3:3)*

*"Nicodemus saith unto him, How can a man be **born** when he is old? can he enter the second time into his mother's womb, and be born?" (John 3:4)*

*"But as many as received him, to them gave he power to become the **sons of God**, even to them that **<u>believe</u>** on his name": (John1:12)*

*"That which is **<u>born</u> of the <u>flesh</u> is flesh**; and that which is **<u>born</u> of the <u>Spirit</u> is spirit**". (John 3:6)*

*"Marvel not that I said unto thee, Ye must be **<u>born again</u>**". (John 3:7)*

*"Nicodemus answered and said unto him, **How can these things be?**" (John 3:9)*

"I have told you earthly things, and ye believe not, how shall ye believe, if I tell you of heavenly things?" (John 3:12)

"And as Moses lifted up the serpent in the wilderness, even so must the Son of man be lifted up:" (John 3:14)

This event with Moses in the Wilderness involves **Rebellion** against God and the authority that He gave to Moses. It involved a **conspiracy** against Moses and his Ethiopian wife. The conspirators attempted to use her **Race** to create division and rebellion against Moses and the authority which God had openly given to Moses!

*Satan was using unbelieving men to stir up a **rebellion** against Moses and they did it by **"playing the Race Card"** against an Ethiopian woman who just happened to be Moses Wife. God responded by sending a plague of poisonous snakes to bite the unbelievers and bigots of Israel.*

The rebels that realized that this was a punishment from God, appealed to Moses. They asked him to appeal to God on their behalf because they were afraid and ashamed to ask for themselves! The Children of Israel requested for God to get rid of these horrible, biting and poisonous snakes.

*But God would not remove the snakes. Instead of removing them, God provided an antidote for the venom. All that God required of them to be saved from the snake venom was to **just take <u>one</u> little look** at a brass likeness of a serpent that was held up high on a pole for **<u>all to see</u>**. Nevertheless, some refused to even take a look! Not even a peek.*

Even after all of the miracles that God showed to them, some just refused to believe till the end. They would not believe and admit that they were wrong. They were too proud to take one look at the brass serpent held up on the pole for all to see. They may have felt that the whole idea of looking to a brass likeness of the snake that was biting them may have been a bit annoying to some who instigated the rebellion. When they looked at the brass serpent on a pole, they were acknowledging their past sin, their past punishment

*"For **rebellion** is as the sin of **witchcraft**, and **stubbornness** is as iniquity and idolatry". (1 Samuel 15:23)*

"And as Moses lifted up the serpent in the wilderness, even so must the Son of man be lifted up:" (John 3:14)

*"That whosoever **believeth** in him should not perish, but **<u>have</u> <u>eternal</u>** life". (John 3:15)*

*Jesus continues to explain to **Nicodemus** about how to be born again in the next verse...*

*"For **God so loved the world**, that **<u>he gave</u>** his only begotten Son, that whosoever **<u>believeth</u>** in him should not perish, but **<u>have</u> <u>everlasting</u>** life". (John 3:16)*

*The moment that you believe the Gospel, you are "born again"; you become a member of God's Family; God becomes your Father; You get a **New Birth** that will never die; you get **sealed with the Holy Spirit** of promise; You are **passed from death unto Life**; you gain **A New Nature**; a new future; . Your citizenship is changed to Heaven!*

*"Whosoever __**believeth**__ that Jesus is the Christ is **born of God**: and every one that loveth him that begat loveth him also that is begotten of him". (1John5:1)*

*God is now your Father! You can never be lost again once you are saved. All relationships in the Spiritual realm are changed, all spiritual matters pertaining to you are new. You become a child of God, a new spiritual Creation of God and all of this is The Gift of God by faith! It's Free! It is not of works, not of yourself, it is by Just **"__Faith__" Alone!***

2.3 Don't Look For An Outward Sign Or Feeling.

Don't expect an outward sign to confirm a person's inward belief on Jesus as Savior. There is no guarantee that there will be an outward sign to confirm that anything took place inside someone's mind. The heart is a pumping organ, it does not think. The term "heart" is used in the Greek to express an emotional belief. Not everyone expresses emotions the same, God accepts a "head belief" just the same as a "heart belief".

The spiritual things that occur the moment that someone believes the Gospel cannot be detected outwardly. Just because you receive a New Nature when you believe, does not suggest that the Old Nature suddenly goes away. Even though the Holy Spirit comes into the new believer immediately upon believing, the Spirit can be quenched! God warns us...

...*"**Quench not the Spirit**"*. (1Thessalonians 5:19)

The Spirit can also be grieved...

...*"And **grieve not The Holy Spirit** of God, whereby ye are sealed unto the day of redemption"*. (Ephesians 4:30)

The new believer may have a situation in his mind to compete with the desire to tell someone about his free Gift of Eternal Life. A believer is capable of selfish motives. Outwardly it is difficult to tell what is going on inside someone's mind. You just can't tell.

When Jesus explained this He said…

"The wind bloweth where it listeth, and thou hearest the sound thereof, but ***canst not tell whence it cometh****, and whither it goeth:* ***so is every one that is born of the Spirit"*** (John 3:8)

It says right there ***"canst not tell"***. When it says ***"canst not tell"*** it is saying that you just can't tell. There may or may not be outward signs of a new birth ever. We all want to know if someone is indeed saved and we should look for outward signs, but don't rest your faith in that persons eternal salvation until you have heard from their own lips that they **understand** and **believe** the Gospel.

Many people are doing good works because they wrongly think that it is helping them get to heaven. Often people just go along with the crowd without understanding and believing. As believers, it is our responsibility to witness to others about the Gospel and find out what they are trusting in. If they are trusting in the wrong thing they are headed for hell!

Just because someone says that he is a Christian and appears to be living for the Lord doesn't mean that the person is even saved at all! Outward appearances can be deceiving. Share the Gospel with your friends and family, ask them questions about their faith. Ask them what they think is necessary to gain Eternal Life. God pays a wage for Christian service. God's rewards are eternal! At the time of this writing, the Judgement Seat of Christ is just ahead for the believer! I believe that it will be an unpleasant event for most believers.

"Wherefore we labour, that, whether present or absent, we may be accepted of him". (2Corinthians5:9)

"For we must all appear before ***The Judgment Seat Of Christ****; that every one may receive the things done in his body, according to that he hath done, whether it be good or bad"* (2Corinthians5:10)

God is going to reward believers for their faithfulness.

*"**Knowing therefore the <u>terror</u> of the Lord, we persuade men**; but we are made manifest unto God; and I trust also are made manifest in your consciences"* (2Corinthians5:11)

This passage is talking about **The Judgement Seat Of Christ**. I believe that will be the first thing that a believer will see after being Raptured! Also notice the adjective **"terror"**. It will be infinitely worse for unbelievers! Not all believers produce fruit. Those who did not produce fruit will suffer loss.

"If any man's work shall be burned, he shall suffer loss: but he himself shall be saved; yet so as by fire". (1Corinthians3:15)

You don't need any works to be saved, it is by Just Faith!

2.4 Confession Before Men Is Not Necessary For Salvation.

Some wrongly say that you must confess with your mouth that you are a Christian to be saved. This is a work, it is something of yourself. Here is the passage that is often used to support that wrong view…

*"That if thou shalt confess with thy mouth the Lord Jesus, and shalt **believe** in thine heart that God hath raised him from the dead, thou shalt be saved".* (Romans 10:9)

*"For with the heart man **believeth unto righteousness**; and with the mouth confession is made unto salvation"* (Romans 10:10)

This passage is not saying that you must confess Jesus out loud before men to be saved. This passage is making a distinction between outwardly saying that you are a believer and actually being one. With the mouth we have an **outward appearance** of salvation, but **Salvation actually occurs the moment that we believe with our mind.**

There are people who will say that they believe when they really don't. They just want to please the crowd. And there are others who will not admit that they are saved for the same reason. But they are saved if they have believed. Confession to God is required but not confession to men. You can talk to God any time, just remember to confess your sins and ask for forgiveness first!

Inward belief must come before outward appearance or else we are talking about "*dead works*". Any works done as an effort to earn or deserve Eternal Life is an abomination to God! According to (Isaiah 64: 6)the best that we can do is like *"filthy rags"* to God.

Works done for appearance sake can be misleading, outward appearances often are. God never promises that every believer will produce fruit to confirm that he is saved. **Not all believers produce any fruit at all,** but they are still saved. We would all love to see new believers taking up their crosses which is their reasonable service, but that is rare. Outward appearances are often deceiving.

"But the LORD said unto Samuel, Look not on his countenance, or on the height of his stature; because I have refused him: ***for the LORD seeth not as man seeth; for man looketh on the outward appearance, but <u>The LORD Looketh On The Heart</u>***". (1Samuel16:7)

The simple message of the Gospel is that we are all less perfect than God. Our imperfections or corruption separates us from our Creator. Our Creator Created us for himself so he loves us and wants to be reconciled to the human race. A sacrifice must be made that only He could make. Our Creator came to the earth as a man. He was born to a virgin and became God in human flesh.

The name of this God/Man is Jesus in several languages. In Hebrew it is Yeshua. That Name means that God is my Salvation. Jesus/Yeshua made an infinite payment on the Cross of Calvary for all of mankind. That is what the Name means and God gives eternal life to everyone who believes on that Name and what it means!

If you are trusting in Christ plus your works or anything else, then you are believing in another Jesus, not the Jesus of the Bible. You need to **change**

your mind about your *"**dead works**"* and wrong beliefs and **believe** the **simple Gospel** that Jesus Christ paid it all! God Loves us and Gave Himself for us, as a substitutionary sacrifice for our sins. Eternal Life is **<u>The</u> <u>Gift</u> of God** (Romans 6:23)(Ephesians 2:8,9) If we **believe** it, then we receive it, at that same moment and forever after.

God divides up the whole human race into two groups! One group is made up of those who have **believed** on Christ; they have been born again by faith; they possess Eternal Life; they have been passed from death unto Life and they are not condemned nor can they ever be! The other group have not believed; they are spiritually dead; they are condemned to the Lake of Fire if they die without Jesus!

*"He that **believeth** on him is **<u>not</u> condemned**: but he that **believeth <u>not</u>** is condemned already, **<u>because</u> he hath <u>not</u> believed** in the name of the only begotten Son of God"* (John 3:18)

God commands all to believe the Gospel. "***<u>believe</u> on the Lord Jesus Christ and thou shalt be saved***". (Acts 16:31)

The word **"*believe*"** is translated from **Strong's #G4100 - *pisteuō*** which is also translated as "**trust**".

Trust Jesus alone to save you. Not you're church; not your favorite "Saint"; or anything else in addition to Jesus alone! Eternal life is a **Gift**, Jesus paid for it in full! Receive it! **It is free!** Just faith. **You can't be trusting and trying at the same time!** They are **Opposites!** It is either one or the other. It can't be both at the same time! Either he paid it all or he didn't. Jesus said that he did!

Eternal Life is <u>not</u> a reward for something that we do for God. A reward is something that you earn. Eternal Life is ***The Gift of God***. It can't be earned! When we realize that we have nothing to offer God in exchange for his righteousness and **we trust in Jesus <u>alone</u> as the only hope of salvation,** then we are born again. We have a new nature; we are passed from death unto life; our citizenship is in Heaven; our mansion is in Heaven and we are God's Ambassadors as long as we are here on this Earth.

Believers who are faithful are rewarded. Believers who are not faithful will lose out on rewards, but they will still be saved.

"Then said they unto him, What shall we do, that we might work the works of God?"(John 6: 28)

*"Jesus answered and said unto them, This is the work of God, that ye **believe** **on him whom he hath sent"*** (John 6:29)

The Gospel says to believe God's record of his Son! (1John 5:10)

These people ***"know not God"*** because they are unbelievers. They did not believe the Gospel.

"*Hear ye therefore the parable of the sower*". (Matthew 13:18)

*"When any one heareth the word of the kingdom, and **understandeth it not**, then cometh the wicked one, and catcheth away that which was sown in his heart. This is he which received seed by the way side"* (Matthew 13:19)

How can you believe something you can't understand? The Gospel is simple. A child can understand it. Satan complicates things with lies.

Satan takes away the truth that you have with lies. Satan is the father of lies. (John 8:44) **Satan is the god of this world**. He blinds the minds of unbelievers with lies so that people do not understand the **simple** Gospel.

*"In whom the god of this world hath **blinded the minds** of them which **believe not**, lest the light of the glorious gospel of Christ, who is the image of God, should shine unto them"* (2Corinthians 4:4)

"And it shall come to pass, that whosoever shall call on the name of the Lord shall be saved"! (Acts 2:21)

2.5 You Can't Be Saved If You Are Trying To Be Saved!

Eternal Life is God's Gift. He does not save those who are trying to be saved, only those who are trusting to be saved!

*"**Now to him that worketh is the reward not reckoned of grace, but of debt**"*. (Romans 4:4)

This verse is saying that if you are working to be saved you won't get grace but instead, you will get debt. In other words, no grace means you will come up short at meeting God's requirement of absolute perfection and that means no salvation! That verse is worth repeating.

"***Now to him that worketh is the reward not reckoned of grace, but of debt***" (Romans 4:4, 5)

"*But to him that **worketh not**, but **believeth** on him that justifieth the ungodly, **his faith is counted for righteousness***". (ROMANS 4:5)

You need God's righteousness and that only comes by grace through faith and not of yourselves.

"*And if by grace, then is it no more of works: otherwise grace is no more grace. But if it be of works, then is it no more grace: otherwise work is no more work.*" (Romans 11:6)(Romans 11:6)

In other words, the moment you add any effort of yourself to the grace formula, you no longer have grace, but instead you have "*dead works*". "*Dead works*" is any work or effort done to help earn your salvation. "Works can't save you! If you think that your works can help you get saved or stay saved, then you are lost and your works are called "*dead works*"; "*filthy rags*" ;and "*iniquity*". Any works done to earn salvation is called "sin" by God.

(Hebrews 6:1) *"Therefore leaving the principles of the doctrine of Christ, let us go on unto perfection; not laying again the foundation of **repentance** from **dead works**, and of faith toward God,"*

"Repentance from dead works" means you should "change your mind" about your works and believe what God has said about your works.

Works and grace are opposites. **You can't have both at the same time.** It is either one or the other. The moment you add **any** works, no matter how seemingly slight, you no longer have grace! You have nullified God's terms for salvation! You have ***another gospel***, an accursed message that cannot save!

"I marvel that ye are so soon removed from him that called you into the grace of Christ unto ***another gospel****"*: (Galatians 1:6)

"Which is not another; but there be some that trouble you, and would pervert the gospel of Christ" (Galatians 1:7)

"For if he that cometh preacheth ***another Jesus****, whom we have not preached, or if ye receive* ***another spirit****, which ye have not received, or* ***another gospel****, which ye have not accepted, ye might well bear with him.* (2 Corinthians 11:4)

"I do not frustrate the grace of God: for if righteousness come by the law, then Christ is dead in vain" (Galatians 2:21)

God only saves those who are **trusting** to be saved, **NOT** those who are trying or working to be saved!

"For they being ignorant of God's righteousness, and going about to establish their own righteousness, have not submitted themselves unto the righteousness of God". (Romans 3:10)

"*Therefore by the deeds of the law there shall no flesh be justified* ***in his sight****: for by the law is the knowledge of sin*". (Romans 3:20)

Trying to establish your own righteousness will not work! If Satan can get people working to be saved, he has won! They won't be saved!

"*Now to him that worketh is the reward* ***not reckoned of grace****, but of* ***debt****"* (Romans 4:4)

"*But to him that **worketh not**, but **believeth** on him that justifieth the ungodly, **his faith is counted for righteousness***". (Romans 4:5)

"*In flaming fire taking vengeance on them that **know not God**, and that **obey not the gospel of our Lord Jesus Christ***" (2 Thessalonians 1:8)

If you are **not trusting** in Jesus **alone**, then you are **not obeying the Gospel**. The gospel says to trust or believe in Jesus **alone** to be saved. The words **"Trust"** and **"Believe"** mean **the same thing.** ***The "Gospel" is the "Power of God" unto salvation*"(Romans1:16)** It is the message that says that **..."The "*Gift of God*" is "*Eternal Life*"**! ; "*it is **not** of yourselves*"! ; It is ***"Free"*** to all those who ***believe*** God's Record of his son.

"*For mine own sake, even for mine own sake, will I do it: for how should my name be polluted? and **I will not give my glory unto another***".!!! (Isaiah 48:11)

God does not share **His Glory** with anyone else! He is God and only He deserves praise! Not men! Men Generally praise men, but only a few give God his praise. Everyone is looking for a hero. So why don't more people just take a little look to Jesus? He says that only he can heal you from within and he claims to be God! Everyone needs a Savior, so do you and I! We all need a Savior and I have found him. My Savior's Name is Jesus!

"*Who can say, I have made my heart clean, I am pure from my sin?*" (Proverbs 20:9)

If you are **trying to be saved**, then you are **not trusting** Jesus to be saved. You can't do both at the same time. If you are trying to be saved; it is because you are **not trusting to be saved**, therefore you have not **understood** and **believed** the Gospel; you have **not obeyed the Gospel**. You... ***have not submitted themselves unto the righteousness of God*"**(Romans 3:10) ... by trying to establish your own righteousness. Don't try to be saved, trust Jesus to be saved. He did all the work! You can't do both at the same time!

God will not share credit with **anyone!**

"I am the LORD: that is my name: and my glory will I not give to another, neither my praise to graven images". (Isaiah 42:8)

God calls our own righteousness "***filthy rags***" when compared to himself, and that is the standard! Don't insult the Creator of the Universe by comparing what you have to offer to what God has already paid! God says that he has met all the requirements to purchase our salvation so that we could receive His Righteousness as a Free Gift! God is offering it for Free!

"*But we are all as an unclean thing, and **all our righteousnesses** are as filthy rags*"…; (Isaiah 64:6)

Our own righteousness is of the law and God says that is not going to help you earn your way into Heaven! You can't earn your way or help out in any way. Salvation is a Gift. It's not of yourself. It's Free!

"***Not by works of righteousness which we have done**, but according to **his mercy he saved us**, by the washing of regeneration, and renewing of the Holy Ghost*"; (Titus 3:5)

There is nothing that anyone can offer to God for their salvation! Everything that you have was given by God. God offers his Righteousness and Eternal Life as a **Free Gift** to anyone who will **believe** the Gospel, that is put your trust in Jesus **alone** to save you, Just Faith!

*"And be found in him, not having mine own righteousness, which is of the law, but that which is through the **faith** of Christ, the righteousness which is of God by **faith**"*: (Philippians 3:9)

If you believe it, then you have it. It is that simple.

2.6 Eternal Life Can't Be Earned And Can't Be Lost!

*"He that **believeth** on the Son of God hath the witness in himself: he that **believeth not** God hath made him a liar; **because he believeth not the record that God gave of his Son**"*. (1John 5:10)

Right here, the Holy Spirit spoke through Saint John and said that the ***"witness"*** is inside of everyone who believes God' record of his Son at the moment of **Belief!** The ***"witness"*** spoken of here is the **Holy Spirit**!

*"In whom ye also trusted, after that ye heard the word of truth, the gospel of your salvation: in whom also after that **ye believed, ye were <u>sealed with that holy Spirit</u> of promise**"*, (Ephesians 1:13)

*"**Which is the <u>earnest</u>** of our inheritance until the redemption of the purchased possession, unto the praise of his glory"(*Ephesians 1:14)

Upon believing during the Church Age, all believers are "***sealed with that holy Spirit of promise***". The Holy Spirit becomes the down payment for the believer until we are redeemed at the Rapture. The believer can't go to hell, he is "***sealed with that holy Spirit of promise.***" and he says that he will not leave you.

"Which is the earnest of our inheritance until the redemption of the purchased possession, unto the praise of his glory" (Ephesians 1:14)

During the Age of Law, believers had to pray and ask God for his Holy Spirit. But not everyone who asked for it got it. This is what makes the Church so special!

This passage (1John 5:10) also says that **everyone must either believe God's <u>Record</u> about his Son or you are calling the Holy Spirit a liar!** Calling God a liar is foolish, it is also Blasphemy! If you don't believe the ***<u>record</u>*** that God gave of his Son, then you are **<u>guilty of unbelief</u>**, you are calling God a liar and you are condemned by God!

*"He that **believeth on him** is not condemned: but he that **believeth not is condemned already**, because he hath not believed in the name of the only begotten Son of God"* **(John 3:18)**

The only difference between being condemned and not being condemned is either **belief or unbelief** in the Gospel or God's Record. It is that simple! Jesus said…

*"Verily, verily, I say unto you, He that **believeth** on me **hath everlasting** life."* (John 6:47)

The moment that we **believe**, we **have *everlasting*** Life! And we can **KNOW** it because God said so!

Our passage in (1John 5) goes on in verse eleven to say…

*"**And this is the record**, that **God hath given to us eternal life**, and this life is in his Son"*. (1John 5:11)

Notice here that the **record** that everyone must **believe** is that **God gives** to those who **believe** his **record**, **Eternal Life as a Gift. Eternal Life is the Gift of God through Jesus Christ! It is Free!** That is it, that is "The Record" that you must **believe!** God says that it is **Eternal**, it is a FREE Gift and you get it the moment you believe God's record about his Son. And that is part of the record. A record is a list of facts.

*"He that **hath the Son** hath life; and he that **hath not the Son of God hath not life**"*. (1John 5:12)

So far we are told that a believer is indwelt by the Holy Spirit the moment that one believes! Now we also find that we receive the Son at the same time! That is all a part of receiving Eternal Life. All of that happens at the same time plus your name is announced in Heaven with rejoicing the moment that you **believe God's Record of his Son.**

"These things have I written unto *you that **believe** on the **name** of the Son of God; that ye may **know** that ye **have eternal life**, and that ye may **believe** on the name of the Son of God"*. (1John 5:13)

God says that you can *Know* that you *Have Eternal Life"* the moment that you believe!"

Eternal Life is never referred to as a reward, it is the Gift of God.

Eternal Life is **Eternal**; it is **Free**; Jesus Paid it All! It is a **Gift**! You can't pay for something that is free. If you tried to, it would no longer be free. A gift is not a gift if you pay something for it. It is either a gift or not. It can't be both at the same time. It might be a good deal, but if you feel that you are obligated in any way to have done anything in the past, present or Future, then it is really not a gift. It is that simple. If you think that you deserve it, then you need to change your mind about your dead works and believe God's Record or you will be lost.

*"All that the Father giveth me shall come to me; and him that cometh to me **I will in no wise cast out**"*. (John 6:37)

God says that **He will not cast out** a believer for any reason. That is why it is called ***"Everlasting Life"***! It is Everlasting and you get it the moment that you Believe God's Gospel of Grace through Faith in God's only begotten Son, Jesus Christ!

The "Gospel of Grace" is another way of saying God's ***record*** of his only begotten Son.

*"And this is the Father's will which hath sent me, that of all which he hath given me **I should lose nothing**, but should raise it up again at the last day"* (John 6:39)

God says that **He will not lose anyone**! You are saved forever the Moment that you **believe! … Just Faith! It is "*Everlasting*"!**

2.7 Change Your Mind About Dead Works

Works done to earn salvation or to impress men are called "***dead works***" In other words, they are worthless and deceptive. Dead works are deceptive in the sense that they are a dead imitation of something that is alive.

*"Therefore leaving the principles of the doctrine of Christ, let us go on unto perfection; not laying again the foundation of **repentance from dead works**, and of **faith toward God**"*, (Hebrews 6:1)

"And he said unto them, ***Ye are they which justify yourselves before men;*** *but God knoweth your hearts: for that which is highly esteemed among men is abomination in the sight of God"* (Luke 16:15)

"For he hath made him to be sin for us, who knew no sin; that we might be made ***the righteousness of God*** *in him"* (2Corinthians 5:21)

The only way to receive God's righteousness is by believing God's message of salvation!

"And be found in him, not having mine own righteousness, which is of the law, but that which is through the ***faith*** *of Christ,* ***<u>the righteousness which is of God by faith</u>***": (Philippians 3:9)

Without God's righteousness, you are condemned, but you receive it as a free gift the moment you believe and it will never be taken back! Even when a believer sins, God's righteousness covers all. God sees you as righteous.

*"**<u>Not by works of righteousness</u> which we have done, but according to his mercy he saved us**, by the washing of regeneration, and renewing of the Holy Ghost"* (Titus 3:5)

Being better than your peers in your own eyes is not the same as having the Righteousness of God given to you as "*The Gift Of God"!*

*"**And this is the will of him that sent me,** that every one which seeth the Son, and **believeth** on him, may **have everlasting life**: and I will raise him up at the last day"*.(John6:40)

The Gospel commands us to believe the record that God gave of his Son. That record says that Jesus paid for all of our sins on the Cross of Calvary so that we could be with him! There is nothing that we can add to what God has already done on our behalf. It is a Gift! It is Free! Jesus Christ **paid for it <u>all</u>!**

"Many will say to me in that day, Lord, Lord, have we not prophesied in thy name? and in thy name have cast out devils? and ***in thy name done many wonderful works?"*** (Matthew 7:22)

"And then will I profess unto them, ***I <u>never</u> knew you****: depart from me, ye that work iniquity"* (Matthew 7:23)

This person doesn't mention what Jesus Christ did on the Cross, only what he calls "***many wonderful works***". He was trusting in his own righteousness and his religious piety. He was too proud for grace. He turned it down. You can't be saved if you are trying to be saved. You must **Trust Jesus** to be saved. God will not share the glory and praise that he rightfully deserves with you or anyone or anything else. Once you are saved, you are always saved! The Righteousness of God is a Gift received by believing in the Jesus Christ of the Bible!

"I am the LORD: that is my name: and ***my glory will I not give to another, neither my praise*** *to graven images"*(Isaiah42:8).

Either you believe what God has said and accept it as a free gift or else you will have to pay for your own sin in the Lake of Fire forever. It is either grace or works. It can't be both at the same time. It is either one or the other. They are opposites! Satan knows that and blinds the minds of men with Lies to stop them from believing.

"*For what saith the scripture? Abraham* ***believed*** *God, and it was* ***counted unto him for <u>righteousness</u>***". (Romans 4:3)

Satan knows that if he can get people to work for their salvation they will end up in the Lake of Fire. It is the Gift of God, it is free! You can't earn it! **Eternal Life is the <u>Gift</u> of God, <u>not</u> a reward!**

"Now ***to him that worketh*** *is the reward* ***not reckoned of grace****, but of* ***<u>debt</u>***". (Romans 4:4)

OK, that is what you will get at the Great White Throne Judgement of the lost if you are trying to be saved. You will fall short of God's Righteousness and be cast into the Lake of Fire. You must trust Jesus alone to receive God's Righteousness!

"*But to him that* ***worketh not****, but* ***believeth*** *on him that justifieth the ungodly, his* ***faith is counted for righteousness***". (Romans 4:5)

It says it right there, that if you are working for it you won't make it. But if you believe what Jesus said on the Cross, **"*it is finished*"**, or in other words he said that He just paid for all of our sins, past, present and future", if you believe that and are trusting in that alone, then you receive **God's righteousness** as a **gift** from God the moment that you **believe.** Just Faith, Simple Faith! Either you believe it or you don't. From that point forward God looks on the believer as a Righteous Child of God.

"*And if by grace, then is it no more of works: otherwise grace is no more grace.* ***But if it be of works, then is it no more grace****: otherwise work is no more work*". (Romans 11:6)

It can't be grace and works at the same time. The Bible says that it is either one way or the other way. It can't be both ways at the same time. They are opposites, like the east is from the west!

There are some that wrongly think that you must do something of yourself to be saved. But God says not so…

"For by grace are ye saved through ***faith****; and that* ***not of yourselves****: it is* ***the gift of God: Not of works****, lest any man should boast.*" (Ephesians 2:8,9)

God says that salvation is "***not of yourselves***" and "***not of works***", he calls it the Gift of God! It is not a gift if you are obligated in any way. You can't have grace and obligation at the same time.

When you add any works to the salvation equation, you no longer have grace! (Romans 4:4,5; 11:6) A false gospel adds something to God's **simple plan of salvation** and does not save because it denies the simple message that…"***the Gift of God is Eternal Life!"*** (Romans 6:23)

*"For they **being ignorant of God's righteousness**, and going about to establish their own righteousness, **have not submitted themselves unto the righteousness of God**"*. (Romans 10:3)

People who are working to be saved are comparing their deeds to Gods! That is Blasphemy! They… *"have not submitted themselves unto the Righteousness of God"* (Romans 10:3). They are working, not Trusting Jesus to be Saved! When you trust Jesus alone to be Saved, then you have **obeyed the Gospel**, you have a **New Birth**; ***"the righteousness of God"***; a **New Nature; a New set of rules** and a **New Heavenly Father who will enforce those rules and discipline you!**

2.8 Obey The Gospel! Unbelief Is Disobedience To The Gospel!

The Gospel tells us to **believe the record** that God gave of his Son; (1John 5:10-13) It also says that if you don't believe God's record; or If you believe someone else's record instead… then you are calling God a liar! Unbelief is rebellion against God! People just need to realize that not believing **God's Record of his Son** is the only thing that can send you to hell!

(2Thesalonians 1:8) *"In flaming fire taking vengeance on them that know not God, and that **obey not the gospel** of our Lord Jesus Christ:"*

This is what God commands all men to do to be saved…

…*"Believe on the Lord Jesus Christ and thou shalt be saved"* (Acts 16:31)…

If you **believe** the Gospel, then you have **obeyed** the Gospel because we are all commanded to **believe the Gospel.**J

If you do not believe God's Gospel, then you have not obeyed it.

*"For the time is come that judgment must begin at the house of God: and if it first begin at us, what shall the end be of them **that obey not the gospel** of God?"* (1Peter4:17) L

*"In flaming fire taking vengeance on them that **__know not God__**, and **that obey not the gospel** of our Lord Jesus Christ*"L (2Thessalonians1:8)

If you have **believed the gospel**, then you have "***Obeyed the gospel***".

*"But they have **not all obeyed the gospel**. For Esaias saith, Lord, who hath **__believed__** our **report**?"* (Romans 10:16)

He that believes is not condemned and he that does not believe is already condemned because he has not believed yet. He is in a state of rebellion against God.

*"He that **believeth** on him is not condemned: but he that **believeth __not__** is condemned already, __because__ **he hath not believed** in the name of the only begotten Son of God."* (John 3:18)

Just Faith!

2.9 Do It Now! Tell God That You Believe That Jesus Paid It All!!

If you haven't believed yet, don't put it off any longer! Just talk to the true and living God in the quietness of your own heart. In your own thoughts, talk to God. He is listening! Start out by admitting to God that you are a sinner. He already knows it but you need to admit that you are guilty and accept his everlasting and eternal forgiveness forever. Just tell God that you are putting your trust in what Jesus did for you on the Cross of Calvary 2,000 years ago. Next, why don't you just thank him for what he has done for you, when he died for all of your sins? Ask for God's protection and pray for a good doctrinally sound Church to join. Pray for good Godly friends, for strong believers to fellowship with. Ask God to protect and nurture you so that you can grow as a new believer.

Jesus said that you receive Eternal Life the **moment** that you **believe** on him. **Eternal Life is eternal**, that means that you can't lose it. Notice that Jesus said that **you __receive__ it __when__ you __believe__ it! Receive when you Believe!**

You can know you **__have, Eternal__** Life the moment you ***__believe__***.

God says that we are Sealed with the Holy Spirit when we understand and **believe** the essential elements of the Gospel. (John 3:16)

*"In whom ye also **trusted**, after that ye heard the **word of truth**, the **gospel of your salvation**: in whom also after that **ye __believed__, ye were sealed** with that Holy Spirit of Promise"*, (Ephesians 1:13)

The Bible says that believers are sealed with the Holy Spirit of Promise until the day of redemption! The day of redemption is the Rapture of the Church.

"And grieve not The Holy Spirit of God, whereby ye are sealed unto the day of redemption". (Ephesians 4:30)

God gives us a new birth with a new nature that will live forever, but the old nature is still with us and it will continue to be with us until either we die or we get Raptured out just before the Tribulation begins. The **__old__ nature** is in conflict with the **__new__ nature.** The nature that you feed and exercise is the one that will dominate your life and produce it's corresponding fruit.

"Behold**, I shew you a mystery; We shall not all sleep**, but we shall all be **changed**, (1Corinthians 15:51)

"***In a moment****, in the* ***twinkling of an eye****, at the* ***__last trump__****: for the trumpet shall sound, and the dead shall be raised incorruptible, and* ***we shall be changed***" (1Corinthians 15:52).

This passage is describing an event that will occur at the end of the Feast of Trumpets on some year in the near future. It is called the Rapture of the Church. After that I believe that the next thing that we will be seeing will be the Judgement Seat of Christ. That is when all believers in the Church will be judged to determine their degree of reward in Heaven. Eternal Life is Free, but God offers rewards for obedience and service! This is when believers who served will get rewarded.

2.10 God Disciplines His Children So We Will Bear Fruit

Although Eternal Life is free and you can't lose it, you can lose peace, joy and happiness. You can lose rewards and if you disobey God he will get involved in your life in ways that will not be pleasant. But God loves his children and he disciplines them when they need it, and for their benefit.

*"For whom the Lord loveth he chasteneth, and scourgeth **every** son whom he receiveth".* (Hebrews 12:6)

"***I will be his father****, and he shall be* ***my son****. If he commit iniquity,* ***I will chasten him*** *with the rod of men, and with the stripes of the children of men:* (2Samuel 7:14)

Seven (7) commands for believers.

God commands Believers to ***Pray; Study***; ***Fellowship*** with other believers; we are also told to **Memorize**, ***Meditate***, ***rightly divide***, and **share** God's word with others.

We are told to be ready always with an answer.

2.11 Chastening, Rewards And The Judgement Seat Of Christ

"Therefore we are always confident, knowing that, whilst we are at home in the body, we are absent from the Lord:" (2Corinthians 5:6)

(For we walk by faith, not by sight:) (2Corinthians 5:7)

"We are confident, *I say,* and willing rather to be absent from the body, and to be present with the Lord". (2Corinthians 5:8)

"Wherefore we labour, that, whether present or absent, we may be accepted of him". (2Corinthians 5:9)

"For we must all appear before the judgment seat of Christ; *that every one may receive the things done in his body, according to that he hath done, whether it be good or bad"*. (2Corinthians 5:10)

"Knowing therefore the ***terror of the Lord,*** *we persuade men; but we are made manifest unto God; and I trust also are made manifest in your consciences* (2Corinthians 5:11)

This passage is talking about the Judgement seat of Christ when the Church will be judged.

"For we commend not ourselves again unto you, but give you occasion to glory on our behalf, ***that ye may have somewhat to answer them which glory in appearance, and not in heart"*** (2Corinthians 5:12)

People who are eye pleasers will have their true motives brought to light.

2.12 Many Believers Will Not Produce Any Fruit At All.

A Person with "dead faith" is still saved! Outwardly they may not look like it but inwardly they have a New Birth! That is the New Nature that will live forever. The works produced by the New Nature will live on forever! The New Nature cannot sin because it is the sinless Nature of God that will live on forever!

The old nature is still there in a Believers life until death or The Rapture. The Old Nature doesn't just die and go away after someone Believes!

The Holy Spirit spoke through the Apostle James about dealing with this Old Nature after one has Believed. The Book of James is talking to Believers who have already been saved. I don't see anything in this Book that is talking to unbelievers. If someone finds a place in the Book of James that is talking to the lost, unbelievers please bring it to my attention. (Endtimesbiblehandbook.org) Well if it isn't talking to Lost Unbelievers about how to get saved, who is it talking to?

The Book of James refers to "brethren" or "brothers" nineteen times in 5 chapters. If you have not been Genuinely ***born again*** then the Book of James is not for you.

A Believer with dead faith and no fruit is the subject of chapter 2. **The New Birth** occurs when a person **believes** God's record, in their heart or mind. Before anyone else has a chance to find out, **<u>you know</u>** that you are saved. It is over and done with in an instant.

<u>Before</u> a person has a chance to tell anyone or say anything to anyone about their salvation experience, they are **already saved!** People get saved before they have a chance to tell anyone, confession has nothing to do with getting saved or staying saved! You are saved at the **instant** that you believe.

We all became "***children of God"*** the moment that we **believed** the Gospel!

"*For ye are all the **<u>children of God</u>** by **<u>faith</u>** in Christ Jesus."* (Galatians 3:26)

I know plenty of people who never indicated to anyone that they believed, but much later it is discovered. I know of believers who denied that they were believers, even denied Christ, but yet they really did believe. I know of plenty of people like that, I was one of them for a while … **but I was still saved!** God was still my Father and I was still his son and he disciplined me severely!

If there was a sign of the Believer, it could only be **The Sign of Chastening**. The **"<u>sign of chastening</u>"** is only to the individual believer about him or herself. "The sign of chastening" is not intended to be a sign to anyone other than the one it is directed toward. That would be called the **"Sign of a Believer"! In the book of Hebrews, chapter 12, the Bible says…**

"For whom the Lord loveth he chasteneth, and scourgeth ***every son*** *whom he receiveth"* (Hebrews 12:6)

God says that **every Son** is chastised.

. ***"for what son is he whom the father chasteneth not?*** (Hebrews 12:7)

"But if ye be ***without chastisement****, whereof* ***all*** *are partakers, then are ye bastards, and* ***not sons****."* (Hebrews 12:8)

If you are a believer, God is going to get involved in your life, he says so right here! This is a warning to new believers. God chastises his children and punishes disobedient sons!

2.13 Earn Eternal Rewards!

Although Eternal Life is a Gift and it is free, God offers rewards to those who will serve him. This is where God is looking for works and he is willing to pay handsomely for it. Believers can earn rewards in Heaven by their obedience on earth. But their failure to do service will in no way jeopardize their salvation! Salvation is a Gift, it is free and it is Eternal! You get it when you believe and it can't be changed!

Once you believe, you have the privilege to serve God and earn rewards! Earning rewards as a Believer is not the same as working for your salvation. You can't work for your salvation, it can't be earned. If you try to serve to be saved, you will be lost. Salvation and Service are opposites, they are separate, not the same. **Salvation is a Gift**, no strings attached it is **free**, but service requires work. The more you work, the more God pays. But God only pays believers. First **you must be "*Born Again*"! The New Birth is the foundation that you build on.** How a Believer spends their time since they believed is how God will reward his Children.

"*For other **foundation** can no man lay than that is laid, which is Jesus Christ*". (1Corinthians 3:11)

The ***foundation*** spoken of here is our saving faith in Jesus Christ and the New Birth. This is a list of six kinds of building materials to represent the different kinds of works that a believer can have. The **first three are valuable** and are not damaged by fire. Gold and silver are purified with fire. The **last three kinds** of building material are basically **garbage** and are consumed by fire. There will be nothing left of those sorts of works.

"*Now if any man build upon this foundation **gold, silver, precious stones, wood, hay, stubble***" (1Corinthians 3:12)

*"Every man's work shall be made manifest: for the day shall declare it, because it shall be revealed by fire; and the fire shall try every man's work of **what sort it is***". (1Corinthians 3:13)

"*If any man's work abide which he hath built thereupon, **he shall receive a reward***". (1Corinthians 3:14)

When a born again believer invests his or her time, talents and treasures for the Lord then he or she will receive a handsome, eternal reward!

"***If any man's work shall be burned**, he shall suffer loss: but **he himself shall be saved**; yet so as by fire*". (1Corinthians 3:15)

It says here that even if all of your works are burned, you will suffer loss of reward but you will still be saved. The foundation of Christ will remain!

The "day of redemption" is when all of the Church is in Heaven together for the first time. Everyone is present and about to be accounted for! The first thing on the agenda for Raptured Believers is to get judged and receive a new body. It will last forever and never become old or tired!

Getting used to a new body with new powers and abilities should be interesting. Scripture indicates that our bodies will be similar to Christ's resurrection body. I assume that we may be able to do some of the things that he did or will do. The Bible says that God will give us a body as it pleases him. That is part of our reward. The Resurrected Believers New body is part of his/her eternal reward and so is a new home in Heaven! How well you are rewarded in the areas of physical or spiritual powers or attributes of your new body along with your eternal accommodations aboard the New Jerusalem, would I believe, be directly proportionate to how well you have pleased your Savior.

We will live in **The New Jerusalem** which measures 15,000 miles on every side. A cube that size is a little bigger than our own moon.

The whole world including the unbelievers will see this giant cube that produces its own special light coming from space toward the Earth.

I would expect for the New Jerusalem to orbit our planet in its own special orbit. I believe that the New Jerusalem will remain parked in that orbit for the entire Millennial Reign of Christ. After that, it will leave and take us along to see new worlds.

We will have new abilities including a perfect memory. Old memories won't go away during that thousand years! After the thousand years are finished, we will be glad to see old memories erased! At the end of the Millennium, Jesus and his New Jerusalem will pull away from this earth to head for a New Earth with no sea and a New Heaven. That is when all of the uncomfortable memories of former things will be forgotten.

2.14 "The Fruit Of The Righteous Is A Tree Of Life"...

"The heart of the righteous studieth to answer". (Proverbs 15:28)

If we are faithful with the truth that God gives us, he will give us more. God commands us to bear fruit that will endure for eternity. He will reward us if we do and discipline us like sons if we don't.

"*The **fruit of the righteous** is a tree of life; and he that winneth **souls** is wise*". (Proverbs 11:30)

"And they that be wise shall shine as the brightness of the firmament; and ***they that turn many to righteousness as the stars for ever and ever".*** (Daniel 12:3)

*"**Herein is my Father glorified**, that ye bear **much fruit**; so shall ye be my disciples".* (John 15:8)

It says here that God is glorified when Believers bear **much Fruit.** The fruit of believers is new believers. The more believers that are won, the more God is Glorified! It is that simple! Reproduce the Faith!

Notice that it doesn't say anything here about God being glorified by someone singing praises or by acting like a rock n roll star on stage while singing about "Jesus". It says that God is glorified every time a New Believer is born and God commands us to win new believers. It is that simple.

*"Now I would not have you ignorant, brethren, that oftentimes I purposed to come unto you, (but was let hitherto,) that I might have some **fruit** among you also, even as among other Gentiles".* (Romans 1:13)

The **fruit of a believer is another believer**. It is that simple.

In the parable of the sower, it is clear that the fruit of the believer is more believers…

… "But he that received the seed into stony places, the same is he that heareth the word, and anon with joy receiveth it;

Yet hath he not root in himself, but dureth for a while: for when tribulation or persecution ariseth because of the word, by and by he is offended" (Matthew 13:20)

"Yet hath he not root in himself, but dureth for a while: for when tribulation or persecution ariseth because of the word, by and by he is offended" (Matthew 13:21)

"He also that received seed among the thorns is he that heareth the word; and the care of this world, and the deceitfulness of riches, choke the word, and he becometh unfruitful". (Matthew 13:22)

Not all believers produce fruit.

"If any man's work shall be burned, he shall suffer loss: but he himself shall be saved; yet so as by fire". (1Corinthians 3:15)

2.15 "The Fruit Of The Spirit Is Love, Joy, Peace...

The fruit of the Spirit should not be confused with Fruit of the Believer. If the Holy Spirit does not have fruit in your life as a believer, then you will not produce any fruit of your own. The Holy Spirit has his own fruit.

"But the fruit of the Spirit is love, joy, peace, longsuffering, Gentleness, goodness, faith", meekness, temperance: against such there is no law. (Galatians 5: 22, 23)

If the Believer has the fruits of the Spirit in their life, then the Believer will also have his own fruit, which is other believers.

The fruit of the believer is other believers, it is a Tree of Life. (Proverbs 11:30). The fruits of the Spirit is listed above. They are not the same as the believer's fruit, they are different. The Spirit must first have fruit in the life of the believer, for the believer to produce fruit of his own.

2.16 "Separated Unto The Gospel Of God"

Paul said that he was ***"separated unto the gospel of God"***. (Romans 1:1)

There are many gospels to choose from but there is only one gospel that saves! We are told to separate ourselves from all false teaching and distinguish ourselves by our accurate, clear and simple gospel message!

"*I marvel that ye are so soon removed from him that called you into the grace of Christ unto **another gospel***" (Galatians 1:6)

"*Which is not another; but there be some that trouble you, and **would pervert the gospel of Christ***". (Galatians 1:7)

"*But though we, or an angel from heaven, preach **any other gospel** unto you than that which we have preached unto you, **let him be accursed***" (Galatians 1:8).

False teachers deliberately look for ways to conceal what they really believe. Often they first want to find out what you believe and pretend to agree. After they have gained your confidence and befriended you, they gradually and subtlety add their false teachings for their own purposes. "Fruit" examination according to scripture means to inquire about someone's beliefs (doctrine). Learn how to apply the grace test. More about that later.

"*Preach the word; be instant in season, out of season; **reprove**, **rebuke**, **exhort** with all longsuffering and **doctrine***" (2Timothy 4:2)

"*For **the time will come** when they will not endure sound doctrine; but after their own lusts shall they heap to themselves teachers, having **itching ears***"; (2Timothy 4:3)

"*And they shall turn away their ears from the truth, and shall be turned unto fables*" (2Timothy 4:4)

"*But **watch** thou in all things, endure afflictions, do the work of an evangelist, make full proof of thy ministry*" (2Timothy 4:5)

Teachers with "***itching ears***" are listening to find out what you want them to say. When they know what you want to hear, that is what they tell you to gain your confidence and support. After that is accomplished, then they look for opportunities to slip in what they want to teach for their own purposes which ultimately gives them control. Ultimately, a false gospel of works and self-righteousness are tools to raise the dough or money.

"*Therefore come out from among them, and be ye separate, saith the Lord, and touch not the unclean thing; and I will receive you*",(2 Corinthians 6:17)

"*Enter ye in at the strait gate: for wide is the gate, and broad is the way, that leadeth to destruction, and* ***many*** *there be which go in thereat*": (Matthew 7:13)

"*Because strait is the gate, and* ***narrow is the way****, which leadeth unto life, and* ***few*** *there be that find it*". (Matthew 7:14)

This passage is saying that there are **many** false gospels, therefore **many** will be lost. Furthermore, there is only one way that leads to eternal life and **few** find it. According to that passage many are the majority that choose the wrong way and **few are the minority** which trust in Christ alone!

"*Jesus saith unto him,* ***I am the way****, the truth, and the life: no man cometh unto the Father, but by me*" (John 14:6)

This passage said that many take the wrong path because there are many ways to choose from but only one way will save.

(John 10:1) "*Verily, verily, I say unto you, He that entereth not by the door into the sheepfold, but climbeth up some other way, the same is a thief and a robber*"

Jesus alone, is the way! Jesus plus nothing equals grace which equals eternal life! Jesus plus anything equals works and does not equal grace. Works will not save and trusting in them will condemn you!

Many is the majority and few is the minority.

"***Many*** *will say to me in that day, Lord, Lord, have we not prophesied in thy name? and in thy name have cast out devils? and in thy name done many wonderful works?*"(Matthew 7:22)

"*And then will I profess unto them,* ***I never knew you****: depart from me, ye that work iniquity.*" (Matthew 7:23)

God says that he **never** knew them even though they preached in his name. Notice that these false teachers call Jesus **"Lord" twice**! This sounds like "Lordship salvation"! Furthermore, he calls their "good works" iniquity. That sounds a lot like Isaiah 64:6

"*But we are all as an unclean thing, and **all our righteousnesses are as filthy rags***" (Isaiah 64:6)

This is what God is telling us to watch for in the last days. That is the majority of so called "Christianity" in the last days. Watch out for deception and false doctrine!

"*And Jesus answered and said unto them, Take heed that no man deceive you*" (Matthew 24:4)

Jesus is answering the Apostles questions about the last days and the first thing he says is ***…"Take heed that no man deceive you"…***

"*For **many** shall come in my name, saying, I am Christ; and shall deceive **many***". (Matthew 24:5)

Jesus is doing the talking here. He is saying that in the last days false teachers will come in his name. Furthermore, they will even admit that Jesus is the Christ or Messiah. In other words, they will say that they are "Christian". Thereby gaining peoples trust so that they can deceive them.

False teachers don't have 666 tattooed on their forehead. They are not wearing a tee-shirt that says "False Teacher" on it. They want to fit in with Christians, look like Christians, and talk like Christians and pretend to be Christians. That is the evil nature of deception. They twist scripture and teach and do things contrary to Scripture to satisfy their own lusts. (2 Peter 2:17, 18) Whatever that may be.

"Narrow is the way" because there is only one way. Jesus said that He is the only way…

"Jesus saith unto him, I am ***the way****, the truth, and the life: no man cometh unto the Father, but by me."* (John 14:6)

"Many" will be left behind at the Rapture who called themselves "Christians" but were not saved because they rejected the truth and chose a counterfeit of the Truth.

They were trusting in their own efforts to save them. Calling themselves "Christians" and playing church along the way, from time to time to maintain outward appearances. Hopefully they will somehow get saved during the Tribulation. There will be greater deception during that time.

2.17 Jesus Is Coming Back!

Jesus returns for the Church to mark the end of the Church age at the near future event called the Rapture. Jesus is only seen and heard by the true believers.

"After this I looked, and, behold, a door was opened in heaven: and the first voice which I heard was as it were of a ***trumpet*** *talking with me; which said, Come up hither,"* (Revelation 4:1)

*"**And <u>immediately</u> I was in the spirit:** and, behold, a throne was set in heaven, and one sat on the throne" (Revelation 4:2)*

As believers, the next thing that we will see is the Judgement Seat of Christ where we will receive our rewards for anything that we have done for Christ. That is the only thing that will be rewarded.

"Let not your heart be troubled: ye believe in God, believe also in me". (John 14:1)

"In my Father's house are many mansions: if it were not so, I would have told you. I go to prepare a place for you". (John 14:2)

"And if I go and prepare a place for you, ***I will come again, and receive you unto myself; that where I am, there ye may be also"****.* (John 14:3)

The Rapture is not explained here, but this is talking about the Rapture. This passage gets explained in 59A.D. by the Apostle Paul in (1Corinthians 15:51-58) (ETBH chapters 33-40)

2.18 "In A Moment, In The Twinkling Of An Eye"

"Behold, I shew you a mystery; We shall not all sleep, but we shall all be changed", (1 Corinthians 15:51)

Ok, this is a mystery being revealed for the very first time. In other words, God hasn't discussed this with any prophets until the time of this writing in A.D. 59. This *"mystery"* involves a future Generation of believers that will never die. The Apostle Paul revealed this for the very first time in A.D. 59. Nobody knew about it until then.

"In a moment, in the twinkling of an eye, at the last trump: for the trumpet shall sound, and the dead shall be raised incorruptible, and ***we shall be changed****"*. (1Corinthians 15:52)

"For this corruptible must put on incorruption, and this mortal must put on immortality". (1 Corinthians 15:53)

Notice that this event will be a changing from mortal believers to immortal believers without experiencing physical death. All Church Age believers will receive their new bodies at this time.

This event occurs faster than a moment. It occurs in the very *"twinkling of an eye"*. I take that to mean the time that it takes for light to reflect off a moving eyeball. About the speed of light or at least beyond our ability to see and notice. Instantaneous!

This has never happened before except for Enoch.

*"And all the days of Enoch were **three hundred sixty and five years**:"* (Genesis 5:23)

*"And Enoch walked with God: and he was not; for God **took** him"* (Genesis 5:24)

The Hebrew word used here for "***took***" is equivalent in meaning to the Greek word "harpazo" used in (1Thessalonians 4:17)

This event is called the Rapture and it is near. (See Chapters 31-40)

Whatever year it happens, the Rapture will occur *"at the sound of the last trump"* of the fifth feast. The fifth feast is called the Feast of Trumpets or Rosh Hashanah. The "last trump" of Rosh Hashanah occurs at the end of the feast at sundown.

The Rapture marks the end of the Church Age and the beginning of the "Day of the Lord". The Day of the Lord should not be confused with the 70th week of Daniel. The Day of the Lord begins at the Rapture and includes the 70th week of Daniel and the time till the end of the 1,000 year Millennium. However, the 70th week of Daniel doesn't begin until the end of Yom Kippur, 10 days after Rosh Hashanah begins or at the end of the "**ten days of awe**".

Jews observe the Ten Days of Awe from the beginning of Rosh Hashanah until sundown at the end of Yom Kippur. I believe that whatever year it is that the rapture happens this ten days will be used by the false prophet and the antichrist to deceive the world with miracles, signs and wonders when the Jews begin rebuilding their Temple.

2.19 The Stage Is Set.

"A man may imagine things that are false, but he can only understand things that are true." **(Sir Isaac Newton)**

Our current Technology, Geopolitical Alliances and world population numbers did not even seem possible to those living 2,000 years ago. Yet the Bible accurately describes the times we are living in. The technologies and events that characterize the times that we are living in right now are described in the Bible thousands of years ago!

Throughout the Ages many have scorned the Bible, saying that the things that it spoke of in the latter days were **impossible fantasies** and could never happen.

Yet today we realize that the world stage is now set for those events described in the Bible thousands of years ago to occur in the near future.

Israel is back in their land on the very next day after their national punishment was completed. The national punishment that was prophesied by Moses in (Leviticus 26) and completed on May 13, 1948 is discussed in (Chapters 30 & 31)

The Catholic Church did not believe the Bible prophecies about Israel returning back to their land in the Last Days. The Catholic Church decided that all of those land promises that God made to Abraham were just going to waist, so the "infallible" Pope claimed the land for himself. Nobody ever thought that it could happen, but Israel is back in their land. Israel returned to their land on the day that God said they would thousands of years in advance!

Current human technology and population numbers can produce events that were described in the Bible thousands of years ago. **Until recently, Bible prophecy seemed to be impossible!** Today even unbelievers can look at current End Time Bible Prophecy and recognize the probability of such events actually happening in our lifetime.

Two **thousand** years ago, men could not imagine a single army that could consist of 200,000,000 Soldiers. Such a thing was unthinkable two **thousand** years ago when the Bible recorded that prophecy. **Two thousand years ago,** there wasn't half that many people on the whole planet!

Since the time those passages were written 2,000 years ago, there have been scoffers like Voltaire. However, sence the 60's the Chinese Army has boasted to have a 200,000,000 million man army. **That is the very same number used in the Bible 2,000 years ago** to describe the army from the East (China), that crosses the dried up Euphrates river to fight against the Antichrist at Armageddon. (Revelation 9: 14-16 & 16:12)

I recently read an article about how many of the eight, major, hydro-electric dams on the river Euphrates were currently under some sort of rebel control. George Soros says that he has "little trusts" set up for rebel groups. Some rebels want to control the water in the Middle-East. Whoever controls the water controls the area. Water is also a source of hydro-electric power. When you control the dams, you control the electricity and the water and the people. I recently came across this article about the Tigris and Euphrates rivers…

Isis Islamic rebels now control most of the key upper reaches of the Tigris and Euphrates; According to the Guardian; Wednesday, July 2, 2014,

Never before, in human history has the possibility existed of the Euphrates River drying up to prepare the way for a 200,000,000 man army, but it does now!

Today End Times Bible Prophecies not only seem to be possible but they appear to be probable. We are definitely heading in the direction of what the Bible has described thousands of years ago.

The Septuagint was translated about 400 B.C. At that time the Hebrew Scriptures were already considered to be ancient. Yet people are still ignoring the evidence, still skeptical and still scorning God. They say "where is the evidence"? Yet they ignore the evidence! People want to argue about anything and ignore what God has said!

God has given extraordinary warnings and provided us with extraordinary evidence and extraordinary signs to match. Evidence that guys like Physicist Michio Kaku and Stephen Hawkins seemed to ignore!

I don't believe that an honest Skeptic can look at the evidence and still be both. Either you will cease from being a skeptic or you are not being honest. The evidence is overwhelming and confirms what the Bible has said; that is why this book was written.

2.20 The 70th Week Of Daniel Is At Hand!

The Church Age will end at the end of the 5th Feast called Rosh Hashanah, at sundown. The 70th week of Daniel will begin at the end of the 6th Feast called Yom Kippur at sundown. The 70th Week of Daniel begins with "***sudden destruction***". (1Thessalonians 5:3)

The Rapture will mark the end of the Church Age and the beginning of the "70th Week of Daniel" which is also known as the Tribulation.

God will send his two witnesses, **Moses and Elijah** to Israel and they will do real miracles.

*"And I will give power unto my **two witnesses**, and they shall prophesy a thousand two hundred and threescore days, clothed in sackcloth.* (Revelation 11:3)

A *"thousand two hundred and threescore days"* is exactly **3 ½ years.**

There will be **28 Judgements** of God spread out over **a period of 7 years**. The **first four judgements**, also known as the "**Four Horsemen of the Apocalypse"** cover the first 3 ½ years of the Tribulation and **wipe out over ¼ of the World's Population**! The **Four Horseman** are also considered to be the first **"Four Seal Judgements".** During the tribulation, the Antichrist will take over the Middle-East.

Jerusalem is betrayed and taken over in **the middle of the Tribulation**. That is the event that Daniel calls the **"Abomination of Desolations". That is when the Jews in Jerusalem are told to run to Petra to escape the Antichrist.** Two thirds of the Jews in Israel will be killed, nine tenths worldwide!

The same thing will happen to the Catholics in the middle of the Tribulation. After making a deal with the Catholics which will put them in control of a one world Church, the Antichrist will betray the Catholics as well. So in the Middle of the Tribulation, the Antichrist will take over Jerusalem and the Temple Site. **The Antichrist will claim to be God** and will demand to be worshiped. This is the 5th seal judgement in the middle of the Tribulation. This is when the killing, torture and terror gets much worse!

The "Tribulation" will last for 7 Hebrew years which consist of 360 days each. The Tribulation is also called the 70th week of Daniel. The 70th Week of Daniel is divided into two half's. The first half is called "*the beginning of sorrows*" and is 3 ½ Hebrew years or 42 months long. The second half of the Tribulation is called "***great tribulation***". That means that things get much worse during the last half of the Tribulation!

During the first half of the Tribulation, twenty five percent of the world's population will be slaughtered by the four horseman. Current population estimates indicate a world population of about 7.8 billion. One quarter of that is about 1.95 billion, which comes out to be about one and a half million deaths per day, every day for three and a half years. I believe that a disproportionately large portion of that number will come from the United States and Europe. After that, things get much worse! The second half of the Tribulation is called "***great tribulation***".

"For then shall be ***great tribulation****, such as was not since the beginning of the world to this time, no, nor ever shall be"* (Matthew 24:21

The second half of the tribulation is when satan is in control, but God fights back with seven Thunder Judgements; seven Trumpet Judgements and seven Bowel Judgements.

3

A Time Of Evil, Deception, Betrayal, Terror, And Genocide!

3.1 Deception!

The Bible warns us about people who will say that they are Christian and "***deceive many"***. (Matthew 24:5) People will smile and pretend to be a "Christian" to gain people's trust so they can use them or gain their vote. It is all about manipulating people for personal gain. God calls them ***"Wolves in sheep's clothing"***

I remember when I saw Presidential candidate George W. Bush look into the camera while apparently struggling to maintain a straight face and saying that he had a personal relationship with Jesus Christ.

At the time I knew that he, his father, George H.W. Bush, and grandfather Prescott Bush were all members of an occult secret society at Yale University called the Skull and Bones Society. Of course, I never considered him to be a believer, but today many people still do. Christians and Muslims consider Bush to be a "Christian" and leader of a "Christian Nation". I don't believe that he is on God's side at all! He has set up America for a big fall!

The skull and bones emblem has been the sign for assassins for centuries. It is also the emblem for Yale's fraternal Order of the Skull and Bones Society in America. During the Nazi rule of Germany during World War 2, it was the emblem for the Nazi SS. There seems to be some connections between these two organizations.

Russ Baker, an investigative journalist documents the alarming history and background of the Bush Family in his book, "Family of Secrets".

Prescott Bush had financial ties to the Nazi's before, during and since World War 2. Both the Nazi S.S. and the Skull and Bones society at Yale have occult beliefs, practices, emblems, oaths and rituals.

George H.W. Bush, Prescott's son, secretly worked for the C.I.A. and headed a team of assassins during the time of the Bay of Pigs. According to witnesses, Bush and his team were in Dallas at the time of the Kennedy assassination, which he denied. When asked, Bush said that he could not remember where he was on November 22, 1963 and denied involvement with the CIA at the time.

Bush ran the C.I.A. for over a decade. He started wars with false pretenses and took the oil revenue. He sold drugs for guns to presumably supply the Contras in South America.

George H.W. Bush married Barbra Pearce, the third child of her mother, Pauline Pierce. Who was one of four persons who were closely and personally associated with Aleister Crowley in 1924. when he underwent a pagan sex ritual that requires four assistances. Eight months later Barbra was born.

There are some who claim that Aleister Crowley is the father of Barbra Bush. Whether or not he is her father, the fact that he could be and the fact that Pauline Pierce was associated with that sort of thing is both disturbing and enlightening about both sides of the family.

His son **George W. Bush** was President during the nine eleven attacks. Many people are beginning to doubt or just disbelieve the media and the government story about the 9/11 tragedy. A retired army intelligence officer by the name of Major General Albert Newton Stubblebine 3rd, U.S. Army retired, is one of many experts speaking out about the 9/11 deception. He has prepared a video on YouTube. Here is a link:

https://www.facebook.com/watch/?v=669117876801651

In the minds of the Muslims, George Bush is the Christian leader of the Christian Nation that has been bombing them into the ground and shooting them from the skies. We started the Middle-East conflict on the basis of lies used by both Bush Administrations to promote the establishment of the N.W.O.

We could have been successful in the Middle-East, but our soldiers were betrayed by their leaders. Those same leaders have been allowing and encouraging terrorists to enter our country illegally.

We haven't accomplished anything in the Middle-East accept for making enemies; destabilizing their governments; creating and supplying I.S.I.S.; stirring-up hatred against the West and creating untraceable oil revenue for the C.I.A. to use to create and supply more international terrorism.

"Ordo Ab Chao" is the Masonic doctrine for establishing the New World Order that George H.W. Bush publicly mentioned in his speeches more than two hundred times. I think the plan is to establish the N.W.O. in the middle of all of the chaos they create. That is what the C.I.A. has been doing all over the mid-east and South America.

Our government secretly organized ISIS and gave them money and weapons to attack defenseless civilians. We pulled out and left our friends who trusted us behind. We promised to protect those who worked with us, but they became the first victims of ISIS when we left. They were betrayed just like in Benghazi and Libya!

Pictures of Muslim prisoners being abused by what they believe are "Christians" in The Guantanamo Bay Detention Camp have been circulated all over the Middle East. This happened during the Bush Jr. administration. So now, according to their religion they are required to return the evil. I believe that Obama let them into this country just for that purpose.

America has been set-up and betrayed by its leaders! Now Bush is changing his story about Jesus. Now Bush is saying that we all worship the same god. He was referring to Alah. That is what Muslims say and that is what the

Free Masons say as well. Billy Graham said the same thing in interviews with Larry King and on other occasions as well. Some of these interviews are still available on YouTube.

Billy Graham was a 33rd degree Mason and a false teacher who misled anyone who listened to him. He preached works for salvation. He promoted ecumenicalism, the Catholic church and the masonic agenda because he was one of them. Billy Graham has misled "mainstream Christianity". Billy Graham has redefined Biblical terms and ignored Biblical definitions and context.

I learned about Free Masons because my father was a 32nd degree Mason and a member of the Egypt Temple. I remember a book I read, written by a former 33rd degree Mason who became saved and wanted to warn the brethren about the deception of Free Masonry. Hear what he said…

"The 33rd degree is an honorary degree bestowed upon especially worthy masons who have accomplished outstanding work in such fields as religion and politics."

In his book, the author described his initiation into the 33rd degree. He strongly implied by description and everything else **<u>except</u>** by name, that Billy Graham was among a number of world famous and influential 33rd degree Masons present at his initiation ceremony in Washington D.C.

Because of oaths taken and fear of retaliation, he said that he could not divulge the names of anyone present. Although his name was not mentioned, he was described so thoroughly that there was no question about who he was talking about. However some years later, as I was driving home to Florida through Charlotte, on the interstate, I noticed a very large sign on interstate 85 which read **"Exit <u>33</u>**, Billy Graham headquarters".

Billy Graham taught works for salvation and worked for a one world religion. After years of pretending that he believes just like we do, we find out that he doesn't. He was a deceiver like the Bible warned us about.

Our illegal Marxist/Muslim administration is covertly bringing into our country an army of radical Muslims to spread terror and anarchy into this once free Christian nation. The (N.W.O.) system has planned to take control of our government by martial law when we are in anarchy and that is where the Muslims come in.

They designed and planned this Anarchy to the very day. This Masonic doctrine is called **"Ordo Ab Chao"**, that means "Order out of Chaos". Google it and see what comes up. 9/11 was planned to justify an unjust war and create laws to lead to a totalitarian N. W. O. Check out the following links…

https://www.youtube.com/watch?v=l0Q5eZhCPuc

https://www.youtube.com/watch?v=CdE1Cwnymzc

Now ISIS is saying that they are coming over here after us! We created, trained and funded ISIS. They have murdered and tortured Christians and they are openly attacking Europeans in Europe. Americans and Europeans are being betrayed by their trusted leaders.

The Bush election was a fraud. Ballots were tampered with!

"The people who cast the votes decide nothing. The people who count the votes decide everything". — **(Joseph Stalin)**

Vote fraud is so common now that it is being done openly and nobody seems to care. During the first Obama election back in 2008, votes were reportedly going for as cheap as $10. People bragged on camera about how many times they voted and how much they made at $10 each.

The voter I.D. requirement in North Carolina has been challenged in court. Other states use electronic ballots without any means of verification or auditing or backup. Voter fraud is becoming easier.

With Obama we had a Marxist Muslim in the Whitehouse, and he is brought Terrorists into this country much faster than anyone wants to believe. I believe that the beheadings that we saw in the Middle-East and spoken of in Revelation are closer to home than anyone wants to admit. People just don't want to upset their comfort zone with unhappy thoughts. So they turn on the TV, listen to Joel Osteen and trust the media instead of facing the truth. They are being misled and betrayed like dumb sheep being led to the slaughter!

"*I saw the souls of them that were* ***beheaded*** *for the witness of Jesus, and for the word of God, and which had not worshipped the beast, neither his image, neither had received his mark upon their foreheads, or in their hands*"... (Revelation 20: 4)

3.2 Satan Blinds People's Minds With Lies.

It is important that believers get their stories straight about what the Bible says! God bases his creditability to man on the accuracy of his record about everything starting with... ***"in the Beginning"*** and on into the future! God's word is important and should be reverenced, not understated; not overstated; not misrepresented or left out! That means rightly dividing his word so that believers can all speak the same message and not be spreaders of confusion! Satan is the author of confusion, don't support his efforts!

"*For* ***God is <u>not</u> the author of confusion****, but of peace, as in all churches of the saints*". (Corinthians 14:44)

"*Finally,* ***be ye all of <u>one mind</u>****, having compassion one of another, love as brethren, be pitiful, be courteous:*" (1Peter 3:8)

"*Fulfil ye my joy, that ye be* ***<u>likeminded</u>****, having the same love,* ***being of <u>one accord</u>, of <u>one mind</u>****.*" (Philippians 2:2) "*Only let your conversation be as it becometh the gospel of Christ: that whether I come and see you, or else be absent, I may hear of your affairs, that* ***ye stand fast in <u>one spirit</u>****, with* ***<u>one mind</u> striving <u>together</u>*** *for the faith of the gospel*"(Philippians 1:27)

"*Finally, brethren, farewell. Be perfect, be of good comfort, be **of one mind**, live in peace; and the God of love and peace shall be with you*" (2Corinthians 13:11)

"*That ye may with **one mind** and **one mouth** glorify God, even the Father of our Lord Jesus Christ.*" (Romans 15:6)

"*That we henceforth be no more children, tossed to and fro, and carried about with every wind of doctrine, by the sleight of men, and cunning craftiness, whereby they lie in wait to deceive;*" (Ephesians 4:14)

If you don't have the story of Creation right, the Bible says you are "***willingly** ignorant*". (2Peter 3:5-7)

Joe Atheist is operating on misinformation! Satan sets up false flags and has people reacting to them.

The Bible is not a religious book, it claims to be the word of God and that makes it science! It is mathematically accurate, scientifically accurate, historically accurate, and prophetically accurate. We are living at a time when we can see prophecy being fulfilled daily.

"*Behold, the former things are come to pass, and **new things** do I declare: **before** they spring forth I tell you of them*". (Isaiah 42:9)

The Bible has pre-told the future and we are seeing it happen on the news every day.

3.3 Erasmus Darwin, Occultism And Secret Societies

Erasmus Darwin Lived from December 12, 1731 to April 18, 1802 and was Charles Darwin's grandfather. He lived at the same time as **Carolus Linnaeus** (May 23, 1707– January 10, 1778) and was born just after **Sir Isaac Newton …**(1642-1727) …died**.**

Erasmus Darwin was a English poet, philosopher, botanist, inventor, physician, pantheist, pagan and A "Naturalist" and much published author.. He was a member of the Royal Society of England. He was politically well-connected at the highest levels in France, England and the newly forming U.S. just to mention a few.

Erasmus Darwin, was a 33rd degree Mason. He was the Master of the famous Canongate Kilwinning Masonic Lodge in Edinburgh, Scotland, who influenced both the American (1776) and French (1789-1799) Revolutions.

"The Temple of Nature", {1802} … Published posthumously by **Erasmus Darwin's** minions, was a collection of poems and miscellaneous writings by **Erasmus Darwin** during the French Revolution. It contained his version of Naturalism, biological Evolution and other miscellaneous subjects. He also authored "**Zoonomia**", or the Laws of Organic Life (1794), which was a renewal of the ancient pantheist belief that nature has creative powers to drive evolution.

Erasmus Darwin grew up in a post-Newton world with a reason explosion in progress! Newton's **"Mathematical Principles of <u>Natural</u> Philosophy"**, (Principia)1687 was one of the most important single works in the history of modern science and it was very popular during the time that Erasmus Darwin was growing up. You could say that Erasmus grew up in the middle of a "reason explosion" caused by Newton's "**Principia**".

Newton's famous book gave God the credit for all of the order in the Universe. Another famous scientist that gave God the credit is **Carolus Linnaeus.** He developed a system of studying, classifying and cataloguing all of God's creation.

"The *species* and the *genus* are always the work of **nature [i.e. specially created]**; the *variety* mostly that of circumstance; the *class* and the *order* are the work of nature and art." (Carolus Linnaeus)

Carolus Linnaeus also gave God the credit. **<u>"God created, Linnaeus organized"</u>** …was his slogan.

Erasmus was from a family of pagan/pantheists who were not friendly to Christian values. Pagan's found Christian rules against sexual immorality to be an attack on the Pagan lifestyle, which included incest, sodomy and Pagan sex rituals.

The Roman Catholic Church would kill Pagans if discovered. Since Roman times, pagans have continued their practices in small secret, networked groups around the world.

It seems like Erasmus Darwin was trying to redefine the term "natural" as it was used by Newton in his book **"Mathematical Principles of <u>Natural</u> Philosophy"**, (Principia)1687. Perhaps this was because Newton gave God the credit but Erasmus would not.

Newton equated "Natural" to the understanding that God Created the physical laws of order and that is what is "Natural". Through **<u>reason and prayer</u>** Newton sought to understand his Creator by understanding God's Creation. That did not sit well with Erasmus.

An acquaintance of Erasmus Darwin once said **…"Erasmus was strongly anti-Christian, and included Credulity, Superstitious Hope, and the Fear of Hell in his catalogue of diseases"**. (Unknown)

The pantheistic concept of "Nature" being god and having mystical creative powers is the essence of Erasmus Darwin's version of "Naturalism" and is implied in every aspect of modern "Evolutionary Theory". It is a very old idea that has nothing to do with real science. It is all about New Age Fantasy religion which is not "new", but is in fact, quite old.

The Darwin family was involved in secret societies and the occult for generations. These secret societies were generally pantheist/pagan in nature and loathed Biblical morality. They married into families with the same beliefs and were often related to their spouses. There was a lot of marriages between close relatives and even incest. Many different groups have continued to secretly meet during the full moons since Roman times. .

Erasmus was the founder of quite a few organizations, some may have been secret but some were not so much, such as **The Famous Lunar Society** that met on every full moon. Erasmus was founder and served as leader from 1755-1765 when the Society met in his home in Lichfield, England. Members insisted that they met on the full moon only for better lighting on the trip home.

No membership list has ever been found, only letters and other correspondence. **Benjamin Franklin**, **Thomas Jefferson** and a long list of Freemasons from America, France, England and Europe attended meetings. Another famous member of the Lunar Society was **Josiah Wedgwood I (1730-95)** who was a wealthy industrialist, inventor, pantheist/pagan/mason and grandfather of Charles Darwin's future wife and first cousin Emma.

The "members" called *themselves* "lunar-ticks" because they met on every full moon. They went on to become known as the "Luna Ticks". Benjamin Franklin and Thomas Jefferson both attended meetings and became leaders of the American Revolution who were among many "Luna Ticks" who were mentored by Erasmus Darwin. Likewise, the Lunar Society also influenced the French Revolution as well.

Erasmus Darwin was from a family of anti-Christian pagan/pantheists on all sides of the family. Erasmus's son Robert and famous grandson Charles Darwin were also high ranking Freemasons.

Charles Darwin's in-laws, the Wedgewoods were also members of the Lunar Society and were known to have practiced incest for generations. Incest, fornication, adultery and sodomy were not ever considered immoral by those who were secretly pagan, even today.

The Wedgewoods and the Darwins were related and Charles Darwins children suffered from genetic problems resulting from inbreeding. Charles Darwin married his mother's brother's daughter, or his 1st cousin on his mother's side. Her name was Emma Wedgwood (1808-1896).

It is interesting and ironic that Charles Darwin married into a family that he was related to and already known for incest going back for generations and yet he is arguing that random change can produce a genetic improvement. While the Darwins were preaching disorder to order, God demonstrates that the opposite is true. Things tend toward greater randomness, entropy!

The Wedgwood family did not regard the Judeo/Christian, Biblical teaching regarding inbreeding within families, neither did the Darwins. At the time, unbelieving pantheists/pagans thought nothing of marrying first cousins and sometimes even siblings. Nevertheless, out of the 10 children that Charles and Emma had together, 3 died as children. Of the 7 that reached adulthood, 3 were infertile and could not have children.. In his later years, Charles began to realize that inbreeding was a bad idea, just like the Bible said. But remember, that God said it first.

Another organization that Erasmus founded to assist in the dissemination of New Age Naturalism was the **Derby Philosophical Society** in (1783); this private organization consisted primarily of Freemasons, the "Lunar Ticks" and potential Masonic recruits. Their philosophy was Masonic/ New Age and Erasmus Darwin was their organizer, founder, leader and chief philosopher.

After his death, the **Derby Philosophical Society** continued to function and supported the promotion of evolution as a scientific theory. A century later, this organization would become one of the most passionate supporters of Charles Darwin's work.

Evolution is the foundational doctrine of the New Age Religion, they both came from Freemasonry which continues their support today.

In America, just prior to the War of Independence, Erasmus Darwin's minions established another organization that was patterned after the **Derby Philosophical Society** that Erasmus founded (1783) in England. It was called the **American Philosophical Society**. It was founded and led by members of the "Lunar Tic" Society and members of the Derby Philosophical Society in England. These groups consisted largely of Erasmus Darwin's minions and protégés.

Right after the American Revolution was over, the American Philosophical Society held a formal election and elected Erasmus Darwin, (an Englishman) as a member. This is evidence that Erasmus Darwin, an Englishmen, was in secret contact with American Freemasons and Revolutionary Rebels before, during and after the American Revolutionary War of Independence. This was Erasmus Darwin's first revolution.

If any Englishman was suspected of collaborating with the American Rebels during that time, they would be hanged. In the case of Erasmus, at least on one occasion where others were jailed for the same offense, Erasmus was overlooked by the crown because he was their physician.

3.4 The Rise Of The Cults In These "Last Days".

Most of the cults rose to prominence over the last 150 years or so. The "false tongues" movement is recent and demonic in nature and has spread like a cancer. Likewise, so have other forms of deception. A small sampling of some of the most influential cults that have risen over approximately the last 150 years are…

New Age Evolution(Erasmus Darwin), Watch Tower Track Society (Charles Taze Russel); Christian Science (Mary Baker Eddy); Mormonism (Joseph Smith); Scientology (L. Ron Hubbard) Just to mention a few.

Islam has also become more aggressive in the last fifty years. Islam is the only "religion" that I know of that requires it's followers to wage war with everyone that doesn't agree with the Koran. In fact, according to Islamic teaching in the Koran and the Hadiths, the only way to guarantee a place in "paradise" is to "kill or be killed" in the fight against infidels.

Christians and Jews are specifically mentioned in the Koran and the Hadiths as inferior, evil and worthy of subjugation, enslavement or death. "Painful chastisement" for Christians professing to believe as most Christians do. That means torture! The goal of Islam is to subjugate the entire world under Sharia law and make all unbelievers <u>feel</u> subdued. According to the Koran and the Hadiths it is the Muslim's duty to punish and terrorize non-Muslims. Some of the ways for Muslims to express their piety is deceiving, terrorizing and punishing non-Muslims.

The Koran says in chapter 9 verse 11…"god has indeed purchased from the believer their lives and wealth in exchange for Paradise. They fight in the cause of god and kill and or are killed. This is a true promise binding on him in the Torah, the Gospel, and the Quran. And whose promise is truer than god's? So rejoice in the exchange you have made with Him. That is truly" the ultimate triumph". **Quran 9:11**.

Do ya think that is the passage that the Muslim Jihadists were following on September 11, 2001?!! Unfortunately, the Bible warns about a seven year period where God will deal with mankind through twenty eight judgements. During the first three and a half years, also known as "*the beginning of sorrows*", twenty five percent of the world's population will be killed.

The last half or the last three and a half years is called the worst time the world has ever seen or ever will see. Billions will die, very few will survive! The Bible describes Islam taking over the world under the leadership of the Antichrist who is known to the Muslims as the twelfth Imam. Only the real Jesus can stop the Antichrist!

Satan is stepping up his attack on God's people and the truth that they bring, because he knows his time is about to run out. Of all these cults popping up in the last one hundred and fifty years, New Age Evolution is one of the most pervasive and far reaching challenges to God's word.

3.5 The Piltdown Hoax. (Dawson), The First "Missing Link,"

The first "missing link" was a hoax and so has everyone since! The missing Link" was claimed to be found by Charles Dawson, an attorney and wannabe anthropologist. For about fifty years the hoax was promoted by

Free Masons. Charles Dawson and those who participated in the Piltdown Hoax became rich and famous over night!

"But they that will be rich fall into temptation and a snare, and into many foolish and hurtful lusts, which drown men in destruction and perdition. (1 Timothy 6:9)

"For the love of money is the root of all evil: which while some coveted after, they have erred from the faith, and pierced themselves through with many sorrows". (1 Timothy6:10)

At a meeting of the Geological Society of London in December 1912, the fossil remains of what was claimed to be a new type of early human, *Eoanthropus Dawsoni* or 'Piltdown Man', were unveiled to the world. It was claimed by "experts" that irrefutable evidence had at last been found for the much sought after 'missing link' between man and ape.

It was not until the 1950s that Piltdown Man was proven to be a forgery. Members of the Natural History Museum (previously the British Museum (Natural History)), the Geological Society, and the British Geological Survey (previously H.M. Geological Survey) were involved with Piltdown in various ways, from its discovery to its unmasking. Some have been implicated in the forgery itself

The staff of the Natural History Museum (previously the British Museum (Natural History)), the Geological Society, and the British Geological Survey (previously H.M. Geological Survey) were all involved with Piltdown — from discovery to unmasking. Some have been implicated in the forgery itself. Paleontologist Sir Arthur Smith Woodward was involved in the hoax.

The Piltdown Man, is the first of a long line of hoaxes and lies cooked up to promote this New Age Religion called "Biological Evolution". During the time following the meeting of the Geological Society of London in December 1912, numerous other fossilized "missing links" were discovered. For about forty years this went on with many hoaxes popping up around the world. These "missing link" discoveries were becoming almost commonplace.

After testing methods were developed to expose the Piltdown Hoax, all of the other claims about missing link discoveries vanished. We only remember the Piltdown Man because of it's fame which was associated with its key conspirators.

But during those forty some years when The Piltdown Man was considered irrefutable evidence, there were articles written, "Scientific" illustrations and diagrams drawn which were reused to illustrate the Theory of evolution again elsewhere. The artist's rendition of the Piltdown man was simply renamed and reused.

<u>The most remarkable proof of a Creator is His Creation</u>!

3.6 Atheism Is An Emotional Response To Extreme Stupidity

I believe that most Atheists would agree with that statement. But sometimes it is more than just stupidity in the name of Jesus! There are opportunists, without conscience who will betray people's trust for personal gain. Hearing about those kinds of stories just makes people angry.

That sort of thing is almost everywhere. It is represented in every religion, every political party and almost every category and subcategory of people working together.

If a person has ever felt betrayed by someone claiming to be sent by God, then that would provoke an emotional response. Whether it is greed, lust, deceit, treachery or just providing misinformation, all of this is being done in the name of Jesus somewhere right now!

The Bible warns us about Wolves in Sheep's Clothing that prey on the trust of others. Sadly, much of this happens in the name of Jesus.

Why are pedophile and sodomite Priests so common? Priests were tested for HIV on one occasion to determine which groups of men were most likely to contract HIV. The Roman Catholic Priests were more likely to be HIV positive than most other groups. This kind of hypocrisy is likely to provoke an emotional response.

3.7 Evolution Is New Age Fantasy Pretending To Be Science

Charles Darwin is falsely credited as being the originator of the "theory" of Evolution. An important thing to know about the "theory" of evolution is its origin. Charles Darwin's book titled "**The Origin of Species**" (1859), is **not** the origin of the "theory". So where did this so called "scientific theory" come from… and why? Well a good place to start looking is **Erasmus Darwin (1731-1802),** the grandfather of **Charles Darwin (1809-1882)**. Erasmus died before Charles was born.

The connection of Erasmus Darwin to Free Masonry, the occult, paganism and pantheism is hard to hide but that is where this New Age Evolution Lie came from.

You don't hear much about Erasmus Darwin when it comes to the topic of Evolution. That connection, in my opinion, was deliberately kept as a secret, but that is where **Charles Darwin** got his information from. It also seems that Charles Darwin's connection to the Free Masons has been obscured.

The idea of **Biological Evolution** was the result of Erasmus Darwin's desire to promote his Masonic, New Age religion and give it a scientific appearance. At the same time, he was attacking the foundation of the Christian faith, which he hated.

Erasmus Darwin influenced a lot of people. His minions have carried on his work in every country of the world, including the United States, England, France and Germany. Erasmus Darwin took his Masonic/

Pantheistic fantasy and he called it "Natural", but it wasn't really "natural", it was the opposite!

When **Newton and Carolus Linnaeus** talked about "Natural" they were expressing it in mathematically precise, scientific terms. Newton said that he prayed to the God of Creation for his discoveries. Newton and Carolus Linnaeus gave God the credit for the Creation and the laws that govern it and that is what they called "Natural".

What Newton and Carolus Linnaeus called "natural", Erasmus Darwin called unnatural and according to Dr. Immanuel Velikovsky, Darwin forced his own unnatural view on "nature" and tried to cook up support for his belief. He said…

"Most controversial is the evolutionary question. I have done a great deal of work on Darwin and can say with some assurance that Darwin did not derive his theory from nature but rather superimposed a certain philosophical viewpoint on nature and then spent twenty years trying to gather the facts to make it stick" **(Dr. Immanuel Velikovsky)**

Dr. Immanuel Velikovsky was a Russian author who wrote about astrophysics and the history of our solar system. He wrote "Worlds in collision" (1950) ; "Ages in Chaos" (1952) and "Earth In Upheaval" (1955) just to mention a few.

Darwinian Evolution is at the core of this New Age Mystical Religion which is based on New Age fantasy, lies and deception. It is the product of Freemasonry Pantheism. There is nothing scientific about it. The "Theory of Evolution" is not a scientific theory because it has not been observed, ever! It doesn't even rate as a scientific hypothesis. It is New Age Fantasy pretending to be science.

The acceptance of this New Age Fantasy as science is based on a history of scientific hoaxes. People with respectable credentials were profiting from promoting hoaxes to support evolution.

After looking for intermediate species and finding none, Charles Darwin said that there must be evidence in the fossil record.

The imaginations of Darwin's followers got busy and cooked up all kinds of phony "evidence". The hoaxes began after **The Origin of Species**" was published in (1859). Ten years later, we had our first evolution hoax.

The **"Ontogeny recapitulates phylogeny" Hoax** was perpetrated by **Ernst Haeckel**. Haeckel produced faked drawings displaying a series of mini-evolutions occurring in human embryos in (1869).

Evolutionist Stephen Gould wrote the following regarding Ernst Haeckel's work in a March 2000 issue of Natural History:

"Haeckel's … books appeared in all major languages and surely exerted more influence than the works of any other scientist, including Darwin…in convincing people throughout the world about the validity of evolution".. (**Stephen Gould**)

The Java Man Hoax (1891) was reconstructed from two teeth, a thighbone and a piece of skullcap by Dr. Eugene DuBois. It was assumed that the bones came from the same individual. Evidence later indicated otherwise. About twenty six years later, DuBois confessed that Java Man was probably a giant gibbon.

The Piltdown Man was a hoax that lasted for 41 years, from 1912-1953. Perpetrated by Charles Dawson who had accomplices such as Arthur Smith Woodward, keeper of the British Museum's Paleontology department. **Arthur Conan Doyle** the famous author of the Sherlock Holmes novels lived not far from the dig site and knew the other suspects. Famous people were involved, people who were trusted because of their extraordinary credentials.

The Nebraska Man Hoax (1922) was reconstructed from one tooth. People were gullible enough to believe this "missing link" story until it was discovered to be a pig's tooth.

People are being convinced repeatedly with different lies from trusted "experts" who prostituted their credentials and capitalized on promoting the evolutionary lie. There is money to be made by promoting the evolution lie. Where is the money coming from?

I don't believe that biological evolution is a scientific theory at all. Its origin is based in New Age Religion working toward a New World Order. An order that calls Satan by his pre-fallen name which was Lucifer.

Morals and Dogma (of Free Masonry) **By Albert Pike** calls satan by his pre-fallen name… "Lucifer, the Light-bearer!" (Grand Pontiff XIX; page 321)

Pantheism, Free Masonry, Secret societies and the occult make up the New Age Religion of the New World Order! The New Age Doctrine of biological evolution is based on Freemasonry, Pantheism and satanic symbolism with a touch of paganism. A history of deceit, fraud, perjury and occultism follows the origin and promotion of what is called the… "Theory" of Evolution.

It is not a theory, it is a religious doctrine based on lies and deceit. To be a theory it must be observable. No one has ever observed evolution. They are not looking for a missing link, because the whole chain is missing. It just doesn't exist. The evidence confirms the Biblical record.

Charles Darwin got credit for his grandfather's work even though he did not know much about biology or Genetics. Charles Darwin's grandfather, **Erasmus Darwin,** copied Carolus Linnaeus's writings about species and his grandson, Charles Darwin, used Carolus Linnaeus's work about a hundred years later.

People Generally don't know where Charles Darwin got his ideas and information from. It came from reading his grandfather's books. Most of the work was already done for him by his grandfather, Erasmus Darwin.

An earlier and famous publication of Erasmus Darwin was titled "**the Botanic Garden**" in two parts, 1789 and 1791 and was .published throughout the 1790's.

Erasmus Darwin was a Medical Doctor, inventor and was a well published author. Among his publications was a best seller titled "**Zoonomia**"; or the **Laws of Organic Life (1794)** which is a two-volume medical book dealing with anatomy, physiology, pathology and his philosophy about evolution, reproduction and morality.

Just about every topic and example used in Zoonomia reappeared in Charles's Darwin's Origin of Species. Charles's copies of Zoonomia were marked with annotations in every chapter. Other publications by Erasmus that Charles had copies of were also marked and found in the "**Origin of Species**" by Charles Darwin (1859).

"**The Temple of Nature**" was published the year following his death, in (1803). The Temple of Nature was a collection of his writings during the French Revolution. That is the period that he did most of his writing. **Erasmus died just after the French Revolution ended**. Exhaustion and disappointment may have been contributing factors.

Erasmus had a son named Robert. Robert followed in his father's footsteps into medicine and also achieved a high rank in Free Masonry. Erasmus wanted his son Robert to complete his work on biological evolution but it was **Roberts's son Charles** who was also a Mason, **the grandson of Erasmus Darwin** that would complete the task.

Erasmus Darwin is the source of almost all of Charles Darwin's ideas. Charles refined his grandfather's work and presented it to a larger audience without ever mentioning his grandfather's involvement. Charles Darwin took full credit for his grandfather's ideas of "Natural Selection"; "survival of the fittest"; "biological evolution" and the work of **Carolus Linnaeus.**

Erasmus Darwin founded the **Lichfield Botanical Society** for the sole purpose of translating the writings of the **Swedish botanist Carolus Linnaeus** from Latin into English. The task took seven years.

Carolus Linnaeus (1707–1778) is known as the "father of Taxonomy. **Carl Linnaeus is the man who first began the task of studying, measuring, and systematically cataloging all life on our planet**. Carl Linnaeus received his medical degree in (1735) and immediately began publishing his works.

Carl Linnaeus published "**The System of Nature**" (1735) which first organized the systematic study, classification and cataloguing of all life on our planet. He spent his life studying and cataloguing the different life forms of earth. Carolus Linnaeus had followers helping him who shared in his vision as well. He observed no living intermediate links between any species.

"Species", according to Carl Linnaeus, were similar in form because they descended from the same parents created by God …"*in the beginning*"(Genesis 1:1). Carl Linnaeus saw distinct species Created by God, but he saw no intermediate transmutations, ever!

The words… "***After his kind***"… in "(Genesis 1:24) is confirmed by what we see living and in the fossil record.

Carolus Linnaeus went on several research expeditions cataloguing every species that he could find. In 1732 Carolus Linnaeus went on a research expedition to Lapland and again in1734 he went to Falun. There were others. A century later, Charles Darwin would embark on his famous journey.

After spending a lifetime of studying and cataloguing every life form he could find. Carolus Linnaeus insisted that the Creation had to be the work of a Devine Creator. **"God created, Linnaeus organized"** …was his slogan.

Carolus Linnaeus separated all life forms into separate and distinct species, no intermediates were ever found. The cataloguing methods developed by Carolus Linnaeus were accepted and used by scientists all over the world. A form of that system, along with some modifications, is still in use today. A century later, Charles Darwin would use that system.

Carolus Linnaeus's writings included a sophisticated system for classifying, cataloguing, identifying and naming different species that he studied. Carolus Linnaeus did the work, the research and he developed his system to honor his Creator. Remember his motto, "God created, Linnaeus organized".

The **Lichfield Botanical Society** is another one of many organizations founded by Erasmus Darwin. The "Society" consisted of three men who were committed to translating the works of **Carl Linnaeus (1707 – 1778)** the Famous Swedish botanist into English. The Society accomplished their task in seven years and Erasmus would use the information to compare different species and prepare his arguments for Evolution.

Erasmus Darwin raised funds to pay for the translation and used Carl Linnaeus's work of biological classification to predict and then fake what was predicted, but he slightly modified it. His grandson Charles used the system in his work as well. The point is that this system for analyzing, organizing and cataloguing life forms, did not come from the Darwins, It came from Carl Linnaeus (1707-1778), a Christian committed to understanding God's Creation. The system developed by Carl Linnaeus and used by the Darwins, is still being used by scientists today.

A century after Carolus Linnaeus began his research with a scientific voyage abroad (1731), Charles Darwin set out on his famous 5 year voyage aboard the "Beagle.(1831-1836) This was the beginning of his research. This was an attempt to find scientific support for his families pantheistic, and pagan beliefs.

In 1831 Charles Darwin began his famous journey to South America **in his research vessel called the "Beagle".** This trip would last 5 years, but Darwin spent his life looking for living, intermediate species and found none! Darwin's work was supported by Masonic Lodges, Masonic Organizations and individual Masons including his family. Charles's grandfather, Erasmus Darwin established an organization of Masons that would later support Charles's work.

Darwin proved that there isn't a bunch of intermediate species, alive and running around out there because he looked for them and he couldn't find any and neither could anyone else. Darwin insisted that the evidence would be found in the fossil record and there has been hoaxes ever since.

Darwin proved that there were no intermediate species or links running around commonplace, **one hundred years after Carl Linnaeus did the same thing. They are called missing links because they are missing, there is none.** The whole chain is missing! But this is not just one chain but many, many, all of them.

Charles Darwin was looking for entire chains of living intermediate species but he found none! Not even a single link. They said they were looking for a missing link, but it was really entire chains for every species that was missing. They are called "missing links" because there is none, they don't exist, they never did, and that is why they are missing.

You would think that would settle the matter but evolution is a New Age Religious doctrine of the Luciferians that defies all reason. They talk about reason but they do the opposite.

No intermediate species have been found, living or fossilized.

The theory of Evolution is New Age Religion masquerading as science. It is Pseudoscience with a history of fraud, treachery, betrayal and deceit!

3.8 The Danger, Deception And Hypocrisy Of Evolution

The idea of Evolution can be traced back to about the time of the Babylonian Empire, a form of it is found in the ancient teachings of the Hindu's. In the first century the Bible warns about such religious beliefs.

"For the invisible things of him from the creation of the world are clearly seen, being understood by the things that are made, even his eternal power and Godhead; so that they are without excuse" (Romans 1: 20)

"*Because that, when they knew God, they glorified him not as God, neither were thankful; but became* ***vain in their imaginations****, and their foolish heart was darkened*". (Romans 1: 21)

"***Professing themselves to be wise, they became fools***", (Romans 1: 22)

"*And changed the glory of the uncorruptible God into an image made like to corruptible man, and to birds, and fourfooted beasts, and creeping things*". (Romans 1: 23)

That sums up Evolutionary dogma but the consequences of following such false dogma is described in the remainder of the chapter. (Romans 1: 24-32) Fornication, sodomy, disease, violence, hatred and Genocide all stem from such beliefs.

Staling, Hitler, the Khmer Rouge regime and the New World Order all share beliefs in Evolutionary doctrine. They accept human Genocide as a legitimate solution to disagreement about who is in control. The point being that Evolutionary dogma is not only foolish, but it is dangerous!

Since the French Revolution, countries that have embraced evolution and new age religion also tend to commit acts of terror and Genocide. God warns us…

"*And have no fellowship with the unfruitful works of darkness, but rather reprove them.*"(Ephesians 5:11)

"*For it is a shame even to speak of those things which are done of them* ***in secret***" (Ephesians 5:12)

"*For God shall bring every work into judgment, with every secret thing, whether it be good, or whether it be evil* "(Ecclesiastes 12:14)

"*For nothing* ***is secret****, that shall not be made manifest; neither any thing hid, that shall not be known and come abroad*".(Luke 8:17)

God will judge the participants of the **secret societies** and their **secret conspiracies** and deeds.

Biological Evolution is a New Age Fantasy "Religion" based on lies and cooked up and promoted by 33rd degree Free Mason's and other members of the occult. The Masons are a **secret, occultist society** that is promoting the N.W.O.! Biological evolution is a product of Free Masonry from the very beginning till now.

The Piltdown Hoax is an example of Evolutionists committing fraud to promote their New Age religion. Evolutionists have been caught tampering with the evidence for personal gain many times.

God pleads with his Creation to listen to reason!

"*Come now, and let us reason together, saith the LORD*" (Isaiah 1:18)

The whole point of New Age doctrine of Evolution seems to be about changing over to a New Age Morality that denies the God of the Bible. Erasmus wrote about morality during the French Revolution.

The justification through pseudo-science was the task Charles Darwin was charged with later. But he never found any missing links or any of the predicted evidence that he was looking for so his followers cooked up some "evidence".

The idea of Natural Selection was refined somewhat by Charles Darwin but for the most part the ideas of Darwinian Evolution came from Erasmus before Charles was born and at least 65 years before Charles Darwin's book titled "Origin of Species" was published. (November 24, 1859).

The "Theory" of Biological Evolution is not a Scientific Theory at all. It is Masonic New Age Pantheistic Religion masquerading as science. Charles Darwin's Grandfather was a 33rd Mason that wrote about his ideas on evolution before his grandson, Charles was born.

Erasmus created an argument against believing in Biblical Creation. He led the fight to remove God from the morality equation and Evolution is how he did it.

Erasmus was often the mastermind of some of these organizations or at least an active participant. Erasmus networked with other Masonic groups and had close ties with groups such as **Midlands Enlightenment**; **Lunar Society of Birmingham, AKA "Lunaticks"; the Jacobin Masons** and the **Illuminati.** The Illuminati was founded by Adam Weishaupt in **1776**, it was a Masonic group with a N.W.O. philosophy. **Weishaupt,** was considered to be the 'patriarch of the Jacobins and the Jacobin Club.

In his book "Proofs of a Conspiracy" by James Robison, he republishes letters of correspondence between members of the **Illuminati** where their goals and plans are outlined. Founder Adam Weishaupt, known in correspondence as "Spartacus", describes how the mainly academic and political **Illuminati was structured in the image of a mystic sect.** Weishaupt envisioned himself as the creator of a new religion. "The holy faith of Reason".

The Jacobin Club was a democratic club established in Paris in 1789. The **Jacobins were** the most radical and ruthless of the political groups formed in the wake of the **French Revolution**, and in association with Robespierre they instituted the Terror of 1793–94.

The Festival of Liberty and Reason. (November 10, 1793)

Robespierre, conspires and betrays fellow Masons. He uses the **Committee of Public Safety** to kill his competition. The Jacobin Club denounced the Hébertists and Dantonists on framed-up charges. The Hebertists were the antichristian "left" and the Dantonists were the "right". They execute all the popular leaders. Robespierre becomes virtually the dictator (March 24, 1794)

Robespierre decreed the new religion of the Supreme Being. (May 18, 1794)

Robespierre is inaugurated as The Supreme Being (June 8, 1794) The "Terror" begins. It included beatings, assassinations, wars and Genocide.

These groups represent a Secret Masonic System of influence that spawned and controlled the chaos that history calls the French Revolution. They were the chief secret conspirators, the architects of the Anarchy, terror and carnage that characterizes the French revolution!

A religious war was planned, provoked and faked. A response was also planned. All of the resulting calamity would be blamed on religion. Indeed it was! Many Christians lost their heads because of their testimony! There was Genocide in the name of reason. The streets ran red with blood.

Secret groups with a secret religion of illumination were behind the French Revolution. Their cornerstone doctrine was Darwinian Biological Evolution. Key players in the French Revolution were high-ranking Free Masons such as Adam Weishaupt, Napoleon Bonaparte and **Erasmus Darwin.**

Erasmus Darwin was just as much involved in the French Revolution as any of its main players who were influenced by his writings.

At the center of both the American and French Revolutions were the members of **the Lunar Society of Birmingham** who were also known as **"The Lunaticks"**. Erasmus Darwin was the founder, he started it in his house in Lichfield.

The Lunar Society of Birmingham, England was a dinner club and informal, educated society of important figures in the Midlands Enlightenment. At first called the Lunar Circle, "Lunar Society" became the formal name by 1775. The name arose because the society would meet during the full moon, as the extra light made the journey home easier. The members cheerfully referred to themselves as "lunar-ticks", a pun on lunatics. Thomas Jefferson and Benjamin Franklin attended some of those meetings. (They were mostly Masons)

Napoleon Bonaparte was a Free Mason. (1769-1821) He was initiated into Army Philadelphia Lodge in 1798. His brothers, Joseph, Lucian, Louis and Jerome, were also Freemasons. Five of the six members of Napoleon's Grand Council of the Empire were Freemasons, as were six of the nine Imperial Officers and 22 of the 30 Marshals of France.

(November 9, 1799) (*18th Brumaire*) Napoleon Bonaparte named "First Consul," now the effective dictator. (1804) Napoleon was consecrated as Emperor.

The French Revolution embraced Erasmus Darwin's revolutionary idea of Evolution and just look at what happened. The architects of the French Revolution planed on creating a new morality apart from the God of the Bible and then they became Psychopaths. Killing large numbers of innocent human beings with no remorse!

They rejected their Creator, his ways and anyone who believed in him. This led to cultural "cleansing" that began by gradually and strategically targeting the most competent opposition.

The Christian leadership was targeted first. They gradually worked on their opposition priority list following a system of priority levels and easy targets of opportunity that could be concealed. The French Revolution was a colossal conspiracy involving deception; betrayal; assassinations and finally, when the bad guys had control, **Genocide!**

With Evolution comes a new morality, a morality of rebellion against God! That is some of what Freemasonry and the French Revolution had in common. When you attack the Biblical Creation record, you are attacking the credibility of the Bible and the foundation of our faith!

Freemasons were often anti-monarchy and anti-Catholic and attracted people who witnessed the tyranny of the Catholic Church. Generally, people do not distinguish between the Catholic Church and true Christianity.

Erasmus was quite an international networker. He was either the founder, leader or member of quite a large number of organizations internationally.

Erasmus stayed in contact with Masonic Lodges and their front organizations in England, America, France and other European countries and he influenced them all.

Erasmus Darwin was a medical doctor with a successful practice, who actually turned down a job as Royal Physician for the British Royal Family who were among some of his famous patients. Erasmus had connections in high places for sure!

Freemasonry is an international **Secret Society** that is based on **Secret Oaths** and **Deception about Intent**. Freemasonry is based on secrecy and has its own secret god/s; its own secret rituals; its own secret emblems with secret meanings; its own secret philosophies; its own secret morality; its own secret handshakes; it's own secret high days; its own secret code words and phrases and **<u>its own secret agendas</u>**. That is why it is called a "**Secret Society**".

Erasmus Darwin was English but he also secretly supported the American and French Revolutions. It is said that the first Masonic Lodge in France was established by Englishmen.

The French Revolution was started by wealthy agitators who convinced the poor ignorant masses that the King was raising taxes on them, but it was a lie. The King actually raised taxes only on the wealthy and just enough to raise money for the American Revolution that the King of France was supporting. It was believed at the time by many that this deception was another masonic conspiracy. The masons called the charges "**conspiracy theories**".

After stirring up anger against the King which was based on lies, the wealthy agitators led the poor, misinformed people to capture King Louis XVI. The Freemasons of France were accused of this deception which they denied. Any accusations against Freemasons were dubbed as **"conspiracy theories"**. This event marks the beginning of the "radical time of the revolution". Eventually King Louis XVl was beheaded by guillotine. It was a bloody Revolution, full of violence, murder, terror and guillotines.

The last few years of the French Revolution was called the "**Reign of Terror**". **The Committee of Public Safety** was formed. **Maximilien Robespierre was inaugurated as the "Supreme Being"** and the Reign of Terror began. It was characterized by summary executions of suspects without trial. Many Christians were beheaded.

I think that it is important to mention now that Erasmus Darwin was a bit of an international Revolutionary leader. His associates stirred up rebellion in people against the establishment first, then in the case of the French Revolution, against Christianity next, after that anyone could be next.

Erasmus Darwin led a Rebellion against God to promote his fantasy based New Age religion. Evolution is the cornerstone doctrine based on fantasy, lies and deception. From the very beginning evolutionists have faked the evidence like the Piltdown Hoax.

Denying Creation opened the door to a new morality without God, a new order based on Darwin's view that <u>he</u> called "Reason". As a result, two new revolutionary schools of thought about morality developed in France. Utopia was promised but anarchy, death and terror was delivered because Satan always does the opposite!

The two religious Cults of the French Revolution were **The Cult of the Supreme Being** and **The "Cult of Reason"**. One was Atheistic and one was deism. Although these two cults did not agree on everything, they did agree on one thing. Both cults were anti-Christian and anti-Catholic.

Maximilien Robespierre murdered his competition by guillotine and in June 1793 **was inaugurated as "The Supreme Being"** and the reign of terror began. About 14 Months later he was beheaded by guillotine and the reign of terror ended in July 1794. **The Cult of the Supreme Being** disappeared but the "Cult of Reason" continued. Free Masons claim that there are no atheist Masons but that is just another deception.

It was Generally believed by the contemporaries of that time that the French Revolution was engineered by the Free Masons of Europe. The Masons denied involvement and called it a conspiracy theory.

I think that the term “Conspiracy Theory” may have been a phrase that was thought up and discussed in secret meetings of secret societies long ago. The term “**Conspiracy Theory**” when it is used in a condescending manner seems to be a thought strangling ploy used to silence those who want to expose Masonic conspiracies.

The term “**conspiracy theory**” is an easy way for a Mason to raise doubt. By pretending to be a condescending skeptic, a Mason can subtly attack those who attempt to expose free masonry. This aids in concealing a matter without revealing that one is a Mason.

France was to become the Masonic “New Order”. Governing bodies were formed, committees were created such as the National Assembly, the Committee of Public Safety etc. It is interesting to note the Committee of Public Safety became dominate by use of assassination and terror during the “Reign of Terror”.

The lie about safety and security to gain control is the same tactic used by the U.N. and the N.W.O. in America and Europe today. This is where it came from and Erasmus Darwin’s teachings are a common link behind the violent revolutions of his time and communist or Fascists revolutions since.

Erasmus was one of the most internationally famous and influential writers in Europe during the time of the **French Revolution, (1789-1799)** that is when he did most of his publishing.

Erasmus died in 1802, just after the French Revolution ended.

All of the denial about the Masonic conspiracy to overthrow the monarchy doesn’t erase the fact that all of the major players were Masons. Their plan to establish a secular, Godless, New Order and destroy the Catholic Church and Christianity in General was called a conspiracy theory until a Catholic Jesuit Priest decided to research and document the facts.

Although much had already been said by many about the Masonic involvement, **Abbe Augustin Barruel (1741 –1820)**, a French Jesuit priest and publicist was the first to do a thorough documentation of their involvement.

In his book, originally titled "Mémoires pour servir à l'Histoire du Jacobinisme" he documents the creation of the **Bavarian Illuminati** and **the Jacobins** in his book, the new title in English is**... "Memoirs Illustrating the History of Jacobinism" published in 1797.** Barruel documents the involvement of the Bavarian Illuminati, the Jacobins and other secret societies in the French Revolution. In his book he meticulously documents the conspirators, their associations, statements and activities planned and executed by the secret societies.

The conspirators called it conspiracy theories in a most condescending manner. People became afraid to talk about these things because of fear of being ridiculed for it. Now look at how far things have gone because good people would not speak up or look in to things a little more.

Just before the French Revolution got started, Erasmus Darwin published "**The Botanical Gardens**" in two parts. "**The Loves of the Plants"** was published in 1789 and that is the year that the French Revolution broke out. In 1789 the "National Assembly" was formed and the Bastille was stormed.

"**The Economy of Vegetation**" was published in 1791, and Part one of "Zoonomia" was published in 1792. **Zoonomia was Erasmus Darwin's most famous book** that included a discussion about his theory of Evolution. It took the Creator out of their view of morality and everything. **That is the year that the Palace was stormed. (1792)**

The year after Zoonomia was published, on January 21, 1793 King Louis XVI was beheaded.

in June, 1793 Robespierre was inaugurated as "**The Supreme Being"** and the "Reign of Terror" began. **The "Reign of Terror" began June 1793** and ended with the fall of Robespierre on **July 1794**. In 13 months there were 16,594 official death sentences in France, but the death toll was much higher.

The French Revolution was supposed to be based on science and reason. Darwinian Evolution is neither. They were the result of Free Masonry, Pantheistic Paganism, occultism and lies! They cooked up their own religion to justify doing anything for personal gain.

The printing press made rapid, mass communication possible. The revolutionary ideas of the French Revolution could be distributed rapidly by the written word. The printing press made it all possible.

You don't hear much about Erasmus when it comes to Evolution or the French Revolution, but he seems to be a key figure who has concealed his involvement in many things.

When you consider all of the violent revolutions by groups who have embraced the New Age Religion and its main doctrine of Darwinian Evolution, at some point you must realize the connection.

Evolution leads to Atheism which is an open door to psychopathic behavior. They believe that everything is an accident so they have made up their own rules, to suit themselves. Life no longer has any meaning for them.

Millions of people were slaughtered during the Russian Communist Revolution, 20-60,000,000; The Chinese Communist Revolution slaughtered 60,000,000; The Nazis slaughtered about 17,000,000 in war crimes that involved the systematic extermination of entire races, cultures, religions and anyone they didn't like. Nazis embraced New Age Occult religions and rejected the Bible. They believed in Evolution, they all believe that life is the result of a whole lot of accidents consistently happening at the right time and place. They don't want to know about God and they don't want anyone else to know either!

First let's consider Goebbels' Principles of Propaganda.

These principles were compiled Jowett and O'Donnell.

Avoid abstract ideas; appeal to the emotions; Constantly repeat just a few ideas; Use stereotyped phrases; Give only one side of the argument;

Continuously criticize your opponents; Pick out one special "enemy" for special vilification.

Erasmus Darwin was outspoken against Christianity. He cooked up the idea of biological evolution to support his pantheistic beliefs. He challenged the Biblical Creation story and the belief in the God of the Bible. His grandson took credit for his work and never mentioned his grandfather's contributions.

Freemasons are Generally accepted as the architects of the **French Revolution**, the **New Age Movement**, **Darwinian evolution**, Secret Societies and more. Erasmus Darwin was a 33rd degree Mason that was a major contributor to all of those just mentioned and more.

3.9 Designs and Probabilities, or Fantasy, it is your choice.

It is unreasonable to <u>assume</u> that disorder can organize itself or anything! Order cannot exist without an Organizer! The Organizer comes before His Creation. God always presents himself as present tense, Omniscience, Omnipotent, masculine, and plural. A Trinity! Order requires an organizer with the magnitude of organization and power required to Create and maintain the Creation. It is that simple, yet there is a **new age religion** that teaches otherwise.

Sir Isaac Newton published his most famous book titled, **"The Mathematical Principles of Natural Philosophy" (1687)**

It is considered by the Royal Society of England, to be **<u>the</u> most influential science book <u>ever</u> published!**

From real science, I know that things tend toward greater randomness... Everything gets "old" or wears-out; gets broken or lost. That describes what everyone observes in everyday life. That is the nature of things. This is also what the Bible teaches…

"Of old hast thou laid the foundation of the earth: and the heavens are the work of thy hands" (Psalms 102: 25)

"*They shall perish, but thou shalt endure: yea,* ***all of them shall wax <u>old</u> like a garment****; as a vesture shalt thou change them, and they shall be changed:"* (Psalms 102: 26)

Things just tend to wear out.

As a believer, I have often considered how life tends to adapt to its environment. We can lose genetic information but we can't gain new information without **an intelligent designer**. Without intelligent intervention, everything decays and becomes less organized, less efficient and dies. That is the nature of things.

Even the aging process is genetically **<u>designed</u>** <u>and</u> **<u>controlled</u> by what can only be known as <u>a personality</u>**. A Personality that knows everything and must be all powerful. A Personality that Created everything for Himself. He is Perfect and Created us to have fellowship with Him, but an enemy corrupted everything which ended human perfection and the fellowship that we had with God. We are not perfect and now we age.

Every living thing has **preprogramed obsolescence** built into their DNA. The reason that we all grow old and die the way that we do is because of the **design** of our genetic code.

After puberty our genes cause a steady decline in our immune system and our free radical control systems. Also controlled by our DNA, is the decline in our repair systems and brain function as well. The fact is that all of our hormones begin to decline after puberty and that is what controls the aging process. The aging process is by **God's design**. It sends us the message that sin has a consequence and we have a limited amount of time to deal with the problem.

Adaption requires already existing systems that are designed to become activated or enhanced in some way by environmental stress. We are talking about different genetic traits such as color of eyes, skin, hair, hand size, sight, hearing, smell, immunity, intelligence, strength etc. In finches it could be characteristics such as beak size or shape.

Genes are sections of DNA code that carry the **design** for specific traits. Regardless of the genetic characteristic, genes designate the trait and variations of that trait are made possible by the alleles. Some alleles are dominant and some are recessive. When a dominant allele corresponds to a recessive allele, the dominant allele will be expressed.

When a recessive allele corresponds to another recessive allele, something has to determine which of the two or more recessive alleles will be active. Environmental stresses play a role in making that determination.

Alleles designate the form that the gene takes. The different variations of any trait are produced by stored alleles. Stored recessive alleles could remain dormant for generations. The activation of the recessive alleles is more than just random chance. There seems to be a biofeedback system in place that is designed to translate the environmental stress put on the organism and it's corresponding systems and the alleles. Alleles are the source of adaption within a species.

Allele, also called **allelomorph**, is any one of two or more genes that may occur alternatively at a particular site (locus) in a chromosome.

Alleles may occur in pairs, or there may be multiple alleles affecting the expression (phenotype) of a particular trait If the paired alleles are the same, the organism is said to be homozygous for that trait; if they are different, the organism is heterozygous. A dominant allele will override the traits of a corresponding recessive allele in a heterozygous pairing. In some traits, however, alleles may be codominant and neither acts as dominant or recessive.

Nearly all blood types come from combinations of just six alleles. In the human ABO blood system, those with type AB blood have one allele for A and one allele for B. Those without either are type O.

Most traits are determined by more than two alleles. Multiple forms of the allele may exist and be dormant for generations before being expressed. It is currently not fully understood how the alleles are affected by environmental factors. There could be any number of recessive alleles hanging around, dormant just waiting to become activated by some kind of stress or something else.

All life is the result of a design there is nothing left to chance.

The probabilities of a simple, nonliving, fully functional protein molecule just all coming together by chance, under the best fantasy circumstances could only be expressed with a zero or some negative digit. It can't happen, ever.

Biological Evolution is not an observable scientific theory as it is so deceptively misrepresented as being. Biological Evolution is the central doctrine of the New Age Religion of Freemasonry. The Freemasons publish a magazine called the New Age Magazine that promotes their religious and political agendas. Evolution is a masonic creation.

Immanuel Velikovsky once said…

…"Most controversial is the evolutionary question. I have done a great deal of work on Darwin and can say with some assurance that Darwin did not derive his theory from nature but rather superimposed a certain philosophical viewpoint on nature and then spent twenty years trying to gather the facts to make it stick" **(Immanuel Velikovsky)**

True science indicates that things tend toward greater randomness. Things break, wear out and die. That is the **nature** and **law of our universe** and that is what the Bible teaches. This also applies to biological life including the human race. Contrary to the followers of Darwin, the human race is not getting more Genetically sophisticated, but quite the opposite. We are becoming weaker, more frail. This may be why twins are not so common any more.

Genetic defects reflect how order can become less and less orderly and that is the fundamental issue. The Bible, Newton and Carolus Linnaeus all taught that our world requires an intelligent and all-powerful Creator and that things tend to wear out and become disorganized. On the other hand, Darwinian evolution insists the opposite. Darwinian evolution teaches that random chaos somehow has become organized by chance. Impossible!

A horse has 64 chromosomes and a donkey has 62. This is a classic example of the loss of Genetic information. When a horse and a donkey mate, an infertile offspring is produced called a mule. The mule has Genetic problems that include reproduction.

Similarly in humans we have Genetic defects that are the result of Genetic malfunctions. Downs Syndrome is one that involves a sticking chromosome during mitosis. Such an error doesn't produce new, superior Genetic offspring. It uses damaged D.N.A. to produce an offspring with health problems, reduced abilities and a reduced lifespan; an offspring that still needs to learn about Jesus, just like everyone else.

Every living cell in your body has a computer that is programed with a code that guides our growth, development and health. This code is determined by the order of the four nucleotide bases that make up DNA, adenine, cytosine, guanine and thymine, A, C, G and T for short. Within our DNA is a unique chemical code for each of us.

DNA has a twisted structure in the shape of a double helix. Single strands of DNA are coiled up into structures called chromosomes. Your forth six chromosomes are located within the nucleus in each cell. Within our chromosomes, sections of DNA are "read" together to form Genes. Genes control different characteristics such as height and eye color.

All living things have a unique Genome. The human Genome is made of 3.2 billion bases of DNA but other organisms have different Genome sizes. If printed out the 3.2 billion letters in your Genome would: Fill a stack of paperback books 200 feet (61 m) high; Fill 200 500-page telephone directories and Take a century to recite, if we recited at one letter per second for 24 hours a day. If you think that just happened by accident, then you just aren't thinking!

Ever since God fearing men began to take a serious look at the so called "Theory" of Evolution, they have talked about the probability problem. Evolutionists often use probability illustrations that simply do not reflect the reality of the problem.

When you consider the probability of a protein just happening by chance, you need to consider more than just the numbers in this illustration of chance. For the sake of making a point we are going to have to imagine a world that was made up of plenty of all five atoms required for life. So imagine a world made up of an unlimited supply of carbon, oxygen, hydrogen, nitrogen and sulfur. Also imagine that entropy is not a force that opposes anything being made by chance. So we have a perfect, imaginary world for evolution to occur by chance.

Let us consider how much time it would take to make a smaller than average protein consisting of 150 amino acids in this imaginary world. Scientists have calculated that to be about 1 in 10 raised to the **164th power**. If you use imaginary conditions that no one believes could ever exist. But evolutionists insist that it could happen if given enough time.

At the rate of 6,000(thousand), 000,000 (million), 000,000,000, (billion), 000,000,000,000, (trillion), 000,000,000,000, (trillion) attempts per minute there would be enough time to cook up 10^{58th} of failed attempts out of the necessary 10^{164th} attempts required produce one functional protein during the oldest possible age of the earth being 4.6 billion years.

To give a little perspective about the size of these numbers, scientists estimate that there are about 10^{80th} atoms in the whole universe. There simply is not enough time to even produce a single protein by chance. Not even close, not a chance. But it takes a whole lot more than just one protein to start life.

To have life, you need proteins, enzymes, DNA, carbohydrates, sugars, specialized organs and other complete life forms and their systems to turn the waist back into food. The whole eco system had to come into existence at the same time and that is what the fossil record shows.

The complexity of the physical laws that govern our universe is proof of an intelligence beyond human understanding. The complexity of any living thing is currently beyond human comprehension. The complexity of the Genetic code that controls every living thing is currently beyond human comprehension, and yet we are asked to believe that all of this incomprehensible and perfectly brilliant order just happened without the direction from a brilliant, perfect and all-powerful Creator as described in the Bible!

Biological evolution is **a renewal of the ancient pagan belief that "nature" has creative power. It is a** They don't believe that Order requires an organizer! Our universe is governed by physical laws. Those laws define our universe, or describe how it is organized. Those laws did not just happen by themselves because they represent order and order requires an organizer.

Disorder cannot organize itself, an organizer is required! Disorder can destroy order but it cannot create order. Disorder cannot create anything. A Universe with laws of order requires an organizer/Creator!

Carl Sagan was looking for signs of intelligent life from other worlds in the cosmos. He believed that any form of language detected by radio sensors coming from space, no matter how simple, would be proof of intelligent life in other worlds. However, Carl Sagan ignored the complex language in the nucleus of the cells of every living thing on this earth. **What hypocrisy!**

The Genetic language that controls every living thing on this earth is extraordinary proof of an extraordinarily intelligent and powerful Creator! The Bible speaks about men who would believe the lie of evolution…

*"**Professing themselves to be wise, they became fools**"*, (Romans 1: 22)

Not a chance! Evolution can't happen, ever. It is time to move on; leave this new age fantasy behind.

3.10 Two Wrong Assumptions.

"Atheism is so senseless and odious to mankind that it never had many professors". **(Newton)**

When a person says that… "there is no God", they are making **<u>two</u> wrong assumptions!**

The <u>first</u> wrong assumption that Atheist make is that they already know about everything there is to know about including God! Wow! Arrogant Pride can be delusional!

The <u>second</u> wrong assumption that Atheists make is that everything just happened by chance. God calls both of these assumptions "foolish".

Pride can cause a full grown educated man to stumble at something so simple that a child can understand it! But Satan attacked God because of pride. Satan uses our feelings to control us. Those who are controlled by their feelings are easy prey for Satan because feelings can be a way to bypass your reason. Pride is a feeling accompanied by certain thoughts used often because it stims from feelings and desires.

God's gift of eternal life is a gift of love from God himself! Don't make prideful and unrealistic assumptions about your ability to know everything! Accept God's Free Gift of Eternal Life Now! It is Free! Just tell God that you believe. Thank him for giving you Eternal Life! Say something like… **"THANK YOU JESUS!"** and that is it! It can't be undone!

The second thing that Atheists are assuming is that everything just happened by chance. Atheists say that they don't believe that there is any evidence of a Creator! Wow, all of this order and the only thing out of order is the unbeliever! Unbelief is an act of rebellion against God, the Creator! It is willful, prideful, ignorance about God and rebellion against the one who Created you and Loves you!

God calls both of these assumptions "foolish".

To take the attitude that you" know it all" is traditionally considered to be prideful and therefore foolish. To say that disorder organized itself to make up our world is lunacy.

"To the chief Musician, A Psalm of David.]] ***The fool hath said in his heart, There is no God.*** *They are corrupt, they have done abominable works, there is none that doeth good"* (Psalm 14:1)

"DON'T BE A FOOL"

"Surely your turning of things upside down shall be esteemed as the potter's clay: for shall the work say of him that made it, He made me not? or shall the thing framed say of him that framed it, He had no understanding?" (Isaiah 29:16)

"And they said, ***Believe on the Lord Jesus Christ, and thou shalt be saved,*** *and thy house"***.** (Acts 16:31)

Evolutionists Begin With Wrong Assumptions

Evolutionists assume that everything happened by chance and there is no God.

Darwin spent his life looking for living, intermediate species called transmutations or "Missing Links" and found none. While Evolutionists are spending their time explaining why there are no living intermediate species they look to the fossil record to support their claims. Once again, they don't find any "missing links" because there is no chain.

Biological evolution is not possible. The probability of something as complex as life coming into existence by random chance is nonexistent. A design requires a designer. The complex program code that makes up our D.N.A. requires an intelligent and powerful God not random chance!

3.11 A Wrong Choice

*"For the wrath of God is revealed from heaven against all ungodliness and unrighteousness of men, **who hold the truth in unrighteousness**;"* (Romans 1:18)

God has said that he has given every man a certain amount of knowledge about their Creator.

*"Because that which may be known of **God is manifest in them**; for **God hath shewed it unto them**".* (Romans 1:19)

This passage says that God has installed a God awareness inside of every man. There are other passages that say the same thing.

God has given every person a built in awareness of himself and the ability to seek him out. This is an ability that animals just don't have.

Knowing our Creator is the most important thing to us. After all, if we are Created, then everything would be Created relative to the Creator. The Creation would be a reflection of himself and what he wants!

If everything in the Creation is relative to the Creator, then the Creator must be perfect! Anything that is contrary to the Creator is Anarchy, enmity or you can just call it sin. Sin cannot exist with God.

Satan introduced sin into the world and he introduced it to mankind through Adam. The only way that we can have fellowship with God is through trusting in Jesus Christ!

God wants to have a personal relationship with everyone in the human race.

When we believe the Gospel message that God came to the earth; was born of a woman; is 100% God and 100% man; lived on the earth, a perfect life as God in perfect human flesh; offered himself as the perfect, infinite, Human sacrifice for all mankind; Jesus was crucified, buried and rose again from the grave as proof that all sins were paid for if we believe; Just Faith!

His sacrifice is complete, nothing else can be done. Jesus did all of the work and I did none! It is a ***Gift!***

*"For the wages of sin is death; but **the gift of God is eternal life** through Jesus Christ our Lord".* (Romans 6:23)

Eternal Life is "*the Gift of God*"… To suggest otherwise is to be contrary to what God has said. That is sin! Unbelief in what God has said is contrary to God. Unbelief is sin!

"*He that **believeth** on him **is not condemned**: but he that **believeth not** is **condemned already**, **because he hath not believed** in the name of the only begotten Son of God*". (John 3:18)

Not believing what God has revealed about himself and our relationship to him will condemn an unbeliever for eternity if he dies like that! **You need to change your mind!** Quit believing in fairytales and **hear the God of Reason!** You need to believe God's Record!

*"For God so loved the world that he gave his only begotten Son, that **whosoever believeth** in him should not perish, but **have everlasting life**".* (John 3: 16)

Not that is a summation of God's record that everyone must believe. The only way to receive Eternal Life is **by *believing* God's *record* about his Son Jesus!** That is called **"receiving Jesus"**, He is **The Gift! When you believe** God's record about his Son, then you become born again that same instant! God becomes your Father and you become God's child.

*"**But as many as received him**, to them **gave he** power to become the sons of God, even **to them that believe** on his name"*: (John 1:12)

You must ***believe*** to **receive God's *Gift*!** Just Faith! **It is Free!** It is a ***Gift!***

God gives himself so that everyone could be saved. Their sin is already paid for, but they just don't know it! Everyone must be told and they must believe or they are condemned for eternity upon death, not because God is mean and never loved them, but because of the nature of our universe and the nature of its Creator.

Unless someone is a child or mentally retarded, they must understand and believe the simple truth summed up In (John 3: 16). God also says that you receive the Holy Spirit; possess Eternal Life; are born again, of the Spirit the moment that you **believe God's Record**.

When that happens, regardless of whether or not you tell anyone, you can know that you **have Eternal Life** and that your Citizenship is in Heaven because God said so. He did! That settles it! **Just Faith!**

(1John 5:10) *"He that **believeth** on the Son of God **hath** the witness in himself: he that **believeth not** God **hath made him a liar**; because he **believeth not** the record that God gave of his Son".*

This record is a list of facts about God, his Son, sin and the Salvation of humanity. **Eternal Life is the Gift of God through Jesus Christ!** (Romans 6:23) Just Faith! Confession to anyone other than God himself is not part of the deal! You are only required to confess to God Almighty himself and no one else! God says that **when you believe** what he has said, then you are believing God's Record and are saved from that moment and forever after regardless of what you do or do not do.

If you do chose to confess before man, beware! Satan wants to silence you! Choose carefully how you confess and who you confess to! **Pray about it every time! Follow the Spirit.**

To look at this awesome universe, and attempt to understand the mind of the Creator is hopeless outside of the path he has provided. He said so!

Having fellowship with our Creator is the purpose of every human being. That is why we were made! If you ask God for the Truth and respect it when you get it, God will give you more! If you reject truth when it is offered, God will step back and let Satan, the father of lies have you. Satan is a deceiver, he will promise you one thing and do the **opposite**. Satan is committed to the destruction of humanity!

Choosing to remain willingly ignorant and ignore the remarkable evidence of our Creator and his nature and remain in unbelief has serious consequences! When you disbelieve God's record, **you are calling him a liar! (Verse10).** That is a bad idea! That is the **<u>only</u>** thing that can keep you out of Heaven is **<u>unbelief!</u>**

*"And this is the **record**, that God hath given to us **<u>eternal</u>** life, and this **life is in his Son*** (1John 5:11)

"He that hath the Son hath life; and he that hath not the Son of God hath not life". (1John 5:12)

Got Jesus? You can <u>know</u> that you <u>have</u> <u>eternal</u> life!

*"These things have I written unto you that **<u>believe</u>** on the name of the Son of God; that ye may **<u>know</u>** that ye **<u>have</u>** **<u>eternal</u>** life, and that ye may **believe** on the name of the Son of God"*. (1John 5:13)

God says right here that **you can <u>know</u> that you <u>have</u> Eternal Life when you <u>Believe</u>**. Just as sure as you can **know** that you **have Eternal Life <u>when</u>** you **<u>believe,</u>** you can also be sure that you will spend all of eternity in Hell if you don't believe! Not because that God likes to be mean, but because you have chosen to ignore God even though he is pursuing you with truth and the Holy Spirit!

"The same shall drink of the wine of <u>the wrath of God</u>, which is poured out without mixture into the cup of his indignation; and he <u>shall be tormented</u> with fire and brimstone in the presence of the holy angels, and in the presence of the Lamb" (Revelation 14:10)

"And the smoke of <u>their torment</u> ascendeth up for ever and ever: and <u>they have no rest day nor night</u>, who worship the beast and his image, and whosoever receiveth the mark of his name". (Revelation 14:11)

Heaven is real, Hell is real, God is real, Jesus is real and Satan is real. Wake up, there is a spiritual war going on even if you can't see it directly. God's Angels are fighting against Satan and his demons. The fight is over the survival of the human race! Eternal Salvation is free! "***The Gift of God is Eternal Life***", it is **a Gift! It is Free!** No one can get saved without hearing the good news about how God has paid our way to Life Eternal! Paid it in Full! That is what Jesus said as he was hanging on the cross, beat to a pulp! He said it was done, "finished" and he did it alone! Not you or anyone else can contribute to what he did! He said that He paid it all, not you or anyone else.

God is offering to anyone who will take it, the **Free Gift** of God, ***Eternal Life.***

Everyone can see the Creation. Only a perfect, all powerful, all knowing God could Create all of this. If there is a Creator, there will be a judgment.

DON'T CHOSE TO DISBELIEVE! You must **Change Your Mind about what you believe about Jesus** and **believe what God has said!**

3.12 Don't Look For A Hero Before Jesus.

The Bible doesn't say that a real hero, a savior would show up before the real Jesus, but it does say that a false Christ and a false prophet would show up offering peace. But he will bring war, famine, pestilence and Genocide **suddenly! He does the opposite of what he says**! The Antichrist will come in peaceably but he will oppose God's word and God's people.

The Biblical description of the Antichrist matches "The Hadith of the Twelve Successors" and it's description of the 12th Imam or the Mahdi.

Muslims who believe in twelve divinely ordained leaders, known as the twelve Imams are known as "Twelvers". They believe that the last Imam, Muhammad al-Mahdi, lives in occultation and will **reappear** as the promised Mahdi. The Mahdi will have no birth certificate and that is important to the "Twelvers".

Evil And Corruption Will Grow Worse And Worse

***"But evil men and seducers shall wax <u>worse and worse</u>, deceiving, and being deceived".* (2Timmothy3:13)**

The Bible says that things are going to get **worse and worse**, until Jesus comes and destroys the Armies of the Antichrist! Till then, if you want to be thinking scripturally, then you need to be thinking that things are going to get **worse and worse**. I know that sounds pessimistic, but **that is what the Bible says.**

The Antichrist and the False Prophet will deceive the world.

The Muslims are looking for their version of "Jesus" the Muslim Prophet to reveal the Mahdi to the world. Well that sounds like the False Prophet of Revelation, or the first Islamic Pope**. Robert Spencer called Pope Frances 1st, "<u>the first Islamic Pope</u>".** I think that is very significant that he said that.

Robert Spencer is a Catholic Scholar who has studied Islam for most of his life and authored seventeen books about Islam. Robert Spencer's latest book is **"THE HISTORY OF JIHAD FROM MUHAMMAD TO ISIS"**. "The comprehensive history of the role of war and terror in the spread of Islam"**.** Robert Spencer's website is… <u>https://www.jihadwatch.org</u>

The Antichrist will be Islamic and deceive the world with lies. The Muslims call it "**Taqiyya**" or "Religious Deception" We are already seeing that now. The Antichrist will promise one thing and do the opposite.

*"For when they shall say, Peace and safety; then **<u>sudden</u> destruction** cometh upon them, as travail upon a woman with child; and they shall not escape".* **(1Thessalonians 5:3)**

(ETBH chapter 31)

3.13 A Time Of Evil, Deception, Betrayal, Terror, And Darkness!

Israel's punishment away from home ended on May 13, 1948, the next day they were declared to be a sovereign nation for the first time in 2,000 years. Israel is back in their land and we have been living in the "Last Days" ever since.

God describes this time as a time of increased evil...

"But evil men and seducers shall wax worse and worse, deceiving, and being deceived. (2Timmothy 3:13)

A description of the beginning of the Tribulation is found in (Micah 7:1-5).

*"**Woe is me!** For I am as when they have gathered the summer fruits, as the grapegleanings of the vintage: there is no cluster to eat: **my soul desired the firstripe fruit**"*. (Micah 7:1)

Basically this first verse is describing the reaction of some people just after the Rapture. The ***"firstripe fruit"*** spoken of here is **believers** that just got raptured. The ***"grapegleanings"*** spoken of here are the **unbelievers** who just got left behind.

Some will realize that they have been left behind because their Christian friend witnessed to them about the **Rapture**; the **Last Week of Daniel** and hopefully **Salvation** by Just Faith. But they rejected it! But when the Rapture happens the unbelieving friends of Christians that will be left behind may miss their Christian friends and remember. Maybe they were together at the end of Rosh Hashanah, at the sound of the last trump". That is when the Rapture will occur some year in the future, maybe this year.

The one's who were with their Christian Friend at the time of the Rapture are likely to see it for themselves and tell others. Those who did not see it happen, but heard about it 2nd hand will look for missing Christians, but will not find them.

Enoch could not be found when his friends were looking for him (Genesis 5: 24). They knew because he told them like he was supposed to.

The Bible says that the Jews are looking for a sign. Maybe the Rapture of the Church will be that sign. If an unbelieving Jew is with a born again Believer at the end of Rosh Hashanah on the year of the Rapture and sees Believers being Raptured with their own eyes I think that they would believe! The Rapture will be a sign to them

I believe that they are going to miss their Christian friend. ***"My soul desired the firstripe fruit"***. When they realize that he was right about these things and they have been left behind because they did not believe the Gospel. Maybe they will change their minds after the Rapture. Or maybe they will be deceived like the masses.

After the rapture, all Believers will be gone. The restraining power of the Holy Spirit that acted through them is gone! There is nothing to hold back the evil. The "restrainer" is gone. ()

***"The good man is perished** out of the earth: and there is none upright among men: they all lie in wait for blood; they hunt every man his brother with a net"*. (Micah 7:2)

(Micah 7:2) begins by saying …***"the good man is perished"***. This passage is talking about **believers** being raptured **before** the Tribulation begins. The phrase ***"the good man"*** indicates that we are talking about **believers**.

"Perished" is translated from Strong's # **H6 - *'abad* It is** The term ***"perished"*** can mean to **vanish;** to **disappear**; to be completely and utterly removed; **gone completely**;

Notice the next verse indicates that the result of the ***"good man"*** being removed is political corruption.

"***That they may do evil* with both hands earnestly***, the prince asketh, and the judge asketh for a reward; and the great man, he uttereth his mischievous desire: so they wrap it up"* (Micah 7:3)

This sounds like the bad version of "The Art of the Deal".

This passage is similar to (2Thessalonians 2: 6-8). Both passages indicate a spreading of evil as a result of believers being removed. (2Thessalonians 2: 6) is talking about a "*restrainer*" that restrains "*the wicked one*". Perhaps God will raise up special people for that purpose.

The restrainer is identified as the Holy Spirit that seals every believer in the Gospel of Jesus Christ! They are both taken out at the Rapture. (Micah 7:1-6) (Ephesians 1:13) (2Timmothy 2:6-8)

"The best of them is as a brier: the most upright is sharper than a thorn hedge: the day of thy watchmen and thy visitation cometh; now shall be their perplexity". (Micah 7:4)

It continues on to describe how corrupt men will be during the Tribulation. You can't trust a friend, your wife or even a judge. Can't trust a guide either. It sounds like everyone is selling out whoever they can for whatever they can get for themselves. Honor and loyalty are a thing of the past.

*"Trust ye not in a friend, put ye not confidence in a guide: **keep the doors of thy mouth from her that lieth in thy bosom**"*. (Micah 7: 5)

*"For the son dishonoureth the father, the daughter riseth up against her mother, the daughter in law against her mother in law**; a man's enemies are the men of his own house**"*. (Micah 7: 6)

"*Therefore I will look unto the LORD; I will wait for the God of my salvation: my God will hear me"*. (Micah 7: 7)

Some people say that they will believe when they see the rapture for themselves. Many are prepping for what they think will be a big Hurricane party or a scenario like the post-Apocalyptic movie titled "Mad Max and the Thunderdome" starring Mel Gibson. Many are storing food, guns, ammo, gold and silver to help weather the storm. They think that they are ready, because they don't know their Bible. Some say "bring it on, I am ready."

*"**Woe unto you that desire the day of the LORD!** to what end is it for you? the day of the LORD is darkness, and not light."*(Amos 5:18)

The 70th week of Daniel will be a time of darkness and terror! I am expecting widespread power outages which will contribute to the "darkness" spoken of repeatedly. Also the lack of power will be a real health challenge. The lack of refrigeration will cause food to go bad. Following kosher dietary laws during this time will save many lives.

"As if a man did flee from a lion, and a bear met him; or went into the house, and leaned his hand on the wall, and a serpent bit him."(Amos 5:19)

*"Shall not the day of the LORD be **darkness**, and **not light**? even **very dark**, and **no brightness in it**?"*(Amos 5:20)

The only thing the Tribulation has in common with a hurricane party may be the darkness part. If 20% of the Nations power grid goes down, it all goes down except for Texas. Sabotage, Terrorism E.M.P.'s or solar flares could do the job. The Bible speaks of sever darkness and emphasizes it often in relation to The Day of the Lord.

*"That day is **a day of wrath**, a day of trouble and distress, a day of wasteness and desolation, a day of **darkness and gloominess**, a day of **clouds** and **thick darkness**"*,(Zephaniah1:15)

Some of that darkness may be the result of smoke caused by the many judgements that occur. One third of the trees will be burned up and all of the grass. (Revelation 8:7) I suspect that the thunder judgements will be a source of smoke as well.

If the Yellowstone Caldera blows, it could block the suns light all over the world. By the way, if the Yellowstone Caldera blows, it will be the first time in human history that a super volcano has erupted. Scientists say Yellowstone will turn into a giant lake of molten lava. That sounds like a literal Lake of Fire to me. Scientists are saying that it is past due for an eruption and has been showing signs that an eruption may be near. It is good to know that believers will be raptured before any of this happens.

*"**For God hath not appointed us to wrath**, but to obtain salvation by our Lord Jesus Christ"*, (1Thessalonians 5:9)

I am expecting for Terrorists to play a role in shutting down the power grid initially. The result would be more than just darkness. People will starve and have to fight for food. The have nots will outnumber the haves. Everyone will lose!

The Bible says that one fourth of mankind will be killed as a result of the first four judgments known as the "Four Horseman". That will cover the first three and a half years of the Tribulation. Twenty Five Percent of today's population means that about 1.8 billion people will die in the first half of the Tribulation. That is 4 judgements out of 28 total.

3.14 Evil Will Grow Worse And Worse

"This know also, that in the last days perilous times shall come". (2Timothy3:1)

This is talking about the time that we are living in now. The Day of the Lord is about to begin! We are seeing wars and rumors of wars already.

"For men shall be lovers of their own selves, covetous, boasters, proud, blasphemers, disobedient to parents, unthankful, unholy (2Timothy3:2)

"Without natural affection, trucebreakers, false accusers, incontinent, fierce, despisers of those that are good", (2Timothy3:3)

"Traitors, heady, highminded, lovers of pleasures more than lovers of God"; (2Timothy3:4)

"Having a form of godliness, but denying the power thereof: from such turn away". (2Timothy3:5)

Finding a doctrinally sound Church is difficult today. It has been that way for quite some time. If you want a good Church, you need to ask God for help. You are not likely to find a good Church on your own. Ask God to help.

"Yea, and all that will live godly in Christ Jesus shall suffer persecution" (2Timothy3:12)

*"**But evil men and seducers shall wax <u>worse and worse</u>**, deceiving, and being deceived"*. (2Timothy3:13)

3.15 Deception By Our Leaders And News Media.

Our Leaders have secretly sold us out!

American technology is now completely controlled by the most evil and diabolical secret societies on earth. They are called "Luciferians" and they control our country and the world. They are using our banking system and technology to advance the N.W.O. They have secretly taken over and sold us out!

Stories about weapons of mass destruction were lies about Iraq to justify the war. Likewise, stories about Gaddafi abusing his people were also lies. The fact is Libya had free education as high as you wanted to go for men **and** <u>woman</u>; free health care for all; free apartments for newlyweds without a job or means to pay just to help them get started.

Nevertheless America bombed Iraq and Libya into the ground so that Muslim extremist could take over the country. This was done by the "community organizer" who illegally occupied the Whitehouse.

The United States of America is the 11th horn spoken of by Daniel. It is the Nation of the Antichrist (ETBH 28-38) and its technology will be used to deceive the world. I believe that the Miracles, signs and wonders used by the False Prophet, the Antichrist and their false prophet minions spoken of in the book of Revelation are based on human technology.

When Moses was withstood by Pharaohs magicians', they used the technology of their day to mimic the miracles of Moses. Moses miracles however, were superior to the "miracles" performed by Jannes and Jambres.

"Now as Jannes and Jambres withstood Moses, so do these also resist the truth: men of corrupt minds, reprobate concerning the faith" (2Timothy 3:8)

Military grade holographic technology can make you think you are seeing whatever they want you to see. Civilian holographic technology is advanced much further than most Americans can imagine. Military holographic technology is advanced about 50 years beyond that. The technology is kept secret because it is no good if the enemy knows what you can do.

It is interesting how holographic technology is used to entertain the public all over the world, but Americans don't get to see much of that. You can go on Youtube and see the holographic displays enjoyed in Japan, Russia and other countries, but you can't see it in 3-D on Youtube.

American holographic technology is the most advanced in the world, but Americans don't get to enjoy it like people in other countries. I believe that we are being dumbed down so we don't realize when it is being used to deceive us.

I don't see the holographic entertainment in the U.S. that is seen by the public in Asia and Russia and to a lesser degree in parts of Europe. I believe that if Americans realized how far holographic technology has advanced they would not be so skeptical about the belief that holograms could be used to create the images of planes going into buildings on 911.

Cut and paste the following link into the address bar of your browser…

https://www.youtube.com/watch?v=CUoqwUVOxHE

The Bible mentions an image that is placed in the third temple of Israel in the middle of the Tribulation. It is an image of the Antichrist that everyone will be required to worship. The same thing happened in Ancient Babylon.

The late Dr. M.H. (Hank) Lindstrom was a speculator when it came to Bible Prophecy. One of the possibilities he suggested about the image was that it could be a computer Generated hologram or robot look alike. Such an image could give the impression that the Antichrist was always awake and always watching.

"And deceiveth them that dwell on the earth by the means of those miracles which he had power to do in the sight of the beast; saying to them that dwell on the earth, that they should make an ***image*** *to the beast, which had the wound by a sword, and did live".*(Revelation 13:14)

This image is also mentioned in (Revelation 13:15; 14:9, 11; 15:2; 16:2; 19:20; 20:4)

Now consider the passage that mentions the Beast rising up out of the sea.

"And I stood upon the sand of the sea, and ***saw a beast rise up out of the sea,*** *having seven heads and ten horns, and upon his horns ten crowns, and upon his heads the name of blasphemy".* (Revelation 13:1)

Consider the possibility that this passage may also be describing a hologram used by the Antichrist to deceive. Maybe not, but the point is that Jesus described this time as a time of deception. It could be almost anything. (Matthew 24:4, 5) The point is that you cannot believe your eyes, but you can believe what God has written!

Hologram technology might be used to fake a U.F.O. landing. Plastic surgery could be used to fake an Alien visitor. This has been talked about by President Regan and others. I don't think that we will have to wait very long to find out.

3.16 The Stage Is Set.

Until recently, the Ancient nations spoken of in end times Bible prophecies had ceased to exist. The Catholic Church said that God was through with Israel. The end times prophecies looked like an impossibility to most of the world. Sir Isaac Newton disagreed with the "infallible" Pope and said that Israel would return to their land in the last days just like the Bible said. World War One changed all of that.

The Ottoman Turkish Empire that controlled all of those territories lost the war. The British gained control of what used to be the Ottoman Turkish Empire. As a result, independent Nations were reborn. Nations that had ceased to exist for millenniums began to reappear. Now what used to be considered impossible is now clearly on its way to being fulfilled.

Here are the countries spoken of in the last days prophecies that either ceased to exist for a while and recently came to be reborn or in the case of the U.S. born for the first time. The first on this list is The United States of America 1776; Egypt 1922; Turkey 1923; Jordan 1927; Ethiopia 1930; Saudi Arabia 1932; Iraq 1932; Syria 1941; Lebanon 1946; Israel 1948; Libya 1951.

4

Scientists Make Mistakes But God Never Does!

(Jeremiah 9:23, 24)

4.1 If You Want To Know The Truth, God Will Get It To You!

"*And ye shall seek me, and find me, when ye shall search for me with all your heart*". (Jeremiah 29:13)

"*They that know thy name will put their trust in thee: for thou, LORD, hast not forsaken them that seek thee*" (Psalm9:10)

"*That was the true Light, **which lighteth <u>every</u> man that cometh into the world***". (John 1:9)

"*The heavens declare the glory of God; and the firmament sheweth his handywork*". (Psalms 19:1)

"*For the wrath of God is revealed from heaven against all ungodliness and unrighteousness of men, who hold the truth in unrighteousness*"; (Romans 1:18)

"*Because that which may be known of God is manifest in them; for God hath shewed it unto them*". (Romans 1:19)

"*For the invisible things of him from the creation of the world are clearly seen, being understood by the things that are made, even his eternal power and Godhead; **so that they are without excuse***": (Romans 1:20)

God has provided proof of his existence in many ways including prophecy and science but nothing is so obvious, so magnificent and as amazing as God's Creation! The spirit of man itself is another amazing miracle of God!

Yeshua has provided truth to every man which leads to him. He said if you want to know the truth, ask God and he will get it to you. You are getting it now!

"But without faith it is impossible to please him: for he that cometh to God ***must believe*** *that he is, and that he is a rewarder of them that diligently* ***seek him***". (Hebrews 11:6)

"The secret of the LORD is with them that fear him; and ***he will shew them his covenant***". (Psalms 25:14)

This passage talks about God's covenant with man. This is a reference to the Gospel. The Gospel is God's simple plan of Salvation for all mankind. This passage states that God will get that message to anyone who fears and seeks God.

"If any man will do (willeth to do) his will, ***he shall know of the doctrine****, whether it be of God, or whether I speak of myself."* (John 7:17)

God says that if you humble yourself before God and seek him, you "*shall know of the doctrine*".

"Nevertheless he left not himself without witness, in that he did good, and gave us rain from heaven, and fruitful seasons, filling our hearts with food and gladness". (Acts 14:17)

God says that he has always had a witness available for anyone, anywhere, anytime that is seeking to know their Creator.

(Jeremiah 9:23, 24)

4.2 God Confirms The Unseen With What We Can See.

We cannot see the future, but God can. We cannot see the spirit world, but God can. In the Bible, God reveals knowledge about these things and more. He confirms his word with prophecy and scientific knowledge revealed from ancient times that only God could have known!

Only the God described in the Bible could have known about the things revealed to us through the Bible. That is why God tells us so much in advance. That is also why the Bible is laced with so much scientific information that only God could have known. All of this confirms the Bible to be the word of God.

God has given us a written record that is known to exist, in some cases, for over 3700 years. From ancient times, the Bible has also demonstrated a knowledge of the Creation that was far beyond man's understanding at the time. Man's knowledge is still playing catchup with the Bible.

"Where the senses fail us, reason must step in." (Galileo Galilei)

4.3 God Reveals Science Before Man Had The Knowledge

"I believe the more I study science, the more I believe in God."(Isaac Newton)

God reveals science before man had the knowledge. (Isaiah 48:3-6)

The Bible reveals a knowledge of Genetics. (ETBH chapter 18:3) God reveals a knowledge of biology and physiology. God said that all flesh is not the same. There is the blood of different animals and there is the blood of man and they are all different. (1Corinthians 15:39)

In the 1600's blood transfusions between man and animals was proposed and faked. There is a story about a researcher who tried to duplicate the faked experiment on himself. As the story goes, He received blood from a sheep. At the time, it was believed by some that alcohol would protect against any negative effects.

The researcher completed the procedure, walked across the street to the tavern, ordered a drink and began bragging about what he had just proved to the world. As the story goes…the man died suddenly before he could finish his drink. I doubt that he would have been able to make it across the street. The story may have been embellished a bit, but scientists of that day had to find out the hard way that you just can't mix human and animal blood.

The Bible says that all flesh is not the same, neither is the blood. Even though the Bible said "*The life of the flesh is in the blood*" (Leviticus 17:11) Bloodletting was used to treat diseases until the 1800's. George Washington died from bloodletting.

The Bible exhibits a knowledge of geography before man could have known. The Bible talks about when the continents were divided. (ETBH chapter 23)

The Bible describes the geography around Siberia before the prophet Ezekiel could have known about it. There were no reliable maps available. (ETBH chapter 46.1)

The Bible demonstrated a knowledge of astrophysics…

God said the earth is round, suspended in space on nothing. (Isaiah 40:22)

The Bible talks about an empty place in the north. (Job 26:7) Astronomers have found out that such a place does exist.

The Bible describes how far God removes our sins from us and compares it to how far the east is from the west. (Psalms103:12) He doesn't say as far as the north is from the south, but compares it to the east from the west. There is a finite distance between the north and south poles. You can only go north for a limited distance before you are heading back south again, but you can go east for infinity and you will always be going east. The same thing applies to going west. The Bible is demonstrating global knowledge to express infinity before man had such understanding.

This is God's version of "scientific method" I call it **Confirmation by Observation**. He has provided us with plenty of evidence to observe. This is **not "Blind Faith"**, it is the opposite! It is **Confirmation by Observation**. That is a boast that evolutionists can't make!

4.4 God Declared The Beginning And End Of The Universe!

I remember as a child, being told by an older adult that "the Universe has always been here and it always will be". I heard that a lot when I was growing up. If you said otherwise, you might find yourself in an argument. People used to have strong feelings about that, some still do.

"Modern Science" used to teach that matter cannot be created or destroyed. That was called The "Law" of Conservation of Mass. The Bible was criticized for saying otherwise. Now we know that the Bible was right all along. The "Law" was changed to the conservation of Mass and Energy.

Thousands of years before man knew it, God declared that the Universe had a beginning and it will have an end!

"In the beginning God created the heaven and the earth". (Genesis 1:1)

The Bible also teaches that the Universe will have an end. (Revelation 21:1) It will be destroyed by fire.

*"And I saw **a <u>new</u> heaven** and a new earth: for **<u>the first heaven</u> and the <u>first earth were passed away</u>**;* "(Revelation 21:1)

Not only does this passage declare that this heaven and earth will pass away, but it says that there will be another, a new one! This may be an indication that there are not only other worlds, but maybe other universes as well!

*"For, behold, **I create <u>new heavens</u>** and a <u>**new earth**</u>: and the former shall not be remembered, nor come into mind".* (Isaiah 65:17)

The Bible teaches that the Universe is **<u>not</u>** eternal or infinite. God says our Universe had a beginning and he Created it by his ***"power"***! Energy and Power are equivalent terms.

The law of the conservation of mass contradicted this notion! The world belief about the origin of the Universe was that it was eternal, with no beginning or end. The Bible said otherwise and was criticized for it. Today we know that our universe had a beginning which implies an end. The "Law" of the conservation of mass was amended by Einstein. The new version is called the Law of the conservation of Mass **and Energy**.

The Bible states repeatedly that God Created everything by his ***"power"***. Power and energy are equivalent terms. The Bible said a long time ago that matter was Created from energy.

*"He hath made the earth by his **power**, he hath established the world by his wisdom, and hath **stretched out the heaven** by his understanding"* (Jeremiah 51:15).

This same passage also describes an expanding universe. Imagine that, God also said that first as well.

When you consider what the Bible says about the Creation, it stands out from all other beliefs of the past and present. The difference is that the Bible accurately talks about science and the future and that is how God says that he confirms his words.

The Law of the Conservation of Mass was changed to "The Law of conservation of Mass and Energy.

The mystery of what holds the protons in the nucleus of an atom together is also explained in the Bible. Scientists called this force that holds the nucleus together "nuclear glue" for a lack of explanation. Now they call it the "strong force", but still don't understand where it comes from.

*"And **he is <u>before all</u> things**, and by him all things **consist**"*. (Colossians 1:17)

Here is more insight into the nature of the Universe. God says that he holds everything together.

"Consist" is translated from Strong's G4921 – *synistēmi* one of the definitions is to hold or put things together to make one whole thing.

This passage says that Jesus is the Creator and "all things" are held together by him. That sounds like **"nuclear glue"** to me.

4.5 "They All Shall Wax Old As Doth A Garment"

"And, Thou, Lord, in the beginning hast laid the foundation of the earth; and the heavens are the works of thine hands:" (Hebrews 1:10)

*"**<u>They shall perish</u>**; but thou remainest; and **<u>they all shall wax old</u>** as doth a garment;" (Hebrews 1:11)*

"And as a vesture shalt thou fold them up, and they shall be changed: but thou art the same, and thy years shall not fail". (Hebrews 1:12)

Excuse me, but I think that sounds a lot like the 2nd Law of Thermodynamics. ***<u>"they all shall wax old as doth a garment;"</u>***

At a point in time that the Bible calls ***"In the beginning"*** the Creator organized and "wound up everything" and now it is becoming unwound and less organized because of sin. Where there is sin, there is death. (Romans 6:23)

God is describing a Universe or at least a part of our universe that is growing less orderly, more random, decaying and growing old, unwinding. A world that will ultimately be eliminated and replaced with another, a **<u>new</u>** heaven and earth. A heaven and Earth ***<u>"wherein dwelleth righteousness"</u>*** *(2Peter 3:13)*

I think the point here may be that our Universe or our world is polluted by Satan and will eventually be replaced by one that is unpolluted. But that won't happen for **at least** another one thousand years **after** *Armageddon.* This passage may be talking about an entire universe or it may just be referring to a new planet, in a new solar system with a new heavenly view when looking up.

God will always be! The Universe will grow older and be replaced with a new one! *"wherein dwelleth righteousness" (2Peter 3:13)*

"Nevertheless we, according to his promise, look for new heavens and a new earth, ***wherein dwelleth righteousness****".* (2Peter 3:13)

Entropy is a measure of disorder. Sin is also a measure of disorder.

The Universe is wearing out. This is the opposite of what_Evolutionists believe, yet is considered a law of science.

The Bible says the Universe is already old and the earth is also already old, they both had the same beginning and they will have an end, but God is forever!

4.6 Looking For Truth In All The Wrong Places.

"He who thinks half-heartedly will not believe in God; but he who really thinks has to *believe* in God." (Sir Isaac Newton)

If Satan can get you to stop searching and thinking about God, he has won.

"In whom the ***god of this world*** *hath* ***blinded the minds*** *of them which* ***believe not****, lest the light of the glorious gospel of Christ, who is the image of God, should shine unto them".* (2Corinthians 4:4)

Satan blinds the minds of unbelievers through lies and misinformation. Much of this misinformation comes from our churches; some from our schools; some from our Lodges and mass media.

When honest, thinking people hear things that just do not make any sense, it can be offensive. When the same misinformation is coming from a so-called authority claiming to be sent by God and used for personal gain, it can make honorable people angry! As a result, people reject another counterfeit of Christianity. Repeat that process enough times and you have an angry, frustrated atheist who just does not want to talk about it anymore.

Atheism is an emotional response to repetitive disappointment with the clergy. It seems like there is too many wolves in Sheep's clothing and not enough real sheep! Someone must be willing to guard the flock from the wolves, preserve the truth and share it with others.

The Bible has been misrepresented and the atheist argues and wins against "straw men" with wrong information. Meanwhile the real issues and facts are missed.

Galileo, the Italian inventor and Genius was threatened with death by torture by the so called "infallible Pope" for disagreeing with him. The Pope said the earth was the center of the universe and Galileo said not so…

"It vexes me when they would constrain science by the authority of the Scriptures, and yet do not consider themselves bound to answer reason and experiment." **(Galileo Galilei)**

The Catholic Church opposed the average man possessing a copy of Scripture. They liked to read it in Latin, secretly, behind locked doors so there could be no argument about what the "Church" said that it said.

Men who copied or translated or distributed Scripture without authorization by the Roman Catholic Church were in danger of being hunted down and burned alive.

"It is surely harmful to souls to make it a heresy to believe what is proved." (Galileo Galilei)

The Bible agrees with Galileo's statement. The Vatican and the so called Roman Catholic "Church" is nothing but Roman Paganism masquerading as Christianity.

The Roman Catholic Church has burned, tortured and murdered true, born again believing Christians since they had the power to do so. They had elaborate torture chambers with every evil contraption that their perverse, unregenerate, demon possessed Priests, Cardinals and Popes could conceive of.

If you want the truth, ask God, He will get it to you. Maybe you already did as a child. It may be years later and you are no longer an innocent child, but now is the time that the truth got through and **you need to believe** it!

"Verily, verily, I say unto you, He that ***believeth*** *on me* ***hath everlasting*** *life".* (John 6:47)

"He that ***believeth*** *on him is* ***not condemned****: but he that* ***believeth not is condemned already****, because he hath not believed in the name of the only begotten Son of God".* (John 3:18)

You get it when you believe it, and it is **Everlasting!**

4.7 The Bible Has A Verifiable Record From The Beginning

God has declared a record of significant events starting with the Creation of the universe and our planet in particular. God's record of these events includes information that only God could have known. Scientists are just now becoming aware of some of these facts. (ETBH chapters 7-16)

"***I have even from the beginning declared it to thee; before*** *it came to pass I shewed it thee:*" (Isaiah 48:3)

The first and most significant event to record is **"*the beginning*"!** The Bible describes this event literally and in the most absolute sense of the word.

Science has confirmed the Biblical description of the Creation, but there are men who just won't admit it… This is discussed in (ETBH Ch. 7-8).

The Old Testament describes an expanding universe at least 14 times. Job said it twice 3500 years ago, if you are counting. (ETBH Ch. 5:3).

The Bible does <u>not</u> say the earth is young or 6,000 years old, but some men misrepresent what God has said and are creating confusion. The Bible says **the earth is old** and **has a history** long **before man was Created**. (ETBH chapters 7-17)

In the beginning God Created everything… <u>except</u> man. He Created the earth… "***<u>not in vain, he formed it to be inhabited</u>***". (Isaiah 45:18) There is no six days of Creation. Everything was Created in the Beginning with a single command from God. Everything that is, except for man. Man was Created much later. That is where the 6 days came in. But they are not 6 days of Creation. They are **<u>6 days of restoration</u>** after the destruction of the Earth. A prehumen civilization was Created to live on the earth. Lucifer was left in charge of the planet as *"god of this world"*. He ruled over the male occupants which have been changed into their current two forms that we call Angels and demons. (ETBH chapters 7-17)

The only thing Created during those 6 days of restoration were **all new species of air-breathing life and man.** (ETBH chapter 18) Plants and fish were not recreated during those 6 days. The same fish and plants that filled the earth and seas when Lucifer ruled the earth are with us today, but they don't grow as big because the environment has changed.

"*The god of this world*", Lucifer, as he was called back then, is one of earth's original occupants. He declared war against God. He deceived and coerced one third of the Earth's population to join in his rebellion. As a result of Lucifer's war, God judged the earth and all air breathing life was destroyed. There is plenty of evidence and the Bible said it first. (ETBH chapters 12-15)

The Bible says that Lucifer and the Angels were the first occupants of this earth. Lucifer, was and still is *"the god of this world"*. They built cities long before man was Created. They apparently developed their own technology which included intergalactic space travel, time travel and perhaps more. This may all sound like a lot to swallow at once, but that is what the Bible says and God gives proof. (ETBH chapters 4 and 7-16)

The evidence of this pre-human civilization is being discovered at the bottoms of the oceans of the world today. (ETBH chapters 12-16)

3500 years ago the Bible said that before the Tower of Babel was constructed all the continents were connected together. According to the Bible, the continents were divided over a period of 239 years. (ETBH chapter 23) It was a time of continual earthquakes.

Today scientists recognize that the contents were once all connected but they believe that it took 12 million years based on current movement. The Bible said it first and gives the real story that the tectonic plates moved faster for 239 years continually.

One thing that is conspicuously absent from the Bible is the unscientific mistakes that you find throughout all of the ancient civilizations even up to modern times. When man's science disagrees with the Bible, eventually science discovers that the Bible was right and man's contemporary science is changed. Our scriptures are thousands of years old, yet they got it all right the first time!

4.8 God Left Evidence Of Events Recorded In Scripture

All of this is declared in God's word the Bible, long before man understood or knew about it.

"***I have even from the beginning declared it to thee; <u>before</u>*** *it came to pass I shewed it thee: lest thou shouldest say, Mine idol hath done them, and my graven image, and my molten image, hath commanded them*". (Isaiah 48:5)

God has declared the Creation; the history of the Earth (2Peter 3:5-7) and the history of Man. God has also declared their future. God has even declared how the World will end.

If you could demonstrate that the Bible was wrong about something, you would discredit the whole Bible and its claim to be the Word of God.

If God's record has been true about the past, then it is reasonable to consider what it says about the present and future. It would not be reasonable to ignore it.

Only the Bible offers a look into the future from the past to show us details about the present. The present confirms the prophecies of the past because that just happens to be the time that we are living in!

4.9 The Geological Nature Of The Earth, God Said It First!

Traditionally, believers have taught that the earth was hollow. After all, that is what the Bible says. For those of you who don't know about what the Bible says about the heart of the earth, let me just briefly say that according to the Bible, there is a lot going on down there and it doesn't sound like it is all molten lava. (ETBH chapters 19:9, 10)

But that is not all. The Bible talks about all of the continents being connected at one time. (ETBH chapter 23) It sounds like a very large tectonic plate that was partly above sea level. Many Geologists believe that much of what we call dry land today was at one time the bottoms of the oceans. Much of what was once the bottom of the Oceans is now dry land.

Before the big asteroid strikes of the **Cretaceous–Paleogene extinction event,** also known as the (K–Pg boundary), a different world existed. That was probably about 65 million years ago.

I think of a subtropical paradise with greenhouse conditions that could support a thick layer of airborne moisture so that the temperature was almost constant worldwide.

Everything on land grew to be much larger than today because the environment has changed.

N.A.S.A. has said that all life on the planet was destroyed about 65,000,000 years ago, along with the dinosaurs. But N.A.S.A. has changed that assertion down to 70 percent. The new age, evolutionists must have had some influence in that change. The evolutionists maintain that not more than 50 percent of all life was destroyed in the **Cretaceous–Paleogene extinction event.**

The Biblical record indicates that **all** air breathing life was destroyed but non air breathing sea life was not. Before the event that left the Chicxulub crater, things were much different I believe. The days may have been much shorter back then. And that is another factor that may have influenced the growth rates. If the earth's orbit around the sun is the same but there were more days in a year, the annual rate of growth may have been significantly affected. The impact of an asteroid the size and velocity of the one that left the Chicxulub crater would certainly affect earth's rotation. (ETBH chapter 16)

Imagine a perfect temperature all day and night. Any breeze would be Gentle. There may or may not have been any night at all but different shades or colors of light at different times. The earth **before** Adam was different in many ways.

The environment was also different Before Noah's Flood

Before Jesus rose from the dead, the Bible says that all dead went to a place in the heart of the earth called Hades. The Bible taught that the heart of the earth is hollow. Until recently, men believed that.

Recently "Scientists" decided that the earth was not hollow, but had a core of hot liquid rock and iron. The Bible position was attacked. However, new evidence indicates that these "scientists" need to recheck their math. (ETBH chapters 19.10-19.11)

4.10 The Nature Of Our Universe, God Said It First!

"<u>things which are seen</u> were <u>not</u> made of things which do appear". (Hebrews 11: 3)

Now we have the very nature of the Universe being revealed in the Bible. Consider what that passage says… it is saying that what we see is made from what we don't see. That is a scientific statement.

A CNN website features a report by Monica Sarkar titled "Large Hadron Collider: World's biggest physics experiment restarts". (April 5, 2015) In that report she says …

"The burning questions that remain include the origin of mass and why some particles are very heavy, while others have no mass at all; a unified description of all the fundamental forces such as gravity; and uncovering dark matter and dark energy, **<u>since visible matter accounts for only 4 percent of the universe</u>**."

A direct link to the full article… (http://www.cnn.com/2015/04/05/europe/hadron-collider-restart/index.html)

Just remember, God said it first.

4.11 The Expanding Universe, God Said It First!

3500 years ago Job is the first to mention an expanding universe.

(Job 26:7) "*He stretcheth out the north over the empty place, and hangeth the earth upon nothing*".

(Job 26:7) is the oldest known reference to an expanding universe!

Job says that God *"stretcheth out the north over the empty place"*

He says it again in Job 9:8. Altogether, the Bible describes the heavens as expanding 14 times. Here are the other 12 times that use the Hebrew word ***"natah"*** (נָטָה) (Strongs#5186) to describe an expanding universe… (2Samuel 22:10) (Psalms18: 9) (Psalms104: 2) (Psalms144: 5) (Isaiah 40:22) (Isaiah 42:5) (Isaiah 44:24) (Isaiah 45:12) (Isaiah 51:13) (Jeremiah 10:12) (Jeremiah 51:15) (Zechariah 12:1)

When God describes the Creation of the heavens, he consistently uses terms and phrases like "*stretcheth forth the heavens*" or "*stretcheth out the heavens*" or "*spreadeth out the heavens*".

Job was written about 3500 years ago, is the oldest book of the Bible and it said that God expands the Universe! That is more than 3400 years before Sir Edwin Hubble figured it out! <u>God said it first!</u>

God goes on to say that he *"hangeth the earth upon nothing"*. (Job 26:7) I guess that describes space pretty well for a nomadic tribesman who doesn't travel far from home.

About 2700 years ago, the earth is described as a sphere in an expanding Universe! (Isaiah 40:22)

"*It is he that sitteth upon the <u>circle</u> of the earth, and the inhabitants thereof are as grasshoppers; that <u>stretcheth</u> out the heavens as a curtain, and <u>spreadeth</u> them out as a tent to dwell in*": (Isaiah 40:22)

The original Hebrew is clearer than the English translation, of… *"circle"*. (Chuwg) Strong's # H2329 = sphere

Isaiah goes on to say… that God *"stretches"* and *"spreads out"* the heavens!

Not only do we see God describing the Earth as a sphere, but he is also describing an Expanding Universe again!

"*Thus saith the LORD, thy redeemer, and he that formed thee from the womb, I am the LORD that maketh all things; that stretcheth forth the heavens alone; that spreadeth abroad the earth by myself*"; (Isaiah 44:24)

More "*stretching*" and "*spreading*", it seems like God is describing an expanding Universe again!

"*The burden of the word of the LORD for Israel, saith the LORD, which stretcheth forth the heavens, and layeth the foundation of the earth, and formeth the spirit of man within him*". (Zechariah 12:1)

About 2700 years ago, God tells man again that he Created our expanding universe. God uses terms like…"*that stretcheth out the heavens*" and "*spreadeth them out*" (Isaiah 40:22) and "*stretcheth forth the heavens*" (Zechariah 12:1)… to describe the Creation of the heavens.

Don't think of it as the "Big (random) Bang" theory, think of it as … "God's Big perfectly organized Bang"! I call it a "bang" because in terms of time and space, it happened exceedingly fast, began with a point and expanded out. It is not a theory. God said it first and that makes it a fact! You can call it a bang or swoosh or whatever you want, **just don't call it random!** It was ordered and organized by God's command! It was sudden, all at once with perfect precision!

Our universe was Created faster than we could imagine because God operates outside of our dimensions of time and space.

4.12 God Confirms His Word By Declaring The Future!

From ancient times God declared the future of Israel and the very day their messiah would present himself on Palm Sunday. (ETBH chapters 23-25)

The Bible pre-told the very day Christ would ride into Jerusalem on Palm Sunday and the very hour he was crucified!

Over 300 detailed prophesies were fulfilled in 32A.D. by Jesus. The rejection by the Jews and acceptance by the Gentiles was also foretold. (ETBH chapters 24-26)

The world-wide dispersion of Israel, the destruction of the second temple and their regathering the second time in the "last days" was all foretold. (ETBH chapters 23-30)

The Bible is mathematically accurate, it predicts the very day and hour of events, hundreds and even thousands of years in advance! The main reason I believe in the Biblical record is because of Israel. God has done and is doing to Israel exactly what he said he would do. Israel has done and is doing what God said they would do. God will continue and fulfill **all** his promises and predictions!

The accuracy of the Bible is demonstrated by Prophesies that have already been fulfilled. End Time Prophecies are currently being fulfilled and the stage is set for the Tribulation to begin. (ETBH chapters 26-32)

4.13 God Said It First!

The Bible talks about a catastrophic event that occurred long ago. That "wobble" in the Earth's rotation may be left over from what happened about 65 million years ago.

"*Behold, the LORD maketh the earth empty, and maketh it waste, and turneth it upside down, and scattereth abroad the inhabitants thereof*". (Isaiah 24:1)

"***The earth shall reel to and fro like a drunkard****, and shall be removed like a cottage; and the transgression thereof shall be heavy upon it; and it shall fall, and not rise again*". (Isaiah 24:20)

According to NASA, the earth was struck by a very large asteroid about **65 million years ago**. That is when the dinosaurs and all air breathing species disappeared suddenly. This is Known as the **Cretaceous–Paleogene extinction event.**

The asteroid struck the Earth in what is now called the Yucatan Peninsula and has been studied and scrutinized for years. Those studies have revealed that the Earth's axis has changed significantly.

The Polar Regions were once closer to the position of the equator relative to the sun and were more tropical. In other words, the earth's axis has changed significantly just like the Bible said. The contents of that asteroid are being studied further.

Many sudden changes occurred as a result of that asteroid hitting the Earth 65 million years ago. NASA scientists originally reported that the asteroid was responsible for the sudden death of all life on the planet; the ice ages; and changes in the Earth's magnetic field; rotation and alignment. All of this is bad news for the Evolutionists but it is exactly what the Bible describes as the result of a war that involved the Earth. The Bible explains with more details about the events that led to the end of all air breathing life on Earth. (ETBH chapters 10-13)

After some political pressure, NASA has changed their official position on this matter. NASA now says that 70% of all life was destroyed by the asteroid. Some scientists say 75%. The Bible said that **all <u>air breathing</u> life was destroyed**. Only plant seeds and sea life survived. That sounds about like 30% to me. They were underwater like the Bible said.

Man's "science" has consistently made mistakes, often politically motivated, when the Bible was right all along! The Bible said it first, long before man knew about these things, so that we can confirm that it is indeed the word of God!

"Thou hast heard, see all this; and will not ye declare it? I have shewed thee new things from this time, even hidden things, and thou didst not know them." (Isaiah 48:6)

"*Will not ye declare it?" is another way of saying* "won't you admit it?"

Throughout history the imperfect science of Man has been in conflict with <u>true</u> science, but the Bible <u>never</u> has. Man's science is catching up to some things in the Bible but they are still behind. No other book can compare it to!

If modern day physicists would consider what the scriptures say, it would advance science! Biblical faith is based on fact not fantasy!

"*Now faith is the substance of things hoped for, the evidence of things not seen.*" (Heb. 11:1)

The original Greek is clearer. "*Substance*" is best translated as "substantiating". The thought here is not blind faith, it is the opposite!

True Biblical faith is built on the foundation of evidence! Biblical faith is all about substantiating the foundation of evidence and that is what Newton did.

The Bible is God's record of what he has declared about the past and the future and when he declared it. God records the past from the beginning so that we can know that it is truly His Story.

Here is a sampling of some of the scientific statements and teachings in the Bible long before man had such knowledge; **The earth is round**... (Isaiah 40:22); **suspended in space upon nothing** ... (Job 26:7); **there is an empty place in the north** ... (Job 26:7); the **conservation of Mass and Energy** ... (Jeremiah 10:12); the Big bang... (Genesis 1:1); Order and Design require a Designer... (Psalm 19:1; Romans 1:19-20); First time Anesthesiology was used in surgery... (Genesis 2:21); use of running water to prevent infection... (Leviticus 15:11-18); Infectious disease control by quarantine ... (Leviticus 13:46); the optimum time for circumcision, the 8th day ... (Genesis 17:12); clouds formed from distilled water... (Ecclesiastes 1:7); elements crack apart in a nuclear explosion... (II Peter 3:10); God holds the elements/atoms together... (Colossians 1:17). Recently man has confirmed the Bible record about all of this, but it is important to remember that **God Said It First!**

There is plenty more. No other ancient book claims to be the word of God or offers proof like the Bible. Certainly not the Koran!

5

A Biblical Perspective Of Physics.

(Jeremiah) (Ezekiel 8:1-18);

5.1 The Bible Is The First To Mention Time Travel!

The Bible speaks of time travel in both directions,

*"**Declaring the end from the beginning**, and from ancient times the things that are not yet done, saying, My counsel shall stand, and I will do all my pleasure":* (Isaiah 46:10)

..."***From Ancient times***", God has declared: "***the things that are not yet done***!" That includes God's record of the ancient past and the future.

This passage is describing time travel in both directions! First we will look at two events that involved going back in time.

God sent Jeremiah his Prophet <u>back in time</u> to see things that occurred a very long time ago. The Biblical description of Lucifer's war matches the cretaceous-Paleogene extinction event that destroyed the dinosaurs and a whole host of magnificent air-breathing species. According to NASA, the cretaceous-Paleogene extinction event probably occurred about 65 million years ago. That event meets the description of Lucifer's War that killed all of the air-breathing life on the planet!

Jeremiah was taken back in time to see this event and the aftermath of the destruction. (Jeremiah 4:23-28).

(Jeremiah 4:23) *"I beheld the earth, and, lo, it was* ***without form, and void;*** *and the heavens, and they had* ***no light****."*

This passage is telling us how the earth <u>became</u> "*without form and void*" as it is found (Genesis 1:2)

(Jeremiah 4:24) *"I beheld the mountains, and, lo, they trembled, and all the hills moved lightly"*

(Jeremiah 4:25) *"I beheld, and, lo,* ***there was no man****, and all the birds of the heavens were fled."*

This happened before man was Created.

(Jeremiah 4:26) *"I beheld, and, lo, the fruitful place was a wilderness, and all the cities thereof were broken down at the presence of the LORD, and by his fierce anger."*

God is ending all air-breathing life on earth and the civilization that was formerly ruled by Lucifer. Jeramiah is witnessing this first hand! God says that in the last days men would scoff at God's record.

(2Peter 3:3) "*Knowing this first, that there shall come in the* ***<u>last days</u>*** *scoffers, walking after their own lusts"*

(2Peter 3:4)*"And saying, Where is the promise of his coming? for since the fathers fell asleep, all things continue as they were from the beginning of the creation"*

Notice that in the **last days** men would scoff at God's record of the future events because of their lack of understanding about God's record of the Past.

(2Peter 3:5) *"For this they* ***willingly are ignorant*** *of, that by the word of God the heavens were of* ***old****, and the earth standing out of the water and in the water"*

This passage is saying that men are willingly ignorant if they don't know God's record about the earth being old and it was flooded twice. (Genesis 1:2; 7:1)

(2Peter 3:6) *"Whereby the **world that then was**, being overflowed with water, perished:"*

We have another passage talking about another **world**. But this world perished from this planet long before man was Created. This passage is talking about a fully populated planet with plants, animals and a civilization of angels living and working here under Lucifer's rule as "*god of this world*".

(2Peter 3:7) "*But the heavens and the earth, which are now, by the same word are kept in store, reserved unto fire against the day of judgment and perdition of ungodly men.*"

The Bible talks about a distant future event that will burn up the earth. Also mentioned is some unusual solar activity in the not so distant future, during the Tribulation period. God is trying to get men's attention throughout the Tribulation. According to (Revelation 16:8) the sun will scorch men for several days during the Tribulation. That is the sign of an unstable star recorded in the Bible 2,000 years ago.

Scientists have believed that our sun is unstable for decades and have been measuring it's neutrino output for quite some time. No one believes that there is any cause for concern any time soon, but the next stage for our sun's cycle is the red giant stage. Scientists believe that the red giant stage will consume the first three planets.

Most Astrophysicists agree that this will not happen for millions of years. Some boldly say billions of years, but not the Bible. The Bible describes the final destruction of the earth by fire after the thousand year reign of Yeshua has finished. Apparently all believers leave to visit new worlds with Yeshua while the unbelievers stay behind and release satan from the heart of the earth. We are talking about just over a thousand years from now.

(2Peter 3:8) *"But, beloved, be not ignorant of this one thing, that one day is with the Lord as a thousand years, and a thousand years as one day."*

God takes another Prophet, back in time.

The Elders of Judah questioned why God was punishing them. **God took Ezekiel <u>back in time</u>** to show him why he was angry with Israel.

Ezekiel is sitting in his living room with the Elders of Judah near the river Chebar. This is a considerable distance from where God transported him back in time … (Ezekiel 8:1-18);

Dr. Scofield has a note on verse 3 which says…"It was as if he were transported back to Jerusalem and to the time when these things occurred"…

(665)… (Ezekiel 8:1-18)… *"And it came to pass in the **sixth** year, in the **sixth** month, in the **fifth** day of the month, as I sat in mine house, and the elders of Judah sat before me, that the hand of the Lord GOD fell there upon me",* (Ezekiel 8:1)

"Then I beheld, and lo a likeness as the appearance of fire: from the appearance of his loins even downward, fire; and from his loins even upward, as the appearance of brightness, as the colour of amber". (Ezekiel 8:2)

"And he put forth the form of an hand, and took me by a lock of mine head; and the spirit lifted me up between the earth and the heaven, and brought me in the visions of God to Jerusalem, to the door of the inner gate that looketh toward the north; where was the seat of the image of jealousy, which provoketh to jealousy."

"And, behold, the glory of the God of Israel was there, according to the vision that I saw in the plain".

"Then said he unto me, Son of man, lift up thine eyes now the way toward the north. So I lifted up mine eyes the way toward the north, and behold northward at the gate of the altar this image of jealousy in the entry".

"He said furthermore unto me, Son of man, seest thou what they do? even the great abominations that the house of Israel committeth here, that I should go far off from my sanctuary? but turn thee yet again, and thou shalt see greater abominations"

"And he brought me to the door of the court; and when I looked, behold a hole in the wall".

"Then said he unto me, Son of man, dig now in the wall: and when I had digged in the wall, behold a door".

"And he said unto me, Go in, and behold the wicked abominations that they do here".

"So I went in and saw; and behold every form of creeping things, and abominable beasts, and all the idols of the house of Israel, pourtrayed upon the wall round about."

"And there stood before them seventy men of the ancients of the house of Israel, and in the midst of them stood Jaazaniah the son of Shaphan, with every man his censer in his hand; and a thick cloud of incense went up"

"Then said he unto me, Son of man, hast thou seen what the ancients of the house of Israel do in the dark, every man in the chambers of his imagery? for they say, The LORD seeth us not; the LORD hath forsaken the earth".

"He said also unto me, Turn thee yet again, and thou shalt see greater abominations that they do."

"Then he brought me to the door of the gate of the LORD'S house which was toward the north; and, behold, there sat women weeping for Tammuz."

"Then said he unto me, Hast thou seen this, O son of man? turn thee yet again, and thou shalt see greater abominations than these".

"And he brought me into the inner court of the LORD'S house, and, behold, at the door of the temple of the LORD, between the porch and the altar, were

about five and twenty men, with their backs toward the temple of the LORD, and their faces toward the east; and they worshipped the sun toward the east."

"Then he said unto me, Hast thou seen this, O son of man? Is it a light thing to the house of Judah that they commit the abominations which they commit here? for they have filled the land with violence, and have returned to provoke me to anger: and, lo, they put the branch to their nose."

"Therefore will I also deal in fury: mine eye shall not spare, neither will I have pity: and though they cry in mine ears with a loud voice, yet will I not hear them" (Ezekiel 8:1-18)

God took Ezekiel back in time to show him what Israel did to provoke him. They turned their back to the Temple, faced the rising sun like some do on Easter Sunday and they worshiped Tammuz. Easter is a pagan sex goddess and this is a forerunner of our modern day "Easter" celebration. And that is why Christians who are "in the know" don't use the word "Easter" because that is the name of a pagan sex goddess. We call it resurrection day or Feast of First Fruits.

The Bible is proof that time travel exists! The ability of God's Prophets to accurately describe the catastrophic destruction of massive amounts of plant and animal life that produced our coal and fossil fuels is more proof that God's record is true.

5.2 Peter, James, and John Traveled Into The Future

"Verily I say unto you, There be some standing here, which shall not taste of death, till they see the Son of man coming in his kingdom" (Matthew 16:28)

The chapter breaks are not always where they should be. In this case, the chapter ends after the first verse talking about the second return of Yeshua. (Jesus Christ) Much confusion has resulted from people not continuing to read the next chapter.

The chapter and verse divisions are not inspired by God. They were added by men in the 1300th century and often break in the wrong places.

(Matthew 17:1) *"And **after <u>six</u> days** Jesus taketh Peter, James, and John his brother, and bringeth them up into an high mountain apart"*

(Matthew 17:2) *"And was transfigured before them: and his face did shine as the sun, and his raiment was white as the light."*

(Matthew 17:3) *"And, behold, there appeared unto them Moses and Elias talking with him"*

They saw the 2nd return of Christ 2,000 years into the future!

Peter, James and John accessed the future through a portal in Time. This is no joke! **Later on Peter wrote about it**. Peter wrote…

*"For **we** have not followed cunningly devised fables, when **we** made known unto you the power and coming of our Lord Jesus Christ, but <u>**were eyewitnesses of his majesty**</u>"*. (2Pe 1:16)

Peter Said They "***Were Eyewitnesses***" Of Christ's Return

! A portal between the then present with the Second Return of Christ 2000 years into the future is what Peter, James and John were describing.

They saw the spot where they were standing, on the Mt of Olives, 2000 years into the future. They became *"eyewitnesses"* of an event that would occur 2000 years later. Peter, James and John were transported forward in time. They were there when it happened and saw it happening, even though it had not yet happened from their original time reference or yet ours! What a paradox! Theologians call it "The Transfiguration". (Mathew 16:28-17:3); (Mark 9:1-4) Any way you look at it, it is time travel.

5.3 John Also Wrote About The Experience.

John was also there and an eye witness to the Return of Jesus. He talks about it in (Revelation 1: 2, 11)

"*Who bare record of the word of God, and of the testimony of Jesus Christ, and of all things that he* ***saw***". (Verse 2)

"*Saying, I am Alpha and Omega, the first and the last: and, What* ***thou seest****, write in a book, and send it unto the seven churches …*" (verse 11)

Either John got a second look or he is describing in more detail what he saw at the transfiguration.

Next verse…

They were witnessing the second return of Christ from the vantage point where they were standing on the Mount of Olives and saw 2000 years into the future!

The second return of Christ is also described in other places in the Bible. Notice the similarities…

"*And he had in his right hand seven stars: and out of his mouth went a sharp two-edged sword:* ***and his countenance was as the sun shineth in his strength****.*" (Revelation 1:16)

5.4 The Speed Of Light Is Constant.

The distance to the edge of our Universe is believed by some scientists to be at least 4 trillion light years away. Traveling at light speed, it would take you about 8 trillion years to go round trip, but Jesus did it in less than three days. (Ephesians 4:8, 9)

Light travels 186,000 miles per second, that may sound fast, but when you consider the distances involved with intergalactic space travel, the speed of light is too slow! For Jesus to travel to heaven and back within three days there would have to be some kind of extra dimensional portal like the one that Jacob saw in (Genesis 28:12)…

"And he dreamed, and behold a ladder set up on the earth, and the top of it reached to heaven: and behold the angels of God ascending and descending on it" (Genesis 28:12).

Jacob described a portal that went to Heaven…with Angels traveling up and down between Heaven and earth. This is a mass transit system for angels or intergalactic space travel by way of an inter-dimensional portal!

Do you remember the old science fiction series called Star Gate SG1? It is interesting to note how some of the themes from science fiction movies have so many parallels with scripture.

A portal from our dimension to another that connects the two would do just nicely, I think. Traveling round trip to Heaven in less than three days (Ephesians 4:8, 9) requires "extra dimensional" Involvement. (Interacting with dimensions other than the four that we are aware of)

One of the missions of the Large Hadron Collider (LHC), the behemoth accelerator straddling the French-Swiss border, has been to test the possibility of unseen extra dimensions. They said… "Since the discovery of the Higgs Boson (particle / field) in 2012, completing the Standard Model of particle physics, **the idea of looking for other dimensions has become more central"**.

This is not science fiction! This is real. They have just spent a whole bunch of money to discover and understand other dimensions and they are planning to spend more!

5.5 God Created Other Worlds.

According to the Bible, we are not alone. The Bible says…

…*"Through faith we understand that the **worlds** were framed by the word of God, so that things which are seen were not made of things which do appear".* (Hebrews 1:2)

Notice that the word ***"worlds"*** is plural. Other *"worlds"* are mentioned again…

*"Through faith we understand that the **worlds** were framed by the word of God, so that things which are seen were not made of things which do appear".* (Hebrews 11:3)

One of the places that Lucifer wanted to attack was the Mount of the Congregation in the sides of the North. This is not talking about earth. (Isaiah 14:13) And then there is the Mountain of God that he visited. (Ezekiel 28:14) There is only two places named but there probably is more.

Not only is there other worlds, but there are also other intelligent beings. Some are much like ourselves and others are like nothing we could imagine! (ETBH chapters 9; 10; 11)

Just remember… God said it first!

5.6 Who Is The "Congregation"?

I believe that the "congregation" (Isaiah14:13 & 14) was made up of the "Hosts of God" which consists of **at least** two separate species of extra-terrestrials described in the Bible (Psalms 148:2) and called Cherubim (Genesis 3:24) and Seraphim. Notice that "The Hosts of God" are named as a distinctly, different species or group. (Isaiah 6:2, 6) I call these creatures "extra- terrestrials" because there is no mention of them ever living on the Earth except for Lucifer who became satan.

Cherubim may have visited the earth from time to time, but for the most part we only hear about them in heaven. Three Cherubim are mentioned in the Bible. Lucifer was a Covering Cherub, he was given the title of "god Of This World" from the beginning. He still holds that title along with others.

Two other Cherubim are (Genesis 3:24) which temporarily guarded the Garden of Eden after Adam and Eve are kicked out for sinning. But that is all that is mentioned until Armageddon is stopped by Jesus and his armies consisting of Cherubim, Seraphim, Angels and the returning saints. (I'll be there.)

Cherubim are definitely strange looking creatures. I believe that there may be more varieties or species of these "*Hosts of God*" than those mentioned here. There is one reference to a group of "*Watchers*" (Daniel 4:17) but no description is given. Watchers were either a type of Angel or another member of the "Hosts of God" that was assigned the responsibility of watching over us. I think it was probably a type of angel because they watch over us, in a larger scale.

At the moment a person believes the Gospel they are assigned an angel which may be what a watcher is.

There may be a very large number of different kinds or varieties of these Hosts of God, perhaps a very large number of them.

If you saw the Movie "Star Wars", you may recall the scene where Luke Skywalker & Han Solo were in a type of Alien bar with all sorts of creatures carrying on around them. This passage makes me think about that scene. The mount of the congregation seems to be like a place where God's Host's gathered themselves together for fellowship or whatever.

Apparently things weren't working out for Lucifer there because he wanted to go there and take it over also.

Where is the ***"sides of the <u>north</u>"***? What is it? How was he going to get there…?

Job described it as… *"He stretcheth out the **north** over the empty place"* and Lucifer thought about it as *"in the sides of the **north**"*. Based on what God said and what Lucifer thought, there seems to be a corridor in space between Heaven and earth; (Job 26:7); the corridor has *"sides"* (Isaiah14:13 & 14), according to Lucifer's thoughts; it was used for transportation between Heaven and Earth which is what Jacob saw in his dream. (Genesis 28:12)

"And he dreamed, and behold a ladder set up on the earth, and the top of it reached to heaven: and behold the angels of God ascending and descending on it." (Genesis 28:12)

This is like a mass transit system between heaven and earth.

I believe that Satan may have developed the technology to travel on his own through other dimensions, **another way**.

"*Verily, verily, I say unto you, He that entereth not by the door into the sheepfold, **but climbeth up some other way**, the same is a thief and a robber*". (John 10:1)

Some passages seem to have double meanings.

5.7 Scientists Are Researching Other Dimensions. (CERN)

The CERN super collider is 17 miles in circumference, making it the largest machine in human history. The super collider borders on three countries. Over 10,000 top scientists from more than one hundred countries have worked on the project. It took them 30 years to build and it cost over ten billion dollars.

The **Higgs boson** (or **Higgs particle**) was the central theme until it was **discovered on (March 14, 2013)**. Now that the **Higgs particle** has been discovered and the more significant theories have either been confirmed or disproven, CERN has a new focus.

Today Physicists are researching the possibility of other dimensions, even time travel. That is currently the central focus of CERN, the world's biggest and most expensive Science Experiment!

Research is currently being done at the CERN Facility in Europe to better understand things that the Bible discussed 3500 years ago. Things that critiques like Voltaire publicly scorned!

The only way to explain some of the things that the Bible talks about is other dimensions and time travel. They are learning a lot over there right now. Sergio Bertolucci, former Head of Research and Scientific Computing at CERN said that the super collider could open a door to other dimensions or other worlds for a tiny fraction of a second.

In his own words Sergio said that this may be just enough time "*to peer into this open door, either by getting something out of it or sending something into it.*" There is concern among his colleagues about what might come out.

Unfortunately there seems to be an occult connection with CERN. There are ancient demonic statues, ritualistic pagan dancing and emblems surrounding this place. A statue of the CERN mascot, Apollyon the goddess of destruction, stands in front of the building complex. What does all of that have to do with science?

The French town where C.E.R.N. is located is called "Saint-Genus-Poilly". The name Pouilly is Latin for "Appolliacum" . In the Roman era a temple stood there in honor of Apollo. The people who worshiped there believed it is a gateway to the nether world. CERN is built on the same spot.

At this point, I think that it is important to consider (Revelation **9:11**) where Apollyon is mentioned. Apollyon is the angel of the bottomless pit who is in control of the stinging "locusts" that come out of it during the second half of the Tribulation. This is called the "***first woe***". These stinging locusts will torment men for five months, who do not have the seal of God in their foreheads.

Notice in (Revelation 9:1) a star falls from heaven to the earth. Talking about the "***star***" it goes on to say that a "***key of the bottomless pit***" was given to him. Notice that the **"star" is actually a person.** I am going to speculate a little about this because it seems to me that CERN may be the key to the bottomless pit spoken of here. After all it is referred to as a door to other dimensions.

That "star" that fell from heaven may be satan losing his title of "*Prince of the power of the air*". Satan ends up possessing the antichrist. This happens in the middle of the Tribulation. Perhaps (Revelation 9:1-11) is describing a failed attempt by satan to break in to the heart of the earth using human technology, but instead, God's angel guarding the Bottomless Pit sends these flying locusts into our world for five months. Although this first attempt is unsuccessful, it appears to me that this same technology may be developed further and used again after the Millennial reign of Yeshua is completed. That is when satan is released from the bottomless pit.

Physicists are already wanting a bigger and more powerful super collider. It has been proposed to build another collider four times as long and six times more powerful. After the one thousand year rule of Yeshua, all believers will be evacuated before our unstable sun goes into the "Red Giant Stage" and burns up the earth.

After the believers evacuate, only unbelievers will remain. The Bible indicates that satan will be loosed from the bottomless pit at that time. Perhaps the unbelievers will build this larger collider and successfully use it to release satan from the heart of the earth at the end of the Millennium.

There may no longer be an angel to guard the place because they will probably be evacuated too. Of course this is speculation, but it certainly fits the facts. We will just have to see what happens.

5.8 Heaven Is In The North!

Earth is like a giant gyroscope spinning in space with its North Polar Axis always pointing in the same direction. On more than one occasion, the Bible implies that Heaven is in the North.

"For thou hast said in thine heart, I will ascend into heaven, I will exalt my throne above the stars of God: I will sit also upon the mount of the congregation, ***in the sides of the north":*** (Isaiah14:13)

(Isaiah14:14) ***"I will ascend above the heights of the clouds****; I will be like the most High."*

Lucifer was on the earth, below the clouds when he was thinking these thoughts. From this, we can conclude that the earth's axis was pointing toward Heaven at that time.

Apparently this empty corridor between Heaven and Earth has sides. The Boötes void is a curios "empty place in the north" that meets Job's description.

"*He stretcheth out the* ***north*** *over the* ***empty place****, and hangeth the earth upon nothing.*" (Job 26:7)

God is telling us here in Job, that there is an *"empty place"* in the north and that he is stretching out the heavens over it. In other words… the ***"empty place"*** is somewhere between heaven and earth because God is stretching out the north (heavens) **over** it.

The Boötes void was discovered with the 200 inch telescope on Mount Palamar, today other observatories are still studying this space phenomenon in the North.

The Bible teaches that Heaven is in the North and that is why the sacrifices are offered "*northward*".

*"And he shall kill it on the side of the altar **northward** before the LORD: and the priests, Aaron's sons, shall sprinkle his blood round about upon the altar".* (Leviticus. 1:11)

North is the direction that promotion comes from. Notice the indirect way this is expressed. This indirect method only slightly obscures the meaning. Those who want to know the truth will "connect the dots", those who don't will miss it.

*"For promotion cometh neither from the **east**, nor from the **west**, nor from the **south**...But God is the judge: he putteth down one, and setteth up another".* (Psalm 75:6 & 7)

There is only one direction left. God is in the North and Heaven is in the North. Currently our North Axis points to the North Star (Polaris)! I don't believe that our axis points directly toward Heaven, but I think that it used to. Northward means approximately but not exactly north.

North is a direction relative to the earth. As the Earth rotates on its axis, it maintains its position in space like a gyroscope

5.9 An Empty Corridor Between Heaven And Earth

..."*He stretcheth out the north over the empty place, and hangeth the earth upon nothing*". (Job 26:7)

Job 26:7 refers to it as the empty place in the north.

The earth is spinning like a top in space at about 1000 miles per hour, with its axis always pointing toward Heaven (north) with an "*empty place*" in space between.

Because it says that "*He stretcheth out the north over the "empty place*", I believe that the "*empty place*" spoken of here is between Heaven and Earth.

Isaiah records what God tells him...

*"For thou hast said in thine heart, I will ascend into heaven, I will exalt my throne above the stars of God: I will sit also upon the mount of the congregation, **in the sides of the north**:"* (Isaiah 14:13)

The *"sides of the north"* mentioned in (Isaiah 14:13) and the "empty place" in the north mentioned in Job 26:7, seem to be talking about the same thing.

What we have here is an empty corridor between Heaven and earth. I believe that Heaven may be directly behind The Boötes void. That is where I expect Jesus Christ and all of his armies and the New Jerusalem will come from.

5.10 Intergalactic Space Travel Implies Time Travel.

Time travel implies other dimensions and other dimensions implies time travel. The Bible talks about time travel in both directions. Accurate, detailed prophesies confirms God's Word and the fact that there are other dimensions. The Bible gives us clues about how God may use them to tell the future.

Intergalactic space travel requires the ability to access and navigate through other dimensions. Intergalactic space travel is talked about in the Bible. The ability to appear and disappear, pass through walls or produce the sound of a voice out of seemingly thin air are all indications of a technology much higher than our own. A technology that can access other dimensions.

When they lived on the earth a very long time ago, perhaps billions of years ago, I believe that Angels had to learn a lot like we do. I believe that they lived here and built cities and many wonderful things. I believe that some of the ruins of those Ancient Angelic cities are buried under water.

Ancient underwater ruins have been found which could be Angelic or Demonic depending on how you look at it. Their leader, Lucifer, *"the god of this world"* decided to rebel against God and God destroyed the Earth.

Probably about 65,000, 000 years ago if NASA is right. It is called the **Cretaceous–Paleogene Extinction Event.**

Just remember, **the Bible said it first**.

5.11 Satan Was Also A Time Traveler. (LUKE 4: 5)

Satan manipulated time when he tempted Christ on the pinnacle of the Temple and the Bible records that fact for our edification. (Luke 4: 5)

*"And the devil, taking him up into a high mountain, shewed unto him all the kingdoms of the world **in a moment of time.**"* (Luke 4: 5)

If that isn't Time manipulation, then what else could it be? Remember that Lucifer who became Satan left Earth and traveled across the Universe to attack Heaven. To travel that kind of distance requires the use of other dimensions. When the time and distances are too great, other dimensions are required. It is the only way that amount of time and distance can be dealt with.

Satan knew how to use other dimensions. Satan was a Time Traveler and deep-space traveler. After all, he went to the sides of the north and the Mount of the Congregation. God apparently provided transportation in some situations such as Jacob saw…

"And he dreamed, and behold a ladder set up on the earth, and the top of it reached to heaven: and behold the angels of God ascending and descending on it". (Genesis 28:12)

I believe that this dream provided a view through an Interdimensional Portal similar to the one that Peter, James and John saw during the transfiguration of Christ on the Mount of Olives.

I don't believe that satan used public transportation when attacking Heaven. I don't believe that he traveled through normal channels.

"*Verily, verily, I say unto you, He that entereth not by the door into the sheepfold, but climbeth up some other way, the same is a thief and a robber*". (John 10:1)

Some passages have double meanings. More about all this later.

Intergalactic space travel is not possible without time manipulation yet Satan attempted to attack Heaven. Satan and one third of his Angels must have had at one time, the ability to navigate through other dimensions and time. Right now, they seem to be stuck where they are.

5.12 The God Of This World Rebelled Against His Creator!

"*For thou hast said in thine heart, I will ascend into heaven, I will exalt my throne above the stars of God: I will sit also upon the mount of the congregation, in the sides of the **north**:* (Isaiah14:13)

"*I will ascend above the heights of the clouds; I will be like the most High*". (Isaiah14:14)

Satan was on the Earth and beneath the clouds when he was thinking this. The North Polar Axis is always pointing in the same General direction like a gyroscope in space. It is always pointing toward Heaven in the North.

God is revealing Lucifer's thoughts when he was plotting, in his mind, to overthrow God in Heaven! God knew what he was thinking as the thoughts passed through Lucifer's mind.

"*For thou hast said in thine heart*"... **(Isaiah 14:13)**

Since the beginning, Lucifer has been the ***"god of this world"***. (2Cor. 4:4) However his name was changed to satan when he led a rebellion against God. Satan means "enemy of God" or "Adversary" and he will continue to be *"god of this world"* until Jesus takes the title away at Armageddon. (Revelation 20:1-4)

"*In whom the* ***god of this world*** *hath blinded the minds of them which believe not, lest the light of the glorious gospel of Christ, who is the image of God, should shine unto them*". (2Corinthians 4:4)

Lucifer lived and ruled on the earth from the beginning. The earth was Lucifer's planet, his domain and home. To a certain extent satan is still "*god of this world*", this is still his territory and he considers us as inferior, offensive invaders that he despises and wants to destroy us along with everything else God Loves...

"*For thou hast said in thine heart, I will ascend into heaven, I will exalt my throne above the stars of God: I will sit also upon the mount of the congregation,* ***in the sides of the <u>north</u>****: I will ascend above the heights of the clouds; I will be like the most High*". (Isaiah14:13 & 14)

Notice where Lucifer was, he was below the clouds, apparently on Earth.

Lucifer wanted to "*ascend into heaven*". This is a clear indication that he was not in Heaven. He must be somewhere else. If he is called "*the god of this world*" it seems reasonable that he would be at home, on Earth, below the clouds. (Isaiah 14:14) Lucifer said…"*I will ascend above the heights of the clouds*". Lucifer was below the clouds, on the Earth, maybe even at home in his living room, but he wasn't in Heaven because that is where he wanted to go.

Lucifer's throne was below the stars of God and that was not satisfactory to him. He wanted his throne to be exalted… "*Above the stars of God*".

Lucifer wanted to… "*Sit also upon the mount of the congregation, in the sides of the north*" (Isaiah 14:13)

It sounds like Lucifer did not sit upon the mount of the congregation which is located "in the sides of the north", but he wanted to. Did pride cause all of this?

6

God Said It First!

6.1 Why The Bible Was Written

"But these are written, that ye might believe that Jesus is the Christ, the Son of God; and that believing ye might have life through his name. (John 20:1)

"For the Son of man is come to seek and to save that which was lost". (Luke 19:10)

"Believe on the Lord Jesus Christ, and thou shalt be saved" (Acts 16:31)

*"For by grace are ye saved through faith; and that **not of yourselves**: it is the gift of God, **Not of works**, lest any man should boast".* (Ephesians 2:8, 9)

6.2 Why This Book Was Written.

"knowing therefore the terror of the Lord, we persuade men; but we are made manifest unto God; and I trust also are made manifest in your consciences". (2Corinthians 5:11)

The Bible accurately records the past.

In the past, the Bible has discussed events that would happen in the future. The most important parts of those ancient prophecies in the Bible, deal with Jesus. Those prophecies give us the very day that Jesus would present Himself to Israel as their Messiah. (ETBH chapters 27, 28)

God puts his Time Stamp on events in the Bible so that his word is always confirmed.

Events described in the Bible that occur toward the end of the Church Age have already happened, right on schedule even though they seemed impossible at the time that it was written about!

Much of what was spoken of by God's Prophets dealt with the time that we are living in right now! It is just like the Bible said it would be in **The Last Days.** (ETBH chapters 40 through 51) Almost all of the World political alliances described in the Bible, appear to have been made or are in the final process. I can see, through God's word, that this N.W.O. is coming like a giant steam roller and it is upon us right now! According to scripture, the best strategy is to evacuate! All of those who have put their trust in Jesus Christ will be evacuated at the Rapture. Those professing "Christians" who believe that their behavior is a factor in becoming "saved" will be left behind with all of the other unbelievers. The Antichrist will take over the West first.

"***And it was given unto him to make war with the saints****, and* ***to overcome them****: and power was given him over all kindreds, and tongues, and nations".* (Revelation 13:7)

The Antichrist is going to take over. If you fight against him after the Rapture, you will lose. If you think that fighting a fight that God says that you will lose is a good idea, then you can't be reasoned with or you just don't know God! You have been warned. You better not be an unbeliever in the wrong place when the Tribulation begins! It will begin **suddenly** and without notice **to the unbeliever**!

Believers who are faithfully watching for their Lord's return will not be caught off guard. Signs will continue until the Rapture for the Believers. After the Rapture, some unbelievers will recognize that the Rapture has happened and they have been left behind. As a result some may change their mind and believe! The ones who were warned by believers like you and me and those who read this book should have a better chance of recognizing **What Is Going on and What Will Happen Next**.

My grandfather was known to be a big, strong, and tough man with a strong, deep voice that sometimes sounded more like a growl than a voice. He was a real tough guy who has been in some rough places. But, My grandfather's most famous slogan was......."**A good run is better than a bad stand**"!!! Sometimes running is the best thing to do, just escape and save yourself and your loved ones! This is a time to be thinking about escaping what is coming over the horizon!

The Rapture is God's way of providing the way to escape the most awful time that this world will ever endure! All born again believers will be taken off the planet **before** the Tribulation begins. Those who call themselves "Christian" because of how they live their life, are not born again because they are trusting in their works. **They will be <u>left</u> <u>behind</u>**. It will be kind of like **<u>The Parable of the 10 Virgins</u>**...

"*Then shall the kingdom of heaven be likened unto **ten virgins**, which took their lamps, and went forth to meet the bridegroom*". (Matthew 25:1)

"*And **five of them were wise**, and **five were foolish**.*" (Matthew 25:2)

"*They that were **<u>foolish took</u>** their lamps, and took **<u>no oil</u>** with them*": (Matthew 25:3)

"*But **the wise took oil in their vessels with their lamps***" (Matthew 25:4)

"*While the bridegroom tarried, they all slumbered and slept.*" (Matthew 25:5)

"*And at midnight there was a cry made, Behold, the bridegroom cometh; go ye out to meet him.*" (Matthew 25:6)

"*Then all those virgins arose, and trimmed their lamps.*" (Matthew 25:7)

"*And the foolish said unto the wise, Give us of your oil; for our lamps are gone out.*" (Matthew 25:8)

"*But the wise answered, saying, Not so; lest there be not enough for us and you: but go ye rather to them that sell, and buy for yourselves*" (Matthew 25:9)

"And while they went to buy, the bridegroom came; and they that were ready went in with him to the marriage: and ***the door was shut****"*. (Matthew 25:*10*)

"afterward came also the other virgins, saying, ***Lord, Lord****, open to us"* (Matthew 25:*11*)

"But he answered and said, Verily I say unto you, ***I know you not****"*. (Matthew 25:1*2*)

This passage has similarities to (Matthew 7:21) and (Revelation 4:1). In (Matthew7:21) the condemned appeal to Christ for a pardon on the basis of their good works done in Jesus's Name. They also said *"**Lord Lord**"*. Jesus replied *"I **never** knew you"*. These people were unbelievers. They were never saved because they were trusting in their good works and not in the finished work of Jesus on the Cross on behalf of all mankind!

If you believe in a false Gospel then you are an unbeliever and not saved. You will miss the Rapture. If you are a religious person who calls yourself a "Christian" because you think that you are "good enough" to call yourself that. Then you don't understand God's only plan of Salvation. Then you probably aren't really saved. There are people who say that they believe in Jesus but what is it that they believe? You need to believe in God's Record!

A person who is saved knows that eternal life is a gift. Anyone who understands that Eternal Life is not of works should also understand that it is not of yourself! Not of works and not of yourself mean the same thing. Nobody gets saved because they are "good enough"! People only get saved when they realize that what Jesus did on the Cross was good enough and trust in that!

Those who are trying to be saved by making Jesus "Lord of their life" are working for their salvation and don't know it. They are *"blinded"* by Satan. They will be left behind at the Rapture along with almost all Catholics, Presbyterians, etc. If you get left behind this book should help.

Just like the five foolish virgins in the parable. Outwardly they appeared like the five wise virgins. (Outwardly they appeared religious) The only difference was the oil. The oil represents the Spirit of God that we all get when we believe the Gospel.

The whole world is preparing for a climax of evil. They are talking about Armageddon at Halloween, Zombies are talked about all year long. Nero's "Broken Cross" was brought back as a "Peace Sign" by occultists in the 60's. They say that it's not the Broken Cross of Nero that symbolized the destruction of Christianity, but an innocent sign of Nuclear Disarmament. We are condescendingly told that it just happens to look that way and it is no more than a coincidence. If we suggest that it may be more than a coincidence we find ourselves facing scorn from scoffers.

Just because the "Peace Sign" was chosen by occultists that oppose God and his people, we are told not to be prejudice in our view about what they are planning and doing. We are told and expected to believe that what they are doing is innocent. I don't think so!

One of our most popular drinks today has the 6th Hebrew letter written three times in a row, across the front of the container disguised as a monster claw mark. One of the slogans used to promote the drink is **"Unleash the Beast Within!"** Well 666 is the mark of the Beast. Are we supposed to believe that this is another coincidence with no connection? Scoffers and Scorners are the same things, just another term for Satan's minions.

The fact is that these secret societies have been around since Ancient Babylon and have spread around the World and have secretly taken over. Most people don't want to know about it. The Bible warns about this in the "Last Days". We need to pay attention to God's warnings! The purpose of this book is to warn people about what God has said and to establish the reliability of the Holy Scriptures.

6.3 The Bible Demonstrates A Knowledge Of The Future!

*"But **there is a God in heaven that revealeth secrets**, and maketh known to the king Nebuchadnezzar what shall be in the **latter days**. Thy dream, and the visions of thy head upon thy bed, are these"* (Daniel 2:28)

God says …

*"For I know that after my death ye will utterly corrupt yourselves, and turn aside from the way which I have commanded you; and evil will befall you in the **latter days;** because ye will do evil in the sight of the LORD, to provoke him to anger through the work of your hands"*. (Deuteronomy 31:29)

*"And now, behold, I go unto my people: come therefore, and I will advertise thee what this people **shall do** to thy people in the **latter days**"*. (Numbers 24:14)

Here is more about *"the latter days"…*

*"When thou art in tribulation, and all these things are come upon thee, even in the **latter days**, if thou turn to the LORD thy God, and shalt be obedient unto his voice;"* (Deuteronomy 4:30)

God says what will happen and when it will happen…

*"And **he hath confirmed his words**, by bringing upon us a great evil: for under the whole heaven hath not been done as hath been done upon Jerusalem"*. (Daniel 9: 12)

*"Now I am come to make thee understand what shall befall thy people in the **latter days**: for yet the vision is for many days.* (Daniel 10:14)

God did to Israel exactly what he said he would do. He even gave a timeline of events. The Bible is God's record of what he has declared from Ancient times. Some of what God has declared has already been fulfilled just like the Bible said that it would.

God confirms his word with a built-in record of future events and knowledge pre telling the future with accuracy, describing events of the past, **before** men could possibly have known about it.

The Bible reveals the nature of these events and nature of the universe where these events occurred, long before man knew about it. God has revealed the nature of Time, Space and all of his Creation with the accuracy, awareness and perspective that only an infinite, all-powerful and all-knowing Creator could have! God has a track record of 100% accuracy in forecasting the future!

6.4 Alexander, God's Great Time-Stamp

Alexander III of Macedon (356–323 B.C.) conquered the world in a little more than 13 years. He was known for his speed of success and the city that he built to carry his name. All of this happened centuries before Christ was born.

The "Ancient" Hebrew Scriptures were considered to be "Ancient" at the time that Alexander gave the order to have them translated into his language which was Greek. Alexander liked to get things done in a hurry so he commissioned seventy scholars to get the job done. That translation is called "the Septuagint" which is Greek for "seventy scholars".

Included in this Ancient collection of generally accepted as inspired Scriptures were the writings of God's Prophets. The whole Bible existed long before it was translated into Greek and yet it foretold the history of Israel; the Messiah and the world!

The prophets of those scriptures spoke of events that occurred long before man was on the earth. Those scriptures also spoke of events that were happening at the time that they were translated; things that were going to happen in the future which have already been fulfilled and things that are due to be fulfilled in the near future.

God uses events to confirm the presence of his word in a place and time. The story of Alexander the Great demonstrates this. Alexander is a timestamp from God that establishes without question that prophecies about the Messiah coming the first time were written centuries before the events occurred.

The prophecies about Israel returning to their land at the appointed time are real prophecies with details that eliminate the possibility of chance. All of that and more was written long before any of it happened. Nobody can argue about when Alexander conquered the world. They can't argue about when the Ancient Hebrew scriptures were translated into Alexander's language either. The order was given while he was still in Egypt. I believe that those 70 scholars may have been leaders of the first project of the Library. Any way you look at it, the Old Testament or Hebrew Scriptures were around long before Alexander conquered the world. We can prove that the scripture was written long before the events described occurred! God predicted the past with accuracy, so **you can believe what God says about what happens next.**

6.5 Alexander "The Great" Conquered Egypt In 332 B.C.

At that time, Alexander controlled the whole world! Upon arrival in Egypt, Alexander was shown possible sites for the city that was to bear his name. Alexander ordered his architect named Deinocrates to build the city to be called Alexandra. It was to be at the location of a small village named Rakotis located near the Nile.

Alexandra Egypt is where the Septuagint was translated into Alexander's language which was Greek. The word "Septuagint" is the Greek word that means "Seventy Scholars. That is how many scholars were being assigned at the beginning of this project.

Assigning seventy scholars at the beginning of this project is an indication of a sense of urgency to complete the project quickly. Alexander wanted to read those preexisting Hebrew Scriptures for himself and he wanted it in a hurry! That is how he conquered the world; that is how he wanted the city that was to carry his name and that is how he wanted this project done as well! He wanted it NOW! At the time, the Greeks were known for speed.

Seventy scholars were a lot of resources to be devoted to a single project, but it was. Because of the magnitude of the resources committed to this project from the very beginning, it appears that there was a big sense of urgency to get the job completed in a hurry. The World was conquered in a hurry. It took less than 14 years for Alexander III of Macedon to conquer the world. The city of Alexandra Egypt was built for a man that was known for being incredibly fast. The "Septuagint" was most likely done in a hurry. The first known Greek translation of the already recognized as ancient Hebrew scriptures is the Septuagint. It probably was completed before the Library was. The Library of Alexandria was completed and use by 288 B.C.

The Library of Alexandria consisted of an Academy, a Museum, Research Center, and Library. The "Septuagint" A.K.A. the Alexandrian Text is the crowning achievement of that Library and confirms that the Scriptures were already recognized as being Ancient at the time of this translation in the 4th century B.C.

6.6 "There Shall Come In The Last Days Scoffers"

*"Knowing this first, that there shall come **in the last days** scoffers, walking after their own lusts"* (2Peter 3:3)

*"And saying, **Where is the promise of his coming?** for since the fathers fell asleep, all things continue as they were from the beginning of the Creation".* (2Peter 3:4)

*"For this **they willingly are ignorant** of, that by **the word** of God the heavens were of **old**, and the earth standing **out of the water** and **in the water:**"* (2Peter 3:5)

*"Whereby the world that **then was**, being overflowed with water, **perished**":* (2Peter 3:6)

*"But the heavens and the earth, which **are now**, by the **same word** are kept in store, reserved unto fire against the Day of Judgment and perdition of ungodly men".* (2Peter 3:7)

This passage is talking about men who "***willingly are ignorant of***" **three things**:

The first thing mentioned that men "***willingly are ignorant of***" is the fact that the Heaven and Earth are both **old** and were **Created** by the **Word** of God, a very long time ago;

The second thing that God says that men "***willingly are ignorant of***" is **the history of the earth**… "*and the earth standing **out of the water** and **in the water**:*" (2Peter 3:5) This passage is talking about **two floods.** I believe **the first flood** was the result of God's Judgement on Lucifer and his earth. I believe that event probably occurred about 65,000,000 years ago and is referenced in (Genesis 1:2) and (ETBH chapters 8-10). Noah's flood was the **second and last global flood**.

"***The world that then was***" (verse 6) is talking about the **first world** to occupy this planet. That **first world** is the world that was destroyed **by God's First Global Flood Judgement.** That Judgement was against Lucifer as he was called back then. "Lucifer" is the name given to a Cherub that was given the earth as his home. A Cherub is not the same as an Angel. (ETBH chapters 12-18) After Lucifer attempted to forcibly over through God's authority with violence, God changed Lucifer's name to Satan. God still calls him *"god of this world"*; *"Prince of the power of the air"* and the father of lies.

When God judged the Earth for the First time, it was a planet with only one big continent and plenty of water. That big continent was home to a large variety of extinct air-breathing animal species; fish and other non-air breathers which are still in existence today; plants which are also still with us today and a very large population of advanced Humanoids who are also known as Angels. (ETBH chapters 12-18)

Long before man was Created, Angels lived on this earth and Lucifer ruled over them as *"god of this world"*. They were an advanced civilization with the ability to access and use other dimensions for space and time travel. Energy was no problem. The Universe is full of energy and I believe that they learned how to harvest it.

*"Whereby **the world that then was,** being overflowed with water, **perished**":* (2Peter 3:6)

The History of the Earth is the <u>second thing</u> that ***"Men willingly are ignorant of"***

This is talking about **the history of the earth** which is **the 2nd thing** that ***"Men willingly are ignorant of"...*** The Bible talks about **two floods** and **two worlds** that ***"Men willingly are ignorant of"***. The Bible describes the destruction by flooding of a world with only one big continent. (ETBH chapters 10-18)... That is ***"the world that then was"*** that ***"perished"***.

The Earth was a single continent planet with an advanced civilization living on it when it was flooded and destroyed. That is **The First Global Flood** and is described in (Genesis 1:2) and (ETBH Chapters.8-23) The destruction of the pre-human world that was Lucifer's home is the subject of (2Peter3:6)

The Bible goes on to talk about the world that **we live in <u>now</u>** which is the **2nd world.**

*"But the heavens and the earth, **which are <u>now</u>,** by the **<u>same word</u>** are kept in store, reserved unto fire against the day of judgment and perdition of ungodly men"* (2Peter3:7)

<u>The third thing</u> that ***"Men willingly are ignorant of"*** is the fact that by the same power or ***"word"*** that ***Created,* Judged** and **destroyed the former world**, the same God or ***"word"*** that judged the former world is going to judge this world that we live in now. We are waiting for the 70th week of Daniel to begin! After 7 years of the worst time the world has ever seen or ever will see, the Battle of Armageddon will be fought.

The stage is almost set. The Bible tells us the direction that things are going to go until Jesus returns. This is a time of evil, deception, betrayal, violence and terror. Watching for those last few pieces of the puzzle to fall into place is both scary and exciting! The things that the Bible said would happen during the 70th week of Daniel are ready to begin next fall or maybe the fall after that, or after that! My guess is that it is more likely to be sooner than later.

6.7 May 14, 1948, Was The Beginning Of The "Last Days"

The terms ***"Last Days"*** or ***"Latter Days"*** are translated from the same Hebrew words and can be used interchangeably. They mean the same thing. The term ***"Last Days"*** refers to the period of time beginning when Israel returned to the Promised Land on **May 14, 1948**, continuing through the Tribulation, Armageddon and up to the Feast of Tabernacles when Christ sets up his earthly Kingdom. That means that we have been living in **"*the Last Days*"** since **May 14, 1948**!

"And it shall come to pass ***in that day****, that the Lord shall set his hand* ***again the second time*** *to recover the remnant of his people, which shall be left, from Assyria, and from Egypt, and from Pathros, and from Cush, and from Elam, and from Shinar, and from Hamath, and from the islands of the sea".* (Isaiah 11:11)

That is exactly what has happened! God has brought back Israel to the Promised Land ***"again the second time"***, just like he said he would do in the ***"latter days"***. Since **May 14, 1948** we have been living in this exciting time that the Bible calls "The "***Last"*** or "***Latter***" Days!

The **"*latter days*"** is described often in the Bible. It is not an obscure teaching in scripture but more of a central theme that is discussed repeatedly.

"Yet will I ***bring again the captivity of Moab*** *in the* ***latter days,*** *saith the LORD. Thus far is the judgment of Moab".* (Jeremiah 48:47)

"But it shall come to pass in the ***latter days****, that I will* ***bring again the captivity of Elam****, saith the LORD".* (Jeremiah 49:39)

So you may wonder about what is so special about this time called ***"the Last Days"?*** As we have said, The ***"Last Days"*** began on **May 14, 1948** but it continues through to the end of the "***70th week of Daniel"***.

6.8 The 70th Week Of Daniel! (Tribulation!)

The **"70th week of Daniel"** is the final "week" of Daniel's Prophecy concerning Israel.

"Seventy weeks are determined upon thy people and upon thy holy city" (Daniel 9:24)

A "week" is 7 of anything, but in the case of (Daniel 9:24-27) the term means **a week of years.** The final week of the prophecy involves Israel in "***the Last Days"*** (ETBH chapters 33-49) The "Last Days" builds up to a climax during the final 7 years of the Age of Law, which is also known as the 70th week of Daniel (Daniel 9:24) Some just call it the "Tribulation".

"The Tribulation" and "The 70th Week of Daniel" have the same meaning. The Bible describes this time in detail. Right now it is all coming together just like God said it would! God has given us plenty of warning about what happens next and about when this time is upon us. If anyone wants to know the Truth, God will get it to them.

*This know also, that in the **last days** perilous times shall come.* (2Timmothy 3:1)

The **70th week of Daniel** has been described by God as the worst time the world has ever seen or ever will see. Many "preppers" think that they are going to be prepared. How can you be prepared if you don't understand what is about to happen? God warns the world about what is upon them. Gold and silver will become illegal.

*"Your gold and silver is cankered; and the rust of them shall be a witness against you, and shall eat your flesh as it were fire. Ye have heaped treasure together for the **last days**"* (James 5:3)

"the anger of the LORD shall not return, until he have executed, and till he have performed the thoughts of his heart: in the ***latter days*** ***ye shall consider it perfectly****"*. (Jeremiah 23:20)

God's Prophet's foretold the future of Israel as proof that they were speaking for God. God foretold Israel's rebellion against God and he foretold Israel's punishment, to the day! The amount of punishment due was in terms of their years of offence. Historically we have start and stop dates so that we can do the math and find out that it equals the exact number of days of punishment prophesied by God's Prophets thousands of years **before it happened!** (ETBH chapter 28-30)

The Bible foretold that Israel would rebel against God and that God would punish them for it! God said that he would scatter Israel among the Nations for two thousand years. He did! God said that Israel would be scattered among the Nations without a Priesthood, without a Temple or sacrifice for over two thousand years and then He would bring them back the second time in the ***Last Days.*** During this time Israel would also be without household gods. (ETBH chapter 28-30)

According to the Bible record, Israel had a bad habit of praying to false gods. The classic example is the golden calf that was made right after God had delivered them from Egypt. That made God jealous! Nevertheless God said that He would bring unbelieving Israel back to their land in the ***"latter days"***. The Catholics said it would never happen but **God did it anyway**, just like he said he would! (ETBH chapter 28-30)

"Afterward shall the children of Israel return"...

"... and seek the LORD their God, and David their king; and shall fear the LORD and his goodness in the ***latter days****"* (Hosea 3:5) (ETBH chapters 30-49)

Israel was reborn in May 14, 1948. The birth of Israel was exactly when the punishment that was prophesied expired. For about 2,000 Hebrew years, the Jews have been scattered among the Nations. Nobody believed that they could return.

6.9 The Nile River Would Dry Up In The Last Days.

"And they shall turn the rivers far away ;"(Isaiah 19:5, 6)

Ethiopia's Grand Renaissance Dam will divert the waters of the Nile away from Egypt just like the Bible said. Today, the risk of drying up the river is real.

The biggest events in human history are about to unfold just as prophesied!

I am talking about the 70th week of Daniel. Twenty eight judgments of God spread out over Seven years of the worst time the world has ever known or ever will know. Twenty five percent of the world's population will be destroyed by the first four judgments alone!

Armageddon is fought at the end of those 7 years. The Battle lasts for only 1 hour and destroys a third of the world's population. Dr. Lindstrom estimated that there may not be more than a hundred million survivors after Armageddon!

6.10 The Ultimate Survival Guide!

The Bible is the Ultimate Survival Guide! The Bible tells you how to be saved eternally and how to avoid all of the judgements that are about to come on the earth. If you don't get the first escape message, there may be a second chance. God will raise up 144,000 young, unmarried, Jewish Missionary Men who will have information for survival!

Considering how many the Bible says will die during the Tribulation it might be helpful to have a program guide to help you get to the right place at the right time or not be in the wrong place at all! **A chronological outline of events that occur during the Tribulation would be the <u>Ultimate Survival Guide!</u>** I believe that the One **Hundred and Forty Four Thousand** Jewish Missionaries From Israel will provide information about **<u>when</u>** and **<u>where</u>** these **<u>28 Judgements will occur</u>**. This information will likely be supplied to them by Moses and Elijah. The 144,000 spoken of in (Revelation 7:4-8; 14:1-6) are Moses and Elijah's Disciples.

"And I looked, and, lo, a Lamb stood on the mount Sion, and with him an **hundred forty and four thousand**, having his Father's name written in their foreheads". (Revelation 14:1)

Moses, Elijah and their 144,000 Disciples are all killed by the Antichrist by the Middle of the Tribulation. (ETBH chapters 44-48)

The Antichrist wants to take over **all** of the Middle-East. He is looking at Ethiopia and Libya to be next, but … *"But tidings out of the east and out of the north shall trouble him":* (Daniel 11:44)

That would be China and Russia. The bad news from China and Russia is that if you try to go any further, we have put together a coalition to stop you. That is the kind of news that it will take to cause him to stop and proceed no further.

"But tidings out of the east and out of the north shall trouble him: therefore he shall go forth with great fury to destroy, and utterly to make away many". (Daniel 11:44)

Jordan will escape the Antichrist.

Consider that the Antichrist is given a Kingdom; he is the 11th horn and the U.S. is also called the 11th Horn. (ETBH chapters 40-42)

6.11 Jordan Will Escape The Antichrist!

Jordan may be the Safest Place on Earth during the Tribulation! The Bible says that Jordan will escape the Antichrist.

"He shall enter also into the glorious land, and many countries shall be overthrown: but ***these shall escape out of his hand, even Edom, and Moab, and the chief of the children of Ammon****".* (Daniel 11:41)

"Edom, Moab, and Ammon" will escape the Antichrist. That is talking about modern day Jordan! How would you like to be King over that nice little piece of real-estate! Well King **Abdullah II of Jordan** may not know it yet but he is in control of what may be the best piece of real-estate on earth! He has natural resources that he doesn't know about. Safety will be a resource in great demand.

6.12 Siberia Will Survive Armageddon.

Today Russia is fighting to protect Syria from Muslim Rebels who want to kill everyone who is not a Muslim. When Russia steps in to defend Christians from Rebel groups who are organized, trained, funded and controlled from powers within the U.S. that is when rolls have been reversed. How come Russia is doing what the U.S. used to do and the U.S, is doing what the Communists do?

The U.S. has rejected God and is ready to take on the roll as the Nation of the Antichrist. This country is calling Good, "evil" and they are calling evil, "good".

Putin has been taking the moral high ground against America in the Middle-East. He has been showing videos released by Muslim Rebels to spread terror to their enemies. These videos showed torture, mutilation and cannibalism. The rebels were spawned, organized and supplied by the U.S.

Putin has tried to get this information to Americans but the American media ignores the story about ISIS. For now there is plenty of information on youtube or the internet if you are looking. Most Americans are caring less because they are losing trust in our government. They have always trusted the government and now they are hearing much about government conspiracies; corruption and cover-ups. Americans are confused and they know it. They don't know who to trust or believe.

Russia and China lead the resistance against the N.W.O. but the leadership of both is primarily unbelieving. Siberia in Northern Russia where everyone dreads to be sent is the only part of Russia that will survive Armageddon which is the final battle at the end of the Tribulation! Siberia will survive, the rest of Russia will be destroyed at the end of the 7 years. (ETBH chapters 48 & 49)

Poor Egypt, Poor Israel! They both are going to have it rough! (ETBH chapters 45-50) God speaks to us through the Bible about what will happen; approximately when it will happen and why. **God wants us to understand**.

God has **accurately foretold the past**, our present and He has told us... **What Happens Next**?

When God says that a country will escape, then it will escape.

6.13 Armageddon, 1/3 OF Humanity Killed In An Hour!

God says **who will start the Battle of Armageddon.**

God even says about when it will start, but not the day or hour.

The battle will last for only one hour! The Battle of Armageddon destroys one third of the worlds remaining population.

The battle would destroy all life if Jesus did not return and stop it.

The Battle of Armageddon will be **Nuclear!**

Armageddon is located in the valley of Megiddo in Israel

According to the Bible, the River Euphrates would dry up to prepare the way for the 200,000,000 man army of China to march into the Middle-East. (Revelation 9:14-16; 16:12)

China has maintained a 200,000,000 man and woman army since the sixties. That is the exact number mentioned in the Bible! (Revelation 9:16)

This was written at a time when there were not that many people in the world. The technology to build giant dams like we have today was unimaginable back then. To do the things mentioned in the Bible in the "Last Days" was considered by most to be unimaginable at the time it was written! Today these things are quite possible. The stage is actually set for events to unfold that were described thousands of years ago with great detail.

There are five dams along the river which can shut off the water with the push of a button.

God's enemy's will part his land (Israel) and they will be Judged and punished for parting God's land and for persecuting Israel.

"*I will also gather **all nations**, and will bring them down into the valley of Jehoshaphat, and will plead with them there for my people and for my heritage Israel, whom they have scattered among the nations, and parted **my land**.*" (Joel 3:2)

A gathering of the Nations against Israel has been building up. The stage is set.

In the last days a wall throughout the land is mentioned. Today they have it built. That has never happened before.

*"And he shall besiege thee **in all thy gates**, until thy high and fenced **walls come down**, wherein thou trustedst, throughout all thy land: and he shall besiege thee in all thy gates throughout all thy land, which the LORD thy God hath given thee"*. (Deuteronomy 28:52)

6.14 Another Holocaust Is Coming!

There will be another, final dispersion of Israel. It will occur in the middle of the Tribulation. That would be 3 and one half years from the beginning. That is when the *"**Abomination of Desolations**"* takes place. God says for believers living in Judea to run to the rock city called Petra which is in the Kingdom of Jordan. I believe that this is believing Jews, mostly. Some Gentile believers who are in the right place at the right time may be able to mix in with escaping Jews to Petra and Jordan. Some say that the city could accommodate over a million people. God will feed them and protect them for 3 and ½ years.

Jordan will escape the Antichrist. God will protect it on behalf of his people. Jordan may become the safest place on earth during the Tribulation. Property values should skyrocket!

Upon finding out about the ***Abomination of Desolations,*** believers are told to just escape, don't take anything with you that may slow you down or indicate that you are leaving for an extended stay! The Bible says to run immediately and don't think about taking anything with you. Just come empty-handed. That is when the Antichrist takes over Jerusalem; takes up residency in the Temple of God; pretends to the world that he is God; and demands worship.

From that point on there will be no buying or selling without his mark. Worldwide slaughter of Jews will begin immediately, suddenly, without expectation. The warnings were ignored by seemingly everyone.

Nine tenths of the Jews will be killed, only 1/10th will survive worldwide. About this same time, the Catholics will find that their agreement was also disannulled. **All deals are off**, in the middle of the Tribulation the Antichrist is indwelt by Satan himself. This is called the **Abomination of Desolations.**

Here are passages that describe the event and even gives the name of the man that he will indwell.

"And the seventy returned again with joy, saying, Lord, ***even the devils are subject unto us through thy name***". (Luke 10: 17)

This conversation is about casting out evil spirits. Jesus had just sent out 70 disciples to do miracles and to cast out demons. Now they are coming back together to talk about it. They were impressed by the power to cast out demons.

"And he said unto them, **I beheld <u>Satan</u> as <u>lightning</u> fall from <u>heaven</u>"**. (Luke 10: 18)

Jesus responds to their excitement about casting out demons by reminding them that Satan will possess the Antichrist in the middle of the Tribulation. After Armageddon Jesus will have to cast Satan out of the body of the Antichrist. The Antichrist and the False Prophet will both be cast alive into the Lake of Fire. Satan will be cast into the bottomless pit where he will remain for 1,000 years.

Verse 18 is Jesus responding to his disciples in verse 17 who are talking about casting out demons. I believe that verse 18 is **<u>another description of the Abomination of Desolations</u>** when Jesus stops the abomination by seizing the Antichrist. He gives the name of the man who will be indwelt by Satan and called by believers the "Antichrist". (ETBH chapters 47, 48)

There will be worldwide Genocide, like never before and never will be again in human history. Two thirds of the Jews in Israel will die horrible deaths at the hands of the most evil people on earth. But this will be worldwide, on the same day, it will be done **suddenly** and **unexpected**. Luciferians are masters of deception and betrayal and that is exactly what is going to happen.

As a result of this **sudden** and **unexpected <u>betrayal</u>**, Jerusalem is taken by the Antichrist in the middle of the Tribulation.

Survival Secret Number 2!

The Antichrist consistently does the opposite of what he says; that is how he is able to besiege the gates; break down the walls of Israel; invade and take over Jerusalem. First he gains trust and then he uses that trust as a weapon against his victims. Betrayal is how it is done.

6.15 The Stage Is Set.

Signs in the Earth.

Global arms race

*"And ye shall hear of wars and rumors of wars: see that ye be not troubled: for all these things must come to pass, but **the end is not yet"***. (Matthew 24:6)

Small Nations around the world are investing large sums of money on National Defense. Superpower Nations around the world are doing the same.

*"Beat your plowshares into swords, and your pruninghooks into spears: **let the weak say, I am strong**"*. (Joel 3:10)

There is an unspoken contest to see who can be the most intimidating threat. It seems like almost every week some country is releasing scary new information about what they can do to an adversary.

*"**For nation shall rise against nation**, and kingdom against kingdom: and there shall be **famines**, and **pestilences**, and **earthquakes**, in divers place"*. (Matthew 24: 7)

The increased earthquake and volcanic activity over the last century is undeniable, but yet there are well organized and well-funded deniers. Who is funding all of these organizations?

The word ***"nation"*** as it is used in (verse 7) **is talking about race** or ethnic group. **Strong's #G1484 – *ethnos,*** is the Greek word used in the original text. This passage is talking about world-wide ethnic cleansing on a global scale! It will be unlike anything that has happened in human history or ever will in the future! Unprecedented, Global Anarchy for 7 years!

Genocide is coming and it is important to understand where it is coming from. It is coming from Satan, and Jesus will personally come back and stop it before Satan can destroy the whole world!

"Ye must be born again", Believe on (Jesus), Eternal Life is Free, it is the Gift of God, you get when you believe it! The New Birth can't be earned and it can't be lost, **<u>ever</u>!**

God has revealed to us the history and nature of our world and its future long before man had such knowledge. Those facts confirm his word as true.

Right here, God is saying to the Nations that the Land of Israel belongs to Him. God is the one who says who gets to live there. The Nations that have plotted to divide up God's land will be judged and punished for that.

(Joel 3:2)

"*Beat your plowshares into swords, and your pruninghooks into spears: let the weak say, I am strong*" (Joel 3:10)

"Proclaim ye this among the Gentiles; <u>Prepare war</u>, *wake up the mighty men, let all the men of war draw near; let them come up:"* (Joel 3:9)

We are experiencing an **unprecedented global arms race**! I believe that we are watching the fulfillment of this prophecy every time we checkout the news.

As it was "***in the days of Noah"*** and Sodom and Gomorrah, or Pompeii in 79A.D. It is interesting to note that there is an active super volcano located underneath Yellowstone National Park that is past due on its eruption. If it does erupt, it could destroy the entire Midwest up to the east coast. All of the farm land will be gone! That alone could cause famine and pestilence, but there will be more, plenty more!

6.16 Jesus Returns!

"Put ye in the sickle, for the harvest is ripe: come, get you down; for the press is full, the fats overflow; for their wickedness is great"(Joel 3:13).

The Bible says that Jesus will return one hour after Armageddon begins. Jesus will do that to stop the total destruction of all life on the planet.

The Bible says that no one knows the day or the hour except the Godhead. But **you can know approximately when it will happen**, **within 13 days**. It will happen between the Feast of Yom Kippur and the Feast of Tabernacles. No one knows the exact day or the hour.

That is a total of 13 days to choose from. It can't happen before the first day and it can't happen after the 13th day either.

I believe that most believers will be looking for Jesus at the second return. Otherwise how could they have survived through all of that without knowing something about when, where and what? They must be seeking God to have gotten saved in the first place.

When Christ returns, he will be visable to everyone. They will see the New Jerusalem coming toward earth from, I believe, the North.

The New Jerusalem is a cube that measures 1500 miles on all sides. That is just a little bigger than the moon. The Vatacin has a very large telescope it is called the "**Vatican Advanced Technology Telescope".** It is a giant infared telescope that is called "L.U.C.I.F.E.R. They reciently changed it to LUCI for short.

Here is a Wicapedia link.. https://en.wikipedia.org/wiki/Large_Binocular_Telescope

The Vaticin, headed up by the False Prophet, will prbabilly spot **The New Jerusalem** first and will throw all of it's spiritual athority behind the decloration that the earth is under attack by alian forces.

To those who have been brainwashed by Hollywood (A.K.A. "Hellywood") will see the New Jerusalem coming! When they see an enormus, structure the size of the moon and shaped like a cube coming from the North, the unbelievers will be tererfied! They may think of the "Borg" from Strar Trek. After all, both are extremly large flying cities shped like a cube. There are other similarities that make me think that Satan has many planed deceptions against God's signs for man kind.

The ability to attack an incomming extraterrestial army from space did not exist until receintly.

NASA's Astroid Redirect program will give the United States the ability to attack incomming astroids.

This technology meets the Biblical discription of an attack against Christ as he returns. We now have the ability to track and attack incomming targets. According to the Bible, that target will be Jesus, his armies and the New Jerusalem!

As you may be beginning to see, the world is preparing for these things spoken of in the Bible 2,000 – 3,000 years ago.

Now is a good time to trust Christ as your Savior! When Christ comes back at the end of the Tribulation to stop the Battle of Armageddon, the Angels will gather all of the unbelievers together and kill them all. He then will judge them and assign them a degree of punishment in the Lake of Fire. (Hell)

Surviving believers will inter into the Millennium. Some estimate that there may be less than 100 million surviving believers. (ETBH chapters

6.17 About 20-40 Million Survivors, Worldwide!

Immediately after Armageddon is stopped, the armies of the world will attack Jesus Christ and his armies. The Angels will gather up all of the unbelievers ***first***. This time the Believers are the ones who are left behind. **All unbelievers will be taken *first***, whether they are carrying a weapon or not. All of the unbelievers will be taken first to the valley of Jehoshaphat. All unbelievers will be judged to determine their degree of punishment in Hell. Every one of them will be cast into the Lake of Fire forever!

Dr. Lindstrom estimated that not more that about 100 million would survive the Tribulation, the Judgment and enter into the Millennium in their mortal human form. Lindstrom did not have a passage that he referred to for his position. He said that 100,000,000 is about the world population of 32A.D. Lindstrom often said this would be the worst time the world has ever seen or ever will see in the future.

"(Joel 3:13) speaks of events taking place in 'the valley of Jehoshaphat,' which seems to be an extended area east of Jerusalem."—J. Dwight Pentecost, Things to Come: A Study in Biblical Eschatology (Grand Rapids, MI: Zondervan Publishing House, 1958), 341.

*"**Put ye in the sickle**, for the harvest is ripe: come, get you down; for the press is full, the fats overflow; for their wickedness is great".* (Joel 3:13)

Jesus Christ and his angels with gather up all of the unbelievers and kill them in The Valley of Jehoshaphat. There is a lot of debate about which valley that is. I don't believe that that valley exists yet. In fact, I believe that is talking about the valley spoken of in (Zechariah 14:4). That is the valley that is created when Jesus first lands his feet on the Mount of Olives. The Mountain will split apart from east to west...

*"And his feet shall stand in that day upon the mount of Olives, which is before Jerusalem on the east, and the mount of Olives shall cleave in the midst thereof toward the east and toward the west, and there shall be a very great valley; **and half of the mountain shall remove toward the north, and half of it toward the south".***

I have spent years wondering about how many people would survive the Tribulation and how they would do it. The best plan is not to be here when it happens. I am looking for the pre-Tribulation Rapture every year at the end of the Feast of Trumpets.

I finally figured out how to estimate the survival rate for the Tribulation and talk about that in detail in chapter (45 ETBH). My estimated number of survivors of the 70th Week of Daniel is roughly about 30-50 million believers to enter into the Millennial reign of Christ in their mortal bodies.

But for those folks that seem to think that they will do just fine because they are so good and sweet, let me just warn you that you must be "***born again***" to be raptured. Being religious and calling yourself a Christian is not a substitute for a new birth which God requires of everyone! Just Faith! Everyone must understand and believe the Gospel! That is all! It is a Gift!

If you get left behind, the first thing that you need to do is change your mind about what you believe about Jesus and believe the Gospel. The next thing to think about is staying alive!

After the Rapture, you have 10 days to get as far from the Antichrist as possible. The U.S. is the 11th little horn and Europe is the ten big horns That Daniel talks about. At sundown on Yom Kippur will come "***<u>sudden</u> destruction***". If you don't get out before that happens at sundown following Yom Kippur, you probably won't!

*"For when they shall say, Peace and safety; then **<u>sudden</u> destruction** cometh upon them, as travail upon a woman with child; and **<u>they shall not escape</u>**"* (1Thessalonians 5:3)

Seven years later, the valley of Jehoshaphat is where the Buck stops! That is where all unbelievers for all of human history to that point in time, will be judged immediately after the 7 year Tribulation is finished and Armageddon has been stopped.

After that event, there will be a river of blood two hundred miles long and deep as a horses bridle. That is the killing of the unbelieving, ruling elite and their unbelieving slaves. The few who managed to escape the Antichrist, will not escape Jesus! They will be pulled out of their holes in the rocks and taken lake all of the other unbelievers.

"And they shall go into the holes of the rocks, and into the caves of the earth, for fear of the LORD, and for the glory of his majesty, when he ariseth to shake terribly the earth". (Isaiah 2:19)

30-50 million survivors is not much for the whole planet! (See chapter 45)

7

"In the beginning God Created"...

EVERYTHING EXCEPT MAN

7.1 The First Hebrew Word Of The Bible..."re'shiyth"

"In the beginning" was translated from that single Hebrew word ..."re'shiyth", רֵאשִׁית

Genesis 1:1 (Genesis 1:1) **"re'shiyth", רֵאשִׁית** the first word of the Hebrew Bible, describes an event that occurred a very long time ago. "The Event" that God is talking about here is... The Creation of the Universe!

(Strong's # H7225 – def.; first, beginning, best, chief)

I believe that this single Hebrew word, in this context, is describing a point of origin for all dimensions in our finite Universe.

God has said that this Universe had a beginning, (Genesis 1:1), and it will have an end, (Revelation 21:1).

Human Science is beginning to accept that fact, but **God said it first!**

This Hebrew word, "re'shiyth", gives us some insight into the nature of God, the nature of our universe and our relationship to him. I am talking about the relationship of our entire Universe to God!

7.2 "In The Beginning", God Was Already There

Before the "*beginning*" was, God was already there! (John 1:1)

Before the word "*beginning*" had any meaning, God was already there! (John 1:1)

Before the word "before" had any meaning, God was already there! (John 1:1)

"*In the beginning **was** the Word, and the Word was with God, and the Word was God*". (John 1:1)

He is before all things in our Universe! (Genesis 1:1) (John 1:1)

God doesn't just live somewhere inside our Universe, God exists before and after our Universe. I believe that our Universe may actually be "inside" of him! I can't wait to find out.

God transcends everything that we know of and then some! You can't talk about where God is, or when God is, or was… because God transcends all of that! He is "outside" of our Time, Space and our entire Universe! He is before and after our Universe (Genesis. 1:1) (John 1:1)

God is outside of Time… (Genesis 1:1) (John 1:1)

God is outside of Space… (Genesis 1:1) (John 1:1)

God is outside of all of the dimensions that make up our Universe and at the same time has access to every point in every dimension!

God is Perfect, Eternal, Infinite, omnipotent, omnipresent, All Knowing, Unchanging, and incapable of sin!

7.3 "In The Beginning" Is Not The Same As The First Day

There once was a time when there was no time.

Before *the beginning*, the word "*beginning*" had no meaning.

"*In the Beginning*" is where it all came from. That was **<u>before</u>** there was a day or anyone to even know about the idea except for God.

The First Day of Earth's restoration came much later and is a different event. Don't confuse these two events which are separated in time, perhaps by billions of years. The human race is relatively new to this earth. The earth is very old, perhaps billions of years old. It has been around for much longer than man.

"In the Beginning God Created"... everything <u>except</u> man. (Psalm 148:1-5) Contains a list of seven things that God Created "*in the beginning*". **Everything in the world is represented in that list <u>except man</u>**. (Isaiah 45:18) Describes how God Created the Heavens and the Earth...God says that he did not Create it like you see it in (Genesis 1:2), God Created the Earth Perfect and Complete in the Beginning! (Isaiah 45:18)

God says that the earth is old; He never calls it young and neither should anyone else!

*"For this they willingly are ignorant of, that by the word of God the **heavens were of <u>old</u>**, and the earth standing out of the water and in the water:"* (2Peter3:5)

The Heavens and the Earth were both Created at the same time called**...** ***"In the Beginning"***. (Genesis 1:1) (Isaiah 45:18)

However there are some who profess to be Christian teachers and insist that the earth is young. **Not So!**

7.4 The 2nd Hebrew Word, Announces The "Trinity"

The Hebrew word for "God" used here is "Elohiym". The form of the word used is plural. In Hebrew, there is a singular, a dual and plural form. The plural form used here means 3 or more.

The Third And Fourth Hebrew Words ...Bara' And 'Eth,_

The two Hebrew words translated as ..."*Created*" are...

bara' -Strong's (H1254) to shape, fashion, Create (always with God as subject)

'eth -Strong's (H853) particle, not translated

The word "*Create*" is used very sparingly in the Bible and always with God as the subject! Its meaning is very specific.

Genesis 1:1 combines re'shiyth, (In the beginning) with Elohiym, (plural for God) and Bara' (Create) to say "*In the beginning God Created*".

Notice what is being created. "The Heavens and the Earth". In the Hebrew that is ... shamayim, שָׁמַיִם for (Heaven), and 'erets אֶרֶץ for (Earth).

There are other passages that use the word "Create" or "Bara' " in combination with the Heavens and the Earth. Psalm 148:1-5 and Isaiah 45:18 are both describing this same event.

These are all **<u>Creative</u>** acts done by God, just like we find in (Genesis 1:21, 27; 2:3), but on a much larger scale and at a much earlier time.

7.5 In The Beginning, God Created Everything Except Man.

When God made a complete list of everything that He Created in the Beginning man was not on that list.(Psalm 148:1-5) Man was not Created in the Beginning, he was Created about 6,000 years ago. We are newcomers in a very, very old universe with a history.

On the other hand, Cherubim (Ezekiel 10: 1-11) and Seraphim (Isaiah 6:2, 6) have also been around for a very, very, long time. They are two different species which make up the "Hosts of God" that are mentioned by name and description. (Ezekiel 10: 1-11) and (Isaiah 6:2, 6)

I suppose that the "Congregation" spoken of in (Isaiah. 14:13) consisted of Seraphim, Cherubim and whatever else God Created in the Beginning. The Hosts of God may include other intelligent spirit beings beside those two mentioned. These two, I believe, are mentioned because they have been here before and men will have an encounter with them again at the 2nd return of Christ.

The Hosts of God, I believe are intelligent, extra-terrestrial spirit beings. An Intelligent spirit being is someone that can have an intelligent conversation with God, I think. Extraterrestrial means that they are not of this earth, however Angels are from this earth.

The Mount of the Congregation in the sides of the North is where the Hosts of God went to congregate. This is also a place that Lucifer wanted to attack and take over. (Isaiah 14:13) He had some history there. The Hosts of God are space travelers and they are coming back with Jesus along with the Angels and the Saints of God.

The Mount of the Congregation is a place where God's Extraterrestrials could go and fellowship. And along comes a Cherub named Lucifer, (as he was called then), and crashes the party with violence. It reminds me of the scene in Star Wars where there is an alien bar room full of strange looking, fictitious extraterrestrials and then a laser gunfight breaks out. I am sure that the first real Star wars was different and more intense. When Lucifer attacked the mount of the Congregation he must have been packing extreme destructive power to use against defenseless, unarmed and peaceful folks.

There may be other species of extraterrestrials which make up the Hosts of God, but not mentioned because they are not relevant to us. (No need to know)

Angels are not part of that Group called "Hosts of God". The Hosts of God are extra-terrestrial. Angels are different than the Host of God and these two species for several reasons; they are not from the earth and the Angels are: Furthermore, Angels don't have wings and the Cerebrum and the Seraphim do have wings: Angels look like men, but the Seraphim have reptilian features and the Cerebrum are different in other ways.

THE ORIGINAL CREATION

The Heavens, Earth, Angels, Hosts of God, plants, animals and everything else, except Man was Created from the command (Psalm 148:1-5), given by God, *"in the beginning"* (Genesis 1:1).

"In the beginning (Genesis 1:1) *God Created*" all dimensions in our Universe.

7.6 God Created The Visible And The Invisible, All At Once

"*For the invisible things of him from the creation of the world are clearly seen, being understood by the things that are made, even his eternal power and Godhead; so that they are without excuse*:" (Romans 1:20)

The Bible teaches that there is an unseen world. Scientists are beginning to recognize that our Universe probably consists of more dimensions than we are aware of. Subatomic particles that suggest there may be more dimensions have been found.

Researchers are creating forces to produce reactions that are measured to reveal the secrets of our Universe. Using a $6 billion particle accelerator straddling the French-Swiss border, CERN physicists have collided protons at an energy level unmatched at any other physics lab.

Using the Large Hadron Collider (LHC) physicists are searching for evidence of other dimensions. That sure is a lot of money to spend on researching something that doesn't exist. Enough scientific experts have convinced enough people with enough money to pursue knowledge about dimensions other than the four that we are aware of.

Six Billion Dollars is a lot of money to be spending on anything!

The Bible implies that there are more dimensions by describing events that require more dimensions to even be possible. If scientists would study their Bibles, like Sir Isaac Newton did, it would advance Man's understanding of Science.

7.7 If You Take This Passage Literally...

... And you interpret it in the most absolute since of the word, you have a fascinating situation to consider...

The first Hebrew word of the Bible, רֵאשִׁית ..."re'shiyth" (Strong's # H7225 – def.; first, beginning, best, chief), in this context has some interesting implications.

The nature of this definition and the context that it is in requires it to be interpreted in the most absolute sense of the word.

This is the first word of the first sentence of the first paragraph of the first chapter of the first book of the entire Bible! There is no context to consider before this first Hebrew word. There is nothing before it.

And what is that word saying? It is talking about a single event, the Creation of the Universe by God. The Creation of everything that we know of and more, in a single point in time and space called "*the beginning*".

"*In the beginning*"; (Genesis 1:1), these first three words are God telling us when and where "*the beginning*" was, in terms of the four dimensions that we are aware of and any others that may exist.

7.8 "In The Beginning" God Created Time

When we think about Time, we think linearly, like a time-line. A time-line has a "B.C." and "A.D." without the zero in the middle. But to describe the "Beginning of Time" we would start with a point and draw a straight line to the right ending in an arrow also pointing to the right. That would indicate the Beginning of Time.

...The beginning would be called..."the first point after Time = 0, (t=0). That would express when. When It happened was immediately after Time = 0.

When considering the meaning of "*In the beginning*" as found in the first verse of the Bible, there are more dimensions to consider than just time. The single dimension of time is the easiest to understand, because we can think of it linearly. It begins in a point, or so we currently think... so just after "Time = 0" or "t=0"... God created the heaven and earth!

7.9 "In the beginning" All Dimensions Were Created.

The "*beginning*" is... The first point of existence or the origin of all dimensions that make up our Universe!

When you consider the three dimensions of space, it gets complicated very, fast! Our Universe consists of one dimension of Time and three Dimensions of Space that we know of for sure.

We call ourselves "four dimensional beings" but the Bible implies that there is more.

7.10 "The Beginning" The Point Of Origin Of Our Universe!

Space was created just after time equal Zero (T=0)!

Let height, width and depth be represented by the three coordinates (x, y, z,) for the three dimensions of space.

Because we are discussing an event which occurred at a point in time called "Time = 0" or (t=0) to indicate the beginning, we also set values for all other dimensions at zero to indicate that this is their origin as well. This is the common point of origin for all dimensions in our Universe, not just Time. God calls it "*the beginning*". Therefore we have…x=0, y=0, z=0 at t=0 to represent the four dimensions that we know of "*In the beginning*".

All dimensions were created…"*In the beginning*," when God commanded (Psalm 148:1-5). Genesis Chapter one, verse one is the beginning of all dimensions in our universe!

The Old Testament describes an expanding universe at least 14 times. Job said it twice 3500 years ago if you are counting. (See Ch. 5:3) It all started from a point just like astrophysicists believe today.

7.11 Biblical Teachings Imply Other Dimensions.

Angels and demons must occupy dimensions that we cannot yet detect. They occupy the same time and space as us, but are undetectable to us at this time. The Bible records unusual events that implies more dimensions. Sudden Appearances and walking through walls just to name a few. These phenomenon and plenty more like them, imply that there are more dimensions than our four. That explains how Angels can occupy our same space and not be detected.

Intergalactic Space Travel, is not possible in our four dimensions. To travel from Earth to Heaven, Lucifer and his rebel Angels had to move a lot faster than the speed of light! That is not possible in our four dimensions. The ability to navigate through other dimensions is necessary for intergalactic space travel.

7.12 A Beginning Implies An End.

"Lift up your eyes to the heavens, and look upon the earth beneath: for the heavens shall vanish away like smoke, and the earth shall wax old like a garment, and they that dwell therein shall die in like manner: but my salvation shall be forever, and my righteousness shall not be abolished." (Isaiah 51:6)

"And I saw a new heaven and ***a new earth: for the first heaven and the first earth were passed away****; and there was no more sea."* (Revelation. 21:1)

The new Earth will not have any more seas! Now that is interesting! The new earth sounds like an interesting planet.

Other worlds are mentioned and new ones to come.

8

"He Created it not in vain, He formed it to be inhabited"

8.1 "Not In Vain" Means Not Like You See It In (Genesis 1:2)

"*For thus saith the LORD that* ***Created*** *the heavens; God himself that formed the earth and made it; he hath established it, he* ***Created*** *it* ***not in vain, he formed it to be inhabited****: I am the LORD; and there is none else.*" (Isaiah 45:18).

God is telling us here in (Isaiah 45:18) that he did not Create the earth as seen in (Genesis 1:2), **but in contrast to that**, in (Genesis 1:1), ***"he formed it to be inhabited"*** or "***not in vain***".

The word translated into ***"not in vain"*** (Isaiah 45:18) is the same word that is used in (Genesis 1:2) and translated into ***"without form".*** That word is… "tôhûw, to'-hoo;" Strong's #H8414 which means… "from an unused root meaning to lie waste; a desolation (of surface), i.e. desert; figuratively, a worthless thing; adverbially, in vain:—confusion, empty place, without form, nothing, (thing of) naught, vain, vanity, waste, wilderness." (Strong's Dictionary)

God is talking about the original Creative act, that Created the heaven and the earth from that single command (Psalm 148:5)… "*In the " beginning*" (Genesis 1:1()… at Time = 0.

God speaking through Isaiah tells us that when God Created the earth "***he Created it not in vain****, he formed it to be inhabited:*"! (Isaiah 45:18)

In other words he Created it perfect and complete.

The command that was given in (Psalm 148:5) resulted in the Creation of a fully inhabited earth as well as everything mentioned in (Isaiah 45:18), (Psalm 148:1-5) and (Genesis 1:1)!

"*And the earth was without form, and void; and darkness was upon the face of the deep. And the Spirit of God moved upon the face of the waters.*" (Genesis1:2)

The Hebrew word translated as "***was***" can also be translated as "***became***". It is translated as "***became***" in (Deuteronomy 27:9)

"*And Moses and the priests the Levites spake unto all Israel, saying, Take heed, and hearken, O Israel; this day thou art **become** (H1961) the people of the LORD thy God*". (Deuteronomy 27:9)

"***he formed it to be inhabited***" … by Lucifer and his Angels. (Isaiah 45:18)

They were the original inhabitants of the Earth, but that was a very long time ago.

Lucifer was a Cherub and is one of the original inhabitants of the Earth.

"*In the beginning*" Lucifer was made the "*god of this world*" (2Corinthians 4:4). He lived here on the Earth and ruled over its inhabitants for a very long time before there was a problem. This is long before Man was Created.

The Angels were the original inhabitants of the Earth and were ruled over by Lucifer as "*god of this world*". (2Corinthians 4:4)

I believe the Earth was inhabited by a large population of Angels, Lucifer as "*god of this world*" and a very large variety of animal and plant life.

I believe that the earth's environment was more nurturing than it is today and was more densely populated. The Earth supported a vast civilization of Angels; some very large creatures; some very unusual creatures; and perhaps creatures with unusual powers. Most of that is gone now.

That world was destroyed by God's judgment a very long, long time ago. All that is left today are coal deposits, oil, fossils and possible ruins of the pre-human civilization on the bottoms of the oceans.

Angels had plenty of time on Earth, to develop their own technology.

Apparently the Angels lived, learned and worked on planet Earth for a very long time. During that time, they had no wars, no diseases, and no death. But they did have a lot of time to learn things. I believe that their memories were better than ours today. I believe that they learned quickly and advanced in science much, much more quickly than Humans have.

According to scientists this Universe has been around for about 15 to 20 billion years. They may be right. It seems that Angels had plenty of time to develop their own technology and build their own cities. I believe that they accomplished things beyond our imaginations, right here on earth, long before man was Created!

No need for magic, just pure technology that was supposed to be used for God's Glory. The things that they could do back then would even surprise the folks at D.A.R.P.A.

Angels and the Hosts of God were Created at the same time as everything else **except man**. (Psalm 148:1-5; Isaiah 45:18; Genesis 1:1) **Everything** was **Created** "In the Beginning" **except** for Man.

"*For thus saith the LORD that* ***Created*** *the heavens; God himself that formed the earth and made it; he hath established it, he* ***Created*** *it* ***not in vain****, he formed it* ***to be inhabited****: I am the LORD; and there is none else*". (Isaiah 45:18)

There is no mention of Man in the original Creation. (Genesis. 1:1)

The Bible talks about the **Creation** of… ***"the "heavens" and the earth,*** "and says when it happened… It happened ***"In the beginning"***. That would be the same as T=0… but **there is no mention of Man.** (Genesis 1:1, Isaiah 45:18 & Psalm 148:1-5)

Comparing Genesis 1:1 with Isaiah 45:18 and Psalm 148:1-5 …we discover…

Genesis 1:1 says… *"In the beginning God **Created** the heaven and the earth".*

Isaiah 45:18 says…*"For thus saith the LORD that **Created** the heavens; God himself that formed the earth and made it; he hath established it, he **Created** **it <u>not</u> in vain**, he formed it **to be inhabited**: I am the LORD; and there is none else".*

According to (Isaiah 45::18)…this event where the Heaven and Earth were Created resulted in a perfect and completely inhabited earth. This all happened ***"In the beginning"*** because that is when the Heaven and Earth were Created. (Genesis 1:1 & Isaiah 45:18) There is no six days mentioned but yet we have an inhabited Earth.

Now consider what (Psalm 148:1-5) has to say about it. This is **a list of 7 things that were <u>Created</u>** *"in **the Beginning**".*

Once again, we know when this event occurred, because the Heaven and Earth are both mentioned in this list. There is only one command mentioned, therefore we believe that all 7 items in this list were **Created** ***"In the Beginning"***. The Bible says…

*"Praise ye the LORD. Praise ye the LORD from the heavens: praise him in the heights". "Praise ye him, all his **angels**: praise ye him, all his **hosts**." "Praise ye him, **sun** and **moon**: praise him, all ye **stars** of light" "Praise him, ye heavens of **heavens**, and ye **waters** that be above the heavens." "Let them praise the name of the LORD: for he commanded, and they were **Created**".* (Psalm 148:1-5)

This passage mentions two different groups of intelligent, spirit beings being Created but there is no mention of Man.

All three of these passages talk about the **Creation** of the **heavens** and the **Earth**. That is because they are all talking about **the same event**. During that original event called "***the Beginning***" several things are specifically mentioned as being **Created** at that point in time; "*the Heavens; The Earth; the Angels; the Hosts of God; the Sun; the Moon; and ye **waters** that be above the heavens.*"

All of this happened at the same point in time, the time known as... "***in the beginning***" (Genesis 1:1), or if we go to Psalm we find ... "***when God commanded***" (Psalm 148: 1-5).

"In the beginning" we have a fully formed and inhabited Earth but there is **no mention of Man.**

We have **three different passages** describing an event that they all identify as the **Creation.** They identify the time of the event as ***"in the Beginning"***. The three descriptions of the event called the **"*Creation*"**, all match up. It is the same event. We have **three** witnesses.

The Angels were Created at the same time as the Hosts of God, Heaven (Psalm 148:1-5), Earth and everything else **except man**, as the result of ONE COMMAND given by God (Psalm 148:5) ..."*for **he commanded** and **they were Created***" (verse 5); *"In the beginning"* (Genesis 1:1); at time equals 0, **(t=0).**

One command was given and the Creation of everything mentioned in (Genesis 1:1), (Isaiah 45:18) and (Psalm 148:1-5) resulted. It all happened ***"In the beginning"***. (Genesis 1:1)

The phrase ***"In the Beginning"*** is talking about more than just a Point in Time and Space. A Perfect description for a **Point** in **Time**, **Space** and **all other dimensions** of this Universe; **The Origin Of Everything in this Universe** that God Created is ...***"In the Beginning"***. He made it perfect and complete! He made it a long time ago, long **Before Man Was Created**.

God <u>Never</u> calls the Earth "<u>Young</u>"! God calls the Earth "*Old*". Anyone who calls the Earth young is arguing against God! Anyone who calls the Earth Young when God calls it Old is a good person to avoid when searching for Biblical Truth.

God commands us to "*reprove, rebuke, exhort with all longsuffering and doctrine.*" (2Timmothy 4:2)

8.2 God Created The Earth Perfect And Complete.

(Isaiah 45:18) "*For thus saith the LORD that Created the heavens; God himself that formed the earth and made it; he hath established it, **he Created it not in vain**, he formed it to be inhabited: I am the LORD; and there is none else.*" (Isaiah 45:18()

In (Isaiah 45:18), when God Created the Earth, the Bible says…"*he Created it **not in vain**, he formed it to be inhabited*". The Hebrew word translated here as "***not in vain***" is "tohuw", (Strong's H8414)…and is the same Hebrew word translated as "*void*" in (Genesis 1:2).

The Bible says that God did not Create the Earth like you see it in verse 2 of Genesis chapter one, "***he formed it to be inhabited***". (Isaiah 45:18)

There is no mention of Man in this passage either. All three passages are talking about the Creation of the heavens and a fully inhabited Earth at the same point in time called "*the beginning*" (Genesis 1:1); Psalm 148:1-5); (Isaiah 45:18), but there is no mention of man in any of them. Man was not Created in the beginning, he was Created much later.

8.3 An Ancient, Advanced Pre-Human Civilization!

I believe those cities were occupied by Lucifer and his Angels. The ruins of those cities may still be at the bottoms of the oceans today. Portions and fragments of those incredible cities may be popping up all over the world. There are videos of very large, very deep, underwater structures of unnatural origin. They are finding these things all over the world. Some of these may be the remains of those Angelic cities.

Much of the dry land, even tall mountains were once submerged under the sea.

(Jeremiah 4:27); "*For thus hath the LORD said, The whole land shall be desolate; yet will I not make a full end.*"

All air-breathing life on the Earth was destroyed, but not the fish. God was not through with the Earth. God still had future plans for the Earth, but until then it would remain in darkness. If you cut off the Sun's light from the Earth, the surface would freeze very quickly.

"*For this shall the earth mourn, and the heavens above be black: because I have spoken it, I have purposed it, and will not repent, neither will I turn back from it.*" (Jeremiah 4:28);

8.4 God Judges Satan. His Planet, The Earth Was Destroyed!

"*Behold, the LORD maketh the earth empty, and maketh it waste, and turneth it upside down, and scattereth abroad the inhabitants thereof*". (Hebrews 11:3) (Isaiah 24:1)

"Empty" בָּקַק, baqaq, (Strong's H1238) - baqaq to empty, lay waste, to make void (fig.)

"*maketh it waste*", בָּלַק, - balaq, (Strong's H1110) to waste, lay waste, devastate, (Poel) to make waste, (Pual) devastated (participle)

"*turneth*" Strong's H5753 - `avah, to bend, twist, distort

And that is how the Earth ***<u>became</u>*** "***without form and void***" in (Genesis 1:2).

Notice that it says that God is doing something to the Earth causing it to become that way. The perfectly Created and inhabited Earth is being destroyed by God. Nothing like that has happened in human history. **This happened <u>before</u> Man was Created.**

8.5 A Time Traveler's View Of The Earths Destruction!

Jeremiah actually went back in time to become an eye witness. He said…

"***<u>I beheld the earth</u>**, and, lo, it was **<u>without form</u>**, and **<u>void</u>**; and the heavens, and they had **<u>no light</u>***". (Jeremiah 4:23)

"***<u>Without form and void</u>***" is also used in Genesis 1:2. This is the same event. (Jeremiah. 4:23) & (Genesis. 1:2)

(Jeremiah 4:23) and (Genesis 1:2) makes it clear that the Suns light was not reaching the Earth. This is the same event.

Jeremiah is looking at the Earth. I believe he is arriving just after the Earth is struck by something big, like the asteroid that made the Chicxulub crater in The Yucatán Peninsula. That asteroid is believed to have destroyed 70% of all life on the planet. Originally N.A.S.A. stated that it was 100%. Today there are evolutionists that want to argue and say that it was only 50%.

But let us remember that N.A.S.A. originally said that after the sun's light was cut off. There was a type of dust cloud surrounding the planet causing a type of "nuclear winter". The dust cloud cut off the sun's light for an indefinite period of time. There is plenty of Evidence to indicate that the earth was indeed flash frozen at the time of the Mass-Extension Period also known as the **Cretaceous–Paleogene boundary.** (K-Pg) or (K-T)

The Iridium anomaly discovered in the sediment layers between the Cretaceous period and the Paleogene period is also called the Cretaceous–Paleogene boundary. This sediment layer indicates multiple asteroid strikes. At least one of these is much larger than the strike discovered at Chicxulub, on the Yucatan Peninsula in Mexico.

Iridium is an interesting element. It is very hard. It resists oxidation even at high temperatures. It is also used as a superconductor. This stuff is associated with Asteroids and Meteors. Tracing this stuff is how scientists discovered the connection between these Asteroid Impacts and the Mass Extinctions of the Cretaceous–Paleogene boundary.

I think that it is important to remember that this is the time when Lucifer and his fallen Angels were both cast back down to the earth. Perhaps there is a connection. Did Lucifer and his rebel Angels build military space craft out of Iridium?

I believe that the whole surface was flash frozen and remained that way until 6,000 years ago. That is when Global Warming **really** began.

Scientists either believe that the earth quickly recovered from this Deep Impact Mass Extension Event or it gradually recuperated over a period of time.

Those extreme optimists who believe that the earth recovered quickly from this event also call themselves evolutionists. Only an evolutionist could believe such a thing. Evolutionists need to believe in a quick recovery from a mild event otherwise they have got a lot of explaining to do!

The earth remained in a deep freeze for an undetermined period of time. Not years, but many, many millenniums. Nothing could have survived. According to N.A.S.A.'s original statement, the earth was a dead planet.

The fossil record supports the notion of a mass extension of very large reptiles and mammals that occurred about 65,000,000 years ago. I believe that was the time of earth's destruction by God. He just used a large mass traveling at just the right velocity that struck the earth in just the right location, with the right trajectory, which caused the earth to turn upside down and wobble like a drunkard.

"*I beheld the mountains, and, lo, they trembled, and all the hills moved lightly*" (Jeremiah 4:24);

Jeremiah was viewing the Earth just after the Mass-Extension Impact. The Earth is still vibrating from the impact. If the earth rang like a bell then that would be an indication of a hollow earth. If planets are hollow, then they would resonate when struck.

"*I beheld, and, lo,* ***there was no man****, and all the birds of the heavens were fled*". (Jeremiah 4:25);

This is before Man was Created. This is "***the World that then was***" (2 Peter 3:6)

"*I beheld, and, lo, the* ***fruitful place was a wilderness****, and all the* ***cities*** *thereof were broken down at the presence of the LORD, and by his fierce anger*". (Jeremiah 4:26);

"The Presence of the LORD is something that is very rare and very destructive when it comes to sin. No sin can exist in the Presence of the LORD! These passages demonstrate the magnitude of destructive Power of God's wrath against sin!

In verse 26 we have the fruitful place ***becoming*** a wilderness. As you can see here**, it wasn't "*<u>Created</u>*" that way, it "*<u>became</u>*" that way**.

If this destruction took place **<u>before</u>** man was Created**, who built those cities** that were destroyed? The Bible describes the destruction of more than just an ecosystem. The Bible also describes the destruction of an ancient and advanced civilization. A civilization that Lucifer led in a rebellion against God! I believe that we may already be discovering their remains on the bottoms of the oceans.

8.6 God Gives Full Account Of Time And Events.

God gives a full accounting of Time. That is why it is called <u>His</u> Story. History.

God has his own Time Accounting System! It is called the Bible. In the Bible we can find God's Account of what happened, where it happened, when it happened and why it happened.

We can also find out in the Bible, about what is happening in our world right now! The Bible tells us what is happening, where it is happening and it tells us why it is happening! We can be sure that the Bible is God's Record and that God's Record is true because of the Bible's Record of Accuracy about the past and present!

Now consider what God says about **What Happens Next!**

9

A List Of 7 Things God Created, "In The Beginning"

(Psalms 148:1-5) & (Genesis 1:1).

9.1 God Uses David To Describe … "The Beginning".

"*Praise ye the LORD. Praise ye the LORD from the heavens: praise him in the heights*". (Psalm 148:1)

*"Praise ye him, all his **angels**: praise ye him, all his **hosts**.*" (Psalm 148:2)

"*Praise ye him, **sun** and **moon**: praise him, all ye **stars** of light*" (Psalm 148:3)

*"Praise him, ye heavens of **heavens**, and ye **waters** that be above the heavens."* (Psalm 148:4)

*"Let them praise the name of the LORD: for **he commanded**, and **they** were **Created**".* (Psalm 148:5)

Man is not in this list!

9.2 God Completed The Creation With One Command

Notice that the past tense verb *"commanded"* is singular and the object, "*they*" is plural.

"*Let them praise the name of the LORD: for he commanded, and they were Created*".

Everything mentioned in (Genesis. 1:1) and (Psalm 148:1-5() was Created at the same time, with the same command!

If it was all Created at the same point in time, there could only have been one command. The point in time we are talking about is…"*the beginning*". (Genesis. 1:1)

At time = 0, (t=0), God gave the command that Created the Heaven and the Earth, including Angels, the Hosts of God and everything else mentioned in (Psalm 148:1-5)!

God spoke through David (Psalm 148:1-5) to say that the "*heavens*", the "*Angels*", all of "*God's hosts*", the "*Sun*", the "*Moon*", the "*stars of light*" and the "*waters above the heavens*" were all "Created" at the same point in time called ***"the Beginning"*** with ONE COMMAND! (Psalm 148:1-5)

Psalm 148:1-5 gives us additional details about what happened and when it happened. So when you combine Genesis. 1:1 with Psalm 148:1-5, you realize that the command was given "*in the beginning*" and all of Heaven, Earth, Angels, Hosts of God etc. were all Created at Time=0! Genesis says it happened in the beginning and Psalm says it happened when God commanded it to happen. It all happened "*In the beginning*" when God gave the command at Time=0!

9.3 God Commanded And 7 Things Were Created.

"*All his **angels***" (verse. 2); "*All his **hosts**.*" (verse. 2); "*Praise ye him, **sun***" (verse. 3);… "*And **moon**;*" (verse. 3); "*All ye **stars** of light*" (verse. 3); "*Praise him, ye **heavens** of heavens,*" (verse. 4); "*Ye **waters** that be above the heavens.*" (verse. 4)

But no mention of man.

So when do you think God commanded? "*In the Beginning God Created the heaven and the Earth*" (Genesis. 1:1). Heaven and Earth includes everything in our Universe!

Heaven is the first item in that list of 7 things that God Created when he commanded (Psalm 148:1-5), it is also the first thing mentioned to be Created in Genesis 1:1. We know from Genesis 1:1 that the Heavens were Created in the beginning and so was the Earth.

God has associated these 7 things together for a reason.

The first item in that list, (heaven) includes everything else in the Creation; and Genesis 1:1 says that heaven was Created *in the Beginning*, Therefore I believe that this must be a list of 7 things that were Created *in the beginning*.

The other 6 things in that list were also Created *"In the beginning"*, along with everything else **<u>except Man</u>**. (Genesis. 1:1) (Psalm 148:1-5)

That list includes two groups of intelligent, spirit beings, but once again, Man is not included in the list. I believe those two groups include a large variety of intelligent spirit beings, but man is not there. He is not included in the list.

"In the beginning", God commanded *"and they were Created"*. Everything in that list was Created at the same point in time because it is a list of things Created *in the beginning*.

By leaving Man off of this list, **God is saying that he did not Create Man in the Beginning. Man was Created later**.

Either all 7 things were Created in the beginning with one command or all of the commands happened at the same point in time called *"the beginning"*. Either way, from our point of view it all means the same thing because that is when time was Created. However, the word "commanded" is singular.

From the Bible, I believe that *"the beginning"* was the original and smallest unit of time, space and other dimensions that is possible, a point!

Remember that in *"the beginning"* **God was <u>already there</u>!**

"In the beginning <u>was</u> the Word" (John 1:1() God is before; after; inside of and outside of that point called ***"The Beginning"***.

Time is not a boundary for God and neither is space!

9.4 This List Of 7 Includes Everything Except Man.

When we compare (Genesis 1:1) with (Psalm 148:1-5). In Genesis 1:1 we have God Creating the heavens and the Earth *"in the Beginning"*. Now here in (Psalm 148:1-5) we have a list of seven things that *"were Created when God commanded"*.

Included in that list is the Heavens and things associated with the Earth. That list also includes two different groups of intelligent, spirit beings, which are identified as *"all his angels": and "all his hosts."* (Vs. 2) but there is no mention of Man. According to Genesis 1:1 the heavens and the earth were both Created …" *in the beginning*". Both are part of that list that were all Created when he commanded.

"In the beginning God Created" (Genesis 1:1) **… everything <u>except</u> Man. (<u>Psalm 148:1-5</u>)**

Genesis 1:1 tells us when and where the Creation of the heavens and the earth was. It was *"in the beginning"*.

(Psalm 148:1-5) tells us what was Created when God commanded. It was a list of 7 things that included the heavens; the earth; Angels; the Hosts of God and everything else <u>except</u> man. Man is not on the list of things God Created *"in the beginning"* but Angels and the Hosts of God are. **Man is not on the list. (<u>Psalm 148:1-5</u>)**

"*For he commanded, and they were Created*". (Psalm 148:5)

This command spoken of in Psalm 148:5 is the command that Created everything mentioned in (Genesis 1:1) (Isaiah 45:18) <u>and</u> (Psalm 148:1-5).

This was the command that was given "*in the beginning*".

"*In the beginning*" describes a point in time.

A point in time, implies one command.

Man's absence from the Creation list in Psalm 148:1-5 implies that Man was Created later.

I have heard it said… that "7 is the number of perfect completion".

9.5 Heavenly Hosts And Angels Are On The List

The Angels were Created at the same time as ***"all his hosts."***; the Heavens; the Earth and everything else mentioned in (Psalm 148:1-5) and (Genesis 1:1); but are mentioned separately.

Angels are not included in the group called "God's Hosts", they are not the same.

The Bible mentions the Creation of two groups of spiritual beings in the beginning. (Psalm 148:2) (Genesis. 1:1)

"*Praise ye him, all his angels: praise ye him, all his hosts.*" (Psalm 148:2)

God's Hosts and Angels are the two separate groups of Spiritual beings Created "*In the Beginning*". Man is conspicuously missing from that list.

In other places the Bible describes Seraphim and Cherubim which are included in the group called "God's Hosts" I believe there is more.

God's Hosts are extraterrestrial. Angels are not. There are no references to Angels being in heaven prior to Satan's rebellion.

There are three references of Cherubim on earth. The first was Satan the Covering Cherub; the second was when two were guarding the Garden of

Eden when Adam and Eve were evicted (); and the third is at the Battle of Armageddon.

Seraphim must be members of that group called "The Hosts of God" and are mentioned at the Battle of Armageddon. ()

Later on "*Watchers*" are mentioned (Daniel 4:17) but not described. Watchers are probably a special group of Angels because Angels are associated with the earth.

If there are more species other than Seraphim and Cherubim, I don't believe that mankind will know of them this side of the Millennium, otherwise they would be mentioned. However the **Bible does speak of other worlds.** (Hebrews 1:2; 11:3)

9.6 No Mention Of Man's Creation "In The Beginning"

The Creation of Man is conspicuously missing in this list of seven. There is no mention of the Creation of Man in any of these passages. (Genesis 1:1) (Psalm 148:1-5)(Isaiah 45:18) And yet these passages are talking about the original Creation of this Universe "*In the Beginning*".

According to the Bible, Mankind will have a close encounter of the third kind with the Seraphim and Cherubim at the end of the Battle of Armageddon. I believe that is why they are mentioned and described in the Bible.

Seraphim and Cherubim are both Genuine, authentic Extraterrestrial's (not from earth) and they will be here, along with the Angels, the Saints and the Lord of Glory, Jesus leading them at the end of Armageddon. (Revelation 19:11-14)

God's Hosts may consist of many different kinds of Extra-terrestrials which we may encounter later.

Angels are currently in both Heaven and Earth, Angels are not extraterrestrial, I believe because they originally lived here on the Earth with Lucifer.

"*God's Hosts*" may consist of many different species other than the Cherubim and Seraphim. Lucifer was a Cherub and now is called Satan. Satan will continue to be "*the god of this world*" until Jesus returns to defeat him at the end of Armageddon. (Revelation 19: 16-20)

"*God's Hosts*" are associated with Heavenly duties; "*the sides of the north*"; and the "*Mount of the Congregation*".

This is why I refer to them as "Extra Terrestrial".

Angels are associated with the Earth. Angels, I believe are the original inhabitants of Earth with Lucifer as their leader. After Lucifer's fall, a lot of things changed.

The command spoken of in (Psalm 148:5) is the command that Created the heavens and the earth as spoken of in Genesis chapter 1:1 and everything else except Man.

No mention of the Creation of Man in either passage.

No mention of "6 days of Creation" in either passage.

Only ONE COMMAND and the heavens; earth; Angels; hosts of God; all came into existence from a single point of origin called "*The Beginning*"!

9.7 "In The Beginning" God Also Created Other Worlds.

The Bible mentions ***"worlds"*** **plura**l that were Created ***"In the beginning"***

"*Hath in these last days spoken unto us by his Son, whom he hath appointed heir of all things, by whom also he made the* ***worlds***"; (Hebrews 1:2)

"*Through faith we understand that the* ***worlds*** *were framed* ***by the word of God****, so that things which are seen were not made of things which do appear*". (Hebrews 11:3()

Contrary to popular beliefs about the Bible, the Bible says that God Created ***"worlds"***, that is ***"worlds"*** **plural**, as in more than one. It doesn't say how many "worlds" were Created, but it does mention places located in "the sides of the north". God doesn't give us all of the details but He says that He Created other worlds". I don't think that God told us everything in the Bible. He told us what we need to know to prepare us for the greatest events in human history which is either death or the return of Jesus to the earth. Either way, we will discover the rest latter!

10

Two Kinds Of Extraterrestrial Species Are Mentioned.

10.1 We Are Not Alone

Scientists are busy looking for signs of intelligent life in space while God is looking for intelligent life on Earth. The Bible talks about other worlds and other forms of intelligent life in space. Carl Sagan said that any form of language is proof of intelligent life. The DNA found in every living cell makes up the language of life that is more complex than anything that man can produce but that fact is being ignored.

10.2 The Hosts Of God Are Not From The Earth.

"Hosts of God" is a term that applies to spirit beings other than Angels. (Psalm 148:1-5)

The only other two spirit beings mentioned in the Bible other than Men and Angels is Cherubim and Seraphim. There may be more!

The activities of Cherubim and Seraphim are associated with the "sides of the North"; Heaven; the Mountain of God and Armageddon. Therefore, I call them "Extraterrestrials" because their activities seem to be associated primarily with the throne of God and "*the sides of the north*".

Satan is a Cherub and his home has been the earth since before the 6 days of restoration and the Creation of all air breathing animals and man.

Since the fall, I believe that Satan has lost some of his power and cannot travel to the degree that he once did. Satan's' titles include "Prince of the Power of the Air" (Ephesians 2:2) and "*god of this World*". (2Corinthians 4:4)

I believe that Satan currently can travel in the space around the earth. That is why he is still called "The Prince of the Power of the Air". He will lose that ability in the middle of the Tribulation and will be confined to the surface of the earth for the last 3 ½ Years. At Armageddon he will lose the title of "God of this world" and be imprisoned in the heart of the earth for a thousand years.

10.3 God Created… "Extraterrestrials".

They just don't live here they are from somewhere else, but they are coming back at the end of Armageddon.

When Jesus, (Yeshua) returns, those two members of the Hosts of God that are mentioned, shall be a part of his army. That may be why they are mentioned.

These are real space invaders and they are part of God's Army! They along with the Angels and returning saints will defeat the Antichrist and all of the armies of the world. They will gather all of the unbelievers of the world into one place

10.4 Seraphim Are Some Very Strange-Looking Creatures.

Seraphim will be present when Christ returns, to stop the Battle of Armageddon.

Seraphim will assist, along with the Angels and the resurrected Church, "*Saints*", in destroying the armies of the Antichrist.

The Seraphim will also assist in gathering together the Antichrist's followers and anyone else that has not yet believed God's Message of salvation. (John 3:18)

"He that believeth on him is not condemned: but he that believeth not is condemned already, because he hath not believed in the name of the only begotten Son of God". (John 3:18)

Seraphim are Majestic beings with serpent and human features

(Seraphim: Strong's H8314) def. … Majestic beings with serpent features, 6 wings, human hands and voices, in attendance upon God.

"Above it stood the seraphim: each one had six wings; with twain he covered his face, and with twain he covered his feet, and with twain he did fly". (Isaiah 6:2)

"then flew one of the seraphim unto me, having a live coal in his hand, which he had taken with the tongs from off the altar:" (Isaiah 6:6)

That doesn't sound like anything that I have seen. It doesn't sound like anything I have ever heard of, even in Science fiction movies. This is about as "Alien" in appearance as one could imagine!

When saraph (Strong's H8314) – is used twice in a row, it is also translated as flying fiery serpent…

(Isaiah 14:29) *"Rejoice not thou, whole Palestina, because the rod of him that smote thee is broken: for out of the serpent's root shall come forth a cockatrice, and his fruit shall be a fiery (H8314) flying serpent"*. (H8314) (That is the same word being used twice in a row and being translated as *"fiery flying serpent"*.)

"The burden of the beasts of the south: into the land of trouble and anguish, from whence come the young and old lion, the viper and fiery (H8314) flying serpent, (H8314) they will carry their riches upon the shoulders of young asses,

and their treasures upon the bunches of camels, to a people that shall not profit them". (Isaiah 30:6)

The late Dr. Hank Lindstrom said that he often wondered if the frequency that the number "thirteen" was associated with Satan or the Antichrist was more than just a coincidence. (Isaiah 14:13) (Revelation. 13:13)

10.5 Cherubs Are Unlike Anything That We Have Ever Seen

"Also out of the midst thereof came the likeness of four living creatures. And this was their appearance; they had the likeness of a man". (Ezekiel 1:5)

"And every one had four faces, and every one had four wings". (Ezekiel 1:6)

"And their feet were straight feet; and the sole of their feet was like the sole of a calf's foot: and they sparkled like the colour of burnished brass". (Ezekiel 1:7)

"And they had the hands of a man under their wings on their four sides; and they four had their faces and their wings". (Ezekiel 1:8)

"Their wings were joined one to another; they turned not when they went; they went every one straight forward". (Ezekiel 1:9)

"As for the likeness of their faces, they four had the face of a man, and the face of a lion, on the right side: and they four had the face of an ox on the left side; they four also had the face of an eagle". (Ezekiel 1:10)

"Thus were their faces: and their wings were stretched upward; two wings of every one were joined one to another, and two covered their bodies". (Ezekiel 1:11)

"And they went every one straight forward: whither the spirit was to go, they went; and they turned not when they went". (Ezekiel 1:12)

"As for the likeness of the living creatures, their appearance was like burning coals of fire, and like the appearance of lamps: it went up and down among the living creatures; and the fire was bright, and out of the fire went forth lightning". (Ezekiel 1:13)

"And the living creatures ran and returned as the appearance of a flash of lightning". (Ezekiel 1:14)

10.6 Cherubim Are Members Of The "Hosts Of God".

"*Hosts of God*" are Extra Terrestrial in origin. I say that because there is no mention of them here on Earth except for 2 Cherubs guarding the entrance on the East side of the Garden of Eden when Adam and Eve were cast out.

"So he drove out the man; and he placed at the east of the garden of Eden Cherubim, and a flaming sword which turned every way, to keep the way of the tree of life". (Genesis 3: 24)

Lucifer who is now called Satan and is "*god of this world*" is a Cherub. (2Corinthians 4:4)

10.7 Cherubs Are Mentioned As Being Near God's Throne.

They may also be part of the Mount of the Congregation, located in the "*Sides of the North*". (Isaiah 14:13) The "*sides of the north*" appears to be an empty place in the North (Job 26:7). A very big corridor that leads to heaven like the one Jacob saw. (Genesis 28:12)

KJV Strong's H3742 matches the Hebrew כְּרוּב (kĕruwb).

"Cherubims" AND "H3742"Cherubs have wings.

(Exodus 25:20)"*And the cherubims H3742 shall stretch forth their wings on high, covering the mercy seat with their wings, and their faces shall look one to another; toward the mercy seat shall the faces of the cherubims* (H3742) *be*".

A Cherub is another strange looking creature. (Ezekiel 10:14) *"And every one had four faces: the first face was the face of a cherub, and the second face was the face of a man, and the third the face of a lion, and the fourth the face of an eagle"*. (Ezekiel 10:14)

Cherubs are first referenced in the Bible as guarding the Garden of Eden. (Genesis 3:24)

"So he drove out the man; and he placed at the east of the garden of Eden Cherubims, and a flaming sword which turned every way, to keep the way of the tree of life". (Genesis 3:24)

(Ezekiel 10:16) *"And when the cherubim's went, the wheels went by them: and when the cherubims lifted up their wings to mount up from the earth, the same wheels also turned not from beside them"*.

As a Cherub, before he sinned, Lucifer had access to "*the Holy Mountain of God"* and probably the Mount of the Congregation.

"*Thou art the anointed Cherub that covereth; and I have set thee so: thou wast upon the holy mountain of God; thou hast walked up and down in the midst of the stones of fire*" (Ezekiel 28:14)

It must have been an interesting place, but Lucifer was not satisfied with his position in all of this. It also sounds like he was heat resistant. Verse 14 says…

"thou hast walked up and down in the midst of the stones of fire". I wonder if any Cherub could do that or if it was just certain ones"? (Ezekiel 28:14)

Lucifer was Created perfect. He chose to become evil.

"Thou wast perfect in thy ways from the day that thou wast Created, till iniquity was found in thee". (Ezekiel 28:15)

10.8 Satan Counterfeits The Truth

Satan "*blinds the minds*" of unbelievers with lies! People are seeing so many counterfeits that they also reject the real thing or they just become frustrated and stop looking!

Some people are so tired of all the garbage being spread out in the name of Jesus that they don't want to hear anything remotely related to religion or Jesus. Many have already made up their minds based on commonly accepted lies.

There is so much misinformation about so many things regarding the Biblical Record of Creation; Bible Prophecy; Eternal Life; God and his Bible that it all sounds like nonsense!

With all of the misinformation and misunderstandings about what God has said or not said, it is easy to make the choice that Satan wants you to make.

Satan is working overtime and the Church is asleep! Satan's lies are being spread by the "Churches" as well as secular sources and not being corrected by the members of the true Church.

That is the sort of things that causes Atheism.

"In whom the god of this world hath blinded the minds of them which believe not, lest the light of the glorious gospel of Christ, who is the image of God, should shine unto them". (2 Corinthians 4:4)

10.9 Satan Masquerades As An Angel Of Light

Satan is a Cherub, not an angel and not an angel of light.

"By the multitude of thy merchandise they have filled the midst of thee with violence, and thou hast sinned: therefore I will cast thee as profane out of the mountain of God: and I will destroy thee, ***O covering cherub****, from the midst of the stones of fire."* (Ezekiel 28:16)

Satan masquerades as an Angel of light but he is neither. (2Cor. 11:13-15)

11

Angelology 101; Myths And Facts.

11.1 The "First Estate" Of The Angels.

I believe, the "*first estate*" of the Angels was living here on Earth in the four dimensions that we are all familiar with.

The Earth was their original habitation. Lucifer, as he was called then, ruled over them as "*the god of this world*".

I believe that the Demons that followed Satan used their own technology to access other dimensions. They used that technology to travel to the other galaxies and apparently are stuck in that estate or dimension here on earth until they are Judged by God. They have lost much of their previous abilities. They can occupy our same space without detection.

11.2 Earth Was First Inhabited By Satan And His Angels.

I believe the Angels lived and worked here for a long time before there was a problem. (Ezekiel 28:13)

"*Thou hast been in Eden the garden of God; every precious stone was thy covering, the sardius, topaz, and the diamond, the beryl, the onyx, and the jasper, the sapphire, the emerald, and the carbuncle, and gold: the workmanship of thy tabrets and of thy pipes was prepared in thee in the day that thou wast Created*" (Ezekiel 28:13)

Satan was a musician. He played tabrets and pipes and was the original occupant of Eden. (Ezekiel 28:13)

Angels Are The Original Inhabitants Of The Earth. The Fallen Angels *"left their own habitation"* (Jude 6), They left their homes on Earth to attack God in Heaven.

"And the angels which kept not their first estate, but left their own habitation, he hath reserved in everlasting chains under darkness unto the judgment of the great day." (Jude 6)

The Angels were not at any of the locations that Lucifer wanted to lead them to. They were with him on the Earth because he is the *"god of this world"* (2Corinthians 4:4) and they were his subjects.

"And the angels which kept not their first estate, but left their own habitation, he hath reserved in everlasting chains under darkness unto the judgment of the great day". (Jude 6)

"In whom the god of this world hath blinded the minds of them which believe not, lest the light of the glorious gospel of Christ, who is the image of God, should shine unto them." (2Corinthians 4:4)

11.3 Angels Attend To Matters Of The Earth And Man.

"And he dreamed, and behold a ladder set up on the earth, and the top of it reached to heaven: and behold the angels of God ascending and descending on it;" (Genesis 28:12)

Angels travel between Heaven and Earth through a portal that Jacob got to see. This portal through *"the sides of the north"* to Heaven.

"For thou hast said in thine heart, I will ascend into heaven, I will exalt my throne above the stars of God: I will sit also upon the mount of the congregation, ***in the sides of the north****:"* (Isaiah 14:13)

Angels tend to the Earth and Man.

There is a group called "Watchers" which most likely are a type of Angel, but not much is said about them.

"*This matter is by the decree of the watchers, and the demand by the word of the holy ones: to the intent that the living may know that the most High ruleth in the kingdom of men, and giveth it to whomsoever he will, and setteth up over it the basest of men*". (Daniel 4:17)

No Bible passage says that Angels ever sang anything.

11.4 Angel's Always Appear As Men And Don't Have Wings.

Angel's appear as men and are mistaken for travelers and strangers.

"*Be not forgetful to entertain strangers: for thereby some have entertained angels unawares.*" (Hebrews 13:2)

Abraham had three strangers for dinner but did not know at first that two of the "men" were actually Angels that would be destroying Sodom and Gomorra. The third "man" was the Lord himself. (Genesis. 18:1-7)

(Genesis. 18:1) "*And the LORD appeared unto him in the plains of Mamre: and he sat in the tent door in the heat of the day;*"

"*And he lift up his eyes and looked, and, lo, three men stood by him: and when he saw them, he ran to meet them from the tent door, and bowed himself toward the ground,*" (Genesis. 18:2)

"*And said, My Lord, if now I have found favour in thy sight, pass not away, I pray thee, from thy servant:*" (Genesis. 18:3)

"*Let a little water, I pray you, be fetched, and wash your feet, and rest yourselves under the tree:*" (Genesis. 18:4)

"And I will fetch a morsel of bread, and comfort ye your hearts; after that ye shall pass on: for therefore are ye come to your servant. And they said, So do, as thou hast said."(Genesis. 18:5)

"And Abraham hastened into the tent unto Sarah, and said, Make ready quickly three measures of fine meal, knead it, and make cakes upon the hearth." (Genesis. 18:6)

"And Abraham ran unto the herd, and fetch a calf tender and good, and gave it unto a young man; and he hasted to dress it." (Genesis. 18:7)

11.5 Sodomites Recognized Angels As Men And So Did Lot.

"And there came two angels to Sodom at even; and Lot sat in the gate of Sodom: and Lot seeing them rose up to meet them; and he bowed himself with his face toward the ground"; (Genesis. 19:1)

"And he said, Behold now, my lords, turn in, I pray you, into your servant's house, and tarry all night, and wash your feet, and ye shall rise up early, and go on your ways. And they said, Nay; but we will abide in the street all night". (Genesis. 19:2)

"And he pressed upon them greatly; and they turned in unto him, and entered into his house; and he made them a feast, and did bake unleavened bread, and they did eat". (Genesis. 19:3)

"But before they lay down, the men of the city, even the men of Sodom, compassed the house round, both old and young, all the people from every quarter": (Genesis. 19:4)

"And they called unto Lot, and said unto him, Where are the men which came in to thee this night? bring them out unto us, that we may know them" (Genesis. 19:5)

"And Lot went out at the door unto them, and shut the door after him", (Genesis. 19:6)

"And said, I pray you, brethren, do not so wickedly". (Genesis. 19: 7)

11.6 God Never Called An Angel His "Son". (Hebrews 1:5)

"For unto which of the angels said he at any time, Thou art my Son, this day have I begotten thee? And again, I will be to him a Father, and he shall be to me a Son?" (Hebrews 1:5)

Sons of God is a term used exclusively for believers... (John 1:12) (Job 38:7) (1John 5:1)

"But as many as received him, to them gave he power to become the sons of God, to them that believe on his name" (John 1:12)

"When the morning stars sang together, and all the sons of God shouted for joy"? (Job 38:7)

"Whosoever believeth that Jesus is the Christ is born of God: and every one that loveth him that begat loveth him also that is begotten of him". (1 John 5:1)

Daughters of men or Sons of men and Children of men are terms used exclusively for unbelievers. (Ecclesiastes 3:10) (2Samuel 7:14) (Psalms 145:12)

"I have seen the travail, which God hath given to the sons of men to be exercised in it." (Ecclesiastes 3:10)

"To make known to the sons of men his mighty acts, and the glorious majesty of his kingdom". (Psalms 145:12)

"I will be his father, and he shall be my son. If he commit iniquity, I will chasten him with the rod of men, and with the stripes of the children of men": (Psalms 145:12)

God says that once you become his son, he will chasten you when you need it. God chastens his sons *"with the rod of men, and with the stripes of the children of men"*. In other words God will use unsaved, evil men to carry out his judgements on believers!

11.7 Angels Don't Breed (Matthew. 22:30)

There are no baby Angels, no female Angels and no marriages among Angels. (Matthew. 22:30)

"For in the resurrection they neither marry, nor are given in marriage, but are as the angels of God in heaven". (Matthew. 22:30)

11.8 The Lady With An "Angel" Collection.

Dr. Lindstrom once told of a story of lady that collected pictures and statues of what she called Angels…

The lady was showing off her collection of "Angel" statues to Dr. Lindstrom.

The statue collection consisted of what is commonly seen in the shopping malls. You know, beautiful women with wings, halos and long flowing white gowns.

Of course no "Angel collection" is complete without baby "Angels". You know, the cute little, fat chubby artist rendition of what they think an Angel should look like. They included tiny little wings and little bows and arrows in their renditions.

Dr. Lindstrom had to inform the lady the truth about her statues and what real Angels look like.

Contrary to popular belief, Angels are always seen as men. Have you ever seen a statue or picture of a male Angel?

All of the "Angel" statues and pictures that I have seen are blond haired, blue eyed beautiful women.

Have you ever seen a picture or statue of an "Angel" with red hair or freckles? Well I am offended. My eyes are not blue either. How about a dark-skinned Angel, or an Angel with an "Afro" style hairdo? Nope.

Well Dr. Lindstrom had to tell the Lady that those were not pictures and statues of likenesses of real Angels.

Angel's don't have wings... (Hebrews 13:2) (Genesis. 18:1-7) (Genesis. 19:1-7)

No passage says Angels ever sang anything. The passage used at Christmas to say that Angels sang is (Luke 2: 13, 14). The passage says that the Angels said it, not sang it.

Angels were the original inhabitants of the earth. ()

She was more than surprised. She was horrified!

The point of this story is.... Better pay close attention to what you believe about the Scriptures, because Satan is misrepresenting everything!

11.9 Angels Should Not Be Sought After Or Prayed To

Angels should not be sought after, prayed to or worshiped.

"Let no man beguile you of your reward in a voluntary humility and worshipping of angels, intruding into those things which he hath not seen, vainly puffed up by his fleshly mind", (Colossians 2:18)

11.10 Angels Will Be Judged By Believers. (1Corinthians 6:3)

"*Know ye not that we shall judge angels? how much more things that pertain to this life*"? (1Corinthians 6:3)

11.11 Satan Coerced And Deceived One Third Of The Angels.

…to "*ascend into heaven*" and follow him in his attack against "The Mount of the Congregation, in the sides of the north:" (Isaiah 14:13); and against God himself!

That means that they were not in heaven or any of those other places, but that is where he wanted to go.

"*North*" is a reference relative to the Earth, therefore Satan and his Angels were on the Earth. North is the direction of heaven relative to the earth. The North Star is a good General direction pointer.

They were with Lucifer/Satan in his domain, beneath the clouds on the surface of the earth. (Isaiah 14:14)

"*I will ascend above the heights of the clouds; I will be like the most High.*" (Isaiah 14:14)

Satan probably did not have all the facts or it seems to me that he would not have attacked God! He was also blinded by his own pride.

11.12 Satan Is Not An Angel Of Light.

Satan is <u>not</u> an Angel of Light, but he <u>masquerades as one</u>.

Just as Satan masquerades as an Angel of Light so does his false apostles masquerade as apostles of Christ. (2Cor. 11:13-15)

"*For such are false apostles, deceitful workers, transforming themselves into the apostles of Christ*" (2Cor. 11:13).

"*And no marvel; for Satan himself is transformed into an angel of light.*" (2Cor. 11:14)

"*Therefore it is no great thing if his ministers also be transformed as the ministers of righteousness; whose end shall be according to their works.*" (2Cor. 11:15)

"*Transformed*" Strong's G3345 – metaschēmatizō – to change oneself into something else; Pretend to be something else; masquerade;

Satan's ministers masquerade or pretend to appear as holy, good, loving and kind by acting too sweet to be real. A mature Christian should recognize it as probably fake and put up their guard by asking the right questions. But carnal Christians along with the lost are blinded by Satan's lies and they are seduced.

11.13 Satan Is A Fallen Cherub.

"*Thou art the anointed Cherub that covereth; and I have set thee so: thou wast upon the holy mountain of God; thou hast walked up and down in the midst of the stones of fire.*" (Ezekiel 28:14)

"*Thou wast perfect in thy ways from the day that thou wast created, till iniquity was found in thee*". (Ezekiel 28:15)

"*By the multitude of thy merchandise they have filled the midst of thee with violence, and thou hast sinned: therefore I will cast thee as profane out of the mountain of God: and I will destroy thee, O covering cherub, from the midst of the stones of fire*". (Ezekiel 28: 16)

11.14 Company Is On Its Way!

They are coming back soon! Jesus will come with the New Jerusalem which is a bright shiny cube with 12 gates and measuring 1500 miles on each side. That is big! A sphere big enough to hold the New Jerusalem will be bigger than the moon.

The New Jerusalem will produce its own light and will be visible from earth as it approaches. As Jesus Christ and his armies approach the earth they will be attacked, probably as invading aliens. They may call the New Jerusalem the Borg or something similar. That is the name of some bad guys from the Star Trek T.V. series.

I believe that the New Jerusalem will orbit the earth during the Millennium.

12

Men… "Willingly Are Ignorant Of"… 3 Things and 2 floods.

(2Pet. 3:5)

12.1 "Men Willingly Are Ignorant Of"… The Creation;

Men are willingly ignorant about the Creation of the Universe; the age of the Universe and the history of the Earth. (2Pet. 3:5)

"*…by the word of God the heavens were…*" (2 Peter 3:5)

God Created everything by his word! He spoke it! (2Pet. 3:5)

That is the first thing God says that people are willingly ignorant of... "*that by the word of God the heavens* ***__were__***". (Created)

It sounds to me like God spoke this Universe into existence with a single command! Compare (Psalms 148:1-5) with (2Pet. 3:5). "*Word*" as it is used here may be singular as well. That would make it a one word command! All I can say is that must have been a really big word. God's language is more than we can understand!

…"*for* ***he commanded****, and* ***they*** *were Created*". (Psalms 148:1-5)

To me, that sounds like a single command!

Any way, it doesn't sound like a long discussion. "**he commanded**, and **they** were Created".

"For this they willingly are ignorant of, that by the word of God the heavens were of old, and the earth standing out of the water and in the water:" (2Pet. 3:5)

12.2 "Willingly Are Ignorant Of"... The Age Of The Universe.

God said that the heavens and the earth were both Created *"in the beginning"..."the heavens were of* ***old****"*. God never calls the earth young!

In (2Pet. 3:5) God is saying that the heavens are **old**.

"For this they willingly are ignorant of, that by the word of God the heavens were ***of old****,"* ... (2 Peter 3:5)

"***Of old*** *hast thou laid the foundation of the earth: and the heavens are the work of thy hands*" (Psalms 102:25).

I know that "*old*" is a relative term but nevertheless God chose to use it here. I believe there is a reason for that. Young is also a relative term, but he chose not to use that term. Since he used the word "*old*" let's try to think of it that way and encourage others to do the same. If God calls the Heavens and the Earth old, I would advise against calling them young!

Scientists, who say that they know how old the Universe is say…

…that it is about 13.8 billion years old and the Earth is about 4.5 billion years old, depending on which scientist you ask. To me, that sounds old.

I think that this World may be getting close to the end of its usefulness to God. (Hebrews 1:10, 11)

"And, Thou, Lord, in the beginning hast laid the foundation of the earth; and the heavens are the works of thine hands:" (Hebrews 1:10)

"They shall perish; but thou remainest; and they all shall wax old as doth a garment;" (Hebrews 1:11)

Sometime after the Millennial reign of Christ, God will destroy the earth with fire and there will be a new Heavens and Earth. I take that to mean that our sun will go into what astronomers call the "red giant phase". The three inner planets will all be burned up. We will be moved to a new planet with a different heavenly view…

"And I saw a new heaven and a new earth: for the first heaven and the first earth were passed away; and there was no more sea." (Revelation.21:1)

Wow! A planet without a sea! I wonder what kind of ecosystem it will have?

12.3 "Willingly Are Ignorant Of"… The History Of The Earth.

"In the beginning God Created" … everything except Man. In (Genesis. 1:1), "*the heavens and Earth*" are Created but man is not mentioned.

In (Psalms 148:1-5), just like (Genesis 1:1) the heavens are Created, along with Angels, the Hosts of God, and 4 other things mentioned by name, but **there is no mention of Man.**

All of this happened *"in the beginning"*, at the same time that everything else was Created. **Everything except man.** Man is a recently Created being along with air breathing animals. Everything else is left over from… "*the world that then was,*" …

"Whereby ***the world that then was****, being overflowed with water, perished:" (2 Peter 3:6)*

This is talking about two worlds and two floods.

12.4 The Pre-human History Of Our Planet.

Cosmic History 101... There are many Bible teachers who wrongly teach that earth is young. Most popular among these Teachers is Ken Ham. Ken Ham promotes a belief in a young earth, but the Bible never calls the earth young. God always calls the earth old.

Satan's original name was "Lucifer" which means "light bearer" or "morning star". When he was still called "Lucifer", he led a rebellion against God.

Lucifer became the first anarchist and led one third of the Angels that he ruled over in a rebellion against God. Imagine that, a rebellion against his own Creator.

I don't believe that he had all of the facts or maybe he was so blinded by pride that he ignored the facts or some of both. Either way, God knew his thoughts before he did! Lucifer and his army was defeated by God and God changed the name of the god of this world from Lucifer to "Satan". **Satan is still the "*god of this world*"**, that is the title God gave him. (2Corinthians 4:4)

Satan means "adversary of God". Satan created Sin and rebellion against God. That is what God calls him, but he has many aliases. As God's adversary, Satan started a rebellion against God's perfect order and we call that rebellion sin. Satan is the original Anarchist.

Satan's minions are blinded by Satan's lies. They have believed comfortable lies instead of the Truth. Satan and his minions opposes God. Believing or acting on Satan's lies will lead to damnation! Satan's minions declare that good is evil and evil is good. God says beware of them and their damnation!

Satan introduced sin into the Universe. He attacked Heaven and God defeated him. This is the first known intergalactic war! (ETBH chapters 10-13)

Satan was cast back down to the Earth. As a result of the conflict, all air breathing life on **Earth was destroyed**. The physical evidence of Earth's history is everywhere. **(2Peter 3:5,6)** (ETBH chapters 10-16)

Violence, pain, suffering and death are the result of Sin, and Satan is the one that introduced it into the Universe. (Ezekiel 28:16)

There was war in Heaven! (Isaiah 14:13-14; 24:1) (JEREMIAH 4:23-26)

"*I beheld the earth, and, lo, it was without form, and void; and the heavens, and they had no light*" (JEREMIAH 4: 23).

"*I beheld the mountains, and, lo, they trembled, and all the hills moved lightly*". (JEREMIAH 4: 24).

"*I beheld, and, lo, **there was no man**, and all the birds of the heavens were fled*". (Jeremiah 4: 25).

"*I beheld, and, lo, the **fruitful place was a wilderness**, and **all the cities thereof were broken down** at the presence of the LORD, and by his fierce anger*". (Jeremiah 4:26)

If there were no men, whose cities were they?

I believe they were Angelic cities ruled over by Lucifer a long time before Man was Created.

Those cities were destroyed when God defeated Satan and destroyed his home planet. (ETBH chapters 10-16)

12.5 Before Peleg There Was Only One Continent.

God began dividing the continents when Peleg was born.

Peleg's name means "earthquake" because "for in his days was the earth divided" (Genesis 10:25).

Peleg lived for 239 years and that is how much time it took for the continents to get where they are today. After Peleg died the earthquakes became scarce and continental drift slowed down to where it is today.

"*And unto Eber were born two sons: the name of one was Peleg;* ***for in his days was the earth divided****; and his brother's name was Joktan*". (Genesis 10:25)

Strong's H6389 – Peleg (פֶּלֶג) defined as divided.

Strong's Definitions: פֶּלֶג *Peleg, peh'-leg; the same as* H6388*; earthquake; Peleg, a son of Shem:—Peleg.*

"*These are the families of the sons of Noah, after their Generations, in their nations: and by these were the nations divided in the earth after the flood.*" (Genesis 10:32)

The tower of Babel was located in what is modern day Iraq. Men made up their own religion there and left God out. The tower was the focal point of their religion. God was angry! (Genesis 11:1-9)

"*And the LORD said, Behold, the people are one, and they have all one language; and this they begin to do: and now* ***nothing will be restrained from them, which they have imagined to do***." (Genesis 11:6)

Daniel said that in the last days we would reconnect with each other and knowledge shall be increased.

"*But thou, O Daniel, shut up the words, and seal the book, even to the time of the end: many shall run to and fro, and knowledge shall be increased*" (Daniel 12:4).

Rapid travel, interconnectivity and increased knowledge about everything is basically what Daniel was describing and that is exactly what we have going on today.

Before Armageddon, countries will arm themselves with the most advanced super weapons ever produced by man. The Antichrist will control and use our technology against Believers during the Tribulation, he will betray Israel in the middle of the 7 years (70th Week of Daniel) and the Vatican as well.

12.6 Pangaea

The Bible has always taught that all of the continents were at one time just one big continent. (Genesis 10: 25) Today scientists are beginning to recognize that the contents all fit together as if they were once all connected.

A widely accepted theory concerning Continental Drift states that our continents were once all connected as one big continent that Scientists call Pangaea (Pan·gae·a).

The Bible also teaches us that the continents were once all connected and it was after the flood, after the Tower of Babel incident that God divided the land. The Bible even tells us details about the event so that we can verify that it happened. According to the Bible, the continents were created over the life time of Peleg. It began when Peleg was born, 4243 years ago and completed 4004 years ago. (Genesis 11:18, 19) The process began about 1757 and continued till 1996 (Hebrew calendar Adam = year 0). That was 239 years of earthquakes.

If the continents had to move 1500 miles to get to where they are today, then they would be moving at a rate of about 6.29 miles per year or about 28 meters/day or 1.7 meters per hour.

The Bible tells us when it happened, it tells us how fast it happened and it tells us why it happened.

"*And unto Eber were born two sons: the name of the one was Peleg; because in his days the earth was divided: and his brother's name was Joktan*". (1Chronicles 1:19)

Scientists believe that the contents have been moving for the last 12 million years at the same rate that they are today, that would be the starting point. The last point in time when all the land was in one big continent.

This 12 million years is based on the assumption that everything was always moving at the same slow rate that it is today.

Continental drift was first mentioned in the Bible about 2600 years ago. The Bible says that the continents were divided more recently and at a faster rate than scientists currently are aware of. Scientists are catching up with what the Bible has already declared.

The Bible said the earth was divided after the flood, about 4,000 years ago and during the life time of Peleg. He lived for 239 years. So who do you think is right, modern day scientists or our 2600 year old scriptures? The Bible has never been wrong but scientists are continually changing their beliefs about something as they learn more. The Bible doesn't ever do that!

Here is a link to Wikipedia's page about Pangaea… http://en.wikipedia.org/wiki/Pangaea

Another video link https://www.youtube.com/watch?v=3HDb9Ijynfo

Remember, God Said It First!

12.7 The Atlantis Myth.

The similarities between some myths and what the Bible says about the history of the earth may be only coincidental but they reveal some interesting possibilities.

According to the Bible, before Peleg's time there was only one continent. (Genesis 10:25, 32)

Satan, who was and still is the "*god of this world*" started a war against God. God defeated Satan and cast him and his Angels back down to the earth.

According to NASA, an event like the one described in the Bible occurred about 65 million years ago. God judged Satan and the earth, as a result the earth was flooded and all air-breathing life was destroyed.

Many Bible scholars believe that Angels were the original occupants of the earth, Dr. Lindstrom was one of them.

Lindstrom taught that Angels built cities and probably developed technology just like man has done. The only difference is that the Angels lived here for millions and perhaps billions of years. He believed they developed technology far beyond anything that we can imagine.

Many scientists believe that what used to be dry land is mostly submerged under water or rock. Most of the dry land that we occupy used to be underwater. It seems that they got reversed.

The remains of Angelic cities may be under the oceans and perhaps we have already stumbled across some of those ruins. (https://www.youtube.com/watch?v=4-AxXrjBdys)

The Atlantis myth may be a demonic version of what happened. The *"god of this world"* has *"blinded the minds of them which believe not"* … (2Corinthians 4:4)

Satan "*blinds the minds*" of the lost with lies and misinformation. The myth of Atlantis has some interesting similarities with the Biblical record but is wrong, the Bible version is correct!

When the first flood covered the Angelic cities, there was only one continent and that is where the similarity of the truth and the myth ends.

Apparently demons can use information they have about past and future events to deceive men. They use correct information to gain trust and then comes the lies!

Jules Verne's novel also predicted that the United States would be the first to send a mission to the Moon. He also predicted that there would be a dispute between Florida and Texas to host the launch site. That dispute did happen but was settled by the U.S. congress in the 1960s. Florida would be the launch site and Huston would be the mission control center.

The approximate size, weight and dimensions of the spacecraft were also mentioned. Weightlessness was discussed; the use of retro rockets and a splash down in the Pacific Ocean in the same vicinity as actually occurred on the Apollo 11 splash down.

Jules Verne's novel, "From the Earth to the Moon" predicted that there would be a three man crew. The giant cannon that shot the spacecraft into space was called the Columbiad. The name of the actual craft used by NASA was the Columbia.

There is too many similarities to be coincidental and too many errors to be from God. It is taught in the Bible that Satan and his demons can and do influence men in many ways. We are told to be cautious, observant and discerning through God's word.

12.8 Lucifer Lived On The Earth Before Man Was Created.

"*Thou hast been in Eden the garden of God; every precious stone was thy covering, the sardius, topaz, and the diamond, the beryl, the onyx, and the jasper, the sapphire, the emerald, and the carbuncle, and gold: the workmanship of thy tabrets and of thy pipes was prepared in thee in the day that thou wast Created*", (Ezekiel 28:13)

Wow! Lucifer wore Jewels and gold for clothes! Talk about bling! He was also a musician. Sounds a lot like a rock star.

"*Thou art the anointed cherub that covereth; and I have set thee so: thou wast upon the holy mountain of God; thou hast walked up and down in the midst of the stones of fire*" (Ezekiel 28:14)

Sounds like he was heat resistant. I wonder if any Cherub could do that or just certain ones.

"*Thou wast perfect in thy ways from the day that thou wast Created, till iniquity was found in the*" (Ezekiel 28:15)

"*By the multitude of thy merchandise they have filled the midst of thee with violence, and thou hast sinned: therefore I will cast thee as profane out of the mountain of God: and I will destroy thee, O covering cherub, from the midst of the stones of fire.*" (Ezekiel 28:16)

"*Thine heart was lifted up because of thy beauty, thou hast corrupted thy wisdom by reason of thy brightness: I will cast thee to the ground, I will lay thee before kings, that they may behold thee.*" (Ezekiel 28:17)

This is where God defeated Satan and threw him back down to Earth. (Ezekiel 28:13-17) Isaiah talks about the same event.

"*How art thou fallen from heaven, O Lucifer, son of the morning! how art thou cut down to the ground, which didst weaken the nations*"! (Isaiah 14:12)

12.9 Lucifer Visited The… "Mount Of The Congregation".

Satan was on the Earth when he was plotting this out in his mind. God knew his thoughts before he did!

"*For thou hast said in thine heart, I will ascend into heaven, I will exalt my throne above the stars of God: I will sit also upon the mount of the congregation, in the sides of the north:*" (Isaiah 14:13)

Lucifer plots from Earth. (vs13) "*North*" is a term that describes direction relative to the Earth.

The mount of the congregation was on Lucifer's to do list. I suppose that the mount of the congregation was a place where the Hosts of God congregated. Lucifer wanted to go there and conquer it as well. Lucifer also had some history in the *"Mount of God"* and probably even heaven.

"*I will ascend above the heights of the clouds; I will be like the most High*". (Isaiah 14:14)

Sounds like he was below the clouds, on the Earth when he was thinking this.

12.10 Lucifer Plots War Against God.

God defeats him and casts him back down to the Earth. (Isaiah 14:12-14)

"*How art thou fallen from heaven, O Lucifer, son of the morning! how art thou cut down to the ground, which didst weaken the nations*"! (Isaiah 14:12)

"For thou hast said in thine heart, I will ascend into heaven, I will exalt my throne above the stars of God: I will sit also upon the mount of the congregation, in the sides of the north:" (Isaiah 14:13)

"*I will ascend above the heights of the clouds; I will be like the most High.*" (Isaiah 14:14)

It definitely sounds like Lucifer was below the clouds, on the Earth, when he was thinking this.

North is a direction that is relative to the Earth. By referring to the North, God is referencing the Earth.

This is another reference to Heaven's location relative to the Earth. (Isaiah 14:13) Heaven is in the north!

12.11 God Knew Lucifer's Thoughts

Satan's understanding may have been clouded by his pride. It seems that he was very powerful, very smart and very beautiful.

The Bible says that his beauty and music were the source of his pride. I believe that he may have been very beautiful and pleasant to be around. I suppose there was plenty of good music when he was present. He may have used that music to influence the feelings and emotions of those in his presence.

If Satan rationally considered all of the facts that he had available, he would not have rebelled against God. Satan may have not been interested in the facts, he may have been following his feelings instead. I believe he was.

His own feelings and unholy desires may have blinded him from reality, like it does to humans. If you have ever wondered about how Satan persuaded one third of the Angels to follow him he probably used their feelings.

I like to speculate about things like this because it makes me think. Consider how Christians have been persecuted in the past and currently are now.

I believe Satan deceived and persuaded a third of his Earthly subjects, (the Angels) with the same methods used by the Catholics during the Spanish Inquisition and in the dark ages. Islam has been doing the same thing for centuries and is still doing it today.

Deception, Betrayal and Terror is their common tools. Islam and Catholicism have both terrorized, tortured and murdered countless millions of Christians and Jews all over the world and it has been going on for centuries.

Satan probably lied, promised, threatened, tortured and murdered Angels to persuade one third to follow him. It must have been horrific, just like it is going to be during the 7 years tribulation!

He probably told a big lie and then gave the Angels the choice between believing his lies and worshiping him or a painful and terrifying death.

I believe that the two thirds of the Angels that were faithful to God had to be resurrected and relocated to Heaven. Of course this is only speculation. Speculation based on the nature and history of (Satan). I believe that the human race will be tested during the 7 years of Tribulation!

Today Satan's minions are a constant source of terror against God's people. I believe that Satan's minions are using the same methods today as they did when Satan rebelled against God.

Islam routinely and daily, threatens, tortures and beheads Christians, Jews and non-Muslims in the parts of the world that Sharia law governs. Sharia law takes over an area after the United Nations has disarmed the public. The Muslims are supported and armed while the Christians are disarmed and slaughtered. Taking pictures and sending to their friends or selling videos of murder and torture of westerners.

Muslims pray to Alah (Hebrew) for an opportunity to come to America to kill westerners. Obama & Clinton snuck in many thousands, perhaps millions. More than enough to accomplish their goals. They are here now, the U.N. troops are building up and getting ready.

Meanwhile Americans believe that it can't happen here. The fact is that the United States is the 11th Kingdom spoken of by Daniel. I believe that this country is going to be judged first along with Europe. But the U.S. may be judged the most severely.

12.12 Lucifer, Led A Rebellion Against God.

"*In whom the god of this world hath blinded the minds of them which believe not, lest the light of the glorious gospel of Christ, who is the image of God, should shine unto them*". (2Corinthians 4:4)

Satan ruled over the inhabitants of earth as "*god of this world*" and then he led a rebellion against God. (Isaiah 14:13)

"*For thou hast said in thine heart, I will ascend into heaven, I will exalt my throne above the stars of God: I will sit also upon the mount of the congregation, in the sides of the north*": (Isaiah 14:13)

To get to heaven, Satan would have to ascend **northward**. North is a direction relative to Earth. This tells us that Satan was on the Earth when he was thinking this. It also tells us where Heaven is relative to Earth. **Heaven is in the North.**

"*I will ascend above the heights of the clouds; I will be like the most High*". (Isaiah 14:14)

Satan was below the clouds, on the Earth and wanted to ascend into heaven, northward. He ended up taking with him one third of the Angels. (Jude 6)

"*And the angels which kept not their first estate, but left their own habitation, he hath reserved in everlasting chains under darkness unto the judgment of the great day*". (Jude 6)

12.13 This Is A Rebellion Against God's Order

This is a battle of evil against good; Disorder against order and earth is the final battle ground. The Human Race is the object of this battle! This is Satan and his demons against God and his Creation. Satan is using the human race to destroy itself.

Science confirms the Biblical record. God recorded it through men thousands of years before the events happened or men had the knowledge. It is time to believe on Jesus Christ now. Just Faith.

There is Plenty of evidence to substantiate the Bible's claim to be the word of God. The Bible record of the total destruction of the earth and the pre-human civilization that once lived here is supported by plenty of evidence.

Today scientists recognize that there was a catastrophic event that occurred about sixty five million years ago. The earth was struck at least once by an asteroid. The event resulted in the sudden destruction of almost all life on earth.

This was not known until it was discovered by NASA in the mid 1970's. However, it was spoken of in the notes of the Scofield Bible published in 1909. God destroyed the earth, a prehumen civilization and all air breathing animals. That is what the fossil record shows. **Once again, it is proven that God said it first.**

13

Lucifer's War, The Cretaceous–Paleogene Extinction Event.

13.1 The Five "I Will's" Of Lucifer. (Isaiah 14:13)

"*For thou hast said in thine heart, I will ascend into heaven, I will exalt my throne above the stars of God: I will sit also upon the mount of the congregation, in the sides of the north*:" (Isaiah 14:13)

It sounds like Lucifer was on the Earth, below the stars of God, below the heavens.

He wanted to "*sit upon the mount of the congregation*" (Isaiah 14:13). The "*Mount of the Congregation*" is located "*in the sides of the north*". (Isaiah 14:13)

To get to Heaven, Satan would have to ascend northward from where he was.

North is a direction relative to Earth. Therefore, Earth is the place of reference. This not only indicates where Heaven is, but also tells us where Satan was when he was thinking this. He was on Earth.

One third of the inhabitants of Earth joined in Satan's rebellion against God. Those inhabitants of the Earth that followed Satan in his attempt to overthrow God were Angels. Now they are fallen Angels or demons.

Satan and the Angels left their home on the Earth to attack God in Heaven. (Jude 6)

"And the angels which kept not their first estate, but left their own habitation, he hath reserved in everlasting chains under darkness unto the judgment of the great day". (Jude 6)

God judged Satan,

"How art thou fallen from heaven, O Lucifer, son of the morning! how art thou cut down to the ground, which didst weaken the nations"! (Isaiah 14:12)

God cast Satan back down to Earth.

God judged the Angels that followed Satan and also cast them back down to Earth.

13.2 Lucifer Led A Rebellion Against God.

"For thou hast said in thine heart, ***I will ascend into heaven****, I will exalt my throne above the stars of God: I will sit also upon the mount of the congregation,* ***in the sides of the north****"*: (Isaiah 14:13)

Lucifer plotted to overthrow God and he became Satan.

"I will ***ascend above the heights of the clouds****; I will be like the most High"* (Isaiah 14:14)

Satan deceived and coerced one third of the Angels to leave the Earth and follow him. This was the beginning of sin in our universe.

"And the angels which ***kept not their first estate****, but* ***left their own habitation****, he hath reserved in everlasting chains under darkness unto the judgment of the great day"* (Jude 6).

They were defeated…

*"****How art thou fallen from heaven, O Lucifer****, son of the morning! how art thou cut down to the ground, which didst weaken the nations!"* (Isaiah 14:12)

(ETBH chapters 10-16)

God Judged The Earth.

Satan was cast back down to the Earth.

The Angels were also cast back down to the Earth.

"*the Earth* *__was__* ***(became)*** *without form and void*" (Genesis 1:2)

"Behold, the LORD maketh the earth empty, and maketh it waste, and turneth it upside down, and scattereth abroad the inhabitants thereof". (Isaiah 24:1)

(ETBH chapters 10-16)

13.3 Hell Was Made For The "Perps" Of Lucifer's War.

Sometime after Satan's Rebellion, **__Hell was made for Satan and his Angels__** which we will call "Demons" from now on. (Matthew 25:41) Hell wasn't originally made for man.

"*Then shall he say also unto them on the left hand, Depart from me, ye cursed, into everlasting fire,* ***__prepared for the devil and his angels:__"*** (Matthew 25:41)

Hell was Created to contain the sin of Satan and his demons.

Satan has been stuck on the Earth ever since he was defeated. I don't know when Satan found out about Hell, but I Think that he knew he was in trouble when God gave the Gospel to Adam and Eve after they sinned.

Hell is the Lake of Fire or the 2nd death.

"And death and hell were cast into the lake of fire. This is the ***__second death__***". (Revelation 20:14)

Hell is currently empty... (Matthew 25:41)

All the evil in our world will someday be contained in hell...

...until then there will be much more evil and violence.

Hell was Created for Satan and his Angels to contain their sin, but because of Adam's sin, the whole human race also became infected. (Romans 5:12).

People need to understand how mankind became sinful and condemned to the second death.

Most importantly we must know how God has provided a way to save us from the sin problem.

Satan and his demons will someday be judged and kept in Hell for eternity. Satan polluted our Universe with evil, someday hell will contain all of it. Until then we will just have to live with it, but only for a little while longer.

13.4 "Satan" Means "Adversary" Or Enemy Of God.

Satan chose to become the Enemy of God. Today, all unbelieving people are also called Enemies or adversaries of God. (Romans 5:9)

The Riddle of the frozen giants. By Ivan T. Sanderson; Saturday Evening Post; January 18, 1960.

Ivan T. Sanderson was researching the origin of all the frozen, extinct animals discovered permanently frozen in the permafrost of the polar regions of the Earth.

He discovered entire herds of extinct animals that had been flash-frozen. Some were frozen while trying to get up. Others were frozen while eating. Broken bones were not uncommon.

Undigested tropical vegetation was found in the stomachs of the vegetarian species. The stomach is the last part of the body to freeze, but undigested material was found, which supports the belief that these animals were flash frozen.

Sanderson went to the frozen food industry to help analyze samples taken from these animals and discovered something remarkable. These animals were frozen very quickly with very low temperatures.

Flash-freezing occurred. There was no doubt about it, the size of the water crystals determines the rate of freezing. If the rate of freezing is too slow, the water crystals will be too big and cause the cell wall to rupture. When the cell wall ruptures, the cellular contents escape. In the frozen food industry it is considered flash frozen if the cell walls don't rupture. The samples were perfectly preserved.

A Catastrophe caused this event over both frozen Polar Regions of the Earth which is a global indicator.

The Earth's axis changed. Ivan Sanderson knew from geologists also working on Earths history; that the Earths polar axis has changed. This explains how animals living in a tropical to subtropical environment could end up permanently frozen in our polar ice caps.

Impact causalities is the best way to describe animals with broken bones. One hairy mammoth had a broken hip and was frozen while trying to get up. The polar permafrost is full of extinct animals, with broken bones, severed limbs and vegetation mixed in.

The research done by and articles written by Ivan T. Sanderson led others to the belief that a sudden, mass extinction was caused by an asteroid.

The impact would cause worldwide volcanic eruptions. It would be like hitting a whoopee cushion. The pressure would have to spill out somewhere. The volcanoes would act as pressure release openings and spew out volcanic material into the upper atmosphere until the Earths motion in space stabilized.

13.5 Satan Is Still The "God Of This World".

*"In whom the god of this world hath **blinded the minds** of **them which believe not**, lest the light of the glorious gospel of Christ, who is the image of God, should shine unto them*". (2CORINTHIANS 4:4)

Satan apparently built incredible cities on this Earth which were destroyed at this Judgment. (Jeremiah 4:23-26)

A passage describes the coming judgment of men and angels suggests that ruins of those ancient Angelic cities may still be on the bottoms of the oceans. (Revelation 20:13) (To be discussed later)

*"And **the sea gave up the dead which were in it**; and death and hell delivered up the dead which were in them: and they were judged every man according to their works*". (Revelation 20:13)

The phrase...***"And the sea gave up the dead which were in it."*** May be talking about the Angelic bodies, buried in the ruins of those cities. (Revelation 20:13)

I believe that the ruins of those Angelic cities that were killed during Satan's Rebellion may still be on the bottoms of the Oceans. (ETBH chapters 13-17)

Satan Is The "God Of This World" From The Beginning. He will continue to hold that title until Jesus finally defeats him at Armageddon. (Revelation)

"*In whom the god of this world hath blinded the minds of them which believe not, lest the light of the glorious gospel of Christ, who is the image of God, should shine unto them*". (2Corinthians 4:4)

Satan is also called "*The Prince Of The Power Of The Air*". (Ephesians 2:2)

13.6 God Destroyed Satan's Home, The Earth.

"Behold, the LORD maketh the earth empty, and maketh it waste, and turneth it upside down, and scattereth abroad the inhabitants thereof". (Isaiah 24:1)

The Hebrew word for "empty" is… בָּקַק, baqaq, (Strong's # H1238) - baqaq to empty, lay waste, to make void (fig.)

"*maketh it waste*", בָּלַק, - balaq, (Strong's H1110) to waste, lay waste, devastate, (Poel) to make waste, (Pual) devastated (participle)

"*turneth it upside down*" Strong's H5753 - `avah, to bend, twist, distort;

(Isaiah 24:19) "*The earth is utterly broken down, the earth is clean dissolved, the earth is moved exceedingly.*"

What we have here is three witnesses about the same event.

The Earth became without form and void after a perfect Creation, (Genesis 1:1) this is how you find it in the next verse. (vs. 2) The same language is used here in (Isaiah 24:19) as in (Genesis 1:2) and (Jeremiah. 4:23).

The Earth's rotation and orientation in space were affected.

"*The earth shall reel to and fro like a drunkard, and shall be removed like a cottage; and the transgression thereof shall be heavy upon it; and it shall fall, and not rise again*" (Isaiah 24:20)

If the Earth were struck by a large enough asteroid it would cause the Earth to "reel to and fro like a drunkard", as it orbited the Sun just like the Bible describes.

The Earth was destroyed beyond its ability to recover, "*and it shall fall, and not rise again*". (Isaiah 24:20)

This is how the Earth <u>became</u> "*without form and void*" after a perfect Creation.

The Bible says that the Earth… "*The earth shall reel to and fro like a drunkard*" (Isaiah 24:20)

The sun's light was cut off.

"*For this shall the earth mourn, and the heavens above be black: because I have spoken it, I have purposed it, and will not repent, neither will I turn back from it.*" (Jeremiah 4:28)

The destruction of the Earth was so complete that the loyal two thirds had to be relocated. Prior to this destruction of the Earth, there is no reference of Angels in Heaven.

The Earth became…"*without form and void*" (Genesis 1:2); (Isaiah

If N.A.S.A. and all of the other scientists researching the Chicxulub crater are correct, then the earth was struck by a large asteroid that destroyed the earth in a manner consistent with Biblical descriptions about 65 million years ago.

This asteroid event meets the description of Gods Judgment against Satan and his Earth.

I believe that the Chicxulub crater is a monument to Gods first judgment of the Earth. Prior to the impact of that asteroid was the pre-sin era which goes back to Creation.

13.7 God Cast Satan Back To The Earth

The abundance of traffic or trade is associated with Satan's violence, sin and God's judgment. **(Ezekiel 28:16)**

"*By the multitude of thy merchandise they have filled the midst of thee with violence, and thou hast sinned: therefore I will cast thee as profane out of the mountain of God: and I will destroy thee, O covering cherub, from the midst of the stones of fire*" (Ezekiel. 28:16)

Sounds like Lucifer had some history in "*the mountain of God*" and he was about to get thrown out.

"I will cast thee as profane out of "the mountain of God" (Ezekiel. 28:16)

I think this is the point where Lucifer's name was changed to Satan. It sounds to me like Satan was kicked out of the mountain of God, because he was bringing in sin. This is the account of when God took control of the situation. Lucifer who is called Satan from here on, is cast out of the Mountain of God.

"*I will cast thee as profane out of the mountain of God*" (Ezekiel 28:16) "*therefore I will cast thee as profane*" Strong's H2490 – chalal (to profane oneself, defile oneself, pollute oneself)

Satan became proud. Sounds like he thought more of himself than he should have and God set him straight! (Ezekiel 28:17)

"Thine heart was lifted up because of thy beauty, thou hast corrupted thy wisdom by reason of thy brightness: I will cast thee to the ground, I will lay thee before kings, that they may behold thee" (Ezekiel 28:17)

Notice that (Ezekiel 28:17) is talking about Satan being cast to the ground. There are other passages that mention this.

"behold a great red dragon, having seven heads and ten horns, and seven crowns upon his heads". "And his tail drew the third part of the stars of heaven, and did cast them to the earth" (Revelation 12: 3,4)

I believe that this is important. The ground that is mentioned here, I believe, is the earth.

CHICXULUB; LOCAL MAYAN NAME FOR "TAIL OF THE DEVIL"_

(http://www.yucatantoday.com/en/topics/chicxulub)

Chicxulub is pronounced - Cheek-shoo-loob. "For a Mayan village above the center of the crater, that Mayan word means "tail of the devil," which Kring thought was an ideal name for a dinosaur-killing asteroid impact." Inland News Today - full story - http://www.inlandnewstoday.com/story.php?s=13250

Wikipedia says it means… "Chic" = tick or flea and "xulub" = devil

13.8 The Largest Known Asteroid To Have Struck The Earth

The Chicxulub Crater is believed by NASA to have been caused by the largest known asteroid to have ever struck the Earth.

It is believed by NASA to have caused the destruction of 70% of the life on the Earth. This happened about 65 million years ago. Originally, NASA said that all life was extinguished.

"*I beheld, and, lo, there was no man, and all the birds of the heavens were fled*". (Jeremiah 4:25)

This happened before Man was Created.

"*I beheld, and, lo, the fruitful place was a wilderness, and all the cities thereof were broken down at the presence of the LORD, and by his fierce anger*". (Jeremiah 4:26)

When those cities "*were broken down at the presence of the LORD*" Man was not yet Created. (Jeremiah 4:26) I believe that Lucifer and his Angels built and occupied those cities. I believe that some of the remains of those cities may still be at the bottoms of our oceans.

"*The fruitful place was a wilderness*" is a phrase that is saying that the Earth was a fruitful place once, but now it has **become** a wilderness. Jeremiah got to see the "before" and "after" of God's judgment. He was looking at what was left just after God judged the Earth. The fruitful place became a wilderness as the result of God's presence and his "fierce anger".

I believe that the planet was still vibrating from the impact of the largest asteroid known to have ever struck the Earth when Jeremiah was describing what he saw! (Jeremiah 4:24)

The "*presence of the LORD*" is a phrase that is rarely used. Here "*the presence of the LORD*" resulted in the destruction of all air-breathing life on Earth. Sin cannot exist in God's presence! Satan created sin and sin was in the Angels that became demons when they followed him.

"*For thus hath the LORD said, The whole land shall be desolate; yet will I not make a full end*". (Jeremiah 4:27)

God destroyed all air-breathing life on the planet, but he did not completely destroy the planet itself. He had future plans for the Earth.

"*I beheld, and, lo, the fruitful place was a wilderness, and all the cities thereof were broken down at the presence of the LORD, and by his fierce anger*". (Jeremiah 4:26)

This is Satan's home planet being destroyed by God after Satan's failed attempt to attack God in Heaven. Man, talking about a bad idea!

This is rare verbiage in scripture. The "presence of the Lord" is not used often in scripture and is associated with God's Judgment.

13.9 It Struck The Earth About 65 Million Years Ago.

The age of the Chicxulub asteroid impact and the Cretaceous–Paleogene boundary (K–Pg boundary) coincide precisely. (A time of global, mass extensions.)

This is the time of a mass extinction of animals all over the world. This is also referred to as the time the Dinosaurs disappeared. The missing Dinosaurs are only a small part of what disappeared.

"The Cretaceous-Tertiary mass extinction, which wiped out the dinosaurs and more than half of the species on Earth, was caused by an asteroid"... sciencedaily.com - full story - http://www.sciencedaily.com/releases/2010/03/100304142242.htm

"Impact helped result in the extinction of 75 percent of all species on Earth". Space.com (Click link for full story.) http://www.space.com/19681-dinosaur-killing-asteroid-chicxulub-crater.html

The Chicxulub Crater - Additional link http://www.sheppardsoftware.com/Mexicoweb/factfile/Unique-facts-Mexico4.htm

13.10 Jeremiah Traveled Back In Time And Saw This Event...

...he saw, firsthand what happened to the Earth when God judged Lucifer.

(Jeremiah 4:23, 27) is describing an event that happened long ago, before man was Created.

"*I beheld the earth, and, lo, it was without form, and void; and the heavens, and they had no light*". (Jeremiah 4:23)

In (Genesis 1:2), we have... *"and darkness was upon the face of the deep"*; In (Jeremiah 4:23)... "*the heavens, and they had no light*". The Earth was in darkness as a result of this event and remained that way for a long time. If you cut off the sun's light the earth would quickly freeze.

Lucifer, the god of this world, attempted to attack God in Heaven. <u>God cast Satan and his Angels back down to the Earth</u> and destroyed their home planet.

"*I beheld the mountains, and, lo, they trembled, and all the hills moved lightly*". (Jeremiah 4:24)

Now that sounds like the Earth was still "vibrating" after being struck by a big asteroid, like the one that struck the Earth down in the Yucatan peninsula. (See NASA magnetic map of the crater… http://chicxulubcrater.org/) (http://apod.nasa.gov/apod/ap000226.html)

13.11 The Bible Describes This Event. (Isaiah. 24:19-20)

"*The earth is utterly broken down, the earth is clean dissolved, the earth is moved exceedingly.*" (Isaiah 24:19)

Now that sounds like an asteroid impact to me.

"*The earth shall reel to and fro like a drunkard, and shall be removed like a cottage; and the transgression thereof shall be heavy upon it; and it shall fall, and not rise again*" (Isaiah 24:20)

Now that really sounds like an asteroid impact! "The earth shall reel to and fro like a drunkard,"… sounds like an impact wobble from a big asteroid to me. The Earth still has a little bit of that wobble left. But just after the impact the wobble was at its greatest.

If you ever wondered how the Grand Canyon was made, well that wobble spoken of in (Isaiah 24:20) may be involved. All of that water from the oceans would have been sloshing back and forth all over the Earth. There would be continual tsunamis all over the planet until it froze.

Initially the sudden impact of something that big hitting the earth traveling roughly 50,000 miles per hour would vaporize much of the oceans. Like a big whoopee cushion the earth would spew out hot volcanic gases, steam and liquid rock. Most of the hot dust and debris mixing with and vaporizing everything it touches. This hot dust was thrown out in space, some of it continued to orbit the planet and some of it cooled to exceeding low temperatures, condensed and fell back to the earth as giant bubbles of super cold air. This theory has been presented by a researcher looking over the evidence and trying to explain what he sees.

The sun's light would be cut off. The earth would quickly freeze. The Earth would be in a thick cloud of dust and debris indefinitely. Nothing could survive on the surface.

The earth's motion gradually equalized. After 65 million years, that "wobble" is almost undetectable. And that is why I believe that "wobble" is there today.

If somebody wants to talk about "Evolution", just tell them you would rather talk about the "wobble"… and then tell them how it got there and all about the Chicxulub crater. Just make sure you include a simple and clear presentation of the Gospel!

"shall be removed like a cottage" …sounds like an impact for sure.

"*shall fall and not rise again*" all life is extinguished!

This is how the Earth became "without form and void" as described in (Genesis 1:2).

"*he Created it not in vain, he formed it to be inhabited:*" (Isaiah 45:18)

God did not Create it that way! "*The Earth became without form and void*" when God judged Satan.

13.12 The Cretaceous–Paleogene Extinction Event

According to N.A.S.A. the earth was struck by the largest known asteroid in the **Yucatan Peninsula about 65 million years ago**. A crater was discovered there with a small Mayan town built in the center called "**Chicxulub**"

Two minute video about this event, follow this link ... https://www.youtube.com/watch?v=XYJCm6boxjA

(Isaiah 24:1, 19, 20); (Jeremiah 4:23-28)

I believe that event was the same event the Bible describes...

This is how the Earth became without form and void in (Genesis 1:2) (Isaiah 24:1, 19 & 20; Jeremiah 4:23-28)

"Behold, the LORD maketh the earth empty, and maketh it waste, and turneth it upside down, and scattereth abroad the inhabitants thereof." (Isaiah 24:1)

"The earth is utterly broken down, the earth is clean dissolved, the earth is moved exceedingly". (Isaiah 24: 19)

"The earth shall reel to and fro like a drunkard, and shall be removed like a cottage; and the transgression thereof shall be heavy upon it; and it shall fall, and not rise again." (Isaiah 24: 20)

"***I beheld the earth, and, lo, it was without form, and void; and the heavens, and they had no light*"**. (Jeremiah 4:23)

"I beheld the mountains, and, lo, they trembled, and all the hills moved lightly". (Jeremiah 4:24);

*"I beheld, and, lo, **there was no man**, and all the birds of the heavens were fled".* (Jeremiah 4:25);

*"I beheld, and, lo, **the fruitful place was a wilderness**, and all the **cities** thereof **were broken down at the presence of the LORD**, and by his fierce anger".* (Jeremiah 4:26);

"*For thus hath the LORD said*, *The whole land shall be desolate; yet will I not make a full end."* (Jeremiah 4:27);

"For this shall the earth mourn, and the heavens above be black: because I have spoken it, I have purposed it, and will not repent, neither will I turn back from it." (Jeremiah 4:28);

"For thou hast said in thine heart, I will ascend into heaven, I will exalt my throne above the stars of God: I will sit also upon the mount of the congregation, ***in the sides of the north:"*** (Isaiah 14:13)

"I will ascend above the heights of the clouds; I will be like the most High". (Isaiah 14:14)

Once again, it is important to note that Satan was below the clouds, on earth. He wanted to leave the Earth to attack God in the "Mountain of God" (Heaven), and the Mount of the Congregation located in the Sides of the North).

He wanted to sit on the **Mount of the Congregation**. I believe that "***The mount of the Congregation*" is different than "*the mountain of God*"**. The "Mountain of God" is where God is and the "Mount of the Congregation" is in *"the sides of the north"* where the *"Congregation"* gathered themselves.

. The Congregation probably consisted of members of the *"Hosts of God"*. I believe that Lucifer may have been kicked out of the Mount of the Congregation for misconduct, which may have influenced his rebellion.

I believe that Satan attacked the Mountain of God not knowing what he was up against.

13.13 "Chicxulub", Evidence Of A Catastrophic Event

The Chicxulub crater in The Yucatán Peninsula is believed to be made by what may be the largest known asteroid to have ever struck the Earth.

The Chicxulub crater was discovered in the late 1970's by Geologists looking for oil.

Estimates of its actual diameter range from 106 to 186 miles (170 to 300 kilometers), which if proved right could mean it's the biggest.

According to scientists, this crater is one of the oldest and most studied craters on earth.

The Chicxulub crater is believed to be caused by an asteroid traveling at about 50,000 miles per hour.

…"the actual crater is 300 kilometers (186 miles) wide, and the 180 kilometer ring is just an inner wall".

The more this crater is studied the bigger it seems to get.

"Whereby the world that then was, being overflowed with water, perished:" (2 Peter 3:6)

14

There was a War in Heaven! The Earth (Became) Without Form"

(Genesis. 1:2) (Jeremiah 4:23-26) (Isaiah 24:1) (2Peter 3:5 & 6)

14.1 After The War,

The Earth remained dark and frozen until... (Genesis. 1:2)

...Until God started warming things up for Adam. (Genesis. 1:2).

The Earth was restored to accommodate Adam about 6,000 years ago. That is rather recently in astronomical terms.

On the 6th day of Restoration, God put Adam in the Garden of Eden which used to belong to Satan.

Furthermore God put Adam in charge of the planet and even had Adam name all the new animals.

I believe this was done in full view of Satan.

Satan was the "*god of this world*" long before Adam was Created. Satan and the Angels were all Created "In the Beginning". (See chapters 13-18)

14.2 The Earth Was Not Created "Without Form And Void"

"*And the earth was **(became)** without form, and void and darkness was upon the face of the deep*"... (Genesis. 1:2)

He Created it perfect, complete, ***"to be inhabited"***.

"he Created it not in vain, he formed it to be inhabited": (Isaiah 45:18)

There was a war in Heaven! Satan lost the war and the Earth was destroyed.

If NASA is right about the timing of the big asteroid that struck the earth in the Yucatan Peninsula and the mass extinctions then that war probably happened about 65 million years ago.

Both events fit the description of the Biblical account.

14.3 Earth Was Judged, It Became "Without Form And Void"

*"I beheld the earth, and, lo, **it was without form, and void;** and the heavens, and **they had no light**."* (Jeremiah 4:23)

(Genesis 1:2) is talking about the same thing, *"without form and void"*.

These passages are describing what the earth was like after God judged it about 65,000,000 years ago! Here is a reference to the sun's light being cut off. That would freeze things in a hurry! The Biblical record matches with real science and confirms that the Bible is accurate and the word of God!

*"I beheld, and, lo, **the fruitful place was a wilderness**, and all the **cities** thereof were broken down at the presence of the LORD, and by his fierce anger."* (Jeremiah 4:26)

This is before man was Created. These cities were Angelic.

"*Behold, the LORD maketh the earth empty, and maketh it waste, and turneth it upside down, and scattereth abroad the inhabitants thereof.*" (Isaiah 24:1)

"The earth is utterly broken down", (Isaiah 24:19)

This is how the earth ***<u>became</u>*** *without form and void.*

"And the earth was (became) without form, and void; and darkness was upon the face of the deep. And "the Spirit of God moved upon the face of the waters". (Genesis 1:2)

A dark and frozen planet is thawed out and warmed up for Adam.

The Earth is flooded and remains that way until verse 9.

"And God said, Let the ***waters under the heaven be gathered together unto one place, and let the dry land appear:*** *and it was so"*. (Genesis 1:9)

This is the first flood spoken of in (2Peter 3:5, 6)

"For this they willingly are ignorant of, that by the word of God the heavens were of <u>old</u>, and the earth <u>standing out of the water and in the water:</u>" (2Peter 3:5)

"Whereby the world that then was, being overflowed with water, perished": (2Peter 3:6)

This was before Noah and before Adam.

(Genesis 1:2) is where God begins to restore the Earth to a temperature suitable for Man. God starts warming up the earth before counting off six days of restoration. Verse 2 says… *"the Spirit of God **<u>moved</u>** upon the face of the waters"*

"***Moved***" is an old-English term that describes a hen warming her eggs. That is what the Holy Spirit did to the Earth. I believe that the Earth was frozen; in darkness; and had been that way for a very long time and had to be warmed up by God.

If NASA is right about when the big asteroid struck the earth in the Yucatan Peninsula, this probably happened about 65,000,000 years ago.

The earth was warmed up by God before the first day, 6,000 years ago. The Earth was already the right temperature before the first day began.

14.4 Ancient, Underwater Structures Have Been Found

Could it be that they are finding the ruins of those ancient Angelic cities on the bottoms of the oceans? (Jeremiah 4:23-26) (Revelation 20:13) **I don't know, but it could be**.

God destroyed all air-breathing life on the Earth.

*"Behold, **the LORD maketh the earth empty**, and **maketh it waste**, and **turneth it upside down**, and scattereth abroad the **inhabitants** thereof"*. (Isaiah 24:1)

This passage is telling how the earth **became** empty; wasted and turned upside down. God judged Lucifer, defeated him, changed his name to "Satan"; cast him back down to the earth; cast Lucifer's Angels back down to the earth and now they are called demons; turned the earth upside down and destroyed all air-breathing life on the planet. If N.A.S.A.'s calculations are correct, this all happened about 65,000,000 years ago.

If you read this and ask yourself **when** this happened you must realize that this happened **before man was Created**. If this is before man was Created, who are those ***"inhabitants"***

"*The earth is utterly broken down, the earth is clean dissolved, the earth is **moved exceedingly**.*" (Isaiah. 24:19)

"***The earth shall reel to and fro like a drunkard**, and shall be removed like a cottage; and the transgression thereof shall be heavy upon it; and it shall fall, and not rise again*" (Isaiah. 24:20

God struck the earth, made it wobble in space like a drunkard; (Isaiah 24:20), turned it upside down; (Isaiah 24:1) flooded it; (Genesis 1:2) (Jeremiah 4:23-26) cut off the Sun's light; (Jeremiah 4:28) froze it (Genesis

1:2) and left the earth in cold darkness like that with Satan and his demons still on it until God warmed it back up, turned the lights back on and drained the water for Adam 6000 years ago. (Genesis 1:2-9).

According to N.A.S.A., about 65 million years ago the Earth was struck by a very large asteroid in the Yucatan Peninsula which destroyed 70% of all life on the planet. This catastrophic event sounds like the one the Bible describes in (Isaiah 24:1, 19, 20) (Jeremiah 4:23-27) (2Peter 3:5-7) and others.

If N.A.S.A. is right then Lucifer's' War may have taken place about 65 million years ago!

14.5 Jesus Put A 2000 Year Gap Between 2 Words.

Jesus, Demonstrates How To Rightly Divide The Word. Jesus puts a 2,000 Year Gap Between 2 Words in the Middle of a sentence…

…"And he came to Nazareth, where he had been brought up: and, as his custom was, he went into the synagogue on the Sabbath day, and stood up for to read".

"And there was delivered unto him the book of the prophet Esaias. And when he had opened the book, he found the place where it was written,"

"The Spirit of the Lord is upon me, because he hath anointed me to preach the gospel to the poor; he hath sent me to heal the brokenhearted, to preach deliverance to the captives, and recovering of sight to the blind, to set at liberty them that are bruised",

"To preach the acceptable year of the Lord".

"And he closed the book, and he gave it again to the minister, and sat down. And the eyes of all them that were in the synagogue were fastened on him".

"And he began to say unto them, This day is this scripture fulfilled in your ears." (Luke 4:16-21)

Notice that he is reading from Isaiah 61:1, 2 and he stops in the middle of verse 2, in the middle of the sentence. There is a 2000 year gap between the last word he read and the remainder of the sentence! That is between the word "*Lord*" and the word "*and*" in (Isaiah 61:2).

*"To proclaim the acceptable year of the **LORD, and** the day of vengeance of our God; to comfort all that mourn"*; (Isaiah 61: 2)

Jesus stopped reading in the middle of the sentence, closed the book, sat down and said "*This day is this scripture fulfilled in your ears.*" (Luke 4:16-21) The last half of the sentence was 2000 years into the future!

Jesus put a 2000 year Gap between two words in the middle of a sentence. It is all about "*rightly dividing the word*".

"*Study to shew thyself approved unto God, a workman that needeth not to be ashamed, **rightly dividing the word** of truth*" (2Timmothy 2:15)

The gap between Genesis 1:1 and verse two is the first and biggest gap in the Bible that we know of so far. As you can see in Luke 4:16-21 there are other examples of Gaps in time in the Bible. Jesus himself demonstrated that these gaps do exist. (Luke 4:16-21)

The time that elapsed from "***The Beginning***" of Creation, till man was placed here may have been billions of years.

Man was Created quite recently. Man has only been around for about 6,000 years. The Earth is very much older and has an interesting history.

14.6 God's Gap Fact And The Youngster Error...

Ken Ham, the author of "The Answer Book" is one of the leading promoters of the **Young Earth Error**. I call them "Youngsters". If God calls the earth old (Psalm 102:25) (2Peter 3:5), why would **a believer** call it young? God **never** calls the earth young, **He always calls it "Old"**. Let us just call that "**The Youngster Error**".

There seems to be two groups of thought on this matter. First there is God's and then there is the "Youngsters" and they are clearly polar opposites.

Ken Ham may call us "gapists" because we believe that there is a gap between Genesis 1:1 and 1:2. As we just pointed out, Jesus put a 2,000 year gap between two words in the middle of a sentence. Is he going to call Jesus a "gapist" also?

Clarence Larkin wrote a well-researched book titled "**Dispensational Truths**" that discusses these gaps in detail.

14.7 "Create" Is Not The Same As "Made"

The Bible does not say that God Created the Heaven and Earth in six days. It says that God Created the Heaven and the Earth "***In the Beginning***". (Genesis 1:1) There is a big difference!

Satan is the author of confusion. Non-believers, evolutionists and atheists are misled into thinking that the Bible says that God Created the Heaven and Earth in six days. That is not true! God's word is being misrepresented!

There are some that say that God Created the Heavens and earth in 6 days. They use the passage below to support their error. However, this passage does not use the word "*Create*", it uses the word "*made*". The meaning is different. Another thing missing in this passage is the words "*in the beginning*".

God Created everything except man in the beginning. Lucifer and his Angels lived on the earth and Lucifer ruled over them as the god of this world. Lucifer led a rebellion against God and was defeated. He was cast back down to the earth along with a third of his angels. Lucifer's name was changed to Satan and his home planet, earth was destroyed. The Earth remained cold and dark until God warmed things up and prepared it for man. God took six days to restore the earth. The Earth was restored for man about six thousand years ago.

"*For in six days the LORD **made** heaven and earth, the sea, and all that in them is, and rested the seventh day: wherefore the LORD blessed the Sabbath day, and hallowed it"* (Exodus 20:11).

The word "Create" is not used. The beginning is not mentioned either.

The word "Create" is used sparingly in the Bible and always with God as the subject. Its meaning is very narrow and specific.

The word "made" is a common word, often used and has a very broad meaning. It means ..."to do".

*"And God blessed the seventh day, and sanctified it: because that in it he had rested from all his work which God **created** <u>and</u> **made**"* (Genesis 2:3)

Here are both words used in the same sentence which is talking about the six days of restoration. During those six days God "**made**" or reestablished everything except air breathing animals and man. Air breathing animals and man had to be "**Created**" on the fifth and sixth days. That is why the verse says "**Created <u>and</u> made**". Two different words with two different meanings.

The word "*Create*" is used in combination with the words "*Heaven and Earth*" three times in the Bible. The first time is in (Genesis 1:1) where it says when Heaven and Earth was Created. The Heaven and earth was Created in the Beginning.

The second time the word "Create" is used is in the Bible is in (Psalm 148:1-5). Here God gives a list of 7 things that were Created in the Beginning. We know that this is talking about the beginning because the heavens and earth are in the list and it says that they (plural) were Created when He commanded. (Singular) Only one command and everything was Created except man. Angels are in the list, the Host of God are in the list but not Man. Man was not Created in the beginning.

The third time the word Create is used in combination with Heaven and Earth is (Isaiah 45:18). This is the clincher!

"*For thus saith the LORD that Created the heavens; God himself that formed the earth and made it; he hath established it, he* ***Created it <u>not in vain</u>****, he formed it to be inhabited: I am the LORD; and there is none else*". (Isaiah 45:18)

God said that he did not Create the earth like it is described in (Genesis 1:2). As J. Vernon McGee put it… "That settles it". If God Created everything perfect and complete as (Isaiah 45:18) said, then something had to happen between verse one and verse two. That is called the gap. A gap in time between verse one and two of perhaps billions of years. That is where Satan's history on earth took place.

After understanding that God said that he did not Create the earth like you see it in verse two and that God never calls the earth young one should not continue to say or teach otherwise.

14.8 The Earth Remained In Darkness, Flooded And Frozen Until

The period of time the Earth remained **"without form, and void"** after God judged it was until **Genesis 1:2**. That is when God started warming things back up again.

This happened about six thousand years ago.

"And the earth was without form, and void; and ***darkness was upon the face of <u>the deep</u>"*** (Genesis 1:2)

"*<u>the deep</u>*" is a reference to the fact that the whole earth was submerged! That was the first flood that resulted from God's wrath against Satan who attacked his Creator. He was formerly known as "Lucifer" and has human followers alive and on this earth that call themselves "Luciferians".

Luciferians refer to "Satan" by his pre-fallen name. The point being that Satan does not want to admit that he is fallen or defeated. He has deceived his human followers into believing that they are going to be on the winning side of an evil plot to make them, the elite, as gods! Soon the Antichrist will claim to be God and order everyone to worship him or die!

PALEONTOLOGICAL TESTIMONY, The Pleistocene Extinction by R. Cedric Leonard -

The pre-human history of the Earth begins in (Genesis 1:1) and continues till (Genesis 1:2). This period of time is not defined in this context. Pre-human earth history began **"in the beginning", along with everything else <u>except Man</u>.**

"And the earth was **<u>without form</u>**, and **<u>void</u>**; and **<u>darkness</u>** was upon the **<u>face of the deep</u>**. And **the Spirit of God <u>moved</u>** upon the face of the **<u>waters</u>**." (Genesis 1:2)

There is a gap between verse 1 and verse 2.

If NASA is correct, that asteroid struck the Earth about 65 million years ago in the Yucatan Peninsula. According to the Bible, the earth was in darkness, frozen, wobbling in space and without any air breathing life because it was completely flooded. The seeds in the flash frozen ground were preserved for 65,000,000 years.

Finally, about 6,000 years ago, God warmed things up for man. But before man was Created, the surface of the earth was probably nothing but frozen ice sheets that probably covered the entire surface of the earth.

I believe this catastrophic event marks the end of the **former world**. According to scripture, the Earth remained ***"without form and void"***, in **darkness** and **flooded** until Genesis 1:3 when God said **"let there be light";**

15

Six Days Of Restoration;

(Genesis 1:3-2:25) (Exodus 20:11)

15.1 Day 1, Earth's Preparation For Man Begins.

"*And God said, Let there be light: and there was light.*" (Genesis 1:3)

"*And God saw the light, that it was good: and God divided the light from the darkness*". (Genesis 1:4)

Before the first day began in verse 3 God had already warmed things up as stated in verse two. So the first day began with the earth at the perfect temperature. The earth was frozen, flooded and in darkness for about sixty five million years prior to this.

God did not "Create light" on the first day, He allowed the light to be seen on the first day. God removed whatever was blocking the light. It may have been space debris that was still orbiting the earth from the big impact in the **Yucatán Peninsula about sixty five million years ago.**

"*in the beginning*" is not the same as "the first day". That first day may have been when God cleaned up all of that space junk. I believe that this first day may also be when God realigned the Earth and put a new spin on it.

"*And God called the light Day, and the darkness he called Night. And the evening and the morning were the first day*". (Genesis 1:5)

God turns on the lights! The space debris left over from the first war with Satan is cleaned up and sunlight reaches the Earth for the first time in 65 million years.

Notice that the word "Create" is not used. Nothing is said in scripture to be **Created** on The First Day.

15.2 Day 2 The Atmosphere Is Prepared.

"*And God said, Let there be a firmament in the midst of the waters, and let it divide the waters from the waters.*" (Genesis 1:6)

Prior to this, there was probably dense water vapor, fog and mist all the way to the ground. This is the point where all the mist and vapor was separated. The water vapor above was separated from the water below and there became an area in between the waters, where you could see long distances like we do now except those times where it is too foggy. God established a flying area for birds, elevated high enough above the ground to prepare for the birds and other land animals that would be Created on day 5.

"*And God made the firmament, and divided the waters which were under the firmament from the waters which were above the firmament: and it was so*". (Genesis 1:7)

"*And God called the firmament Heaven. And the evening and the morning were the second day*". (Genesis 1:8)

Notice that the word Create is not used. Nothing was Created on the second day.

15.3 Day 3 Begins! Dry Land Appears.

Notice that the waters are gathered together in one place and the dry land appears. This seems to imply one continent! Compare with (Genesis 10:25) and (1 Chronicles 1:19)

"And God said, Let the waters under the heaven be gathered together unto one place, and let the dry land appear: and it was so." (Genesis 1:9)

"And God called the dry land Earth; and the gathering together of the waters called he Seas: and God saw that it was good." (Genesis 1:10)

"And God said, Let the earth bring forth grass, the herb yielding seed, and the fruit tree yielding fruit after his kind, whose seed is in itself, upon the earth: and it was so." (Genesis 1:11)

"And the earth brought forth grass, and herb yielding seed after his kind, and the tree yielding fruit, whose seed was in itself, after his kind: and God saw that it was good" (Genesis 1:12)

It says right here the "seed is in itself, upon the earth" (verse11). God did not Create the seeds on the third day, but he did command the seeds in the Earth to start sprouting after he already warmed the Earth and allowed the Sun's light to reach it on the first day; and make the dry land to appear on the second day.

The word "Create" is not used. Nothing was Created on the third day.

"And the evening and the morning were the third day". (Genesis 1:13)

15.4 Day 4 Begins! God Clears The Sky

"And God said, Let there be lights in the firmament of the heaven to divide the day from the night; and let them be for signs, and for seasons, and for days, and years:" (Genesis 1:14)

"And let them be for lights in the firmament of the heaven to give light upon the earth: and it was so." (Genesis 1:15)

"And God made two great lights; the greater light to rule the day, and the lesser light to rule the night: he made the stars also." (Genesis 1:16)

"And God set them in the firmament of the heaven to give light upon the earth," (Genesis 1:17)

"And to rule over the day and over the night, and to divide the light from the darkness: and God saw that it was good." (Genesis 1:18)

Basically God is clearing the skies so the sun, moon and stars are visible from Earth.

Notice that the word *"Create"* is not used. Nothing was Created on the fourth day.

"And the evening and the morning were the fourth day." (Genesis 1:19)

15.5 Day 5 Begins! The Creation Of Birds And Whales

"And God said, Let the waters bring forth abundantly the moving creature that hath life, and fowl that may fly above the earth in the open firmament of heaven." (Genesis 1:20)

"And God ***Created*** *great whales, and every living creature that moveth, which the waters brought forth abundantly, after their kind, and every winged fowl after his kind: and God saw that it was good."* (Genesis 1:21)

"And God blessed them, saying, Be fruitful, and multiply, and fill the waters in the seas, and let fowl multiply in the earth". (Genesis 1:22)

"And the evening and the morning were the fifth day". (Genesis 1:23)

The word "**Create**" is used for birds, whales and other air-breathing animals. There is no mention of Creating fish. The fish were already there. They survived the cataclysmic Judgment of God by remaining deep underwater. The earth's geothermal energy may have sustained the non-air breathing, gilled sealife during the last 65,000,000 years of earth's deep freeze.

It is interesting that the fossil record indicates that sea life has been around much longer than land animals. They have remained unchanged from the beginning. We may have lost some, but there is nothing new from non-air-breathing sea life. However, air breathing life was completely destroyed and replaced with all new species.

15.6 Day 6! Air-Breathing Land Animals And Man Are Created.

"And God said, Let the earth bring forth the living creature ***after his kind****, cattle, and creeping thing, and beast of the earth* ***after his kind****: and it was so".* (Genesis 1:24)

The words… ***"After his kind"***… in verse 24 is confirmed by what we see living and in the fossil record. **Carolus Linnaeus** (1707–1778) is known as the "father of Taxonomy. He spent his life studying and cataloguing the different life forms of earth. He observed no living intermediate links between any species. After spending a lifetime of studying and cataloguing every life form he could find, he insisted that the Creation had to be the work of a Devine Creator. **"God created, Linnaeus organized"** …was his slogan. But it is important to note that **Carolus Linnaeus** separated all life forms into separate and distinct species, no intermediates were ever found. The cataloguing methods developed by Carolus Linnaeus were accepted and used by scientists all over the world. A form of that system is still in use today along with some modifications.

Erasmus Darwin (Dec. 12, 1731-April 18, 1802) was a **33rd Mason**; a pagan; a nature worshiping pantheist and the grandfather of Charles Darwin. He believed in a "mystical life force" that he called "Nature" and he was the primary source of Charles Darwin's ideas about the origins of life.

Erasmus Darwin lived at the same time as **Carolus Linnaeus** and publicly praised his work and used his methods of cataloguing different life forms according to their physiological characteristics. However he had a different explanation of how life began. Erasmus believed that life somehow made itself from "natural", random processes which are not clearly defined but oppose the Biblical account in every way.

It seemed quite reasonable to Darwin that if there was such a "natural random process" that caused gradual, beneficial, Genetic changes, there should be plenty of living intermediate mutations running around. There should be plenty of both the earliest and latest forms and almost everything in-between. We should be seeing entire chains of intermediate species walking around everywhere we look, but we find none.

Where are all of these trillions and trillions of intermediate species? (Actually Trillions and Trillions doesn't even come close to the real number) It had already been established by Carolus Linnaeus (1707–1778) that they did not exist. **In 1732 Carolus Linnaeus conducted his first expedition** to Lapland in northernmost Scandinavia. At the time, it was uncharted territory for botany. Again Carolus conducted another scientific expedition in 1734 to central Sweden. Carolus Linnaeus described separate and distinct species.

About a hundred years later, Charles Darwin embarked on his famous trip aboard the Beagle in 1831. For five years Darwin traveled all over the world looking for living proof of a chain consisting of many gradual changes between two related species. These supposed changes due to many mutations and transmutations could not be found, not even a single link.

You would think that it would be settled if all evidence was to the contrary, but Darwin would not change his mind. Darwin insisted that there must be evidence in the fossil record of intermediate species. So men began to look there. But before we consider all of that, let's consider something once said by Bill Nye "the Science Guy" in his debate with Ken Ham.

In his debate with Ken Ham, Bill Nye the "Science Guy" said that evolutionists could predict that certain intermediate species would be found in the future. Charles Darwin began that tradition when he predicted that someone in the future would find fossilized bones of intermediate species between man and ape.

Shortly after that public statement was made, the **Piltdown Man was discovered by Charles Dawson.** For over 40 years it was said to be "irrefutable" proof of human evolution until **it was proved to be a hoax.** Technology advanced enough to be able to distinguish between different kinds of bones such as human and animal. That ended that hoax and plenty of others as well.

"And God made the beast of the earth ***after his kind****, and cattle* ***after their kind****, and everything that creepeth upon the earth* ***after his kind****: and God saw that it was good"*. (Genesis 1:25)

"And God said, Let ***us*** *make* ***<u>man</u>*** *in* ***our*** *image, after* ***our*** *likeness: and let them have dominion over the fish of the sea, and over the fowl of the air, and over the cattle, and over all the earth, and over every creeping thing that creepeth upon the earth"*. (Genesis 1:26)

Notice that God is referring to himself in the plural since. This is consistent with the teaching of the three person Godhead, the Trinity.

"And the LORD God said, ***It is not good that the man should be alone****; I will make him an* ***help meet <u>for him</u>****"*. (Genesis 2:18)

"And out of the ground the LORD God formed every beast of the field, and every fowl of the air; and brought them unto Adam to see what he would call them: and whatsoever Adam called every living creature, which was the name thereof". (Genesis 2:19)

"And Adam gave names to all cattle, and to the fowl of the air, and to every beast of the field; but ***for Adam there was not found an help meet for him.****"* (Genesis 2:20)

Man was put in charge of naming animals and caring for the Garden, but he had no companion or helper. *"And the LORD God said,* ***It is not good that the man should be alone****"*;

Eve was Created to be a companion and a helper for Adam.

"And the LORD God caused a deep sleep to fall upon Adam, and he slept: and he took one of his ribs, and closed up the flesh instead thereof;" (Genesis 2:21)

"And the rib, which the LORD God had taken from man, made he a woman, and brought her unto the man". (Genesis 2:22)

"And Adam said, This is now bone of my bones, and flesh of my flesh: she shall be called Woman, because she was taken out of Man"… (Genesis 2:23)

Adam even named Eve. From the very beginning, God placed Adam in charge of the planet and Eve was to be his helper and companion. That was the purpose of the first marriage and the first human institution…

"Therefore shall a man leave his father and his mother, and shall cleave unto his wife: and they shall be one flesh". (Genesis 2:24)

"And they were both naked, the man and his wife, and were not ashamed"… (Genesis 2:25)

"So God ***created*** *man in his own image, in the image of God* ***created*** *he him; male and female* ***created*** *he them"*. (Genesis 1:27)

"And God blessed them, and God said unto them, Be fruitful, and multiply, and ***replenish the earth****, and subdue it: and have dominion over the fish of the sea, and over the fowl of the air, and over every living thing that moveth upon the earth"*. (Genesis 1:28)

"And God said, Behold, I have given you every herb bearing seed, which is upon the face of all the earth, and every tree, in the which is the fruit of a tree yielding seed; to you it shall be for meat." (Genesis 1:29)

"And to every beast of the earth, and to every fowl of the air, and to every thing that creepeth upon the earth, wherein there is life, I have given ***every green herb for meat****: and it was so"*. (Genesis 1:30)

"And God saw everything that he had made, and, behold, it was very good. And the evening and the morning were the sixth day". (Genesis 1:31)

These passages are talking about the six days of Earth's **restoration** after it had been previously destroyed by God.

This completes 6 days of the earth's Restoration, the Creation of all new air breathing animals and Man. The fish and seeds of plants from the pre-human world of Lucifer survived and were already there. I believe they survived about 65,000,000 years of flooded, deep freeze in the dark until God restored the earth as described in (Genesis 1:2). The fossil record confirms this everywhere.

15.7 Day 7! God Rests (Genesis2:2) (Exodus 20:11)

"And on the seventh day God ended his work which he had made; and he rested on the seventh day from all his work which he had made". (Genesis2:2)

"For in six days the LORD made heaven and earth, the sea, and all that in them is, and rested the seventh day: wherefore the LORD blessed the Sabbath day, and hallowed it." (Exodus 20:11)

This completes the **"Week of Restoration"**

16

From Adam To The Tribulation. (Overview)

16.1 Satan Becomes Jealous Of Adam.

The once proud *"god of this world"*, is humbled further and is jealous of Adam. He has gone from proud to jealous.

Satan's original sin was pride which led to everything else he did. God is humbling Satan further by placing Adam, a newcomer to the Universe, in what used to be Satan's garden.

Jealously and hatred sum up what Satan and his demons feel toward the human race. They consider us intruders.

They are doomed and they know that there is nothing that they can do about it.

The only thing they can do to hurt God is to hurt or destroy the human race. One other possibility is to retaliate against God's people in particular. That is the only thing they can do.

They are stuck on the Earth in what is like house arrest. They can't go far from home. They are no longer a threat to the rest of the Universe, only Earth.

16.2 Satan's Fate Announced About 6,000 Years Ago.

Yeshua explains to Adam and Eve his plan of salvation and they both believed. They are sent out of the garden, away from God's presence until God can provide a covering for their sin.

"And I will put enmity between thee and the woman, and between thy seed and her seed; it shall bruise thy head, and thou shalt bruise his heel". (Genesis 3:15)

God would be born of a woman, with no man involved. God would become a man and became a perfect and infinite human sacrifice for sin. All of those that put their trust in that payment will be born again the very instant that they **believe God's record**. Whoever believes God's record will receive a new birth that will never die. (1 John 5:1) (John 6:47)

After the God/man or Messiah paid for the sins of humanity, he will defeat Satan once again at the Battle of Armageddon, at the end of the Tribulation; six thousand years into the future.

About 6,000 years ago Satan found out that he would be defeated by a man born of a virgin. He may have also learned that same day that he only had 6,000 years left, but he definitely knows it by now! **His time is almost up and he knows it**! He is more dangerous than any other time in human history! (See chapters 13-18)

16.3 Satan Intensifies His War Against God And Mankind.

Satan is the Adversary of God. That is what his name means.

From the time that God placed Adam in the Garden of Eden, Satan has plotted against God to destroy the human race. The human race is God's treasure.

Satan hatched a plan which involved the forbidden fruit of the knowledge of good and evil. **Eve was deceived**, Adam was not. Through the sin of disobedience, the human race fell and became infected through the sin of Adam.

Satan tempted and deceived Eve in the Garden of Eden when he questioned God's word. Then he denied what God said and lied to her. Satan told Eve that she would **not** surely die.

Satan is questioning God's word, then denying what God said, actually contradicting him, saying the **opposite**.

Satan's first big lie to the human race is to deny the penalty of sin which is the second death. The second death is also known as the final Hell or the Lake of Fire.

Eve actually died twice. She died spiritually on the spot and later physically. She was going to have to become born again or she would be separated from God and Adam forever.

Adam would not accept separation from Eve. Adam chose to share her fate and eat of the fruit hoping that God would come up with a solution. Adam was not deceived by Satan. He loved Eve and did not want to be separated from her. Now they both need a new birth!

Satan thought that he had won! He destroyed the entire human race when he infected Adam and Eve! *"The soul that sinneth, it shall die"*. (Ezekiel 18:4 and 20) *"The wages of sin is death"*. (Romans 6:23)

We are born into the world as "enemies" of God. (Romans 5:10) Condemned!

We are born spiritually dead, with a sin nature which cannot inherit eternal life. (John 3:6)

We need a new birth! (John 3:3)

We receive a new birth **the moment we believe** the gospel.

"For ye are all the children of God by **faith** in Christ Jesus". (Galatians 3:26)

It happens **when** we believe! It is **free** and cannot be undone!

16.4 Satan's Time Is Almost Up And He Knows It

Apparently Satan hears well. We don't know how much Satan heard at that time, but he probably heard more than is recorded in the Bible. He was put on notice that his clock was ticking. Now he is really motivated to destroy mankind, Jews and Christians in particular! God is not only starting a countdown, but he is going to use a human to destroy Satan. Now he has really got it in for us!

Satan is desperate, deceptive, and dangerous and hates the human race. We are occupying what used to be his fabulous domain. He does not want anyone to understand the Gospel and be saved. He wants to take everyone to Hell with him.

At the time of this writing, satan is still the Prince of the Power of the Air.

"Wherein in time past ye walked according to the course of this world, according to the ***prince of the power of the air****, the spirit that now worketh in the children of disobedience:*" (Ephesians 2:2)

But he will lose that title during the middle of the tribulation.

16.5 Satan Will Be More Dangerous Than Ever Before!

Satan is going to fight to the bitter end to try to destroy humanity!

Satan learned about God's future plans and has been trying to destroy those plans ever since. Satan's hit list consists of Believers first, Jews second then false brethren and anyone else next.

Satan knows that the first 6,000 years of human history prophesied about by God in the Garden of Eden are about done. He is waiting for the 70th week of Daniel to begin. The 70th week of Daniel is also known as the Great Tribulation or the first 7 years of the Day of the Lord, a time of darkness, a time of despair, a time of great deception, betrayal and **a time of harvest**. (Matthew 24: 4, 5, 11, 24)

This is where Satan and his demons square off against God and his people to make their last stand against God. Satan is fighting for the destruction of the souls of men, meanwhile God and his people are fighting and dying trying to warn the lost about the consequences of sin and Christ has paid our way! It is FREE; it is the Gift of God; it is ETERNAL Life by Just Faith!

The Great Tribulation is seven years of God and his people fighting against Satan for the hearts and minds of mankind. Satan will take the human race hostage and use terror and deception against them, but …

"*… whosoever shall call upon the name of the Lord shall be saved*". (Romans 10:13)

16.6 Satan Is The Master Of Lies, Deception And Betrayal.

"*But to him that worketh not, but* ***believeth*** *on him that justifieth the ungodly, his* ***faith is counted for righteousness***". (Romans 4:5)

If Satan can deceive people into working for their salvation through trickery, he has won. No one can get saved any other way than **believing** that Jesus paid it all and trusting in His death, burial and resurrection for the sins of the whole world! It is Free! It is a Gift! **Just** Faith alone! If you haven't done it yet, do it now! You don't have to tell anyone else, just talk to God in the quietness of your own heart, in your thoughts. Talk to Him. Talk to the Righteous Creator of the entire universe in the privacy of your own mind and tell Him that you believe in the God of the Bible who took on a human body and we know Him as Jesus. Tell God that you are trusting in what Jesus did on the cross of Calvary 2,000 years ago to save you and nothing else. Now start talking to God like a son talks to his heavenly Father forever after. Ask God questions and pray for the answers.

You can ask Jesus into your heart and not get saved. You can invite Jesus into your life and not get saved! You can "give your heart to Jesus" and still not get saved! That is because Jesus did not say to do those things to be saved. He said to trust or believe on him and what He did on the Cross of Calvary and he will save you the moment you believe his record about that. Eternal Life is a Gift not a process!

You receive Eternal life the moment you put your trust in Jesus alone for your salvation. Just Faith!

"*For by grace are ye saved through faith; and that not of yourselves: it is the gift of God*" (Ephesians 2:8)

"*Not of works, lest any man should boast*". (Ephesians 2:9)

God only saves those who are trusting to be saved, not those who are trying to be saved! When you trust Jesus to save you, then you are saved forever! Once saved, always saved! It is a done deal!

16.7 You Must Understand The Gospel To Be Saved!

"*In whom the god of this world hath **blinded the minds** of them which believe not, lest the light of the glorious gospel of Christ, who is the image of God, should shine unto them.*" (2Corinthians 4:4)

For every truth in God's word, Satan has cooked up about a thousand lies! Satan blinds people with lies and misinformation.

"*When any one heareth the word of the kingdom, and **understandeth it not**, then cometh the wicked one, and catcheth away that which was sown in his heart. This is he which received seed by the way side*". (Matthew 13:19)

When people hear truth from God, Satan tries to take it away from them with clever lies from false teachers, before they understand and believe.

Nowhere does the Bible tell us to give anything to Jesus or God to be saved. It is ...*"**the Gift of God**"*, he does the giving and we do the receiving it is that simple. **God gives us eternal life as a gift when we believe God's record about his Son.**

You can "give your heart to Jesus" and not get saved because God never said to "give your heart to Jesus" to be saved. He said..."*he that **believes** on me **hath** everlasting life.*"(John 6:47)

16.8 You Can't Be Saved, If You Are Trying To Be Saved!

You can "give your life to Jesus" and not get saved because the Bible never tells you to "give your life to Jesus" to be saved. **Salvation is a gift**, God does the giving and we do the receiving. Satan says the opposite. God says...

"*For by **grace** are ye saved through **faith**; and that **not of yourselves**: it is **the gift of God**": **Not of works**, lest any man should boast.*"(Eph. 2:8,9)

"Grace" is a fancy word that means unmerited favor like a gift. It is something that you don't earn, if it is free, no obligation, someone else did all the work and paid all of the price, then it is called ***"Grace".*** By definition, grace is the absence of merit. Satan knows that, so he has devised subtle ways to deceive people into working for their salvation. It can't be done. They will fail!

You can "turn to Jesus" and not get saved because The Bible says to "***believe***" not turn. Believe God's record about who Jesus is and what He did and be born again. "Turn" can mean a lot of things. God defines his terms in the Bible.

You can try to "turn from your sin" to be saved and not get saved because the Bible doesn't say to "turn from your sin to be saved". The Bible says…

"***Believe** on the Lord Jesus Christ, and thou shalt be saved,*" (Acts 16:31)

Furthermore, you have to ask yourself, how much sin do I have to turn from? Is it certain ones or do I have to turn from all of it? How about the preachers that are saying that. Do they claim that they have turned from all of their sin?

The Bible says …

"If we say that we have no sin, we deceive ourselves, and the truth is not in us". (1John 1:8)

Somebody is lying! Turning from your sin is another way of saying "keep the commandments to be saved", or even do good works. The Bible says…"

"*Therefore we conclude that a* ***man is justified by <u>faith</u> <u>without</u> the deeds of the law.*** " (Romans 3:28)

"*Now to him that worketh is the reward not reckoned of grace, but of debt*" (Romans 4:4)

No grace means no salvation. "Debt" means you have fallen short.

If Satan can get you to work for your salvation, he has won. You can't get saved by working for it. You will end up in Hell! **You can only get saved by <u>trusting Jesus</u> to save you.**

16.9 Change Your Mind About Dead Works!

If you are believing in another gospel, then you need to **change your mind** and **believe** the only true Gospel!

Belief in another gospel is unbelief!

When God reveals the truth from scripture, you need to **change your mind** or beliefs and **believe what God says.**

Trust the God of Reason! (Isaiah 1:18)

"*Come let us Reason together saith the LORD… though your sins be as scarlet, they shall be as white as snow; though they be red like crimson, they shall be as wool*". (Isaiah 1:18)

"*Therefore leaving the principles of the doctrine of Christ, let us go on unto perfection; not laying again the foundation of* ***repentance from dead works****, and of* ***faith toward God****"* (Hebrews 6:1)

Change your mind about your behavior contributing in any way to your salvation. It is not of works! Just Faith! Eternal Life is free to all who believe in Jesus as the only Savior.

"*Jesus saith unto him,* ***I am the way****, the truth, and the life:* ***no man cometh unto the Father, but by me****"* (John 14:6)

16.10 Confession to men Is Not Necessary For Salvation,

I don't believe that confession by mouth or writing to another man is necessary for Salvation, but it is necessary to receive rewards. You need to have a talk with God, in your heart, in your mind where only he can hear you. Talk to him, he is listening, he wants to hear from you. Tell him that you believe and want to believe more and need someone to show you how you can learn more.

This is music to God's ears and his people! There is only one way to Glorify God and that is to tell people about Jesus, who he is, why he died on the Cross and persuade them to believe. When people believe, they receive a new birth, a new life is Created that will never end and you were involved. The people that you help to find Jesus will be your best friends for eternity!

Even the ones that took part in these horrific deeds, could be saved. God will forgive everything except unbelief. (John 3:18)

The tormenters and the executioners could be saved if they changed their minds about who Jesus is and what he did for us. They have until they take their last breath to find out the truth and believe. If they die without Christ Jesus, they will end up in hell with unbelievers and be judged with unbelievers for their works.

The tormenters and the executioners will receive the **maximum** punishment in hell for **not believing and** fighting against God.

16.11 Christians Will Be Beheaded During The Tribulation!

(Matthew 10: 28) *"And **fear not** them which kill the body, but are not able to kill the soul: but rather fear him which is able to destroy both soul and body in hell"*

During the seven years of Tribulation, Christians will be beheaded!

*"And I saw thrones, and they sat upon them, and judgment was given unto them: and I saw the souls of them that were **beheaded for the witness of Jesus**, and for the word of God, and which had not worshipped the beast, neither his image, neither had received his mark upon their foreheads, or in their hands; and they lived and reigned with Christ a thousand years".* (Revelation 20:4)

*"And when he had opened the **fifth seal**, I saw under the altar the souls of them that were slain for the word of God, and for the testimony which they held:"* (Revelation 6:9)

Many, many millions of believers will pay with their lives for not denying Christ! **Confessing Christ is not necessary for salvation**. Salvation is a gift, not of works and not of your selves. These people mentioned here would rather die than deny Christ. They will be rewarded with a Martyr's crown for their testimony in view of all believers. They will receive honor and recognition from Christ for their deeds! But if they had failed instead, they would still be saved and may even still be rewarded for other things they have done for Yeshua.

16.12 Christ Returns At The End Of The Tribulation.

All of the Unbelievers are gathered up by the Angels and resurrected Saints at Christ's return.

Unbelievers are all taken by the Angels to the Valley of Jehoshaphat and killed by Christ.

Lost Men and Fallen Angels are both judged to determine their degree of punishment in the Lake of Fire.

The Antichrist and the False Prophet are the first to be cast into hell! They get the worst punishment.

"And the beast was taken, and with him the false prophet that wrought miracles before him, with which he deceived them that had received the mark of the beast, and them that worshipped his image. These both were ***cast alive*** *into a lake of fire burning with brimstone"*. (Revelation 19:20)

Notice that these two men are not only first in but they are **cast in alive** without dying or being judged. They don't get judged because it is already determined that they are going to get the max!

I believe that these two men are alive, identifiable and in the news now.

"And death and hell were cast into the lake of fire. This is the second death". (Revelation 20:14)

17

The Fall of Man

(Genesis 3:6 & 7)

17.1 The First Human Institution

God Created Adam and the Garden of Eden for himself. God placed Adam in charge of the gardening and naming all of the animals. Adam named the first woman, his wife Eve, after she was taken out of his side.

"And Adam gave names to all cattle, and to the fowl of the air, and to every beast of the field; but ***for Adam there was not found an help meet for him"*** (Genesis 2:20)

God Created Eve for Adam. Marriage is the joining together of a man and a woman for life to honor and glorify God through that union.

"And the LORD God said, it is not good that the man should be alone; I will make ***him an help meet for him****"*. (Genesis 2:18)

Notice why God Created the woman, she was Created because God said that it was not good for a man to be alone. Eve was Created as a companion and **a helper** for the man. The man was Created to do God's bidding on the earth first and then the woman was Created next to help her husband to do as God directs him.

"Neither was the man created for the woman; but the woman for the man". (1Cor 11:9)

Adam was Created **for God** and Eve was Created **for Adam**. Nowhere in Scripture does God ever reverse that order…

"But I would have you know, that the head of every man is Christ; and the head of the woman is the man; and the head of Christ is God". (1Corinthians 11:3)

God says that Eve was Created as a companion and helper for Adam. Just as Adam was supposed to follow Christ, Eve was supposed to follow Adam.

As we shall see, (in chapter 48.1 of this book), Satan will reverse that order in the Last Days, and indeed he has. We see the consequences in our marriages, families and churches today.

Satan is the *"god of this world"* (2Corinthians 4:4) He blinds the minds of unbelievers with lies. He is the *"father of lies"*; the "enemy of God" and the original Anarchist. God's word is always under attack by Satan's minions who teach the **opposite** of what God says for perceived personal gain.

17.2 God Established Gender Roles

Gender roles began in the Garden of Eden with the Creation of Eve. (Genesis. 2-3)(1Cor.11:1-15)(Ephesians 5)

"For a man indeed ought not to cover his head, forasmuch as he is the image and glory of God: but the woman is the glory of the man". (1Corinthians11:7)

"Doth not even nature itself teach you, that, if a man have long hair, it is a shame unto him?" (1Corinthians 11:14)

"But if a woman have long hair, it is a glory to her: for her hair is given her for a covering". (1Corinthians 11:15)

God assigned distinct Gender specific roles, responsibilities and Gender specific appearances.

God assigned the responsibility of caring for the Garden and naming things to Adam. Upon the Creation of Eve, Adam became responsible for her as well. The responsibility of caring for Adam and his needs was assigned to Eve. That was her sole purpose, her responsibility. That was the beginning of Gender roles in marriage.

In (Genesis 3:1-5) Satan approaches Eve. Satan questions God's word; then he entices Eve; Satan denies God's word and finally he tells the big lie, "***Thou shalt not surely die***". **Eve actually died twice!** She died spiritually when she sinned and physically later on.

It has been the teaching of Bible scholars over the years that Eve was alone when this happened because the Bible says that Adam was not deceived but Eve was.

Today modern "Feminists" want to argue against this point. They begin by questioning what God has said.

The passage that one "Feminists" used to support her position is (Genesis 3: 6). At the end of (verse 6) the words **"with her"** was added by the translators without any Hebrew grammatical basis. If you try to use the Strong's concordance to look up the Hebrew basis for those two words… "***with her***" you will find that there is none.

*"And when the woman saw that the tree was good for food, and that it was pleasant to the eyes, and a tree to be desired to make one wise, she took of the fruit thereof, and did eat, and gave also unto her husband **with her**; and he did eat".* (Genesis 3: 6)

The contention that Adam was "***with her***" when this conversation took place and therefore **was** responsible for Eve's bad decision is an attempt to bring into question what God has plainly said and teach the **opposite**.

The "Feminist" that brought up this argument to me accused Adam of "throwing Eve under the bus" when he tried to blame Eve. (Genesis 3:12) This particular "Feminist" claims to be a Christian and probably would deny ever saying this to some people.

When someone pretends to agree to gain the trust of someone else, they may say **"oh yea"** when a verse is quoted. When the **"oh yea"** is followed by a **"but"** that is when either hearsay or another passage is used out of context to refute the original passage. This tactic is deceptive in nature and is used by false brethren who are pretending to be believers. God calls them ***"ravening wolves… in sheep's clothing".***

"Beware of false prophets, which come to you ***in sheep's clothing****, but inwardly they are* ***ravening wolves****"*. (Matthew 7: 15)

In (Genesis 3: 6), **two words were added by the translators.**

Every word in the original Hebrew text has been preserved just like God said it would. All of those original words are found in the Strong's Greek and Hebrew dictionary. God has given a warning to anyone who adds to or takes away from his word. I don't believe that God would give such a warning unless people were going to do that.

The original words that God spoke are still with us today and available to those who are searching. The Old Scofield Reference Bible, Old King James version is, in my opinion, the best place to start in English at this time. If you have an Old Scofield Reference Bible and a Strong's concordance, then you have all that you need to study the Bible in English. If you have questions about the meaning of a passage in any translation, use your Strong's concordance to check out the original language.

The two words ***"with her"*** found at the end of (Genesis 3: 6) do not reflect what was said in the original Hebrew. God's word in the original language that it was written in is inspired and without error. The King James Bible is currently my favorite translation from the original language but it is not without **translation** errors.

We must keep in mind that the original words spoken or written by the Holy Spirit, through the prophets, are without errors. But the translations are human works, not necessarily the Holy Spirit's. God warned the world about changing his words. God has pronounced a special curse on anyone who adds to, or takes away from his word. I don't believe that God would give such warnings if it wasn't going to happen. Well it did happen here. The words ***"with her"*** are not in the original. The King James translators decided to add those words without explanation.

The person who added those two words ***"with her"*** will be held accountable. Furthermore, anyone who knowingly uses this error for their own purpose and knowingly propagates this, will also be held accountable!

17.3 Eve Usurps The Leadership Role And Is Deceived!

Of every section in this book, this is the one that defines the relationship of marriage in God's terms. **Biblical Christian Family Values** come from events which are described in this section. Remember that for every truth found in God's Word, Satan has cooked up about a thousand lies. It doesn't matter when or where you lived because God has said that he has always provided enough truth to lead you to him.

You can't sift through all of Satan's lies in a life time, but if you ask God for truth he will give it to you. If you are faithful with the truth that God gives you, then he will give you more. That includes truth about Marriage.

Satan has cooked up about a thousand lies about Marriage. This chapter will, in some cases, contrast God's Truth to Satan's lies about Marriage. The events that shaped God's decision to establish rules of marriage begin in Genesis.

It is important to remember that God always confirms his word with something! It could be either miracles; signs; wonders; knowledge or Prophecy.

Notice in God's record of the following events that the author of God's word consistently shows a knowledge of the Creation. In this case, the author displays a knowledge of Genetics.

The tree of the knowledge of good and evil is where the forbidden fruit grew. Eating that fruit infected the first man and woman with sin! Now that sounds like a retro virus to me.

The sin Gene is passed down through the man. Probably by the "Y" chromosome that makes him male. Men have an "X" and a "Y" chromosome, but women have two "X" chromosomes. Until recently men did not know this, yet the Bible indicates that only the man can transmit the "sin" Gene. This reflects a knowledge of the "Y" chromosome. Only God could know about the "Y" chromosome at that time.

"And the eyes of them both were opened, and they knew that they were naked; and they sewed fig leaves together, and made themselves aprons." (Genesis 3:7)

Here we have the first religion. All religions have one thing in common, human works are substituted for God's required righteousness.

Satan appealed directly to Eve and she responded independently of Adam. The results were disastrous! Satan presented a pleasant and desirable lie that would make her **feel** good now. **Satan got to her through her feelings.** Adam was not present.

"And Adam was not deceived, but the woman being deceived was in the transgression." (Timothy 2:14)

Aprons made of fig leaves were man's best efforts to hide their sin. This was the first religion and all others followed this pattern and are equally as useless. Adam and Eve would have to **change their mind** about working for their salvation and **believe** the gospel of Grace through faith to become **"*born again*"**.

Only born again believers will enter Heaven. (John 3: 3)

17.4 "Feminism" Is The Opposite Of What The Term Implies.

Satan still attacks with the same tactics; Entice; Question; Doubt; Deceive and Betray. Enticement is associated with feelings and emotions. I believe that Satan uses our feelings and emotions against us. Satan uses our emotions to help him deceive us all, both men and women. But in the case of the woman, he has more feelings to work with.

Satan uses our feelings to influence our decisions. If it feels good enough, or if one thinks that it will, or if the lie comes with a promised reward, then a lie becomes desirable and more likely to be accepted. Women naturally follow their feelings. If they didn't have sensitive feelings, how else could they be good mothers or wives?

Feelings are at the center of a woman's natural function in the family. However, when those feelings are led astray, it can be disastrous! A Godly husband's love and protection can be an effective deterrent to such attacks, if the wife is submissive and obedient.

Satan has a big lie for everyone and every occasion. Feminism is Satan's big lie for women. "Feminism" is Satan's attack aimed at the whole human race through the woman. He has got her right in his sights and with her he can get to the husband, the whole family and the entire human race!

"**Feminism**" is the **opposite** of what the term "feminine" implies. "Feminism" is rebellion against God! It is Perversion, ***"witchcraft and stubbornness"...*** against God!

*"For **rebellion** is as the sin of **witchcraft**, and **stubbornness** is as iniquity and idolatry"*... (1Samuel 15:23)

"Feminism" is an attack on Womanhood, femininity and God himself!

See "**ChristyOMisty**" on line for a Christian, young lady's view about "feminism". The discussion includes a Biblical view contrasted to "Feminism".

https://www.youtube.com/watch?v=IT9jeK30yH8

17.5 Sin And Death Entered Humanity Through Adam's Sin.

Sin and death entered into the human race through Adam's transgression.

*"Therefore, as by one man sin entered into the **world**, and death by sin; and so death passed upon **all men**, for that all have sinned"* (Romans 5:12)

Some try to say that sin entered into the "Universe" through Adam, but that is not what this is saying. The word "***world***" as it is used here means "***all men***" or all of humanity. The Greek word "Kosmos" was translated into the word "***world***" and its meaning is context dependent. It means a group of orderly things. Depending on the context, it could mean all of Creation or it could mean all of mankind. It could also mean all of something else like a system or government or constitution. A complete set of something identified by the context.

*"For God so loved the **world** that he gave his only begotten Son, that whosoever believeth in him should not perish, but have everlasting life"*. (John 3:16)

Here the word "***world***" (Kosmos) is used to describe **who** Christ died for. He died for **all of humanity**, not the Angels or the stars of heaven or the whole Creation or anything else.

*"For God sent not his Son into the **world** to condemn the **world;** but that the **world** through him might be saved"*. (John 3:17)

The word "world" as it is used here is talking about **all of mankind.**

*"And if any man hear my words, and believe not, I judge him not: for I came not to judge the **world**,* (G2889) *but to save the **world**".* (G2889)(John 12:47)

The **"world"** is who Jesus came and died for!

"*Woe unto the* ***world*** (G2889) *because of offences! for it must needs be that offences come; but woe to that man by whom the offence cometh*"! (Matthew 18:7)

Here the word "world" is used to describe the lost, human race.

Satan introduced sin into the former world long before Adam was Created. As a result of Lucifer's sin, there was also death **before** the first man Adam was Created. Sin and death entered the human race at a later date, about 6,000 years ago. They are two different events separated perhaps by millions of years.

The human race has only been around for about 6,000 years but the Earth is old, perhaps billions of years old.

Sin entered the Universe by Satan **long before Adam** was Created.

"For this they ***willingly are ignorant*** *of, that by the word of God the heavens were of* ***old****, and the earth* ***standing out of the water and in the water***" (2Peter 3:5)

That phrase "***the world that then was***" is referring to the former world that is best described as Lucifer's World which lasted until God destroyed it, I believe about 65,000,000 years ago. That is the time described as the he **Cretaceous–Paleogene Extinction Event.**

"Whereby ***the world that then was****, being* ***overflowed with water****, perished."* (2Peter 3:6)

That verse along with others describe the destruction of the earth by God because of Lucifer's rebellion. That is "***the world that then was***" and it was destroyed, flooded, in cold darkness, probably deep frozen until about 6,000 years ago when it was restored for Adam.

"***But the heavens and the earth, which are now****, by the same word are kept in store, reserved unto fire against the day of judgment and perdition of ungodly men*". (2Peter 3:7)

Okay, verse 7 is talking about the "***the heavens and the earth, which are <u>now</u>"***, that is the four dimensional world that we currently live in.

17.6 Why Did Adam Sin?

Adam did not want to be separated from Eve.

Traditionally, I have always been taught that…Adam chose to share Eve's fate, knowing the consequences. He may have been hoping that God would provide a solution. After all, he had a relationship with almighty God for some time before this event occurred.

If Adam believed that God would find a way out of this situation, then he was right! God provided himself as a sacrifice for the whole human race! God offers Eternal Life as a Free Gift to everyone who believes that Jesus is God in the flesh, who paid for the sins of all humanity on the Cross That Day! All who would simply **Believe** that message, can **Know,** that they **Have, <u>Eternal</u>** Life! (1John 5:13)(John 6:47)

*"And **Adam was <u>not</u> deceived**, but the woman being deceived was in the transgression*" (1Timothy 2:14)

That notion… ***"Adam was <u>not</u> deceived"…*** has come under attack by "feminists". One "feminist" said to me that Adam was "**with her**" when Eve was tempted but failed to stop her. She went on to say … "When God questioned Adam about the matter he tried to blame Eve, Adam was ready to just throw Eve under the buss". Such an evil and perverse thing to say.

I believe that Adam loved Eve with perfect love and did not want to be separated from her. Likewise, God loved us and did not want to be separated from us. Kind of like the way that Jesus sacrificed himself for our sins. That whosoever believes in that message, will not go to hell but will have right now, **Eternal Life!** You get it the moment that you believe; it is free; it is Eternal and can't be lost! You become ***"born again"*** the moment that you believe the Gospel! Once born again, always born again! The Bible never talks about becoming "unborn". Once you are born again, you are always born again, it can't be undone, you are stuck. You couldn't go to Hell if you wanted to! But you could go to Heaven a bit early if you are disobedient. Disobedience as a believer will result in Gods intervention by your heavenly Father who loves you! Such intervention or judgment comes with chastening.

"For whom the Lord loveth he chasteneth, and scourgeth every son whom he receiveth". (Hebrews 12:6)

God says that he disciplines all of his children.

"If ye endure chastening, God dealeth with you as with sons; for what son is he whom the father chasteneth not?" (Hebrews 12:7)

Adam loved Eve and he trusted that God would not let them both perish. That is why I believe that Adam chose Eve's fate.

Adam understood the consequences, but he loved Eve and believed that God might find a way for both of them. Adam knowingly took a big risk for Eve!

17.7 God Discusses Gender Roles Again

From the beginning, God intended for the man to *"rule"* over the woman.

"Unto the woman he said, I will greatly multiply thy sorrow and thy conception; in sorrow thou shalt bring forth children; and thy desire shall be to thy husband, and ***he shall rule over thee****"* (Genesis. 3:16)

God has pronounced the Gender roles! "Feminists" want to argue and scorn the Bible, but that doesn't change what God has plainly said…

"For the man is not of the woman; but the woman of the man". (1Corinthians 11:8)

"Neither was the man created for the woman; but the woman for the man". (1Corinthians 11:9)

God describes the chain of command.

"But I would have you know, that the head of every man is Christ; and the head of the woman is the man; and the head of Christ is God". (1Corinthians 11:3)

The woman answers to the man; the man answers to Christ and Christ answers to God the Father. God has made himself clear. Are you going to listen to God or the god of this world?!

17.8 God Commands Men.

"Jesus said unto him, Thou shalt love the Lord thy God with all thy heart, and with all thy soul, and with all thy mind". (Matthew 22:37)

"If ye keep my commandments, ye shall abide in my love; even as I have kept my Father's commandments, and abide in his love". (John 15:10)

"By this we know that we love the children of God, when we love God, and keep his commandments". (1John 5:2)

"For this is the love of God, that we keep his commandments: and his commandments are not grievous". (1John5:3)

"Behold, to obey is better than sacrifice, and to hearken than the fat of rams" (1Samuel 15:22)

"Depart from me, ye evildoers: for I will keep the commandments of my God". (Psalm 119:115)

Obedience is how the believer expresses his love for God.

"If ye love me, keep my commandments" (John14:15)

*"Nevertheless let every one of you in particular **so love his wife even as himself**"*; (Ephesians 5:33)

God commands the man to love God with all of his heart; and he also commands him to love his wife as himself and be fruitful. If you want to obey God, you must understand what he means by *"fruitful"*. Today believers are still commanded to be fruitful and glorify God. So how do you do that? Well it is done through obedience to God. God tells us how to glorify him. All we have to do is obey what he has said and he will bless and reward us.

I'm not talking about material wealth that fades away. I am talking about spiritual wealth that goes on forever!

*"Whereas you do not know what will happen tomorrow. **For what is your life? It is even a vapor that appears for a little time and then vanishes away"***. (James 4:14)

God rewards believers for their degree of Obedience with Eternal Rewards that never end! The Bible says…

"To an inheritance incorruptible, and undefiled, and that fadeth not away, reserved in heaven for you", (1Peter1:4)

God pays a reward, a wage to those believers who chose to obey God and give their time, talent and resources to him! Now remember that we are talking about **Eternal Rewards!**

Everyone that you **help** to win for the Lord becomes your fruit that will live on forever! God is preparing a place in Heaven for all believers, but some may have more authority, more abilities, bigger territory, more responsibility or better accommodations etc. than others. God will give each of us a unique and perfect body. Just think of the reward possibilities that you could associate with your new, perfect eternal body!

*"Herein is my Father glorified, that ye bear **much fruit**; so shall ye be my disciples"*. (John 15:8)

Discipleship requires obedience. God rewards believers for their obedience. Disciples are paid and rewarded lavishly! Those rewards are Heavenly, Eternal and coming soon!

*"**The fruit** of the righteous is a tree of life; and **he that winneth souls is wise**"*. (Proverbs 11:30)

God wants us to be in the soul winning business! Remember, ***"he that winneth souls is wise"*** and God rewards lavishly! The Bible says…

"And they that be wise shall shine as the brightness of the firmament; and they that turn many to righteousness as the stars for ever and ever". (Daniel 12:3)

God is glorified when his word is preached and believed. So he commands us to preach and teach his word. The Bible says…

"Preach *the word; be instant in season, out of season; reprove, rebuke, exhort with all longsuffering and doctrine"* (2Timmothy 4:2)

We are also told to study to have answers to people

*"But sanctify the Lord God in your hearts: and **be ready always to give an answer** to every man that asketh you a reason of the hope that is in you with meekness and fear":* (1Peter 3:15)

When believers preach the word, and people get saved God is glorified! God says that he is glorified when people believe on him. God therefore commands us to study and be ready always with an answer from God's word to any man that asks for ***"a reason of the hope that is in you"***. (1Peter 3:15)

The born again believing man is commanded to share the Gospel with others. That is the believer's purpose and the intended center of the Christian family and Church. Whatever a man does for a living, his purpose is to win souls.

God has given the husband and father the responsibility of making the decision on how to best accomplish this task and how each family member fits into that plan. Any family member is welcome to influence those decisions with advice when it is appropriate. The man's authority comes from God.

"Thy desire *shall be* to thy husband, and **he shall rule over thee**". (Genesis 3:16)

God commands the man to rule his family. It is also the man's responsibility to delegate his authority and responsibilities in his family as he sees fit.

One of the requirements of a Deacon, Elder or Bishop is to **rule** his house well. You can't rule it well if you have an unwilling wife. If the wife is unwilling, the man can't rule and there is anarchy. I believe that has always been an issue to some degree, but not like it is today. One of the signs of the End Times is a divided household.

...*"One that **ruleth** well his own house"* (1Timothy3:4)

*"Let the deacons be the husbands of one wife, **ruling their children and their own houses well**"*. (1Timothy 3:12)

God intends for the man to ***"rule"*** his house. He is also to discern and teach doctrine. These are roles that are intended for the husband. That is something that "feminists" don't like to hear. Biblical marriage is something that Satan attacks. Some of his chief minions in that endeavor are those who call themselves "feminists" but there is nothing feminine about them. They are the opposite of what their name implies. Their intent is to deceive and mislead.

God commands husbands to love their wives as Christ loved the Church... (Ephesians 5:25); and provide for them. (1Timmothy5:8)

"Therefore shall a man leave his father and his mother, and shall cleave unto his wife: and they shall be one flesh". (Genesis 2:24)

17.9 God Commands Women.

The purpose for the Creation of the woman is to be a ***"help meet"*** and a companion for her husband.

"And the LORD God said, It is not good that the man should be alone; I will make him an ***help meet*** *for him".*

The Hebrew word for ***"help meet"*** is assigned the **Strong's # H5828 -** ***`ezer*** and is defined as a female helper, one who provides "succor". "Succor" is an old English term that means to provide assistance, comfort, encouragement, refreshment, companionship.

God says that the wife is to provide her husband with "succor" in whatever form it is needed. That is how God describes her purpose and her function. Therefore, whatever goals her husband has, she should share those goals with him. She can influence him but he decides.

The relationship between a man and his wife is patterned after God's relationship with Israel as his wife and Christ's relationship with the Church as a betrothed bride.

A woman should share the same values and therefore the same goals as her husband. The wife should offer advice and influence in a positive and unselfish way, but her husband decides matters of doctrine, goals and family values. There should never be a cause for division if everyone obeys God!

When a woman marries a man, she is agreeing with and accepting his values, goals and authority. Marriage is about creating a unit called a family. The key word is unity. The opposite of unity is division, opposition and rebellion.

A very, long, long time ago; long, long before man was Created; Lucifer decided that he did not want to do things God's way. He came up with his own version, or should I just say perversion of reality. He rebelled against

God, his fate is sealed and his time is almost up! The "*god of this world*" is about to be dethroned! Satan is the god of sin, rebellion and anarchy.

He lost the fight, but it isn't over yet. He wants to take the whole human race down with him. He has been around a very, very, long, long time! He knows us better than we know ourselves! Wow! If you think of it, Satan has had 6,000 years of dealing with the human race. Six thousand years is nothing in astronomical terms, but that represents a whole bunch of human life times in lengths. Satan wants to destroy the human race at every level starting with the basic unit that we call a family.

The Bible says…

"*And if a house be divided against itself, that house cannot stand*". (Mark 3:25)

"*Wives, submit yourselves unto your own husbands, as unto the Lord.*" (Ephesians 5:22)

"*For the husband is the head of the wife, even as Christ is the head of the church: and he is the savior of the body*" (Ephesians 5:23)

"*Therefore as the church is subject unto Christ, so let the wives be to their own husbands in* ***<u>everything</u>***". (Ephesians 5:23)

If the husband is less spiritual or even lost, God tells the spiritual wife to win her husband's heart for the Lord. God says for her to display her obedience to God in her behavior, particularly mentioned is her submissive attitude toward her husband…

"*Likewise, ye wives,* ***be in subjection to your own husbands****; that, if any obey not the word, they also may without the word be won by the conversation of the wives;*" (1Peter 3:1)

This may sound old fashioned and unpopular to most, but most are either lost or carnal. It is up to the spiritual ones to set the example. If the Bible is really God's word as it claims to be over 3,000 times then we need to do what God says.

"If ye love me, keep my commandments" (John14:15)

We show our love of God by our obedience to him. Likewise, a wife shows her love for her husband the same way. A woman can't obey God while opposing her husband.

The Bible says…

"Every wise woman buildeth her house: but the foolish plucketh it down with her hands". (Proverbs 14:1)

A woman should follow her husband's lead and be submissive to him in all matters, otherwise she would be acting like a rebel. Any form of a woman rebelling against her husband for personal reasons is like witchcraft…

The Bible says…

"For rebellion is as the sin of witchcraft, and stubbornness is as iniquity and idolatry" (1Samuel 15:23)

"Nevertheless let every one of you in particular so love his wife even as himself; and the wife see that she ***reverence*** *her husband."* (Ephesians 5:33)

Today things are getting reversed! Nowhere in scripture does the Bible say for the husband to *"reverence"* his wife. This command is specifically directed to the wife. It directs and defines her attitude toward her husband. If a woman loves God and her husband, then she will obey both and there will tend to be unity, love, peace, harmony, joy and prosperity. That is God's way.

If a woman has a question about doctrine God commands her to ask her husband at home. (1Corinthians14:34, 35)

If she disobeys that command and seeks council elsewhere there will be division. In the Last Days many will depart from the faith, many will claim to come in Jesus name, they will say they are Christian and deceive many. (Matthew 24:5)

"Blessed is the man that walketh not in the counsel of the ungodly" (Psalm 1:1)

Satan will provide plenty of false preachers to give plenty of bad advice which will add to the confusion, but if the woman seeks and obeys God's will, there will be unity. Albert Einstein's wife once said…

…"No, I don't understand my husband's theory of relativity, but I know my husband and I know he can be trusted". **(Elsa Einstein)**

God speaks of a virtuous woman In (Proverbs 31: 10-31) who is the example given by God to the ladies who wish to please him.

17.10 God Commands Children.

Children obey your parents.

"Children, obey your parents in the Lord: for this is right". (Ephesians 6:1)

"Honour thy father and mother; (which is the first commandment with promise;" (Ephesians 6:2)

"That it may be well with thee, and thou mayest live long on the earth". (Ephesians 6:3)

That is pretty simple. Children are commanded to obey their parents and promised blessings if they do what they are supposed to do.

In the Last Days Satan will attack the family unit. He attacks each member where they are most vulnerable.

17.11 Broken Families In The End Times

One of the signs of the End Times is anarchy and the absence of loyalty in the world and in the family. We are in the End Times now and we have been seeing these signs for quite some time.

"Trust ye not in a friend, put ye not confidence in a guide: keep the doors of thy mouth from her that lieth in thy bosom". (Micah7:5)

This passage is talking about the last days. It says that a man will not be able to even trust his wife. Now that sounds like a time of anarchy. But we see that already happening all over the world. When the wife opposes her husband, the house falls! If the wife sets an example of rebellion and disobedience for the children, they are likely to do the same. Maybe against the father at first, but eventually the mother should expect to get some of the same along with any other legitimate authority figures in life. This may direct the children to eventually disobey and rebel against even God.

"For the son dishonoureth the father, the daughter riseth up against her mother, the daughter in law against her mother in law; a man's enemies are the men of his own house." (Micah7:6)

If the wife breaks ranks and divides the household against God's appointed leader, the house will fall.

17.12 A Brief Lesson On Forgiveness.

"Take heed to yourselves: If thy brother trespass against thee, rebuke him; ***and if he repent****, forgive him".* (Luke 17:3)

"And if he trespass against thee seven times in a day, and seven times in a day turn again to thee, saying, ***I repent****; thou shalt forgive him"* (Luke 17:4)

Apparently Zipporah did not have a change of heart. She did not change her mind about what she had done or about what Moses was teaching. In other words, she did not admit guilt and ask for forgiveness. Moses sent her back to her father again for the second time. It seems like a divorce. Without confession, there is no forgiveness from God and no restoration to fellowship.

"If we confess our sins, he is faithful and just to forgive us our sins, and to cleanse us from all unrighteousness. (1John 1:9)

Moses followed God's example of forgiveness. God does not forgive unless the sinner admits guilt and asks for forgiveness. There must be a change of heart. The whole point of forgiveness is to learn from our mistakes. Without confession, sin is not exposed for what it is and nothing has been learned. Furthermore, by forgiving without confession, the sin is accepted and allowed to be spread to others without warning.

"He that covereth his sins shall not prosper: but whoso confesseth and forsaketh them shall have mercy" (Proverbs 28:13)

Apparently Zipporah did not change her mind, confess and ask for forgiveness. Zipporah remained argumentative, stubborn and rebellious even after the separation, which explains why Moses sent her away the second time. God says ... *"For rebellion is as the sin of witchcraft, and stubbornness is as iniquity and idolatry"*. (1Samuel 15:23)

Later on, after Moses led the Israelites out of Egypt, *Miriam,* Moses eldest sister and Aron, his older brother decided to usurp Moses authority. They engaged in seditious activity creating division amongst God's people. They played the race card against Moses because he took an Ethiopian wife.

"And Miriam and Aaron spake against Moses because of the Ethiopian woman whom he had married: for he had married an Ethiopian woman". (Numbers 12:1)

It looks like Moses remarried after he sent Zipporah away the second time. Moses married Zipporah before he led the Israelites out of Egypt. Zipporah was from Median which was south of Jerusalem and shared it's northern border with Jordan and its western border along the northern most part of the Red Sea. The area was inhabited by Semites then and still is today.

I don't believe that it is reasonable to believe that Zipporah, the daughter of a Medianite, could be considered to be an Ethiopian. There is no mention of Zipporah being an Ethiopian and there was no mention of Moses having an Ethiopian wife before this incident with Miriam and Aaron. It seems like the Ethiopian wife was a new thing.

The Ethiopian wife is not mentioned until after Moses led the Israelites and a "*mixed multitude*" out of Egypt. Ethiopia is south of Egypt and there were Ethiopians living in Egypt at the time so the "*mixed multitude*" may have included some Ethiopians.

There was a conflict between what God said to Moses and what the Priest of Midian taught his daughter Zipporah. If Zipporah shared Moses beliefs, values and goals, there would have been no conflict. But she was raised by a man who was a false prophet who profited from his influence. Zipporah and her father were probably not saved.

The grounds for divorce in this case is not because Zipporah did not please Moses. It was because she openly rebelled against God and her husband. God held Moses responsible for her actions and was ready to take Moses life because of her behavior! It was because of the hardness of her hart, not his.

Unlike the Koran, The Bible never condones beating your wife because of rebellion or any other reason. In a case like this, the remedy is to separate or divorce. God Created woman to be a helpmeet for the man because it was not good for the man to be alone. You can be alone and still be married if your spouse.

If divorced, you may have to choose between two evils. If you remarry, it may be considered adultery, if you don't you are alone and that is not good either. In the case of Moses, it seems that God forgave Moses for remarrying. Confess your sin and don't make the same mistake twice.

17.13 Zipporah

Zipporah was probably not saved. She is an example of a rebellious wife. The Bible says…

"Be ye not unequally yoked together with unbelievers: for what fellowship hath righteousness with unrighteousness? and what communion hath light with darkness?" (2Corinthians 6:14)

Moses had marital problems. Moses married Zipporah, the daughter of Jethro, the priest of Midian. Apparently Zipporah's father was a false prophet, probably a pagan.

Zipporah rebelled against Moses at least in the matter of circumcision. (Exodus 4: 24-26) The passages (Exodus18:1-27) tells more of the story.

Moses sent his rebellious wife Zipporah back to her father after that incident described in (Exodus 4: 24-26) Her father brought her back to Moses in an apparent attempt to reunite them. Jethro, the priest of Midian seems like a worldly-wise man. He was smart in the eyes of the world and gave Moses advice which seemed to be good advice when you read about it, but God rejected the advice and replaced it with his own advice. The point being that Jethro was not a priest of God. He was a false prophet who outwardly appeared good but was not. He gave bad advice and mislead those who listened to him.

The name "Zipporah" means "bird". I think that name bears a significant message. In the parable about the mustard seed that grew into a giant tree that represented the Kingdom of God, (Matthew 13: 31, 32) the birds that nested in its branches represented the unbelievers of the world. Just like the birds were attracted to the giant mustard tree, unbelievers are attracted to believers. Although the unbeliever has no spiritual discernment, their carnal minds recognize the benefits of dealing with a Godly person.

It is easy for a believer to be fooled by an unbeliever about their faith. That is why they are called "*false brethren*". An unbeliever may be willing to say they believe just like you, to get the girl or guy they may be attracted to. So don't just believe what is on the surface. You need time to get acquainted well enough to really know a potential spouse. If the person seems just too perfect to be real, stop and ask yourself, "is this person just putting up a front"? These days that is often the case.

17.14 The Gospel Explained To The First Couple. (Genesis 3:15)

*"**And I will put enmity between** **thee** **and the** **woman**, and **between thy** **seed** and **her** **seed**; it shall bruise **thy head**, and thou shalt bruise **his heel**"* (Genesis 3:15)

The sin chromosome would be passed on through the man but not the woman. This is why the coming sin bearer would have to be virgin born.

I suspect that Satan may have already known this. God is telling Satan in terms that he understands how God would be born of a woman without a man and become the perfect sacrifice for all those who believed God's message of salvation.

For God so loved the world that he gave his only begotten Son, that whosoever believeth in him shall not perish but have everlasting life. (John 3:16)

God would undo what Satan has done to the human race by offering himself as a substitute for the sins of mankind. God himself would suffer for the sins of humanity.

God is acknowledging that Satan would declare war on the woman's seed to prevent this from happening and even do some damage to the God-man himself. But the God-man would ultimately destroy Satan.

I don't know if the name Yeshua was revealed at this time, but it was clear that God himself would have to be born of a woman and offer himself as the only sacrifice for sins.

It seems that Adam, Eve, Satan and God were all present at this meeting.

Although not recorded here in its entirety, the complete gospel was explained including details (verse 15) of the sin Gene being passed on by the man.

17.15 Evidence That This Meeting Actually Took Place...

The fact that this meeting actually did take place just like the Bible describes is evidenced by two amazing facts revealed in the context. These facts were reveled long before man had the knowledge.

"*I have shewed thee new things from this time, even hidden things, and thou didst not know them*" (Isaiah 48:6)

The notion of Sin being passed down by the man and not the woman reflects knowledge of Genetics that man has only recently acquired. The woman has two "X" chromosomes and the man has one "X" and one "Y" chromosome. I suppose that the "sin Gene" is located in the "Y" chromosome which only men have.

It is interesting to note that only God could have known this at that time. Not man! This is proof of an intelligent Deity/Creator that exists and actually had that conversation with Adam and Eve. This is our Creator talking to us, not some prophet or some other "god"! This is the one that wrote the code in our DNA. That fact is established by a demonstration of a knowledge of Genetics not known by man at that time. Only in recent history have we understood that DNA even exists!

Satan did attack the seed of the woman starting with Eve's first son Able. The attack continued on each successive Generation as declared in the scripture. Satan's attack continues today against the descendants of Abraham, the physical **and** spiritual descendants. Israel is the physical seed and all believers are considered to be the spiritual descendants of Abraham.

Mankind has always been divided into two groups, those who believe and those who don't. **(John 3:18)**

17.16 Adam Believed God… And His Spiritual Family Begins!

The Family of faith begins. (Genesis 3:20)

"*And Adam called his wife's name Eve; because she was the mother of all living*". (Genesis 3:20)

Adam understood that there would be "**children**", **plural** involved in this promise.

Before a child was born, **Adam believed** what God said. He believed that there would be children and one of those children would be God in a perfect human body. God would sacrifice himself for their sins. This is why Adam called his wife Eve. Adam had already believed the gospel and was born again.

"*Adam believed God*". Just like Abraham did.

"*And he **believed** in the LORD; and he counted it to him for righteousness*". (Genesis 15:6)

You are either "Saved" or "Lost"; you are either a Believer or Unbeliever; not condemned or condemned. The deciding difference is Just Faith!

"*He that **believeth** on him is not condemned: but he that **believeth not** is condemned already, **because he hath not believed** in the name of the only begotten Son of God*" (John 3:18)

Eve Believed God!

"*And Adam knew Eve his wife; and she conceived, and bare Cain, and said, I have gotten a man from the LORD.*" (Genesis 4:1)

The literal translation for … "*from the LORD.*" Is … "*even Jehovah*".

Eve thought that her first child was the Lord himself. **She already believed** but she did not understand that it was not time for the Messiah yet.

One conception and two births implies that Cain and Abel were twins. (Genesis 4:1)

Abel Believed God

Abel demonstrated his faith by offering the proper sacrifice (Genesis 4:4)

Cain did not believe God. He did not offer a blood sacrifice as required. He proudly offered the produce that he grew instead… (Genesis 4:3) and God rejected his sacrifice. (Genesis 4:5-7)

Cain did not believe God, but chose to do things his own way instead. He used his best efforts, but that was not good enough. Cain was condemned by **unbelief!**

The first murder! (Genesis 4:8) The human race got off to a bad start. The first two children were born and the unbelieving one kills the believer. (Genesis 4:8) Looks like Satan wins again.

Seth Believed God

Seth believed God and so did his descendants.

"And to Seth, to him also there was born a son; and he called his name Enos: then began men to call upon the name of the LORD". (Genesis 4:26)

The **believing line of Seth** would be the line of the born again **Sons of God**. That would be the line that the Messiah would come through. But Satan had another plan. He would entice the Sons of God to marry unbelieving women. This would cause division within the family and the gospel message would not be passed down to the children.

;" *and Cainan begat Mahalaleel; and Mahalaleel begat Jared; and Jared begat Enoch; and Enoch begat Methuselah; and Methuselah begat Lamech;*

and Lamech begat Noah; and Noah had three sons Shem, Ham and Japheth". (Genesis 5:1-32)

17.17 All Dead Went To Hades Before The Resurrection of Christ.

The Bible teaches that the heart of the earth is hollow.

"To the end that none of all the trees by the waters exalt themselves for their height, neither shoot up their top among the thick boughs, neither their trees stand up in their height, all that drink water: for they are all delivered unto death, ***"to the nether parts of the earth"****, in the midst of the* ***children of men****, with them* ***that go down to the pit****"*. (Ezekiel 31:14)

This passage is talking about where "*children of men*" go after death. It is described as **a pit leading down "*to the nether parts of the earth*"**. "*children of men*" refers to unbelievers.

The "nether parts of the earth" are also mentioned in (Ezekiel 31:16&18) and (Ezekiel 32:)

"***nether parts*"**. Strong's H8482 – tachtiy; lowest part, heart of the earth.

The Bible teaches that it is where the unbelievers are kept after death **until** their **future judgment.** The judgment to determine their degree of punishment in the Eternal Lake of Fire.

Until Judgment day, the "***children of men***" will go to the heart of the earth. ("*Children of men*" is a term that is used for unbelievers)

"And said, I cried by reason of mine affliction unto the LORD, and he heard me; out of the ***belly of hell (Sheol)*** *cried I, and thou heardest my voice"*. (Jonah2:2)

Jonah descended into the heart of the earth and his spirit remained there for three days and three nights. Jesus called that the sign of Jonah.

"And when the people were gathered thick together, he began to say, This is an evil Generation: they seek a sign; and there shall no sign be given it, but the sign of Jonas the prophet". (Luke 11:29)

Before Christ's resurrection, all of the dead went to the heart of the earth. Both believers and unbelievers were in a place called "Hades" in the New Testament and "Sheol" in the Old Testament. Both terms refer to the same place. One is Hebrew and the other is Greek.

*"For as Jonah was three days and three nights in the belly of the **great fish**, so will the Son of Man be three days and three nights in the **heart of the earth**".* (Matthew 12:40)

The story about Lazarus and the rich man (Luke 16:19-31) describes some of what the heart of the earth was like before the resurrection of Christ. One side was Paradise for the believers and the other side is flame and torment for the unbelievers. The two sides were separated by a big gulf and were visible to each other. Paradise was moved to heaven after the crucifixion of Christ. The unbelievers are still down there waiting for their final judgment before being cast into the Lake of Fire.

Here are some Old Testament references to Sheol. (Genesis 37:35; 42:38) (Numbers 16:30, 33) (Deuteronomy32:22)(1 Samuel 22:6)(Job 7:9; 11:8; 17:13; 21:13)(Psalm 16:10; 30:3 ;)(Proverbs 1:12) (Isaiah 14:9, 11, 15; 28:15)

Here are some New Testament references to Hades. (Matthew 11:23; 16:18) (Luke 10:15) (Acts 2:27; 2:31)(Revelation 1:18; 6:8; 20:13, 14)

17.18 The Hollow Earth Fact.

I have known for years that the Bible taught that the heart of the earth was hollow. I learned it from Dr. Hank Lindstrom. As an electrical engineer/ scientist, he expressed his difficulties at understanding how the earth could be hollow.

When teaching about the rich man and Lazarus, (Luke 16:19- 31), he would just smile, shrug his shoulders and say "I don't know how, but maybe God made a force field down there or something". But he would make it clear that even though he could not understand it now, he knew that it was true because the Bible said so. He taught that the Bible's proven track record confirms that it is the Word of God. Therefore, eventually everything that it says will be confirmed as true science.

According to Luke's record of the rich man and Lazarus, the heart of the earth was divided into two compartments.

One side was for believers and the other for unbelievers. (Luke 16:19-31) The two compartments were visible to each other and were divided by a *"gulf"* in between.

When Christ rose from the dead, he took all of the believers out of the heart of the earth and took them to heaven. (Ephesians 4:8-10)

Christ moved paradise to heaven after his resurrection (2Corinthians 12:1-4)

Now all believers are immediately with the Lord upon death…

"We are confident, I say, and willing rather to be absent from the body, and to be present with the Lord". (2Corinthians 5:8)

The unbelievers are all still there waiting for their judgment. Unbelievers are not mentioned by name after death.

Believers are mentioned by name in the heart of the earth before the resurrection of Christ. Some of them were Samuel and Saul, (1Samuel 15:35; 28:7-19), Abraham and Lazarus (Luke 16:19-31), Jonah. (Jonah 1:17-2:1-10) (Matthew 12:40) and Jesus (Ephesians 4:8-10) (Luke 23:42, 43)

The current accepted theoretical model among geologists is that the earth has a molten, liquid iron core. To them this model seems to fit data collected over the years from a variety of sources including seismographs and changing magnetic fields for example. The currently accepted view seems to conflict with the Biblical description. So who do you think is right?

I have recently discovered that there is a hollow earth theory that is gaining attention and gathering followers.

The Bible teaches that the heart of the earth is hollow… (

https://www.youtube.com/watch?v=ltfuS0WQXuQ)

Most of the information concerning the hollow earth theory is nonsense, but a minority of its proponents seem to actually have good science behind them. I will discuss this further toward the end of this book in the section dealing with the Judgments of the Tribulation.

According to the Bible, there will be interaction between the surface and the bottomless pit during the last half of the tribulation. There is good reasons to consider that the bottomless pit may lead to the hollow center.

18

The Nephilim And A Population Explosion Before The 2nd Flood

18.1 Man Can Accomplish Anything He Can Imagine.

Man can imagine a lot, but God says that man can accomplish anything that he can imagine.

(Genesis 11:6) *"And the LORD said, Behold, the people is one, and they have all one language; and this they begin to do: and now nothing will be restrained from them, which they have imagined to do"*

God is going to slow down that process by doing 3 things…

First is by changing their languages; **second** is by scattering the people in all directions and **third** is by dividing the continents to keep these people separated. (Genesis 11:9)

(Genesis 10:25) *"And unto Eber were born two sons: the name of one was* ***Peleg; for in his days was the earth divided****; and his brother's name was Joktan"*

18.2 Men Lived About 10 Times Longer Before The 2nd Flood.

"And it came to pass, when ***men began to multiply*** *on the face of the earth, and daughters were born unto them"*, (Genesis 6:1)

A longer life meant that they had more reproductive years. The first two children born to Adam and Eve are believed to be twins. There is only one conception mentioned but two births resulted, namely Cain and Able. This may be an indication that twins were common back then.

We are talking about an exponential rate of growth raised to a power that we can only guess at! Such a growth rate would fill up the planet very fast!

The average lifespan for a pre-flood man was almost 900 years. (Genesis 5) Men began to reproduce early in life and continued for many, many years.

I think it is interesting to note that Durk Pearson mentioned in his book "Life Extension A Practical Scientific Approach" in chapter 5, page 83; "If you were to maintain the physiological condition you have in very early adulthood, you might reach a lifespan of over 800 years". Although Durk Pearson does not indicate a belief in the Bible, his numbers match what we find in scripture. Durk believed in evolution.

The women of that time must have been very strong to be able to keep having children for all those years! Women of today could not do that! The human race has devolved since that time.

18.3 Believers Began To Marry Unbelievers; The Nephilim Hoax.

Wikipedia has a wrong definition for "Nephilim" if you search for it. It reflects a wrong definition of **"Sons of God"** in (Genesis 6:2)

However, if you go to **blueletter.org** you find the Strong's definition quite different… Strong's #H5303; Strong's definition; "a bully or tyrant:— giant" Nephilim

Before the second flood there were plenty of giants. It seems that they were not only big but they took full advantage of their superior size, strength and brains to exploit the smaller and weaker. In other words, these giants (Nephilim) had a bad reputation of being bullies and tyrants. The

Nephilim were generally unbelievers for quite some time. At the time of the flood, none of them believed and got on the ark. They were all wiped out by the flood.

(Genesis 6:1-3) Is talking about men. Men are mentioned one time in each verse. No mention of Angels anywhere in the chapter.

"*That the **sons of God** saw the **daughters of <u>men</u>** that they were fair; and they took them **wives** of all which they chose.*" (Genesis 6:2)

There are some that have imagined that this passage is talking about fallen Angels marrying woman and producing giant offspring. **<u>Not so!</u>**

Angels don't marry! (Mark 12:25)

*"For when they shall rise from the dead, they neither marry, nor are given in marriage; **but are <u>as the angels</u>** which are in heaven"***.** (Mark 12:25)

This error came from teaching that the **Sons of God** spoken of in (Genesis 6:2) are fallen angels. This belief comes from a misunderstanding of (Job 1:6 and 7) and not knowing that the term **"Sons of God"** refers <u>exclusively</u> to born again believers throughout the Bible.

"*Now there was a day when the **sons of God** came to present themselves before the LORD, and Satan came also among them*". (Job 1: 6)

"*And the LORD said unto Satan, Whence comest thou? Then Satan answered the LORD, and said, From going to and fro in the **earth**, and from walking up and down in it*". (Job 1: 7)

Because Satan is present and the Lord is also present, someone imagined that this meeting occurred in heaven and the Sons of God are angels. Not so! Heaven is not mentioned in this passage at all, but earth is. Angels are not mentioned either. There is no reference that says that Satan was ever in Heaven. However, before he fell, Satan may have spent some time there, but it is not stated. Since the fall of Satan, he has been an outcast. No sin can enter heaven and that includes Satan! (Revelation 21:27)

When Satan attempted to conquer heaven, he did not get that far.

This meeting mentioned in (Job 1) 0ccurred on earth at the only location designated by God for such meetings and sacrifices. The Sons of God are believers. The Bible states…

"***Then there shall be a place which the LORD your God shall choose*** *to cause his name to dwell there; thither shall ye bring all that I command you; your burnt offerings, and your sacrifices, your tithes, and the heave offering of your hand, and all your choice vows which ye vow unto the LORD:*" (Deuteronomy 12:11)

"*And ye shall rejoice before the LORD your God, ye, and your sons, and your daughters, and your menservants, and your maidservants, and the Levite that is within your gates; forasmuch as he hath no part nor inheritance with you*". (Deuteronomy 12:12)

God warns believers not to offer their burnt offerings anyplace else!

"*Take heed to thyself that thou offer not thy burnt offerings in every place that thou seest*" (Deuteronomy 12:13)

God says that he will meet the believers at his designated place and talk to them there. That is exactly what is being described in (Job1: 6 and 7)The Sons of God are gathering themselves together to commune with God.

"*This shall be a continual burnt offering throughout your Generations at the door of the tabernacle of the congregation before the LORD:* ***where I will meet you, to speak there unto thee*"**. (Exodus 29:42)

When (Job 1:6) says that the "*Sons of God*" came to present themselves to God, it is talking about the place on earth designated by God for such an event. God was present and spoke directly with believers!

The fact that Satan came in among them does not imply that he was visible to them but is an indication that Satan or demons can observe what goes on in our Churches. A possessed person was probably among them. We should always pray for God to keep them away from us.

Some additional passages on the subject of God designating a place where he would meet with believers on earth are…

(Deuteronomy 12:11-14)(Deuteronomy9:10) (Deuteronomy15:20) (Deuteronomy 14:23-26) ()

18.4 God Never Called An Angel His Son

Once again read carefully…

"*That the **sons of God** saw the **daughters of <u>men</u>** that they were fair; and they took them **wives** of all which they chose.*" (Genesis 6:2)

The subject here is *"Men"*, <u>not</u> Angels.

Angels are not mentioned in this chapter at all. Men are mentioned ten times in the first seven verses. The subject here is men, not Angels and not giants either! The subject is believing men marrying unbelieving women. Angels are never called "Sons of God".

"*Being made so much better than the angels, as he hath by inheritance obtained a more excellent name than they*". (Hebrews 1:4)

"*For unto which of the angels said he at any time, Thou art my Son, this day have I begotten thee? And again, I will be to him a Father, and he shall be to me a Son?*" (Hebrews 1:5)

The implied answer is he **never** called an Angel his son. The *"Sons of God"* are human, **believers!**

A big error comes from calling ***"the sons of God"*** Angels. That error leads to more errors.

18.5 Some Biblical Terms Used Exclusively For Believers.

"Sons of God" refers exclusively to believers!

*"But as many as received him, to them gave he power to become the **sons of God**, even to them that **believe** on his name:"* (John 1:12)

*"For ye are all the **children of God** by **faith** in Christ Jesus"*. (Galatians 3:26)

*"The Spirit itself beareth witness with our spirit, that we are the **children of God**"*: (Romans 8:16)

*"Whosoever **believeth** that Jesus is the Christ is **born of God**: and every one that loveth him that begat loveth him also that is begotten of him"*. (1 John 5:1)

*"Behold, what manner of love the Father hath bestowed upon us, that we should be called the **sons of God**: therefore the world knoweth us not, because it knew him not."* (1John 3:1)

*"Beloved, now are we the **sons of God**, and it doth not yet appear what we shall be: but we know that, when he shall appear, we shall be like him; for we shall see him as he is"* (1John 3:2)

*"For whom the Lord loveth he chasteneth, and **scourgeth every son** whom he receiveth"*. (Hebrews 12:6)

Some additional terms for believers are "good", "good man" and "righteous". It is important to understand that no one is considered "good" or "righteous" by God because of their deeds! To be called "good" or "righteous" you must believe God's simple gospel that says that Jesus is God who paid for all sin with his own blood on the Cross of Calvary and Eternal Life is the Gift of God by just faith in what Jesus did. Just Faith, no complicated, manmade substitutes accepted!

18.6 Some Biblical Terms Used Exclusively For Unbelievers.

"*Sons of men*", "*Children of Men*" and "*daughters of men*" are unbelievers!

"*O ye **sons of men**, how long will ye turn my glory into shame? how long will ye love vanity, and seek after leasing? Selah.*" (Psalm 4:2)

Sons of men are unbelievers. They have not been born again.

"*I will be his father, and he shall be my son. If he commit iniquity, I will chasten him with the **rod of men**, and with the stripes of the **children of men**":* (2Samuel 7:14)

God is saying that he uses lost, unbelieving men to chasten his disobedient sons. The terms ***"Children of men"; "Sons of men"*** and "***Daughters of men"*** refer to unbelievers.

"*That is, They which are the **children of the flesh**, these **are not the children of God**: but the children of the promise are counted for the seed"*. (Romans 9:8)

"*And the LORD came down to see the city and the tower, which the **children of men** builded"*. (Genesis 11:5)

"Children of men" are unbelievers. God is the father of born again believers only.

"Daughters of men" refers to unbelieving women.

More Biblical terms for unbelievers include; "*wicked" and "unrighteous"*.

18.7 Before Noah's Flood Men Lived About 900 Years.

(Genesis 5:1-32)

"And the LORD GOD said, My spirit shall not always strive with ***man****, for that he also is flesh: yet his days shall be an hundred and twenty years"*. (Genesis 6:3)

God was going to reduce man's lifespan down to 120 years.

Once again, this is talking about **men.**

The **born again sons of God** were believers who were attracted to unbelieving woman and took them as wives. This caused a division in family values and the transmission of the Gospel to the children was lost. Unbelieving men increased until there were only 8 believers left on the earth. Satan wins again.

The population increased in violence and corruption.

As a result, God would flood the earth; destroy all unbelievers; reduce mans' lifespan down to 120 years and start over with only 8 believers!

18.8 "There Were Giants In The Earth In Those Days"; (Nephilim)

"*There were giants in the earth* ***in those days; and also after that, when*** *the sons of God came in unto the daughters of* ***men****, and they bare children to them, the same became mighty men which were of old,* ***men*** *of renown*" (Genesis 6:4)

God is telling us **when** the Sons of God came in unto the daughters of men. It happened, ***"in those days"*** before the flood **when** giants were plentiful. The flood killed off all unbelievers and apparently there were no believing giants left at the time of the flood.

The Bible says that this happened **when** there were still plenty of giants running around. That means **pre-flood**, and that is all it means!

18.9 The Giants (Nephilim) Were There First.

BEFORE and ***"also after"*** the intermarrying of believers with unbelievers, giants were plentiful. These relationships had nothing to do with a new hybrid race. Giants are mentioned as a reference point in time only. This happened before the flood in the days of Giants. Giants means pre flood. As the rest of the chapter points out the flood was the result, not giants. Read carefully. God describes the order of events…

"There were giants in the earth ***<u>in those days</u>"***…

The giants are mentioned first, they were there first. **Giants were already there!**

…***"and <u>also</u> <u>after</u> that"*** … Now God is telling you what happened next, **<u>after</u>** the giants or "***<u>after</u> that***", period of time (before the flood) when there were plenty of giants.

The giants were **<u>also</u>** present **<u>after</u>** these marriages took place.

*"**<u>When</u>** the sons of God came in unto the daughters of **men"**,* this happened **<u>after</u>** the giants were already there, and giants were ***also*** around when these marriages took place because it was **before** the flood. In other words, this happened in the days of Giants, pre flood.

Before the flood everyone lived longer and the fossil record indicates that things tended to grow bigger. The giant Gene is recessive and apparently affected by the environment.

*"And GOD saw that the wickedness of **<u>man</u>** was great in the earth, and that every imagination of the thoughts of his heart was only evil continually".* (Genesis 6:5)

The wickedness of man spread because of these marriages between believers and unbelievers and God judged the earth! Once again, this is talking about men and the result of their failure to pass down the gospel in their own families.

18.10 Some Interesting Possibilities.

Pre-Flood means giant men, bigger insects, plants and animals. Whether it was Noah's Generation or even before Adam such as the time of Lucifer's rebellion, at some point in time before the flood, air breathing animals, insects and plants grew bigger than today. It is recorded in the fossil record. Carbon dating is subjective and speculative. It makes false assumptions about the past environment of the earth. Zealous Evolutionists have a well-established history of playing with the data. I believe NASA when they say that the earth was destroyed by an asteroid sixty five million years ago. That fits with scripture perfectly. It sounds like Lucifer's war.

When God defeated Satan and destroyed his home planet, the earth, for the **first time,** all air breathing life was destroyed! The earth was struck by at least two significant asteroids from space; turned upside down; darkened; flooded and frozen. (Genesis 1:2)

Noah's flood was the **second** flood and the second time God destroyed air breathing life on earth by flooding. (2Peter 3:5-7)

Before the flood some men carried recessive Genes that caused more growth than others over a longer period of time. With longer lives, larger brains and bigger bodies, these giants were probably dominant. The giants were probably so well off that they did not think they needed God and were lost with the flood. That may partially be why the recessive giant Gene is so rare today.

The flood changed the environment. At the same time, human life span dropped from 900 to 120 years.

Apparently an **environmental factor** changed the lifespan of man after the flood. I believe that environmental factor included much more water in the atmosphere as the Bible describes in (Genesis 1:2, 6, 7; 2:5, 6) and less hydroxyl free radicals in the air to breathe.

Cosmic radiation and free radical exposure affects cell division called "mitosis". The greater the cosmic radiation and free radical exposure, the faster the rate of cell division. Cell division is a defense against genetic damage but there is a limited number of times that a cell can divide.

Free radicals are highly reactive molecules or molecular fragments that cause biological damage associated with aging. The rate of aging is directly proportionate to free radical exposure and damage.

Goliath and his brothers were descendants of Noah so at least some giants were married to Noah's family ancestors.

If this environmental factor was heavier than air and settled in the lowest part of the earth such as the Dead Sea Valley, it might have lingered there a little while longer. It would benefit those living there while it was still dissipating.

If men carrying the giant Gene migrated there they would benefit. That would explain how Goliath and his brothers grew so much bigger than Giants of today. I believe that pre flood giants were much bigger than giants of today.

That may be where the Goliath Family raised their 5 sons, not far from Israel in the Dead Sea Valley.

Size is not the only thing that changed in man, we have devolved, not evolved. That concept is consistent with real science, evolution is not! Scientists have discovered massive amounts of damaged, nonfunctional genetic code that represents lost abilities.

Traces of our original Genetic code that have been damaged beyond their ability to function still remain in our DNA as evidence of the destructive nature of these environmental changes.

18.11 In Conclusion…

The Giants had nothing to do with believers marrying unbelievers.

Sons of God are never called Angels and Angels were never called Sons of God in the Bible. Angels don't breed. The Giants were there **before** the intermarrying took place.

This passage is talking about "*sons of God*", believers. "*Men*" are mentioned 10 times in those first 7 verses. Man is the subject, not Angels. There is no mention of Angels anywhere in this chapter. (Genesis 6:1-5)

The giants mentioned were not the result of these marriages, they were already there. They were mentioned only because this intermarrying with unbelievers occurred while there were still plenty of giants. That means pre flood. The Giants were here first! The result of believers marrying unbelievers was that the gospel did not get taught to the children. The unbelievers corrupted the believers which led to God's judgment and the flood.

Nowhere in the Bible does God call an Angel his son. (Hebrews 1:5)

The terms "Sons of God" or "Children of God" apply to believers only.

"*That is, **They which are the children of the flesh, these are not the children of God**: but the children of the promise are counted for the seed*" (Romans 9:8)

The moment a person **believes** the gospel according to the scriptures he or she is born again or born of God. **Terms like "*Children of God*" and "*Sons of God*" are reserved for believers only**.

Terms like "*Daughters of men*", "*Children of Men*" and "*Sons of Men*" refers to unbelievers.

18.12 Enoch Was Raptured Before The Flood

God warned and evacuated his people **before** Noah's Flood. Noah and his family were the only believers left and they were safe and secure inside the Ark with The Lord himself. Just before the flood, Enoch was raptured…

"*And Enoch **walked with God** after he begat Methuselah three hundred years, and begat sons and daughters*": (Genesis 5:22)

"*And all the days of Enoch were **three hundred sixty and five years**:*" (Genesis 5:23)

That is one year for every day in a solar year. Do you think that is just a coincidence or is God giving us a clue about the Rapture?

Enoch is the first man said to be raptured. The Hebrew words used to describe Enoch's departure are equivalent to the Greek words used to describe the Rapture in the New Testament. I don't think we have much time left to figure out what the clue means! I wonder what a solar year could possibly have to do with the Rapture.

Enoch is the first man to be described as having "***walked with God***".

"*And Enoch walked with God: and he was not; for God **took** him*". (Genesis 5:24)

The Hebrew word translated into "*took*" is Strong's # H3947 – "laqach" (לָקַח). Some of the words and phrases that it was translated into include … "to take; get; fetch; lay hold of; seize; receive; acquire; buy; bring; **marry**; **take a wife**; **snatch and to take away**".

The Church is called the bride of Christ. This Hebrew word is descriptive of New Testament passages of the rapture of the Church.

In this Old Testament passage, Enoch is a Type of the New Testament Church being raptured out of this world. Just before earths Judgement, God's people will also be taken out just like a groom marries and takes his bride.

*"By faith Enoch was **translated** that he should not see death; and was not found, because God had **translated** him: for before his **translation** he had this testimony, **that he pleased God**".* (Hebrews 11:5)

That testimony that Enoch pleased God is an indication that he was a believer.

(Hebrews 11: 6) *"But **without faith it is impossible to please him**: for he that cometh to God must **believe** that he is, and that he is a rewarder of them that diligently seek him"*

Strong's G3346 – metatithēmi is translated into **"*translated*"** here and in other places**..."to transfer or to change".**

*"Behold, I shew you a mystery; We shall not all sleep, but we shall all be **changed**,"* (1Corinthians 15: 51)

*"In a moment, in the twinkling of an eye, at the last trump: for the trumpet shall sound, and the dead shall be raised incorruptible, and we shall be **changed**".* (1Corinthians 15: 52)

Strong's G236 – *allassō* ἀλλάσσω

To change, to exchange one thing for another, to transform

(Hebrews 11:5) is saying the same thing about Enoch being *"**translated**"* that (1Corinthians 15:51 and 52) is saying about the *"blessed hope of the believer"*. They are both *"**changed**"*, transformed, taken away like a bride.

Enoch is a type of the Church being raptured **before** the judgements of the Great Tribulation. In this case it is the judgement of Noah's Flood. It is important to note that Methuselah was Enoch's son and the flood began the same year he died. Methuselah also had a son which died just before the flood. His name was Lamech. Lamech was Noah's father. Other than Noah and his family, Methuselah and Lamech may have been the last of the believers before the flood. At the time the flood began, Noah and his family were the last of the believers still living!

The believers were spared God's judgement Just like Lot and his family had to be removed before God judged Sodom and Gomora.

Enoch is a type of the Church being "Raptured" out of the earth before God's Judgement.

Prior to the flood there was a population explosion. (Genesis 6:1) Believers were intermarrying with unbelievers. (Genesis 6:2) Men were corrupted and thought about evil continually. (Genesis 6:5) God commanded Noah to build an ark. (Genesis 6:14-22)

Noah preached for over 120 years without any converts.

All 8 of the believers left on earth after Satan deceived everyone else got onto the Ark. The Unbelievers were washed away. The flood "***took them all away***". (Matthew 24:37-39)

"But as the days of Noah were, so shall also the coming of the Son of man be". (Matthew 24:37)

"*For as in the days that were before the flood they were eating and drinking, marrying and giving in marriage, until the day that Noah entered into the ark*", (Matthew 24:38)

"*And knew not until the flood came, and **took them all away**; so shall also the coming of the Son of man be*". (Matthew 24:39)

The Genealogy of Adam to Noah (Genesis 5:4-29)

18.13 Believers Don't Always Produce Fruit

Only eight believers left on a well populated earth is proof that believers don't always produce fruit! There are plenty of false teachers that say otherwise.

This is a time when **Believers did not produce fruit** but they were still believers, and therefore they were the "*Sons of God*"! There is no guarantee that a Believer will ever produce any fruit. There is no passage that says otherwise!

These "*Sons of God*" did not produce fruit. They had no fruit but yet they are called the "*Sons of God*". They were saved but never produced any fruit for eternity. They had "*dead faith*" but yet they are called "Sons of God" and therefore are saved.

Dead faith will not save you from God's Dailey chastening or scourging or the judgment that every believer must be judged by, but it will save you from eternal damnation in the Lake of Fire! The Bible makes no distinction between saving faith and dead faith for eternal salvation of the spirit. That is why it is called "Amazing Grace"! Dead faith means no fruit, it also means chastening and scourging from God, loss of reward and even early physical death!

19

Noah's Ark

19.1 God Commands Noah To Build An Ark.

*"Make thee an ark of gopher wood; rooms shalt thou make in the ark, and shalt **pitch** it within and without with **pitch**".* (Genesis 6:14)

*"And this is the fashion which thou shalt make it of: The length of the ark shall be **three hundred cubits**, the breadth of it **fifty cubits**, and the height of it **thirty cubits**".* (Genesis 6:15)

Design Issues

In his debate with Ken Ham, Bill Nye asserted that overcoming the engineering problems associated with constructing an ark of this size out of wood was unrealistic. He said the twisting stresses caused by a structure of that size riding the waves for all that time would cause it to leak. Furthermore, he asserted that it was constructed by unskilled labor with no knowledge of such matters.

Furthermore, Bill Nye asserted there was either not enough room on the ark for enough animals to spawn all the animals that we have today or there isn't enough time for such a small number of animals to give rise to all the animals we have today

Perfectly Designed

Let's not forget who designed the ark, over saw the project and even closed the door (Genesis 7:16)! The Bible said that the boat was covered with pitch inside and out. Pitch is more than just a sealer. It dries like a glue. The barge was glued and I assume that it was also screwed inside and out.

Consider the proportions of the ark. The proportions are perfect for stability.

Bill Nye ignores the Biblical account of God's involvement in the construction this big barge. Once again, it was designed by God! The one who designed the universe and established the laws of physics that govern everything that we know of! God designed life and encoded the program language of life in the nucleus of every living cell.

With God as the designer and project engineer of the Ark construction project, it had to be prefect!

19.2 How Long Did The Ark Have To Float?

How many days does Bill Nye think the ark had to float?

The ark was on dry land until the waters rose enough to lift it off the ground. Then the ark floated until the waters subsided enough for it to rest on the bottom.

The rain and fountains of the deep opened up on the 17th day of the 2nd month of the six hundredth year of Noah's life. (Genesis 17:11) Noah's family entered the ark when it started to rain. (Verse 13) It continued to rain for 40 days and 40 nights. (verse12) but the waters continued to rise.

The Bible says the arc rested on the bottom on the 17th day of the 7th month. (Genesis 8:4)

From the time the waters **began** to rise till the time the water receded enough for the ark to settle on land was 150 days.

The water had to rise enough to lift the ark off the bottom. That may have occurred during the first 40 days or after. Any way you look at it, the ark had to float for less than 150 days. The waters continued to recede for another 150 days after the ark touched bottom.

19.3 How Big Were The Waves?

How big does Bill Nye think the waves were? Were there any waves at all? Remember who calmed the storm in the Sea of Galilee, he was on board! (Mark 4:35-40) This is God's flood, he made the flood, the ark and all of the animals in the ark!

I imagine that there was plenty of wind during those first 40 days and nights when all of the water in the atmosphere fell to the earth, but after that I would expect things to calm down.

With all the moisture gone from the atmosphere there was not much to react with the sun in the atmosphere. I would expect for there to be a temporary state of equilibrium after all of the moisture was gone, a real calm. The fountains of the deep continued to raise the sea level for the remainder of the 150 days, but that would not create waves. The waters would be naturally calm under those conditions. So when did the water level reach the ark? We don't know the starting elevation of the ark and we don't know if the waters lifted it during the first 40 days or some time afterward. It had to float for much less than 150 days, maybe less than 100 days or even less than that.

19.4 How Big Was The Ark?

One Cubit is defined as the length of the forearm from the tip of the longest finger to the elbow. The distance between the middle finger and the elbow can vary on the size of the man and how big he wants others to perceive him to be.

Another factor to consider is that the length of a cubit varies according to how far back in time you go.

The further back in time, the longer the cubit was. Royal cubits are even longer than common cubits. It makes sense to assume that members of a royal family enjoyed superior nutrition and therefore were larger. It seems that men were still bigger just after the flood and began to decline in stature afterward. That may be the explanation for why the size of the cubit declined after the flood.

The cubit measurement used by "Bill Nye the science guy" in his calculations in his debate with Ken Ham was 18 inches. The Roman cubit was also 18" and the most common.

According to his calculations, the ark measurements would be; Length = **300 cubits long** X (1.5) = **450 feet long**; Width = **fifty cubits** X (1.5) = **75 feet wide**; **height =** ***thirty cubits*** X (1.5) = **45 feet high**. That translates into…

(450 feet wide X 75 feet X 45 feet) = **1,518,750 cubic feet of space**. That is big but less than half the size of the real thing.

The **Long Babylonian cubit** which measured 24" is one of the oldest known. The length of a cubit declined from the oldest known post flood measurement of over 25" down to 17.5 inches... At the time of Noah it probably was much bigger, but we just don't know.

If you like working with even numbers, the **Long Babylonian cubit** of 24"fits the bill. With an even number of 2 feet, calculations are easier. We come up with the following dimensions…the ark measurements would be; Length = **300 cubits long** X 2 feet = 600 feet long; Width = **fifty cubits** X 2 feet wide = 100 feet wide; **height =** ***thirty cubits*** X 2 = 60 feet high. That translates into…**3,600,000. Cu/ft. of space**. As you can see, that is more than double the size of Bill Nye's ark.

Sir Isaac Newton is known as the most influential scientist in history. He invented calculus and expressed the laws of physics as mathematical equations that modern technology is founded on.

Newton said the cubit used to build Noah's ark was more than 25". He wrote a dissertation on the matter.

"That the sacred Cubit was very large, appears from the Jewish Calamus or Reed, which contained but six of these Cubits; and from the antiquity of this Cubit, since Noah measured the Ark with it;" –Sir Isaac Newton [1]

[1] Newton, Sir Isaac, "A Dissertation upon the Sacred Cubit of the Jews", Source: Miscellaneous Works of Mr. John Greaves, Professor of Astronomy in the University of Oxford,

"It is agreeable to reason to suppose, that the Jews, when they passed out of Chaldea, carried with them into Syria the Cubit which they had received from their ancestors. This is confirmed both by the dimensions of Noah's ark preserved by tradition in this Cubit, and by the agreement of this Cubit with the two Cubits, which the Talmudists say were engraved on the sides of the city Susan during the empire of the Persians, and that one of them exceeded the sacred Cubit half a Digit, the other a whole Digit … **The Roman Cubit therefore consists of 18 Unciæ, and the sacred Cubit of 25 3/5 Unciæ of the Roman Foot" –Sir Isaac Newton** [2]

Vol. 2, pp. 405-433 (London: 1737). Search the page via CTRL-F for "Noah".
[2] Newton, op cit, Search the page via CTRL-F for "Noah". Note – 25 3/5 unciae equates to 24.83 inches if we take a Roman foot as equal to 0.97 English feet.

If we use Newton's cubit of 24.83 inches then the ark will measure; **Length=** 300 X 24.83 = 7,449 in. =**620.75 feet**; **Width=** 50 X 24.83 = 1,241.5 in. = **103.46 feet**; **Height=** 30 X 24.83 = 744.9 in. = **62.07feet**; Neuton's cubit translates into 3,986,309cubic feet. **That is about four million cubic feet of space**. That is a very big barge and it is more than two and a half times the size of the one calculated by Bill Nye.

19.5 How Many Animals Were In The Ark?

"*Of every clean beast thou shalt take to thee by sevens, the male and his female: and of beasts that are not clean by two, the male and his female*" (Genesis 7:2)

"*Of fowls also of the air by sevens, the male and the female; to keep seed alive upon the face of all the earth.*" (Genesis 7:3)

According to scripture, two of every **kind** of living, air-breathing land animal that could not survive the flood was represented in that ark. (Genesis 7: 14, 15)

This excludes plants; amphibians; aquatic mammals; insects whose eggs could survive the flood; single celled organisms and perhaps others as well.

A **"kind"** is defined as an animal that contains all the Genetic information required to produce all of the species that descended from it that we have today.

For example there are over 300 different species or subspecies of horses today, but only two horses were present in the ark. Those two horses contained all the Genetic information required to produce all of its descendants that we have today. Those two horses represent one "**kind**" and all of the 300 plus species and subspecies that descended from them.

This does not mean that we could repeat the same process again today. Over 4,000 years have passed since the flood. The Genetic integrity of life on this planet has degraded significantly because the environment has changed. Therefore today's horses are not as Genetically complete as they were before the flood.

All species, including humans have devolved significantly since the flood. As already explained, the change in our atmosphere after the flood has allowed more cosmic radiation to reach the earth. Free radicals in our environment has increased dramatically.

I think that it is safe to say that we have a good idea about how many different types of horses there are today.

It took more than 4,000 years for those two horses to develop into more than 300 species or subspecies. They can breed with other types of horses and thereby share each other's Genetic information. This process tends to repair some of the Genetic damage done over the years, but not all. Today you could not take two horses and keep interbreeding them like thoroughbred horse breeders have been doing and expect to have the Gene pool necessary to diversify all over again. The Gene pool has degenerated by natural processes that evolutionists tend to ignore.

This same thing applies to the other animals as well.

The FCI recognizes 339 breeds of dogs which are divided into 10 groups based upon the dog's purpose or function or upon its appearance or size.

Dogs, Coyotes and wolves all have 78 chromosomes and can breed with each other. The list of wolf species is up for debate and ranges from 2 to 30 depending who you are talking to.

There are about 19 subspecies of Coyotes.

When you combine the dogs, wolves and coyotes you get about 388 different species and subspecies from just two baby dogs on the ark. There may be more.

The same thing applies to bears, cats bovine etc. If you took all the known species of animals and divided them up into groups of **"kinds"** like we just did with dogs and horses then you would have a good approximation of how many pairs of kinds were on the ark. With Ken Ham's estimate of 7,000 kinds on board the ark, I think we have a reasonable approximation for the purpose of calculating estimates, but it may have been much more. With about 4,000,000, cubic feet of space in the ark there was plenty of room and plenty of time to have the variations in species that we have today. Therefore we have plenty of room on the ark and plenty of time for genetic losses and separation to produce the animals we have today.

19.6 The Ken Ham / Bill Nye Debate

Ken Ham reportedly has said that there were about 7,000 "***kinds***" of animals on the ark. That would amount to 14,000 pairs plus the "*clean*" animals which were brought in by sevens.

In his debate with Ken Ham, Bill Nye argued against the Biblical flood record. He asserted that the estimated number of animals on the ark could not have given rise to all of the species that he claimed to be living on the earth today.

Bill Nye said that…

…"by the very, very lowest estimate is that there are about 8.7 million species but a much more reasonable estimate is it's 50 or even a hundred million… when you start counting the viruses and the bacteria and all the Beatles that must be extant in the tropical rain forest that we haven't found. So we will take a number which I think is pretty reasonable… 16,000,000 species today."

Bill's Nye's numbers are wrong!

Bill Nye continued on to say that if you take his "much more reasonable" estimate of how many species there are today; (16,000,000.) and divide that by the number of years that have elapsed since the flood; multiplied times the number of days in a year; (4,000 years X 365.25 days) you come up with about 11 new species every day.

In his presentation he made a point that… …"we would expect to find 11 new species every day. So you would go out into your yard,you wouldn't just find a different bird, a new bird, you would find a different kind of bird, a whole new species… a bird… every day, a new species of fish; a new species of organisms you can't see and so on".

Bill Nye went on to ridicule the belief in the Biblical record of the global flood by saying…

"I mean this would be enormous news… the last 4,000 years people would have seen these changes among us. So the Cincinnati Enquirer, I imagine would carry a column right next to the weather report …"today's new species"...and it would list 11 every day, but we see no evidence of that. There is no evidence of these species there just simply isn't enough time"

Bill Nye's presentation misrepresented the facts and wrongly set up his math equation to further mislead his audience so that he would appear to be justified in ridiculing the belief in the Biblical record of Noah's flood.

19.7 My Answer To Bill Nye

Bill Nye cited imaginary estimates cooked up by pseudoscientists trying to rally support for their New Age religion. Evolution is their cornerstone doctrine that we will discuss later in this book.

The so called "Estimates" ranging from 3,000,000 to 100,000,000 "*Species*" on earth are based on fantasy not fact.

Although there are still new "species" or subspecies being discovered from time to time, consider why it has taken so long to discover them. They were just too small to find or so similar to other subspecies in appearance that they were not noticed. In fact, many of these so called new species are not new species at all, but only descendants of close or distant cousins which some call subspecies. It is a lot like human races. We are all human, but from different family trees.

There is a lot of disagreement and confusion about distinguishing species from subspecies. This confusion stems from evolutionists playing with numbers and definitions with deceptive intensions.

Closely related subspecies are related to a common ancestor that had **all** of the original Genetic information of all of its descendants which came to be known as subspecies. The Bible calls this a ***"kind"*** and is consistent with what we observe.

Contrary to New Age evolutionary fantasy, each of the subspecies are lacking Genetic information found in their common ancestors and there is no new Genetic information found in them. This is the opposite of what evolutionists claim but is observable and consistent with scripture.

19.8 The Passenger List Of The Ark

The passenger list of the ark only included the different ***"kinds"*** of air breathing animals. Plants and seeds are not mentioned.

A Swedish scientist named **Carl Linnaeus**, a botanist, physician and zoologist, formalized the modern system of naming organisms called binomial nomenclature. He developed the science and the system for classifying, naming and cataloging animals that is still in use today.

In 1735 the first edition of **Systema Naturae** was printed in the Netherlands. In 1758 the 10th edition was published, by that time it had classified 4,400 species of animals and 7,700 species of plants. In 1758 Carl Linnaeus created and published the system still used today to formally name and describe species.

In the 258 years since, about 1.25 million species have been catalogued. That is about 1 million on land and about 250,000 in the seas. That is what we know for sure because it was catalogued.

Carl Linnaeus did not estimate, he counted, categorized and catalogued to give us real numbers. There is a big difference, especially if you have people doing the estimating that are trying to gain notoriety, grant money and nice jobs by promoting a New Age Religion.

The actual numbers of animals described and catalogued represents a life time of hard work by a real scientist that dedicated himself to finding and establishing the real facts. Carl Linnaeus used his own resources to further man's knowledge of God's Creation because God said that knowledge is the principle thing. Other dedicated scientists have continued his work. Carl Linnaeus was a professing Christian and a real "Science Guy".

19.9 A New Method Of Estimating Numbers Of Species

In August 23, 2011, PLOS Biology published an article by Camilo Mora ; Derek P. Tittensor; Sina Adl; Alastair G.B. Simpson; Boris Worm; titled; "How Many Species Are There on Earth and in the Ocean"?

Published: http://dx.doi.org/10.1371/journal.pbio.1001127

In their article about their new method of estimating titled **"How Many Species Are There on Earth and in the Ocean?"** the authors claim to have a more accurate system of estimating.

The authors of this article claim to have found a pattern that allows them to predict how many different life forms there are on earth. The article goes on to say…

"Assessment of this pattern for all kingdoms of life on Earth predicts ~8.7 million (±1.3 million SE) species globally, of which ~2.2 million (±0.18 million SE) are marine. **Our results suggest that some 86% of the species on Earth, and 91% in the ocean, <u>still await description</u>"**

Now let's consider what is being said here. First they are predicting a total of 8.7 million (±1.3 million SE) life forms on the whole planet including plants and animals in the land and seas. Furthermore, the article states that of that number, 2.2 million are in the seas. So let's subtract that number from the entire predicted number of 8.7 million for the whole planet which leaves us with 6.5 million **predicted** for land.

In his debate, Bill Nye asserted that his estimate of 16,000,000 was "a more reasonable number". It looks like Bill Nye cooked up a number that was **2 and a half times <u>bigger</u>** than the number cooked up by authors of this article. Now let's consider what this article goes on to say…

"Our results suggest that some 86% of the species on Earth, and 91% in the ocean, still await description"…

So that means that only about 14% of that 6,500,000 air-breathing animals are actually known to exist. There are questions about those numbers. Although they are supposed to represent distinct air-breathing species that are currently living on earth today, are any of those "known" living species based on fossils only and not actually living specimens?

Another question to consider is the question of whether any of those "different species" are not actually different species but just a different size or color of the same species. The only way that you can really know for sure is to have a file on every species with DNA information that identifies each "species" to determine if it is indeed a different species or another close relative to a related group with different characteristics.

In other words, of the 6.5 million life forms they are predicting, only 86% have actually been found **according to them**. But for the sake of understanding how extremely ridiculous Bill Nye's math is, we will just use their number for now.

So we multiply 6,500,000 times (.14) and we have 910,000 life forms actually found on land. This number includes amphibians, aquatic birds, plants, animals, bacteria and all kinds of things too small to see.

As you can see, **the real numbers have been grossly exaggerated as evolutionists tend to do** and we still haven't subtracted all of the plants, amphibians and everything else that was not on the ark's passenger list. I would guess that at least half of the life on land is plant life. Bill Nye stated that seeds could not survive the flood, but that simply is not true.

No one knows how many forms of life there are on land that are too small to see, but all of these need to be considered and subtracted from the number also.

When you do all of that, then you have to figure out how many of these different animals are actually related. Then you have a close approximation of the number of ***"Kinds"*** which were on the ark.

Nobody knows exactly how many air breathing, non-amphibious, land animals there are that made the ark's passenger list. Let's just consider who may not be on the list.

The passenger list of animals on the ark did not include animals that could survive the flood, like whales; dolphins; porpoises; sea otters; seals; walruses and other sea mammals.

Aquatic birds that swim and live in and around the water, I suppose would not need to be included either. Seagulls; pelicans; gannets; cormorants; albatrosses; petrels; ducks and penguins are some possible examples to consider, just to name a few.

Sea life was not included in the passenger list either. The list also did not include every bacteria, virus, paramecium, protozoa and insect on earth either. Fish, trees and plants were not included.

I have read Ken Ham's book titled, "The Answer Book" and fail to find any correct answers in it concerning Biblical Creation, however I assume that he has spent more time calculating the number of ***"kinds"*** on the ark than I have, so I will use his number of 7,000 ***"kinds"***.

If you put 2 just weaned babies of every kind of air breathing, land dwelling, non-amphibious species, in the Ark there would be plenty of room left over. The "clean" animals will go in by sevens.

The construction issues brought up by Bill Nye in his debate with Ken Ham basically amounted to the question of… "How could they do that"? Yet when we look at underwater, structures made of stones too big for any crane to move they don't doubt what they see, they just scratch their heads. The answers are in the Bible as we have already discussed.

That is a big barge, no doubt about it! The design was by God. God was inside the ark when he told Noah to come in. Prior to the flood there were no carnivores. They were all vegetarians.

Questions about how all the waist was disposed of and who did the work is irrelevant because God incarnate was on board with them. Questions about how the barge could handle the seas is also irrelevant because God designed the barge and made the flood. The barge only had to float less than 150 days.

Questions about how animals found their way to Australia is answered in the next chapter. All of the continents were connected until after the tower of Babel.

20

"Noah's Flood", The Second And Final Flood;

(Genesis 1:2) (Genesis 6 :)

20.1 When Talking About "The Flood"...

...always consider first which global flood we are talking about.

The First global Flood was when God judged Satan and destroyed the earth. We find in (Genesis 1:2) that the earth had been destroyed; was in darkness, cold and flooded. That was the first global flood.

Noah's flood is the second and last global flood. There won't be another global flood.

"*And I will establish my covenant with you; neither shall all flesh be cut off any more by the waters of a flood; neither shall there anymore be a flood to destroy the earth*" (Genesis 9:11)

20.2 After The Flood God Gives Us A Heavenly Sign.

God put a rainbow in the sky as a sign. (Genesis 9: 9-17)

No more population explosion. No more Giants. The human lifespan declines sharply. We are starting over with only 8 believers. (Genesis 7:1-24) All of the races descended from those eight.

God said, "*be fruitful, and multiply, and* ***replenish*** *the earth*". (Genesis 9:1) That is the same command that God gave to Adam and Eve. God told them to ***"replenish the earth"*** also. (Genesis 1:28)

"Replenish the earth" means to repopulate it.

20.3 After The Flood, Things Were Different.

After the flood, the giants were gone. I believe that the giants of the end of the first millennium were the biggest, just like the dinosaurs.

Some men, I presume, carried a recessive Gene that gave them a longer growth period than others. Because of the environment and extended life spans, the giants grew very big, much bigger than giants of today. (Genesis 6:4)

Apparently the giants did well. Perhaps their confidence in their size and abilities led them away from God. Confidence tends to do that.

By the time Noah built the Ark there were no believing giants left. Consequently, they were not represented in the Ark. Somewhere in Noah's family were some of those "giant Genes". Goliath and his brothers were descendants of Noah.

I believe that the pre-flood environment may have caused the "giant Genes" to be more active than they are in our current environment. The change in the environment affected life on this planet in many ways.

20.4 A New Environment Requires Adaption.

The atmosphere has changed.

Increased free-radical exposure due to more cosmic radiation means a shorter lifespan. Increased free-radical exposure means that cells must divide faster to maintain their Genetic integrity. This increased mitosis increases the need for vitamin B-6, B-12 and protein which are not in abundant supply in the plant world.

God commands man to start eating meat. (Genesis 9:3)

B-12 can only come from animal sources. Meat is an excellent source of B-12, B-6 and protein.

A deficiency of B-12 will lead to brain damage, schizophrenia, mutations and cancer.

The change in the environment that caused the shorter life span is consistent with God making changes to man's diet to compensate for this change and enable man to adapt to what the Bible says happened.

All of the science checks out. This is a true story.

20.5 After The Flood, Lifespan Was Reduced To 120 Years.

The flood changed the environment, human lifespan and more.

Prior to the flood it had never rained. The earth was watered by a mist that came up from out of the ground. (Genesis 2:5, 6) There were no rainbows either. (Genesis 9:11-17)

The Bible describes an environment with a lot more water in the air than we are accustom to. (Genesis 1:6, 7) This would explain the sudden change in human lifespan after the flood.

After the flood, human life span was reduced from about 900 years to 120 years.

It sounds like a thick atmosphere full of moisture. Think of a fine mist of tiny water particles suspended in the air and stacked up to the edge of space. Think of how that would react with incoming cosmic radiation. It must have made a big difference and perhaps the only difference needed to achieve a longer life span.

The geomagnetic field may also have been significantly stronger back then, allowing less cosmic radiation to reach the earth. The Magnetosphere plays a major role in protecting the earth from incoming cosmic radiation. Cosmic radiation generates free radicals in our bodies and in our environment.

Free radicals are a type of highly reactive molecule or molecular fragment that is missing an electron in its outer shell orbital.

Free radicals react with our cells, DNA and all of the bio-chemicals that we are made of. They are generated by cosmic radiation and by chemical reactions in our environment and chemical reactions Generated by our own metabolism.

Free radicals cause oxidation, mutations, cancer and problems associated with aging. Free radicals also affect the **rate** of biological aging as well. The rate of mitosis (cell division/reproduction) of our cells is directly proportionate to the free radical exposure of the organism. Mitosis is part of the cell's defense against free radical damage.

Healthy cells are genetically limited to the number of times they can divide, hence a maximum life span is genetically encoded in our DNA. Everything is genetically programmed to die.

"Ordinary air on a sunny day contains about 1,000,000,000 (one billion) hydroxyl free radicals (the most dangerous type of free radical) per quart of air. That's the bad news. The good news is that certain nutrients are very effective in destroying pathologic free radicals- in fact, that seems to be the only function of vitamin E". (Life Extension by Durk Pearson and Sandy Shaw; chapter 7 pg. 101)

The Bible indicates that human life span will increase during the Millennial reign of Christ. The Bible implies that human lifespans may return back to the pre-flood levels during the Millennium.

"There shall be no more thence an infant of days, nor an old man that hath not filled his days: for ***the child shall die an hundred years old****; but the sinner being an hundred years old shall be accursed."* (Isaiah 65:20)

We were Created with a biological clock built into our DNA. That clock can "tick" at different rates depending on the degree of free radical exposure. The more free radical exposure, the faster the biological clock ticks. This is called the "Free Radical Theory of Aging" simplified.

In 1954 Dr. Denham Harman first implicated free radicals in the ageing process. (http://en.wikipedia.org/wiki/Denham_Harman)

The "Free Radical Theory of Aging" was first introduced by Dr. D. Harman in 1954. The theory was first published in the Journal of Gerontology and has been cited often.

For Wikipedia's full article on Harman's theory go to this direct link… http://en.wikipedia.org/wiki/Denham_Harman

Since then, the theory has been further refined by Harman and others. The updated theory is called "The Mitochondrial Theory of Aging" http://en.wikipedia.org/wiki/Free-radical_theory_of_aging#Mitochondrial_theory_of_aging

He published his new ideas in the April 1972 issue of the Journal of the American Geriatrics Society.

The fact that the Bible has described this and other cataclysmic events with scientific accuracy that is beyond the understanding of any person living at the time, is another indication that the Bible must be the word of God and these events Actually did take place. The Bible's description of the pre and post flood environments is confirming Dr. Harmon's life's work!

To get a better understanding of biological aging, how it works and how it may be ameliorated, I recommend the Book "Life Extension, a Practical Scientific Approach" By Durk Pearson and Sandy Shaw.

The Biblical history of the earth demonstrates that the earth did experience total destruction of all air breathing life on the planet just like the Bible says. The earth has been destroyed, changed and flooded **twice**.

The fossil record indicates that this earth was once densely populated by very large animals which are now extinct. Big animals, big plants, big insects all extinct because of two different, sudden and catastrophic global events.

The fossil fuel that we use today confirms a **cataclysmic** event that buried massive amounts of plant and animal life.

Today you can find fossils of sea life in caves and at the tops of the highest mountains indicating that they were once underwater. Sea life has not changed much but air breathing life has.

Both of those events involved global floods that destroyed all air breathing animals. Just remember, **God said it first!**

21

God Keeps His Promises!

(Genesis 13:14-17) (Genesis 15:1-21) (Genesis 17:7, 8) (Leviticus 26: 40-42)

21.1 God Made Promises To Abraham! (Genesis 12:1-7)

"*Now the LORD had said unto Abram, Get thee out of thy country, and from thy kindred, and from thy father's house, unto a land that I will shew thee*": (Genesis 12:1)

"*And I will make of thee a great nation, and I will bless thee, and make thy name great; and thou shalt be a blessing*:" (Genesis 12:2)

"And I will bless them that bless thee, and curse him that curseth thee: *and in thee shall* ***all*** *families of the earth be blessed*". (Genesis 12:3)

When God said "*in thee shall all families of the earth be blessed*", God was speaking about the Prophets and *Yeshua* the Savior! God promises that the Prophets and the Messiah would come through Abraham's own descendants at a time when he and Sara, his wife were both very old and well beyond their reproductive years. Those promises were literally fulfilled by the Prophets and Yeshua

He is talking about blessings that are coming through Abram's descendants that would be multiplied exceedingly! (vs. 2) The physical Genealogy of Jesus is through Mary only. (Luke 3:27-38)

God promised Abraham land, power, fame and blessings;

"And Abram passed through the land unto the place of Sichem, unto the plain of Moreh. And the Canaanite was then in the land". (Genesis 12:6)

"And the LORD appeared unto Abram, and said, ***Unto thy seed will I give this land:*** *and there builded he an altar unto the LORD, who appeared unto him"*. (Genesis 12:7)

21.2 God Promised Abraham A Savior

God promised Abraham that the Savior would come through his linage! **(Genesis 12:3)**

"And I will bless them that bless thee, and curse him that curseth thee: and ***in thee shall all families of the earth be blessed"***. (Genesis 12:3)

Yeshua coming through Abraham's descendants is central to those promises.

That promise was literally fulfilled by Jesus, the prophets and Apostles.

The physical Genealogy of Jesus can only be traced through Mary's Genealogy in (Luke 3:23-38).

..."*Which was the son of Jacob, which was the son of Isaac, which was the son of Abraham,*" (verse. 34)

21.3 God Promises To Bless Those Who Bless Israel!

If you Curse Israel God says that he will curse you!

"And I will bless them that bless thee, and curse him that curseth thee": (Genesis 12:3)

That is a curse that has been literally carried out by God throughout the past. Some of the prominent ones are the decline of the Spanish Empire after the Edict of Expulsion also known as The Alhambra Decree. This took place when Spain was at the pinnacle of its power. Issued on 31 March 1492, by Isabella I of Castile and Ferdinand II of Aragon ordering the expulsion of Jews from their Kingdoms. 1492 also marks the year Columbus began his search for the New World and the Spanish Empire began its decline!

Look at what happened to Germany. Germany mistreated Jews and made a point of getting rid of them. After the war, the pictures of the destruction and devastation in **Berlin** and across Germany were so bad, it is a wonder that anyone survived.

The Presbyterian Church (U.S.A.) disinvestment from Israel plan involves targeting companies that do business with Israel. They have denounced Israel and denied God's promises to Israel. James Kennedy led the way by posting a letter on his website denouncing Israel as the legitimate occupants of the land of Israel and siding with Israel's enemies. The Presbyterian Church U. S. A. went as far as boycotting companies that did business with Israel. About a week later the Church of Christ followed suit.

Do these people understand that they are asking for God's curse!

21.4 God Repeats The Promise And Says It Is Forever

"*And the LORD said unto Abram, after that Lot was separated from him, Lift up now thine eyes, and look from the place where thou art northward, and southward, and eastward, and westward:*"(Genesis 13:14)

"*For all the land which thou seest, to thee will I give it, and to thy seed <u>for ever</u>.*" (Genesis 13:15)

"*And I will make thy seed as the dust of the earth: so that if a man can number the dust of the earth, then shall thy seed also be numbered.*" (Genesis 13:16)

"*Arise, walk through the land in the length of it and in the breadth of it; for **<u>I will give it unto thee</u>**.*" (Genesis 13:17)

When God says… "*walk through the land in the length of it and in the breadth of it*", that is literal language! (Verse 15) He is standing on the Land and looking at it as God speaks to him! Wow, what an experience that must have been! Can you imagine what that must have felt like!

Anyone that denies that this is literal language, just doesn't believe the Bible.

21.5 The Bible Is Literal, Egypt Is Not A Part Of The Promise!

God also promised that Abraham's descendants would spend 400 years in a land that is **not theirs** and would leave in the fourth Generation. This promise was **literally fulfilled** when they left Egypt. (Genesis 15:13-16)

"*And he said unto Abram, Know of a surety that thy seed shall be a stranger in **a land that is not theirs**, and shall serve them; and they shall afflict them **four hundred years***"; (Genesis 15:13)

Moses was the **fourth Generation**. Notice that God says that Egypt was not included in the promised land. During the past wars Israel could have taken Egypt but did not. The Sinai Peninsula was returned.

The false prophets today are saying that God doesn't mean what he says, but history tells us that God does mean what he says!

God keeps his promises! Everyone needs to believe that fact! Abraham believed God's promises and he was saved by faith.

"*And he **believed** in the LORD; and he **counted it to him for righteousness**.*" (Genesis 15:6)

Abraham believed the promise of a Savior that would come through him and his wife Sarai, even though they were both old and beyond their reproductive years. If you can't believe in Gods' eternal promises to Israel, then how are you going to believe in Gods' promise of Eternal Life?

"*And he* ***believed*** *in the LORD; and* ***he counted it to him for <u>righteousness</u>***". (Genesis 15: 6)

Compare with (Philippians 3:9) Just Faith!

"*And be found in him, not having mine own righteousness, which is of the law, but that which is through the faith of Christ,* ***<u>the righteousness which is of God by faith</u>***": (Philippians 3:9)

21.6 God Repeats The Promises Again. (Genesis 15:1-7)

"*And Abram said, Lord GOD, what wilt thou give me, seeing I go childless, and the steward of my house is this Eliezer of Damascus?*" (Genesis 15:2)

This promise was made when Abram and Saria were both very old and well past normal child bearing age. They had no children of their own. (vs. 2) Yet (vs 6) says … "*Abraham believed God*"

God tells Abraham about a Heavenly Promise. (verse. 5)

"*And he brought him forth abroad, and said, Look now toward heaven, and tell the stars, if thou be able to number them: and he said unto him, So shall thy seed be.*" (Genesis 15:5)

21.7 God Gives Abraham More Land. (Genesis 15:18)

"*In the same day the LORD made a covenant with Abram, saying,* ***Unto thy seed have I given this land****, from the river of Egypt unto the great river, the river Euphrates:*" (Genesis 15:18)

Wow! That is a big piece of real-estate! And prime territory too! Everything between the Nile River and the Euphrates!

Today all Israel wants is a very small piece of that! The land that they have returned to and are now living on. Notice that the land of Egypt is not included in that promise.

Abraham was 99 years old and still childless (Genesis 17:1) when God promised again that he would give all the land of Canaan to the descendants of Abraham for an ***everlasting possession***".

21.8 Abraham Believed God And Was Declared Righteous

*"And he **believed** in the LORD; and he counted it to him for righteousness".* (Genesis 15:6)

Abraham got saved the same way as Adam and Eve and every other human being.

"And be found in him, not having mine own righteousness, which is of the law, but that which is through the faith of Christ, the righteousness which is of God by faith": (Philippians 3:9)

He was also born again when he believed, (1 John 1:12)

A new birth demands a new name! (Genesis 17:4, 5)

"Neither shall thy name any more be called Abram, but thy name shall be Abraham; for a father of many nations have I made thee". (Genesis 17:5)

"*Abraham Believed God*," (Literally)

*"Abraham **believed God**, and it was counted unto him for righteousness".* (Romans 4:3).

*"And he **believed in the LORD**; and he counted it to him for righteousness"* **(Genesis 15:6)**.

God gave us a specific, literal Gospel with literal warnings and Promises of Forgiveness, Salvation and Eternal Life to all of those that **Trust or Believe on Jesus name** according to the scriptures. (Yeshua in Old Testament Hebrew ;)

The salvation message is simple belief!

"For what saith the scripture? Abraham ***believed God****, and it was counted unto him for righteousness*". (Romans 4:3)

Jesus name has meaning! It means that Jesus is God in human form, performing his duties as of a faithful Son.

Jesus obedience to God the Father to pay the sin debt for the whole human race on the Cross of Calvary is what it took to pay for our salvation.

"*For all have sinned and come short of the glory of God.*" (Romans 3:23)

"*the wages of sin is death but* ***the gift of God*** *is eternal life through Jesus Christ our Lord*". (Romans 6:23)

"*For he hath made him to be sin for us, who knew no sin; that we might be made* ***the righteousness of God in him***". (2 Corinthians 5:21).

"*And* ***be found in him****, not having mine own righteousness, which is of the law, but that which is* ***through the faith of Christ****, the righteousness which is of God by* ***faith***": (Philippians 3:9)

"*For God so loved the world, that he gave his only begotten Son, that whosoever* ***believeth*** *in him should not perish, but have everlasting life*". (John 3: 16)

"*Verily, verily, I say unto you, He that* ***believeth*** *on me* ***hath*** *everlasting life.*" (John 6:47)

"*For by grace are ye saved through* ***faith****; and that not of yourselves: it is* ***the gift of God****: Not of works, lest any man should boast.*" (Ephesians 2:8)

Only your **belief** or **trust** in what God has done on Calvary is sufficient, nothing else!

"For what saith the scripture? Abraham ***believed God****, and it was counted unto him for righteousness"*. (Romans 4:3)

"Now to him that worketh is the reward not reckoned of grace, but of debt". (Romans 4:4)

"But to him that worketh not, but ***believeth on him*** *that justifieth the ungodly, his faith is counted for righteousness"*. (Romans 4:5)

Just Faith!

21.9 Abraham Was Justified By Faith, Just Faith Alone!

Either you are **Trusting** to be saved or you are trying to be saved, you can't do both! Only those who are **Trusting** to be saved will be saved. You can't be saved if you are **working** or trying to be Saved!

"*Now* ***to him that worketh*** *is the reward* ***not reckoned of grace, but of debt***". (Romans 4:4)

"*But to him that* ***worketh not****, but* ***believeth*** *on him that justifieth the ungodly,* ***his faith is counted for righteousness***" (Romans 4:5)

God promised Abraham Eternal Life as a gift for just **believing**!

"And he ***believed*** *in the LORD; and he counted it to him for righteousness"*. (Genesis 15:6)

"*Verily, verily, I say unto you, He that* ***believeth*** *on me* ***hath everlasting life***." (John 6:47)

"*These things have I written unto you that* ***believe*** *on the name of the Son of God; that ye may* ***know*** *that ye* ***have*** *eternal life*"? (1John 5:13)

Eternal Life is the **Gift** of God!

"*He that **believeth** on him is not condemned: but he that **believeth not** is condemned already, **because he hath not believed** in the name of the only begotten Son of God.*" (John 3:18)

The only condition here is **belief**.

"*And they said, **Believe** on the Lord Jesus Christ, and thou shalt be saved, and thy house*". (Acts 16:31)

The Gospel is God's Promise of Everlasting Life to all those who put their trust in Gods' sacrifice for our sins on the Cross of Calvary by Yeshua (Jesus).

There is nothing that can be added to that! Belief that your efforts can add to what Jesus did on Calvary is unbelief, it is Blasphemy!

21.10 God Repeats His Promises To Isaac And Jacob.

(Genesis 26:3-5) (Genesis 28:13-15)

God repeats these unconditional, "*everlasting*" and therefore, irrevocable promises again to Abraham and again to Isaac and again to Jacob! (Genesis 17:7, 8) (Genesis 26:3-5) (Genesis 28:13-15)

"*And I will establish my covenant between me and thee and thy seed after thee in their Generations for an **everlasting covenant**, to be a God unto thee, and to thy seed after thee*". (Genesis 17:7)

"*And I will give unto thee, and to thy seed after thee, the land wherein thou art a stranger, all the land of Canaan, for **an everlasting possession;** and I will be their God.*"(Genesis 17:8)

Notice that the land is from an <u>everlasting</u> covenant for an <u>everlasting</u> possession.

The promise is repeated … To Isaac (Genesis 26:3-5); to Jacob (Genesis 28:13-15); and throughout the rest of the Bible.

God says that the land belongs to him and he has given it to Israel forever!

At the end of the Tribulation, Israel will possess all of the promised between the River Euphrates and the Nile. That will occur 7 years after the Rapture, which makes it near future.

22

Israel's Rebellion and Punishment, Prophesied To The Day!

22.1 Blessings Or Consequences?

God Foretold His Rejection By Israel

(Leviticus 26:1-13)

"If ye walk in my statutes, and keep my commandments, and do them"; (Leviticus 26: 3)

"Then I will give you rain in due season, and the land shall yield her increase, and the trees of the field shall yield their fruit" (Leviticus 26: 4)...

...*"And I will walk among you, and will be your God, and ye shall be my people"*. (Leviticus 26: 12)

For many years Israel enjoyed these blessings even though they did not obey ***"all"*** the law. God overlooked Israel's offences and forgave them seventy times seven years. After the upper ten tribes and the lower two tribes had accumulated a total of "7 times 70"= 490 years of not letting the land rest, God judged Israel.

"Then came Peter to him, and said, Lord, how oft shall my brother sin against me, and I forgive him? till seven times?" (John 18:21)

"Jesus saith unto him, I say not unto thee, Until seven times: but, Until ***seventy times seven****"* (John 18:22)

And that is exactly what God did. God overlooked Israel's infraction of not keeping the Sabbatical year until they accumulated a total of 490 years of infractions. That is when God said that it was pay back time. That is exactly one year owed for every seven years that Israel disobeyed. Seven times 70 equals 490 years. That is how long God warned Israel and waited. But the time came when God was through warning and waiting.

22.2 God Promised A Curse If Israel Did Not Obey All Of The Law

(Leviticus 26:13-17)

"*I am the LORD your God, which brought you forth out of the land of Egypt, that ye should not be their bondmen; and I have broken the bands of your yoke, and made you go upright*". (Leviticus 26:13)

"*But if ye will not hearken unto me, and will not do **all** these commandments*"; (Leviticus 26:14)

"*And if ye shall despise my statutes, or if your soul abhor my judgments, so that ye will not do **all** my commandments, but that ye break my covenant:*" (Leviticus 26:15)

"*I also will do this unto you; I will even appoint over you **terror**, consumption, and the burning ague, that shall consume the eyes, and cause **sorrow of heart**: and ye shall sow your seed in vain, for your enemies shall eat it*". (Leviticus 26:16)

"*And **I will set my face against you**, and ye shall be slain before your enemies: they that hate you shall reign over you; and ye shall flee when none pursueth you*". (Leviticus 26:17)

This is describing the change that took place in 606 B.C., on August 11. That day just happens to be the 9th of Av, on the Hebrew calendar. That is the day that Nebuchadnezzar showed up at Jerusalem's front door with all of his armies. That day also marked the beginning of the first 70 year cycle of Israel's punishment under the Leviticus 26 curse and it all began on the 9th of Av on the Hebrew calendar.

God told Israel to let the land rest every 7th year. It is called "The **Sabbatical Year**". Leviticus tells us…

"But in the seventh year shall be a Sabbath of rest unto the land, a Sabbath for the LORD: thou shalt neither sow thy field, nor prune thy vineyard". (Leviticus 25:4)

During the Sabbatical Year, there would be no cultivation of the land. The land was to rest, but Israel did not keep the Sabbatical Year for 490 years and then came payback time!

"And this whole land shall be a desolation, and an astonishment; and these nations shall serve the king of Babylon ***seventy years***" (Jeremiah 25: 11)

If God said that Israel owed him 70 years off the land, then that means that God forgave Israel for 490 years. (7 X 70)

That 490th year ended with the invasion of Nebuchadnezzar in 606 B.C. But God said it centuries before it happened. And it happened on the 9th of Av, according to the Hebrew calendar.

This date of <u>606 B.C</u>. was first established by Usher; then also by Sir Robert Anderson (The Coming Prince, page 227) and in Dr. Constable's Notes on Daniel page 117, just to name a few sources.

22.3 Israel's' Disobedience And Punishment Was Foretold

Centuries before Daniel prophesied the day that the Messiah would present himself to Israel as the King of the Jews, Moses prophesied about Israel's rebellion and disobedience to God.

In the prophecy of (Leviticus 26), God said Israel would not let the land rest every 7th year as required by God's law and there would be a payback.

"Then shall the land enjoy her Sabbaths, as long as it lieth desolate, and ye be in your enemies' land; even then shall the land rest, and enjoy her Sabbaths". (Leviticus 26:34).

"As long as it lieth desolate it shall rest; because it did not rest in your Sabbaths, when ye dwelt upon it" (Leviticus 26:35)

The "payback" will be time away from their land proportionate to the offence, because they did not let the land rest when they were living on it.

Just as God said, Israel chose not to obey God, thereby choosing the curse...

God used Moses to tell what would happen to Israel in the future as a result of their disobedience. This prophecy takes us from the 9th of Av, in the year of 606BC, all the way to the end of Armageddon. During this time Israel would be taken off their land twice and returned twice. All of this was told in advance!

*"**And I will scatter you among the heathen**, and will draw out a sword after you: and your land shall be desolate, and your cities waste".* (Leviticus 26:33)

*"Then shall the land enjoy her Sabbaths, as long as it lieth desolate, **and ye be in your enemies' land**; even then shall the land rest, and enjoy her Sabbaths"* (Leviticus 26:34)

*"As long as it lieth desolate it shall rest; **because it did not rest in your Sabbaths, when ye dwelt upon it**."* (Leviticus 26:35)

God said what Israel would do long before they did it. He also said how he would punish them long before any of it happened. ()

22.4 After 490 Years Of Disobedience, God Judged Israel!

Israel did not keep the Sabbatical Year for 490 (Hebrew) years. Israel did not let the land rest on every 7th year just like God predicted. So we calculate the number of years that the land should have rested …490 / 7 = 70 (Hebrew) years. God overlooked and forgave Israel ***"seventy times seven"*** years. (John 18: 21, 22) But after that…

God said you owe me 70 years for not letting the land rest! God said that Israel would not obey and also said what he would do to them for their disobedience…

*"Behold, I will send and take all the families of the north, saith the LORD, and Nebuchadnezzar the king of Babylon, **my servant**, and will bring them against this land, and against the inhabitants thereof, and against all these nations round about, and will utterly destroy them, and make them an astonishment, and an hissing, and perpetual desolations"...* (Jeremiah 25:9)

(A brief Note about the term "***my servant*** as it is used here.) At the time this was written and fulfilled, Nebuchadnezzar was an unbelieving pagan, yet God calls him his servant. Indeed he was his servant, but he was a wicked servant because he was an unbeliever. Because of Daniel's faithfulness, Nebuchadnezzar eventually believed. (He became Daniel's fruit)

The moment that he believed, he became God's **Righteous** servant. As a new believer, Nebuchadnezzar was still ungodly in behavior and attitude but declared righteous by God because of his faith. As a believer, Nebuchadnezzar was disciplined by God as a father does his son. Eventually Nebuchadnezzar produced fruit as a believer because of God's chastisement.

God has always dealt with believers as a good father does his son. God deals with Jewish Believers the same as Gentile Believers. We are his children.

"I will be his father, and he shall be my son. If he commit iniquity, I will chasten him with the rod of men, and with the stripes of the children of men": (2Samuel 7:14) See also (Hebrews 12: 6)

Being a servant doesn't make you saved because salvation is by grace through believing and not of your works! In the Bible, believers are called righteous because of their faith, not their works. In Contrast unbelievers are called wicked because of their unbelief, not their works. God declares us righteous when we believe the Gospel. Righteous believers can **do** wickedly because they still have their old nature, but they themselves are declared righteous.

"Even as Abraham believed God, and it was accounted to him for righteousness" (Galatians 3:6) see also … (Genesis 15: 6) and (Romans 4:3) and (James 2:23)

Continuing on with Jeremiah 25 …

*…"And this whole land shall be a desolation, and an astonishment; and these nations shall serve the king of Babylon **seventy years**"*. (Jeremiah 25:11)

*"And it shall come to pass, when **seventy years** are accomplished, that I will punish the king of Babylon, and that nation, saith the LORD, for their iniquity, and the land of the Chaldeans, and will make it perpetual desolations"* (Jeremiah 25:12)

After the first curse, then God will forgive Israel **if** they humble themselves before God; confess their sins and the sins of their ancestors; acknowledge that God has justly punished them; and ask God to forgive their sins…

*"**If they shall confess** their iniquity, and the iniquity of their fathers, with their trespass which they trespassed against me, and that also they have walked contrary unto me"*; (Leviticus 26:40)

"And that I also have walked contrary unto them, and have brought them into the land of their enemies; if then their uncircumcised hearts be humbled, and they then accept of the punishment of their iniquity" (Leviticus 26:41)

*"**Then** **will I remember** my covenant with Jacob, and also my covenant with Isaac, and also my covenant with Abraham will I remember; and **I will remember the land**"* (Leviticus 26:42)

If there is no confession, there will be no forgiveness! If they don't confess their guilt to God and the guilt of their ancestors and ask God for forgiveness, there will be no forgiveness! Furthermore, God will punish them **7 times more for their sins**. (Leviticus 26:17)

Notice that it says that when they finally do, then God will also remember his promises to Abraham, Isaac and Jacob concerning the land! That is still future. (Genesis 15:18)

*"I will go and return to my place, **till they acknowledge their offence**, and seek my face: **in their affliction they will seek me early"**.* (Hosea 5:15) 785 B.C. Compare that passage with… *"I go to prepare a place for you"…* (John 14:2, 3)

God says that he will afflict Israel in ***"the last days".*** As a result of God's "***affliction"***, they will finally seek him. This will not happen until the "*last days*", during "*the time of Jacob's trouble*" (Jeremiah 30:7) also known as the 70th week of Daniel.

Jacob was Israel's name before he believed and was born again. The name Israel means "son of God" indicating a new birth. The term "*time of Jacob's trouble"* is another reference to the fact that Israel would return to their land in unbelief. Jacob was Israel's name before he became a believer. A new name for a new birth.

22.5 The 70 Years Begin (9th of Av, 606 B.C.) Phase One!

In 606 B.C., August 11, on the 9th day of the Hebrew month Av, Nebuchadnezzar and his armies showed up at Jerusalem's front door! That was the beginning of the **first curse**. The 70 years of God's chastisement began to tick off.

Exactly one Hebrew year later, **the first deportation began in 605 B.C.** when the Babylonians removed a small sampling of the residents of Jerusalem which consisted of Nobles, Priests and included Daniel (2Kings24:1-4). They also took the Temple Vessels. This also happened on the **9th of Av.** (2Chronicles 36:7) This is when the sacrifices ceased.

The first Babylonian Siege against Jerusalem occurred in 606 B.C., on the 9th of Av and marks the beginning of God's chastisement for Israel. So to count 70 Hebrew/Lunar years from 606 B.C. in the Roman calendar, we first convert 70 Hebrew years into Roman years by multiplying 70 times 360 and dividing that by 365.25 which equals 68.99 Roman years. When rounded off that gives us 69 Roman years. That would bring us to 537 B.C. for the completion of **God's first round of punishment for Israel**. The 70 Hebrew/Lunar years ended on the 8th of Av 537B.C. The first Jews to go back home to rebuild the walls of the city, left on the next day, the 9th of Av.

In 597 B.C. the second deportation began with the removal of the Jewish residents in waves of about 10,000 at a time. (2Kings24:10-17)

22.6 "I Will Set My Face Against You". Ezekiel Fills In The Details;

God warned his people before the Babylonian siege of Jerusalem, but they did not listen. God instructed his prophet Ezekiel to act out the coming siege in the Temple courtyard for everyone to see. God took Ezekiel's voice away because Israel was not listening. This display was a warning and a pronouncement of the number of years of punishment due to the upper 10 tribes of Israel and the number of years of punishment due to the lower 2 tribes of Judah.

"Thou also, son of man, take thee a tile, and lay it before thee, and pourtray upon it the city, even Jerusalem": (Ezekiel 4:1)

"And lay siege against it, and build a fort against it, and cast a mount against it; set the camp also against it, and set battering rams against it round about." (Ezekiel 4:2)

*"Moreover take thou unto thee an iron pan, and set it for a wall of iron between thee and the city: and **set thy face against it**, and it shall be besieged, and thou shalt lay siege against it. This shall be a sign to the house of Israel"*. (Ezekiel 4:3)

Remember in (Leviticus 26:17) God said***..."I will set my face against you"***. Ezekiel is referring to God's original prophecy made 900 years earlier and recorded in (Leviticus 26:17)...

*"And **I will set my face against you**, and ye shall be slain before your enemies: they that hate you shall reign over you; and ye shall flee when none pursueth you"*. (Leviticus 26:17)

Continuing on with Ezekiel…

"*Lie thou also upon thy left side, and lay the iniquity of the house of Israel upon it: according to the number of the days that thou shalt lie upon it thou shalt bear their iniquity*" (Ezekiel 4:4)

God pronounces 390 years of punishment for the upper ten tribes of Israel…

*"For I have laid upon thee the **years** of their iniquity, according to the number of the days, **three hundred and ninety days**: so shalt thou bear the iniquity of the house of Israel"*. (Ezekiel 4: 5)

God pronounces 40 years of punishment for the lower two tribes of Judah...

*"And when thou hast accomplished them, lie again on thy right side, and thou shalt bear the iniquity of the house of Judah **forty days: I have appointed thee each day for a year"***. (Ezekiel 4: 6)

390 years for the upper 10 tribes of Israel, and **40 years** for the lower 2 tribes of Judah. 390 + 40 = **430**

That totals up to 430 years of punishment due for all 12 tribes of Israel.

Subtract 70 years for time spent in Babylon and that leaves **360** years of punishment that was still due!

430-70=360. (Years of punishment due)

22.7 A 2600 Year Old 9/11 Connection With "Alah" (Hebrew)

Compare (Leviticus 26); (Ezekiel 4); (Jeremiah 25) and (Daniel 9)

Near the end of the Babylonian captivity, Daniel was reading (Jeremiah 25: 11, 12) and recognized that the 70 years of punishment, which were prophesied by God's prophets, were almost up. …

"In the first year of his reign I Daniel understood by books the number of the years, whereof the word of the LORD came to Jeremiah the prophet, that he would accomplish ***seventy years in the desolations of Jerusalem****"*. (Daniel 9:2)

Daniel was concerned about what he was reading in the book of Jeremiah. He realized that the 70 years were almost complete. When he compared what he was reading in Jeremiah with Leviticus ("*the law of Moses*") he became even more concerned because Leviticus said …

"And if ye will not yet for all this hearken unto me, ***then I will punish you seven times more for your sins****"*. (Leviticus 26:18)

Daniel realized that Israel had not learned their lesson yet. There was no change of heart and turning back to God and the 70 years were almost completed. Danial said…

"*To the Lord our God belong mercies and forgivenesses, though we have rebelled against him;"* (Daniel 9:9)

"Neither have we obeyed the voice of the LORD our God, to walk in his laws, which he set before us by his servants the prophets". (Daniel 9:10)

"*Yea, all Israel have transgressed thy law, even by departing, that they might not obey thy voice; therefore the* ***curse*** *is poured upon us, and the oath that is written in the* ***law of Moses*** *the servant of God, because we have sinned against him".* (Daniel **9:11**)

It is interesting to note that the word ***"curse"*** used here in Daniel, Chapter **9,** verse **11** is translated from the Hebrew word **Alah** and is equivalent to the Arabic word "Allah". That gives us **a 9/11 connection with Alah that dates back to about 2500 years** ago! Once again, **the Bible said it first!**

"And he hath confirmed his words, which he spake against us, and against our judges that judged us, by bringing upon us a great evil: for under the whole heaven hath not been done as hath been done upon Jerusalem". (Daniel 9:12)

"*As it is written in the law of Moses, all this evil is come upon us:* ***yet made we not our prayer before the LORD our God****, that we might turn from our iniquities, and understand thy truth".* (Daniel 9:13)

Daniel is God's witness that Israel did not have a change of heart toward the Lord. He said that Israel did not confess their sins to God and seek his forgiveness. Daniel makes it clear that Israel was not seeking God. So the verdict given was "***seven times more for your sins***"**!** (Leviticus 26)

23

The Messiah Was Promised And Prophesied To The Day!

(Daniel 9:25)

23.1 Daniel's 69 Weeks Of Years Till The Messiah!

According to (Daniel 9:24, 25), the time interval between the commandment to rebuild and restore Jerusalem, till the Messiah would be 69 weeks of years. That is 69 times 7, which equals 483 Hebrew years.

The 70th week will not be fulfilled yet, that will come later. The 70th week of Daniel will be a future time of God preparing his people for their promises.

The 70th week of Daniel will mark the end of the Age of law. It will be a time when God will make an all-out supernatural attempt to win the hearts and minds of his precious Creation. Satan will make one last attempt to destroy God's Creation before he is defeated. This will be the worst time the world has ever seen or ever will see! Most of the world's population will be destroyed, only a very small portion will survive.

Just like the Bible has accurately foretold events in the past and revealed science before man had the knowledge (Isaiah 48: 3-6) the Bible has also foretold near future events which will also come to pass exactly as described in the Bible.

Just as the Messiah was presented to Israel on the very day prophesied in the Bible; and Israel returned to their land the day after their prophesied punishment off the land was complete, so likewise near-future events will also occur exactly as described in the Bible.

23.2 Palm Sunday Prophesied To The Very Day!

…"***from the going forth of the commandment*** *to restore and to build Jerusalem unto the Messiah the Prince shall be seven weeks, and threescore and two weeks"* … (Daniel 9:25)

That is 69 "weeks" of years. One week equals 7 years, therefore we have 69 X 7 = 483 Hebrew (Lunar) Years. One Hebrew year equals 360 days.

Sir Robert Anderson of Scotland Yard (May 29, 1841 –November 15, 1918) calculated this time interval **to the day** in his classic book titled **"The Coming Prince"**. Link to free pdf file online. Go to… http://whatsaiththescripture.com/Text.Only/pdfs/The Coming Prince Text.pdf .

Anderson established many dates mentioned in the Bible by careful research which included consulting **The Royal Astronomers.** In his book, The Coming Prince, he thanked them …"For this calculation I am indebted to the courtesy of the Astronomer Royal".

According to Anderson, "*the commandment to restore and to build Jerusalem*" (Daniel 9: 25) was the decree issued by **Artaxerxes Longitmanus** in the **twentieth year of his reign**, authorizing Nehemiah to rebuild the fortifications of Jerusalem.

"Now this is the copy of the letter that the king Artaxerxes gave unto Ezra the priest, the scribe, even a scribe of the words of the commandments of the LORD, and of his statutes to Israel." (Ezra 7:11)

The date that the decree was issued by *Artaxerxes* has been established by Sir Robert Anderson to be the **1st day of Nisan (Heb.); March 14, 445 B.C. (Greg.)**: It is interesting to note that **Nisan 1** is also considered by Jewish tradition to be the day of Creation and is used as a Jewish secular New Year for counting months.

Therefore the sixty-nine weeks are to be counted from the 1st day of Nisan (Heb.) = March 14, 445 B.C. (Greg.) … until some special event 483 years in the future which satisfies the words… ***"Unto Messiah the Prince"***.

483 years later, to the very day, counted from Nisan 1 (Heb.), March 14, 445 B.C. (Greg.); Jesus rode into Jerusalem on a donkey on the day known as "Palm Sunday"; as a large crowd threw down palm branches in his path; …

"On the next day much people that were come to the feast, when they heard that Jesus was coming to Jerusalem," (John 12:12)

"Took branches of palm trees, and went forth to meet him, and cried, ***Hosanna: Blessed is the King of Israel that cometh in the name of the Lord."***… (John 12:13)

"And Jesus, when he had found a young ass, sat thereon; as it is written," (John 12:14)

"Fear not, daughter of Sion: behold, thy King cometh, sitting on an ass's colt "(John 12:15)

…Jesus was publicly recognized as the Messiah and thus was the words ***"unto Messiah the Prince"*** (Daniel 9:25) …fulfilled on "**Palm Sunday**. Another prophetic reference to this event is found in (Zechariah 9:9) written 520 years before it happened.

"Rejoice greatly, O daughter of Zion; shout, O daughter of Jerusalem: behold, thy King cometh unto thee: he is just, and having salvation; lowly, and riding upon an ass, and upon a colt the foal of an ass." (Zechariah 9:9)

23.3 The Bible Is Literal And Mathematically Accurate!

Daniel told us the exact year and day that the Messiah would be publicly recognized.

Palm Sunday, began (**Nissan 10**), Julian date; **Sunday morning; April 6, 32 A.D.**

"Know therefore and understand, that from the going forth of the commandment to restore and to build Jerusalem unto the Messiah the Prince shall be seven weeks, and threescore and two weeks: the street shall be built again, and the wall, even in troublous times". (Daniel 9:25)

Remember that 69 weeks is 69 times 7 = 483 (Hebrew) years.

That day can be calculated by counting 483 X 360 = Total Days = 173,880; divide that by 365.25 to convert to our calendar = 476.0574949 years. Therefore… count 476, Roman years from March 14, 445 B.C. (Greg.) … add one year to cross over from "B.C." to "A.D." which takes us to… 32 A.D. the fraction left over can be converted into days by multiplying times 365.25. Therefore … count 21 days from March 14, and that brings us to April 6, 32 A.D. And that is the day that Jesus rode into Jerusalem on a donkey while being hailed as King of the Jews. (John 12:13) Palm Sunday was the day Jesus presented himself as the Prophesied King of Israel to Jerusalem!

All of this was prophesied to the day, hundreds of years in advance!

24

4 Spring Feasts Fulfilled, 4 Divine Appointments Kept

24.1 God's Calendar Is Lunar / Solar

God Defines How A Day Is Counted

"*And God called the light Day, and the darkness he called Night. And the evening and the morning were the first day*" (Genesis 1:5)

In the Hebrew calendar, a day is counted as one revolution of the earth, beginning in the evening at sundown. The night precedes the day. A Hebrew day begins and ends on a sundown.

This counting the night before the day has created some confusion among Gentiles who are accustomed to the Roman/Gregorian calendar. When referencing God's Holy Feast days, the Hebrew date will cover two Roman days which include the night portion of the Roman date preceding the Hebrew date.

Beginning at sundown, the night portion of the preceding Roman day makes up the first half of the Hebrew date which is the night. The second half of the Hebrew date will cover the daylight part of the following Roman day and conclude at sundown.

The days of the Hebrew week are numbered 1-7. Only the seventh day has a name, which is Sabbath. **The Hebrew name for Sabbath is <u>Shabbat</u>.** It means day of rest. All other days are simply numbered without an associated name.

"And God made two great lights; the greater light to rule the day, and the lesser light to rule the night: he made the stars also". (Genesis 1:16)

"Speak thou also unto the children of Israel, saying, Verily ***my Sabbaths ye shall keep****: for it is a* ***sign*** *between me and you throughout your Generations; that ye may know that I am the LORD that doth sanctify you"*. (Exodus 31:13)

The seven feasts were intended to be memorials and descriptions of events that would occur in the future. You could call them ***signs*** in the Sun and Moon because the Hebrew calendar is lunar/solar.

(Isaiah 66:23) *"And it shall come to pass, that from one new moon to another, and from one Sabbath to another, shall all flesh come to worship before me, saith the LORD"*.

When God gave Israel the first Spring Feast called Passover, he said…

"And this day shall be unto you for a ***memorial****; and ye shall keep it a feast to the LORD throughout your Generations; ye shall keep it a feast by an ordinance for ever."* (Exodus 12:14)

The seven Feast days of Israel are Seven Devine Appointments for future generations to watch for.

"And God said, Let there be lights in the firmament of the heaven to divide the day from the night; and let them be for ***signs****, and for* ***seasons****, and for days, and years"* (Genesis 1:14)

This passage refers to the "***lights in the firmament of the heaven***" and declares their Created purpose as being…"***for signs and for seasons.***" This passage connects signs and seasons with the "***lights in the firmament of the heaven***".

"***Seasons***" - Strong's H4150 mow`ed (definition: appointed place, time or meeting; sacred time or feast or in this passage, **Divine Appointment**.

This is a reference to the weekly Sabbath and to the seven Feasts of the Lord in particular. So now we have the Feasts of the Lord connected to "***lights in the firmament of the heaven***".

"***Signs***" are defined in the Strong's Hebrew Dictionary as... ('owth) Strong's # H226; sign, signal; a distinguishing mark; Banner; Remembrance; miraculous sign; **Omen**; **Warning**, token, ensign, standard, miracle, proof.

God says that his "Sabbaths" or Feast days are signs for God's people and so is the heavens. These were signs for what were future events at that time. Christ fulfilled the Spring Feasts when he came the first time. At the time of this writing, we are waiting for the next Feast in line to be fulfilled which is the fifth Feast called Rosh Hashanah or Feast of Trumpets.

24.2 The Months Of The Hebrew Calendar

The New Year for counting months is Nisan.

Just prior to leaving Egypt, God gave Israel their calendar.

"*And the LORD spake unto Moses and Aaron in the land of Egypt, saying*" (Exodus 12:1)

"*This month shall be unto you the **beginning of months**: it shall be the **first month** of the year to you*". (Exodus 12:2)

"*Speak ye unto all the congregation of Israel, saying, In the tenth day of this month they shall take to them every man a lamb, according to the house of their fathers, a lamb for an house:*" (Exodus 12:3)

The first month was later called "***Abib***" by Moses...

"*This day came ye out in the month Abib*". (Exodus 13:4)

"***Abib***" is defined by **Strong's # H24 - *'abiyb* ...**month of ear-forming, of greening of crop, of growing green Abib, month of exodus and Passover (March or April)

Later in the books of Nehemiah and Esther this same month was called "**Nisan**" (Nehemiah 2:1) and (Esther 3:7)

"Nisan" corresponds to Strong's # H5212 – Niycan, which is defined as...

... Literally "their flight"; "the 1st month of the Jewish calendar corresponding to March or April" (Strongs)

It is important to mention here that Sir Robert Anderson, the Author of "The Coming Prince", established the date of ... "*the **commandment to restore and to build Jerusalem***" (Daniel 9: 25); ...was the decree issued by **Artaxerxes Longitmanus** in the twentieth year of his reign, authorizing Nehemiah to rebuild the fortifications of Jerusalem.

That date is 1st day of Nisan (Hebrew.); March 14, 445 B.C. (Greg)

The Hebrew month **Nisan**, similar to our April, but varying from year to year, according to the course of the moon. The name Nisan found only after the time of Ezra, and the return from the captivity of Babylon

Before the Babylonian Exile, only four months had names. All other months were referred to exclusively by number.

Every time Jews refer to the months by their corresponding number they are reminded of their miraculous deliverance from Egypt.

Other than Nisan, only 3 other months had names prior to the Babylonian exile. Those months are:

Ziv is the second month (counting from Nisan) Defined in Strong's Dictionary # H2099 – Ziv = "brightness" name of the 2nd month of the year, corresponding to Apr-May also called (Iyar); this is the month that Solomon laid the foundation for the first Temple. (1Kings 6: 37)

Tishrei is the seventh month of the Hebrew year (counting from Nisan) and corresponds to September and October in the Gregorian calendar. This is the month that Rosh Hashanah and Yom Kippur are celebrated; Rosh Hashanah is The New Year for the purpose of counting years and epochs.

Bul is the eighth month in the Hebrew calendar and corresponds to the months of the Roman calendar beginning from mid-October to mid-November. Bul is the Biblical name for the month that Solomon's Temple was completed. (1Kings 6: 38) Also known as (Cheshvan) The Jews expect for the third Temple to be built in this month but I don't know of a passage where God ever said that it would. We can only speculate about that and then wait to see what happens.

Since the return from Babylon, Israel has been heavily influenced by Babylonian religion. Israel has adopted Babylonian names for the rest of the months that reflects influence from Babylonian religion.

For example, Tammuz is the name of the Babylonian god which Jacob's descendants assigned to the fourth month.

24.3 God Promised Signs

When God restored the earth for man's occupation, he designed the heavens to be used for signs and seasons.

"*…and let them be for **signs**, and for **seasons**…*" (Genesis1:14)

The Hebrew word translated into "*signs*" in (Genesis 1:14) is the same word that was translated as "*token*" in (Exodus 12:13) which is talking about the first Spring Feast of Passover.

Again the same Hebrew word is translated as "*signs*" in (Exodus 31:13) where God says…

*"Verily my Sabbaths ye shall keep: for **it is a <u>sign</u>** between me and you"* (Exodus 31:13)

Notice that it is saying "*Sabbaths*", (plural). This includes all seven "*Feasts of the LORD*".

"*And the LORD spake unto Moses, saying*", (Leviticus 23:1)

"*Speak unto the children of Israel, and say unto them, concerning the feasts of the LORD, which ye shall proclaim to be holy convocations, even these are my feasts*". (Leviticus 23:2)

"*Six days shall work be done: but the seventh day is the Sabbath of rest, an holy convocation; ye shall do no work therein:" it is the Sabbath of the LORD in all your dwellings."* (Leviticus 23:3)

"*These are the feasts of the LORD, even holy convocations, which ye shall proclaim **in their seasons***". (Leviticus 23:4)

God spoke to Moses and commanded him to begin a series of seven annual Holy Days or Holy Feasts. The observance was a prophetic enactment of things God was going to do in the Future. Spring comes before Fall so the **Four Spring Feasts** were fulfilled first.

The Spring Feasts were fulfilled on their appointed days and in their appointed order. We have been waiting two thousand years now for the fifth Feast to be fulfilled. And now that time is at hand.

When Israel observes these Feasts, it acts as an annual reminder of fulfilled prophecies in the past. The first four Feasts remind us that the last three Fall Feasts will also be fulfilled. They are divine appointments!

Not only the days but many other details of the prophesied events were declared well in advance of the events such as the very hour and day of Christ's Crucifixion! Acting out key details of future events during the observance of these seven Feasts was an annual reminder of how God would fulfill his promises in the future.

"*And this day shall be unto you for a **memorial**; and ye shall keep it a feast to the LORD*" (Exodus 12:14)

The 7 Feasts Are Divided Into Spring And Fall,

The first three Spring Feasts represent The **Death**, **Burial** and **Resurrection** of Jesus Christ.

The Fourth Spring Feast marks **the beginning of the Church Age**. It is interesting to note that according to Jewish tradition, the Age of law also began on Pentecost. Jewish tradition holds that this is the day when God gave the law to Moses. Traditionally, Jews have always considered the Feast of Pentecost to represent that day. Pentecost also marked the day that God put the Age of Law on hold at the end of the 69th week of Daniel (Daniel 9: 24, 25) and the Church Age began. (Acts 1:5, Ch. 2) The Church Age will end and the Age of Law will start back up on the fifth Feast.

The first **4 Spring Feasts** collectively are called the "First Advent of Jesus Christ" and they were literally fulfilled on their day, at their appointed hour and in their appointed order. The ceremonial activities described well in advance what would happen in the future fulfillment of each Feast Day. They were all completely fulfilled exactly like the Bible described. (ETBH chapter 25)

The first three Feasts represent the death, burial and resurrection of Jesus Christ. **The fourth Spring Feast**_represents the beginning of the **Dispensation of Grace**, also known as the **Church Age**.

The **3 Fall Feasts** are the last three of the seven feasts. They represent the still future "Second Advent" of Christ, also known as the Day of the Lord.

Feast number five is the Hebrew New Year and is next in line to be fulfilled. Also known as the Feast of Trumpets or as Rosh Hashanah, It begins at sundown on the eve of the first day of Tishri and ends at the next sundown. Tishri is the seventh month of the year but this is the time for counting years and epochs such as the Church Age.

Feast number Five is the Biblical New Year and marks the end of the old year and the beginning of the New Year. The old year ends and the New Year begins ... "*at the last trump*" at sundown (Jerusalem time) which is about 6:30 pm or later.

The time of sunset changes every year. In 2020, on September 20, (Tishri 1) the sun sets at 6:37 pm; in 2021, on September 8 (Tishri 1) it will set at 6:54 pm

Although The Feast of Trumpets is considered to be the civil New Year for counting years and epochs by the Jews, they also have an ecclesiastical New Year on Nisan 1for counting months and holidays in particular.

Every year on the eve of Tishri 1, the Feast day begins at sundown about 6:30pm with a long toot of the Shofar. Throughout the following 24 hours the Shofar is blown in sets of high, low, short and long blasts in a predetermined sequence.

This continues for 24 hours with a total of over 100 blasts of the Shofar during that 24 hours. The Feast ends at sundown, at about 6:30 pm with a final blast of the Shofar and thus the Old Year ends and the New Year begins.

Just like the Church Age began on the 4th Feast, it will end on the 5th Feast,

"*At the sound of the last trump*" (1 Corinthians 15:52).

That final trump of the day marks the end of the previous year and the beginning of the New Year.

Many have quoted... "*That day and hour knoweth no man*" (Matthew 24:36) to prove that you can't know the day and hour of the Rapture. However they are not rightly dividing the word. Christ is speaking these words before his death, burial and resurrection. The Rapture was not yet revealed. The "mystery" of the Rapture was first revealed to the Church after the resurrection in the book of (1 Corinthians 15:51) and later on in (1 Thessalonians 4: 13) because it involves the Church, not Israel.

This passage (Matthew 24:36) is talking to Israel about Israel during the Tribulation. At that time Israel will be looking for Christ's second return at the end of the tribulation.

People confuse the rapture at the beginning of the Tribulation with the second return of Christ at the end of the Tribulation.

Whatever year it happens, the sound of the last trump of Rosh Hashanah will mark the end of the Church Age, the beginning of the Day of the Lord and all believers will be raptured out of this world and judged at the sound of the last trump. (ETBH chapters 29-36) God has declared the Feast day, the hour and the moment but not the year. That is why it is called Imminent.

Yom Kippur begins Ten days after Rosh Hashanah begins. Yom Kippur is also known as the Feast of Atonement. That will be the day that the Antichrist will promise "*Peace and Safety*" (or Peace and Security) and to enforce the "*Treaty with many*" for 7 years. That is the day they will say "*peace and safety*" (1 Thessalonians 5:3). However, at sundown the 70th week of Daniel will begin and … "*then cometh sudden destruction*".

Feast Number Seven is Tabernacles which follows 13 days after Atonement. About Seven years after the Tribulation begins, it will end on Yom Kippur (Atonement) just like it began. Then somewhere between Atonement and Tabernacles, during those 13 days, Russia and its allies will trigger off the Battle of Armageddon which will last for about an hour and be stopped by Jesus himself. Nobody knows the day or the hour of this event.

24.4 Passover, (Pesach), The First Spring Feast

Exodus chapter twelve begins in verse one by stating that the Lord himself is doing the talking.

*"In **the tenth day** of this month they shall take to them every man a lamb"*, (Exodus 12:3)

God describes the Passover Lamb in verse five...

"Your lamb shall be ***without blemish****, a male of the first year: ye shall take it out from the sheep, or from the goats:"* (Exodus 12:5)

Only Jesus is sinless, this is a description of Him! The New Testament Prophets wrote about how Jesus would fulfill these prophecies about seven hundred years later…

"*But with the precious blood of Christ,* ***as of a lamb without blemish*** *and without spot*": (1Peter 1:19)

Jesus Christ is sinless, He is the perfect Lamb of God who was "without blemish and without spot"! He was Crucified as our Passover Sacrifice for the sins of the world.

"The next day John seeth Jesus coming unto him, and saith, ***Behold the Lamb of God****, which taketh away the sin of the world."* (John 1:29)

Getting back to Exodus chapter twelve, we continue on in verse six and find more details about what was then a future Divine appointment.

"*And ye shall keep it up until the fourteenth day of the same month: and the whole assembly of* ***the congregation of Israel shall kill it in the evening***". (3pm in the afternoon by Roman time) (Exodus 12:6)

"And they shall take of the blood, and strike it on the two side posts and on the upper door post of the houses, wherein they shall eat it." (Exodus 12:7) also verses 8-14

Christ was crucified on the first Spring Feast, (Passover) at the very same time of day (3pm) as Passover lambs were slain all over Israel.

"*In the fourteenth day of the first month at even is the LORD'S Passover*". (Leviticus 23:5)

Jesus is our Passover Lamb. Christ was crucified the same hour the Passover Lamb was slain!

The year, month, day and the very hour that Christ was Crucified was prophesied well in advance and declared by God with signs in the heavens; in the earth and in the nations.

"He was oppressed, and he was afflicted, yet he opened not his mouth: he is brought ***as a lamb*** *to the slaughter, and as a sheep before her shearers is dumb, so he openeth not his mouth."* (Isaiah 53:7)

More prophecy fulfilled! Isaiah wrote about this more than 700 years before it happened!

"Purge out therefore the old leaven, that ye may be a new lump, as ye are unleavened. For even Christ our Passover is sacrificed for us" (1 Corinthians 5:7)

Jesus Christ was crucified for our sins at the same hour as the Passover Lambs were being killed all over Israel, at about 3pm that afternoon. Jesus died just before sundown. **Jesus Christ Is The World's Passover Lamb!**

24.5 Jesus Is Betrayed For "30 Pieces Of Silver".

Every detail of Christs Death was foretold!

"And I said unto them, If ye think good, give me my price; and if not, forbear. So they weighed for my price ***thirty pieces of silver****".* (Zechariah 11:12)

"And the LORD said unto me, Cast it unto the potter: a goodly price that I was prised at of them. And I took the thirty pieces of silver, and cast them to the potter in the house of the LORD". (Zechariah 11:13)

Zechariah wrote that 530 years before it actually happened! Judas fulfilled that prophecy 530 years later.

"Then one of the twelve, called Judas Iscariot, went unto the chief priests", (Matthew 26:14)

"And said unto them, What will ye give me, and I will deliver him unto you? And they covenanted with him for ***thirty pieces of silver****"*. (Matthew 26:15)

God described Crucifixion about 1,000 years earlier…

"*They pierced my hands and my feet*". (Psalm 22:16)

God also described it again 530 years earlier…

"And one shall say unto him, What are these wounds in thine hands? Then he shall answer, Those with which I was wounded in the house of my friends" (Zechariah 13:6)

24.6 The Passover Week Prophesied In Detail

Not only does God foretell the day of Palm Sunday (ETBH chapter 25), but all the details of the Passover week beginning with Palm Sunday! Prophecies that were written as far back as 1500 B.C.!

(Zechariah 9:9) was written 520 years before it happened!

"*Rejoice greatly, O daughter of Zion; shout, O daughter of Jerusalem: behold, thy King cometh unto thee: he is just, and having salvation; lowly, and riding upon an ass, and upon a colt the foal of an ass.*" (Zechariah 9:9)

This passage was fulfilled on "Palm Sunday" morning, about 520 years after it was written!

"*Then Jesus six days before the Passover came to Bethany, where Lazarus was which had been dead, whom he raised from the dead*". (John 12:1)

"*On the next day much people that were come to the feast, when they heard that Jesus was coming to Jerusalem*", (John 12:12,)

"Took branches of palm trees, and went forth to meet him, and cried, Hosanna: Blessed is the King of Israel that cometh in the name of the Lord" (John 12: 13)

"And Jesus, when he had found a young ass, sat thereon; as it is written," (John 12: 14)

"Fear not, daughter of Sion: behold, thy King cometh, sitting on an ass's colt." (John 12: 15)

Dr. C.I. Scofield said, "Fulfilled prophecy is proof of inspiration because the Scripture predictions of future events were uttered so long before the events transpired that no merely human sagacity or foresight could have anticipated them, and these predictions are so detailed, minute and specific, as to exclude the possibility that they were mere fortunate guesses."

6pm, Sundown, (Nissan 10) begins, Sunday April 6, 32 A.D. Night comes before the day in the Hebrew Calendar. The next morning is Palm Sunday.

I believe that Jesus entered Jerusalem that day through the Eastern gate at about 6a.m. (Roman time), which is the beginning of the first hour of the day (Hebrew time). I believe that Jerusalem started their day with this event. People got up before the sun, dropped any plans for the day and went to see the one who raised Lazarus from the dead after 4 days. Word had really traveled about that event the previous day. They knew that Lazarus died 4 days earlier because the family was probably still in mourning.

Palm Sunday morning begins in Jerusalem as Jesus rides through the Eastern Gate on a donkey and is hailed as their King. (John 12:1, 12-13) That happens exactly 483 years to the day from the day Artaxerxes gave the command to rebuild Jerusalem just like Daniel (Daniel 9:24, 25) prophesied 600 years earlier! (ETBH chapter 25)

24.7 Hands And Feet Pierced; Lots Cast; No Broken Bones;

About 1,000 years <u>before it happened</u>, Yeshua (Jesus) describes his own crucifixion!

"And I will pour upon the house of David, and upon the inhabitants of Jerusalem, the spirit of grace and of supplications: and ***they shall look upon me whom they have <u>pierced</u>****, and they shall mourn for him, as one mourneth for his only son, and shall be in bitterness for him, as one that is in bitterness for his firstborn.* (Zechariah 12: 10)

Compare with (Psalm 22:16) and remember that crucifixion would not be invented for several more centuries.

"For dogs have compassed me: the assembly of the wicked have inclosed me: ***they pierced my hands and my feet****".* (Psalm 22:16)

About a thousand years latter, the Apostle John confirmed that the Old Testament passage had been fulfilled.

"And again another scripture saith, ***They shall look on him whom they <u>pierced</u>"****.* (John 19: 37)

About 1,500 years before it happened, God said … ***"A bone of him shall not be broken"***

"I may tell all my bones: they look and stare upon me". (Psalm 22:17)

"In one house shall it be eaten; thou shalt not carry forth ought of the flesh abroad out of the house; ***<u>neither shall ye break a bone thereof</u>****".* (Exodus 12:46)

"He keepeth all his bones: not one of them is broken*".* (Psalm 34:20)

"For these things were done, that the scripture should be fulfilled, ***<u>A bone of him shall not be broken</u>****".* (John 19: 36)

"***They part my garments*** *among them, and cast lots upon my vesture*". (Psalm 22:18)

"But be not thou far from me, O LORD: O my strength, haste thee to help me" (Psalm 22:19)

24.8 Unleavened Bread, The Second Spring Feast

The (15th of Nisan) represents the Burial of Christ at sundown 6pm Jerusalem Time!

"And on the fifteenth day of the same month is the feast of unleavened bread unto the LORD:" (Leviticus 23:6)

"In the first day ye shall have an holy convocation: ye shall do no servile work therein." (Leviticus 23:7)

"But ye shall offer an offering made by fire unto the LORD seven days: in the seventh day is an holy convocation: ye shall do no servile work therein." (Leviticus 23:8)

The Feast of Unleavened Bread is observed the day following Passover. Christ was buried on the Feast of Unleavened Bread. (Matthew 27:57-60) (Mark 15:33, 34) (Leviticus 23:6)

Jesus Christ was buried at sundown. Unleavened Bread follows Passover directly. The sundown that marks the end of Passover also marks the beginning of the Feast of Unleavened Bread. At sundown, Jesus was quickly put into his grave at the beginning of the Second Spring Feast, the Feast of Unleavened Bread.

24.9 First Fruits, (Sukkot), The Third Spring Feast

On the seventeenth day of Nisan the first fruits of the barley harvest was gathered and waved before the Lord in celebration.

Jesus Christ rose from the grave at the end of the third Spring Feast, at sundown. That is when The Feast of First Fruits ended, on the seventeenth day of Nisan. (32 A.D.)

Jesus rose from the grave at the end of The Feast of First Fruits (Leviticus 23:9-14)

"And the LORD spake unto Moses, saying" (Leviticus 23:9)

"Speak unto the children of Israel, and say unto them, When ye be come into the land which I give unto you, and shall reap the harvest thereof, then ye shall bring a sheaf of the firstfruits of your harvest unto the priest:" (Leviticus 23:10)

"And he shall wave the sheaf before the LORD, to be accepted for you: on the morrow after the Sabbath the priest shall wave it." (Leviticus 23:11)

"And ye shall offer that day when ye wave the sheaf an he lamb without blemish of the first year for a burnt offering unto the LORD." (Leviticus 23:12)

"And the meat offering thereof shall be two tenth deals of fine flour mingled with oil, an offering made by fire unto the LORD for a sweet savour: and the drink offering thereof shall be of wine, the fourth part of an hin". (Leviticus 23:13)

"And ye shall eat neither bread, nor parched corn, nor green ears, until the selfsame day that ye have brought an offering unto your God: it shall be a statute for ever throughout your Generations in all your dwellings." (Leviticus 23:14)

About fifteen hundred years later Jesus told his disciples about his death, burial and resurrection on the third day.

"But He answered and said to them, "An evil and adulterous generation seeks after a sign, and no sign will be given to it except ***the <u>sign</u> of the prophet Jonah****"*. (Matthew 12:39)

"For as Jonah was three days and three nights in the belly of the ***great fish****, so will the Son of Man be three days and three nights in the heart of the earth".* (Matthew 12:40)

Jesus spent exactly 72 hours in the heart of the earth in a place referred to in the New Testament as "Hades"(Greek). In the Old Testament (Hebrew) it is called "Sheol".

24.10 Pentecost marked the beginning of the Church Age.

Pentecost, **(Shavuot)** the fourth and the last of the Spring Feasts was observed fifty days from the Feast of First Fruits, on the (7th of Sivan)

Pentecost is the only Feast to start on a New Moon (Sunday, 7th of Sivan).

On the day of new moon, the moon rises when the sun rises. It sets when the sun sets. It crosses the sky with the sun during the day and is not visible for a day or two.

Moses was given the law by God on Mount Sinai on the 7th of Sivan.

The last 7 years of the Dispensation of Law has been put on hold for 2,000 years. Also known as the 70th week of Daniel or the Tribulation, it will commence one of these years on the 5th feast. Hosea Chapters 3; 5; and 6 defines that time. That is 2,000 literal Hebrew years which consists of 360 days per year. Therefore we multiply 360 times 2,000 and divide by 365.25 and that gives us the equivalent number of solar years. 2,000 Hebrew years are ***Equivalent*** to 1,971.25 Roman (Solar) years.

God is a date setter and a date keeper! God announces his Divine Appointments in advance and he is always on time!

"And ye shall count unto you from the morrow after the Sabbath, from the day that ye brought the sheaf of the wave offering; seven Sabbaths shall be complete:" (Leviticus 23:15)

*"Even unto the morrow after the **seventh** **Sabbath** shall ye number **fifty days**; and ye shall offer a new meat offering unto the LORD."* (Leviticus 23:16)

"Ye shall bring out of your habitations two wave loaves of two tenth deals: they shall be of fine flour; they shall be baken with leaven; they are the firstfruits unto the LORD." (Leviticus 23:17)

"And ye shall offer with the bread seven lambs without blemish of the first year, and one young bullock, and two rams: they shall be for a burnt offering unto the LORD, with their meat offering, and their drink offerings, even an offering made by fire, of sweet savour unto the LORD." (Leviticus 23:18)

"Then ye shall sacrifice one kid of the goats for a sin offering, and two lambs of the first year for a sacrifice of peace offerings." (Leviticus 23:19)

"And the priest shall wave them with the bread of the firstfruits for a wave offering before the LORD, with the two lambs: they shall be holy to the LORD for the priest." (Leviticus 23:20)

"And ye shall proclaim on the selfsame day, that it may be an holy convocation unto you: ye shall do no servile work therein: it shall be a statute for ever in all your dwellings throughout your Generations." (Leviticus 23:21)

"And when ye reap the harvest of your land, thou shalt not make clean riddance of the corners of thy field when thou reapest, neither shalt thou gather any gleaning of thy harvest: thou shalt leave them unto the poor, and to the stranger: I am the LORD your God." (Leviticus 23:22)

The Church Age began on Sunday, (7th of Sivan) 32 A.D., on the Feast of Pentecost, New Moon.

…A New Dispensation began and the Dispensation of Law is put on hold for about 2,000 (Hebrew) years.

25

May 13, 1948, Israel's Punishment Completed.

25.1 (Phase two) God Said "Seven Times More For Your Sins"

The second phase of the punishment was conditional upon the Jews not having a change of heart after the 70 years spent in Babylon. At the end of the 70 years in Babylon, Daniel said that Israel did not have a change of heart. They still were not listening. (Daniel 9:13)

God said what he would do if Israel did not have a change of heart after the 70 years were completed.

*"And if ye will not yet for all this hearken unto me, then I will punish you **seven times more for your sins**"*. (Leviticus 26:18)

Therefore we multiply **7 times the remaining 360** years of punishment still due (ETBH chapter 27. 6) which brings the number up to **2520 <u>Lunar/Hebrew</u> years** of punishment that was due Israel.

We **convert the 2520 (Hebrew) lunar years into (Roman) solar years** by multiplying 360 (<u>days</u> per **<u>lunar</u>** year) and dividing by 365.25 (<u>days</u> per **<u>solar</u>** year) to equal **2483.78 (Roman) solar years**. That is 2483.78 (Roman) solar years of punishment that Israel would suffer **<u>outside</u>** of their land.

"And I will scatter you among the heathen, and will draw out a sword after you: and your land shall be desolate, and your cities waste." (Leviticus 26:33)

"Then shall the land enjoy her Sabbaths, as long as it lieth desolate, and ye be in your enemies' land; even then shall the land rest, and enjoy her Sabbaths". (Leviticus 26:34)

"As long as it lieth desolate it shall rest; because it did not rest in your Sabbaths, when ye dwelt upon it". (Leviticus 26:35)

The punishment stated was that Israel would be kept off their land because they did not let the land rest during the Sabbatical year.

The 70 Hebrew/Lunar years of punishment due Israel in Babylon ended on the 8th of Av 537B.C... If you count 70 **Hebrew years** from 606B.C. (68.99 **Roman years**) you end up at 537B.C. the 8th of Av. The next day, a group of Jews left for Israel to rebuild the walls on **the 9th of Av!**

To count the time of the second phase of punishment due, start counting on the **9th of Av, 537 B.C**. Av is the 5th month of the Hebrew year which corresponds to July-August on the Gregorian calendar.

The 9th of Av is no ordinary day! It is recognized by the Jews as a day of National tragedies throughout their history.

Subtract 537 years from 2483.78 years and that leaves you with 1946.78 years. Add one year to cross over from B.C. to A.D. (because there is no year 0 on a timeline) and that gives you **1947.78 years.**

To get the number of days represented by ".78", multiply (.78) times (365.25 days in a Roman year) = **285 days**. That is **9** months and **11** days.

The average days in a Roman month is 365.25 divided by 12 = 30.4 days per month. 9 times 30.4 = 273.6. Subtract 273.6 from 285 days and that leaves you with 11.4

Count **9** months and **11** days from, August 3, 1947 (9th of Av) and that brings you to May 13, 1948! That was the day that completed Israel's punishment off their land. The next day Israel was recognized as an independent State by the United Nation. That day was May 14, 1948!

That was the next day after the payment was complete! God keeps track of things! **The Bible is <u>literal</u> and mathematically accurate <u>to the very day</u>!**

I think that it is interesting to note that when we take the fraction left over when converting from the Hebrew calendar to the Roman calendar we have a fraction left over that represents **9** months and **11** days. September is the **9th month** in the Roman calendar and we all know what happened on the **11th day** of that month! Is this more than just another coincidence? Is there another connection with this number? I don't know, but it gives us something to think about.

Here is a list of those calculations written out for anyone interested in checking my math…

The fraction left over when we converted Hebrew years to Roman years is .78, therefore to calculate the total number of days, we multiply ".78 times the number of days in a solar year … (.78 X 365.25 = 284.9days). That rounds up to **285 days.**

To find out how many months that is, we take the average days per Roman month, which equals… (365.25 divided by 12 = 30.44) Now to convert 284.9 days to months, we divide 284.9 by 30.44 which yields **9.36 months.**

Now to convert .36 months to days, we multiply .36 X 30.44 = 10.96 days which equals **11 days** when rounded to significant digits! So .78 times the Roman year equals **9**months and **11**days! I think this may be another hint that there is more to this number than is currently understood.

Wow! Israel returned to the land God gave to them in the last days just like the Bible said they would. And it happened on the very next day of the expiration of the time of God's punishment **away from the Promised Land**! On May 13, 1948 Israel's Punishment required by God in (Leviticus 26: 1-24) (Jeremiah 25: **9-11**) and (Ezekiel 4: 1-6) was completed. **Israel was recognized by the United Nations on the very next day!** God said it thousands of years before it happened!

(Isaiah 66:8) "*Who hath heard such a thing? who hath seen such things? Shall the earth be made to bring forth* ***in one day****? or* ***shall a nation be born at once****? for as soon as Zion travailed, she brought forth her children*"

Since 70 A.D. scoffers have said that Israel was finished. The Catholic Church said that God was through with Israel and claimed all of God's promises to Israel for themselves. Let this be a lesson to the scoffers, God keeps his promises, all of them!

Sir Isaac Newton predicted that Israel would return to their land based entirely on what the Bible said. The unbelievers were wrong and the believers were right.

God's word is confirmed because he does what he says he will do! Every detail was literally fulfilled.

Can a nation be born in a day? (Isaiah 66:8) In May 14, 1948, it was.

25.2 The Identity Of The True Israelites Is Confirmed By God

God said that he would scatter Israel throughout the world and bring them back to their land in the last days after their punishment concerning the sabbatical year is complete. God literally did what he said he would do.

On May 14, 1948, the days required by God for Israel to be off their land had just expired at sundown on the previous day. May 14, 1948 was the day chosen by God, and recorded in the Bible more than 2500 years earlier for Israel to return and take back their land and the Nation would be reborn overnight. (Isaiah 66:8)

Israel's punishment was foretold by Moses, Ezekiel and Jeremiah and it was completed on May 13, 1948 just like the Bible foretold! The people who the world recognized the very next day as Israel on May 14, 1948 was also recognized by God as his people by the fact that this could only happen on the year and day prophesied by God thousands of years in advance!

*"Who hath heard such a thing? who hath seen such things? **Shall the earth be made to bring forth <u>in one day</u>? or <u>shall a nation be born at once</u>?** for as soon as Zion travailed, she brought forth her children".* (Isaiah 66:8)

The people who returned to Israel on May 14, 1948 and are living in Israel today, are authenticated as the true Israelites by God himself with indisputable Bible prophecy. If there is any question about the true identity of the **Israelites**, here is another of many Confirmations that God has provided us with.

If anyone claims to be Israel other than the ones that were recognized on May 14, 1948, and represented in the land at that time, then they are lying! Those who occupy the land today are the real deal! They were reborn as a nation on the right day and place. The day the Bible predicted thousands of years ago.

In spite of all of this irrefutable evidence, the Catholic Church does not recognize Israel to this day. The scoffers will continue to scoff until it is too late. There are a number of groups who erroneously claim to be the "true Israelites", but they are liars.

The Bible is literal, mathematically accurate and God Said It all before it happened!

25.3 God Confirms His Word By Doing To Israel As He Said

God declared the history of the world and Israel in particular, long before it happened. God used Alexander the Great to confirm that fact. **Alexander the Great conquered the world around <u>323 B.C</u>...** It took him about13 years to accomplish this task. In 332 BC Alexander entered Egypt. He named and built a city as a memorial

During that time he gathered 70 scholars together for the task of translating the Ancient Hebrew scriptures into his native language of Greek. That translation is called the "**Septuagint**". Which means "seventy scholars".

That is an extraordinary number of scholars and resources to be gathered together for a single task at any time, but back then it was even more extraordinary! That reflects how important it was to Alexander the Great! This event has a time stamp that cannot be questioned by any rational person with a basic education. **Alexander the Great Timestamp of God!**

At the time of Alexander's translation, the Hebrew Scriptures were already considered to be ancient in all of the known world and recognized by the Greeks. God used Alexander the Great to establish the fact that the Hebrew Scriptures had already been around long before that translation occurred. That translation matches popular modern translations widely used today. No one can deny the antiquity of the Torah. The Old Testament was universally recognized as ancient. We all know when Alexander conquered the world, therefore we know for a certainty that the Hebrew Scriptures and all of those prophecies were written down centuries before the events occurred!

Those Scriptures gave many details about God's punishment against Israel as his son. God said that he would punish Israel like a father. The Messiah was prophesied to the very day; the destruction of the second temple after the Messiah's death was also prophesied; the world wide scattering of Israel and so was Israel's rebirth as a nation and more! God did to Israel what he said he would do and so far they have reacted as God said they would. There should be no doubt that God will continue to do as he has said and Israel will continue to react as God has predicted.

*"And **he hath <u>confirmed his words</u>**, which he spake against us, and against our judges that judged us, by bringing upon us a great evil: for under the whole heaven hath not been done as hath been done upon Jerusalem*". (Daniel 9: 12)

Daniel realized that if Israel did not have a change of heart God said… "***I will punish you <u>seven times more</u> for your sins***". (Leviticus 26:18)

Daniel said that Israel had not changed their heart toward God…

"As it is written in the law of Moses, all this evil is come upon us: yet made we not our prayer before the LORD our God, that we might turn from our iniquities, and understand thy truth" (Daniel 9: 13)

Moses prophesied that Israel would not have a change of mind even after God removed them from their land. (Leviticus 26: 33-35)

*"And if ye will not yet for all this hearken unto me, then I will punish you **seven times** more for your sins".* (Leviticus 26:18)

After all of that (verse 14-17) if they still don't have a change of heart, then God would punish them ***"seven times more"***. (Leviticus 26: 18)

God knew they would not obey and prophesied what would happen after Israel did not respond to God's first Curse. In (Lev. 26:16, 17) God describes what he would do next. After the first punishment if there is no changing of their minds then God would punish them ***"seven times more"***. (Leviticus 26: 18-39)

God said..."*And I will scatter you among the heathen*". (Leviticus 26:33) That happened twice, the last time was in 70 A.D.

"And ye shall perish among the heathen, and the land of your enemies shall eat you up". (Leviticus 26:28)

For the last 2,000 years since Israel was scattered in 70A.D. Jews died and were buried in foreign lands. Just like the Bible said they would.

That part of the curse is complete. Now the worst is at hand!

25.4 The Bible Is Literal, Mathematically And Historically Precise

There are many who say that the Bible is not accurate or not literal. Either way is an attack by Satan's minions on God's word. The fact is that the

Bible is both literal and mathematically accurate. Sir Isaac Newton once said…

…"About the Time of the End, a body of men will be raised up who will turn their attention to the Prophecies, and insist upon their **literal interpretation**, in the midst of much clamor and opposition" **(Sir Isaac Newton)**

Consider the details involved in this "Leviticus 26 Prophecy". The mathematical probabilities of its fulfillment by chance is almost zero! God has used specific dates on the Hebrew calendar to identify events as his Divine Judgment. This is only discovered through the Hebrew calendar.

As already demonstrated, God's Judgments against Israel have all been prerecorded and their fulfillment, carefully documented. This literal precision and accuracy demonstrates how God has used the Hebrew calendar to reveal himself to the world!

Nebuchadnezzar showed up at Jerusalem's front door in 606B.C. on the **9th of Av**. The **walls were breached on the 17th of Tammuz** (605B.C.). Twenty one days later the first deportation took place exactly one Hebrew year after Nebuchadnezzar's siege began. On the **9th of Av (605B.C.)** Nebuchadnezzar took Daniel along with a thousand other captives and all the Temple gold and silver back to Babylon. **The first Temple was burned on the 9th of Av (586 B.C.).** There is a list of catastrophic events that are said to have occurred on the 9th of Av since these first significant events. It turns out that the second Temple was also burned on the 9th of Av (70B.C.)! That one really got the Jews attention!

Modern day Jews all over the world consider the **9th of Av**, (Hebrew calendar), as a time to remember all of the National tragedies that have befallen their ancestors on this day. This Jewish tradition is known today as **Tisha B'Av**, or "**The Fast of the Ninth of Av".** It began after the Jews were taken as captives into Babylon. Av is the fifth month on the Hebrew calendar. (July-August Gregorian calendar)

The first significant event on this day was Nebuchadnezzar and his armies showing up at Jerusalem's front door; the first deportation, **the third event was the destruction of the first Temple.** All of these events occurred on the 9th of Av. The Bible records these events…

"*It came also in the days of Jehoiakim the son of Josiah king of Judah, unto the end of the eleventh year of Zedekiah the son of Josiah king of Judah,* ***unto the carrying away of Jerusalem captive in the fifth month***". (Jeremiah 1:3)

The fifth month is "Av".

After the first deportation, the next big event was the burning of the first Temple (586B.C.). Solomon's Temple…

"*And in the* ***fifth month****, on the* ***seventh day*** *of the month, which is the nineteenth year of king Nebuchadnezzar king of Babylon, came Nebuzaradan, captain of the guard, a servant of the king of Babylon, unto Jerusalem:*" (2Kings25:8)

"*Now in the fifth month, in the* ***tenth day*** *of the month, which was the nineteenth year of Nebuchadnezzar king of Babylon, came Nebuzaradan, captain of the guard, which served the king of Babylon, into Jerusalem*", (Jeremiah 52:12)

"*And burned the house of the LORD, and the king's house; and all the houses of Jerusalem, and all the houses of the great men, burned he with fire*": (Jeremiah 52:13)

The Talmud explains why one passage states the 7th and another says the 10th… (The Talmud is **not** inspired by God.)

"How then are these dates to be reconciled? On the seventh the heathens entered the Temple and ate therein and desecrated it throughout the seventh and eighth and towards dusk of **the ninth they set fire to it and it continued to burn the whole of that day.** ... How will the Rabbis then [explain the choice of **the 9th as the date**]? The beginning of any misfortune [**when the fire was set**] is of greater moment". – (Talmud Ta'anit 29a)

My own observations confirms the Talmud account by comparing these two passages in the following way; First; (2Kings25:8) only tells us what day Nebuchadnezzar's General arrived; Second; (Jeremiah 52:13) tells us when the burning of the Temple was complete. This is consistent with the Talmud account, Josephus's account and Jewish tradition. That is why the 9th of Av is the accepted day for the burning of the first Temple. The fire was set at sundown that begins the 9th of Av, it burned all night and the following day. The following sundown marks the end of the 9th and the beginning of the 10th of Av when the fire was burned out.

The **9th of Av**, known as **Tisha B'Av**, or "**The Fast of the Ninth of Av**". Is believed by many Jews to be a day set aside by God for the chastisement of Israel. Therefore the day is considered to be a day of mourning, meditation and remembrance of **national tragedies** that struck Israel on that day throughout its long and carefully documented History.

A tradition of 3 weeks of mourning the exile of Israel and the destruction of both of their Holy Temples begins on the **17th of the Hebrew month of Tammuz**. The 17th of Tammuz is a day of fasting followed with Three Weeks of gradually-increasing mourning, climaxing on the 9th of Av. The 9th of Av - Tisha B'Av – is the date on which both the First and Second Temples were destroyed, roughly 2,500 and 2,000 years ago, respectively.

It is also believed by Jews to be the day that 10 of the 12 spies sent into the Promised Land returned with a false report. (Numbers 13 and 14)

Joking, Smiling and laughing are not permitted on this mournful, and solemn day.

The 17th of Tammuz is the day on which Nebuchadnezzar and his Babylonian forces breached the walls of Jerusalem after many months of siege. For 21 days the Babylonian armies fought their way through the streets of Jerusalem. The 17th of Tammuz is also the day Moses broke the tablets the 10 commandments were written on when he found the Israelites worshiping a golden calf.

Just like the 9th of Av, the 17th of Tammuz is associated with a number of events. Coincidentally or not, July 4, 1776 also fell on the 17th of Tammuz.

This tradition of three weeks of mourning began after the first exile. After the 70 year Babylonian exile was complete, the Israelites consulted with the priests and Prophets about continuing this tradition of mourning in the 5th month of Av…

"And to speak unto the priests which were in the house of the LORD of hosts, and to the prophets, saying, ***Should I weep in the fifth month****, separating myself, as I have done these so many years"?* (Zechariah 7:3)

Here is the LORD'S answer…

"Then came the word of the LORD of hosts unto me, saying, (Zechariah 7:4)

"Speak unto all the people of the land, and to the priests, saying, When ye fasted and mourned in the fifth and seventh month, even those seventy years, did ye at all fast unto me, even to me?" (Zechariah 7:5)

"And when ye did eat, and when ye did drink, did not ye eat for yourselves, and drink for yourselves?" (Zechariah 7:6)

"Should ye not hear the words which the LORD hath cried by the former prophets, when Jerusalem was inhabited and in prosperity, and the cities thereof round about her, when men inhabited the south and the plain? (Zechariah 7:7)

God is telling Zechariah that Israel did not have their heart right toward God. Their traditions had selfish motivations. There was no acknowledgement of their guilt or remorse. In summation, if their heart is right then the tradition is right. If not, then it is all wrong. For the most part, their hearts were not right. However, other passages tells us that

God will change that with affliction in the End Times. That is definitely near-future.

25.5 The 9th Of Av, "A Day To Recall Israel's National Tragedies"

Jewish tradition assigns the 9th of Av as the day to remember a number of National tragedies throughout their history. Why so many National tragedies have plagued them on this day is a question that we will explore here.

Here is a partial list of significant and tragic events that Jewish Rabbis have assigned to the 9th of Av. This list is compiled from various Jewish websites and is a General representation of Jewish perspective about the 9th of Av. Some of these dates have been questioned because of differences between the Julian calendar and the Gregorian calendar.

Spies return with evil reports of giants in the Promised Land. They said that it could not be conquered. According to Jewish tradition, it was on the evening of the 8th of Av that the spies returned with their evil report. Sundown began the 9th of Av. Out of fear, the Jews give up hope of entering the Promised Land. As a result, they cried all of that night and the following day which was the 9th of Av and rebelled against God. They said that they would rather return to Egypt than be slaughtered by the Canaanites. (Numbers 13, 14) **According to the Talmud and Jewish tradition**, God set aside the 9th of Av to punish Israel because they mourned without cause on that day. (**Note**: The Talmud and Jewish tradition are not inspired by God)

The First Temple, Solomon's Temple was destroyed (586B.C.) on the 9th of Av. Under Nebuchadnezzar, the Babylonians invaded Israel, burned the Temple and many Jews were killed. The remaining population was taken captive to Babylon and Persia.

The Second Temple was also destroyed on the 9th of Av by the Romans (70A.D.). Two and a half million Jews die because of war, famine and pestilence. Over a million more are scattered to all parts of the Roman Empire just like the Bible predicted. Jews were tortured; killed in pagan rituals and celebrations; killed in gladiatorial games and sold as slaves right along with the Christians. At that time the Jews were surprised that the second Temple was destroyed on the same day as the first Temple, the 9th of Av.

The Simon bar Kochba revolt crushed on the 9th of Av (132A.D.). Bar Kokhba took up refuge in the fortress of Betar. After three and a half years of siege and fighting, the Romans finally destroyed it. Over 100,000 were killed. **Simon bar Kochba** was believed by some to be the Messiah. He created an independent Jewish state and forced Jews to join his army or be punished. He ruled about three years. According to Cassius Dio**,** a Roman statesman and historian of Greek origin, overall, about 580,000 Jews were slaughtered by the Romans in the most crewel fashions. According to an account given by Rabbi Yohananby, on one large stone was found the brains of over 300 children. For the most part, the students in the many schools were wrapped up in their own books (scrolls) and burned alive. (Midrash Rabbak, Lamentations Rabba 2:5) The Talmud, refers to Bar Kokhba as "Ben-Kusiba," a derogatory term that implies that he was a false Messiah.

Turnus Rufus ploughs site of Temple on the 9th of Av. (133A.D.) Romans build pagan city of Aelia Capitolina on site of Jerusalem.

First Crusade declared by Pope Urban II. (1095A.D.) 10,000 Jews killed in first month of Crusade. Crusades bring death and destruction to thousands of Jews, totally obliterate many communities in Rhineland and France.

Expulsion of Jews from England, (1290 CE) On July18, King Edward issued the Edict of Expulsion accompanied by massacres and confiscation of books and property. One official reason for the expulsion was that Jews had declined to follow the Statute of Jewry. July 18 was the 9th of Av on the Hebrew calendar.

Inquisition in Spain and Portugal (1492A.D.) culminates in the expulsion of the Jews from the Iberian Peninsula. Families were separated; many die by drowning and there was a massive loss of property. The Jews were given until July 30 to leave the country or face execution.

Important note about the calendar: if you use a Jewish calendar converter to check dates, you will find that some of these dates prior to 1582 are based on the Julian calendar and are a few days before the 9th of Av. These events occurred before Pope Gregory XIII introduced the Gregorian calendar reform. The Gregorian calendar changed the way secular calendars work. Date converters do not take this into account, which causes the discrepancy. Another thing to keep in mind is that a Jewish day is preceded by night which begins at sundown.

Britain and Russia declare war on Germany on the 9th of Av. (1914A.D.) First World War begins. First World War issues unresolved, ultimately causing Second World War and Holocaust. 75% of all Jews in war zones. Jews in armies of all sides - 120,000 Jewish casualties in armies. Over 400 massacres immediately following war in Hungary, Ukraine, Poland and Russia.

Deportations from Warsaw Ghetto to the Treblinka concentration camp begin. (1942A.D.) The deportation began on July 22, at sundown. The Hebrew day, "the 9th of Av" began, but it was still the 22nd (Greg.) until midnight. The Hebrew night begins at sundown. Their night precedes their day. A date converter will show the 22nd as the 8th of Av. At sundown the Jewish victims were on trains and on their way to the Treblinka concentration camp (9th of Av), but it was still the 22nd until midnight on the Gregorian calendar.

Iraq walks out of talks with Kuwait. (1989A.D.)

The Jewish community center in Buenos Aires, Argentina is bombed. Eighty six people were killed and hundreds more wounded. **(9th of Av 1994A.D.)**

This is a list compiled by searching the 9th of Av on the internet. All of my sources are Jewish websites. My favorite sources are as follows;

Judaism 101 http://www.jewfaq.org/holidayd.htm

chabad.org

http://www.chabad.org/library/article_cdo/aid/946703/jewish/What-happened-on-the-Ninth-of-Av.htm

Ohr Somayach International https://ohr.edu/1088

Conspicuously absent from the above compiled list is the day that King Nebuchadnezzar showed up at Israel's front door, **Jerusalem 606B.C., on the 9th of Av!** That is the day that God's prophesied punishment of Israel began!

It seems to me that God is making a connection between the punishment that commenced in 606B.C. on the 9th of Av and the destruction of both Temples which also occurred on the 9th of Av. That connection takes us back to the prophecy of (Leviticus 26) which in turn ties Israel's history to God's prophesied punishment for not keeping the Sabbatical year which began on the 9th of Av, 606B.C!

Israel does not acknowledge their disobedience or God's punishment. At least not yet. That is why I believe they don't consider the dates that involve this prophecy.

The Bible said that God would bring Israel back to their land in unbelief and that is what has happened. Jesus said that he would not return until Israel acknowledges their sins, the sins of their fathers and that God has justly punished them for it. (Leviticus 26:40) As I have already stated that has not happened **yet.** Israel has not acknowledged their sins or God's punishment, but the Bible says that they will during the 70th week of Daniel.

"In their affliction they will seek me early". (Hosea 5:15)

Now that we are talking about the Lord's return again, let's consider what this prophecy reveals about that. First of all, this prophecy tells us when the "Latter Days" begin. In fact, the term is defined in (Hosea 3:5) as when Israel returns to their land. But there seems to be more to this prophecy than just the obvious.

Although this prophecy begins with Nebuchadnezzar showing up **(606B.C.)** on the **9th of Av** and counting 70 Hebrew years (which equals 68.99 Roman years) for the first phase of this judgment which ends in 537B.C., there is also a 70 Hebrew year interval between the first deportation **(605B.C.)** on the **9th of Av** and the beginning of the construction of the second Temple (536B.C.). Furthermore, there is a 70 year interval between the destruction of the first Temple **(586 B.C.)** on the **9th of Av** and the completion of the second Temple (515B.C.) (Ezra 6:15 and Dr. Constable's Notes on Daniel page 117) Also… (The Coming Prince, page 227)

25.6 These Facts Reveal A Distinct 70 Year Pattern.

Keep in mind that 70 Hebrew years equals **69 Roman years.**

In **606 B.C**. on the **9th of Av** the Babylonian siege of Jerusalem began. Seventy Hebrew years later, (537B.C.), the first wave of Jewish captives left Babylon to return to Israel and it also happened on the **9th of Av**! In essence God undoes what he did 70 years earlier.

On the Hebrew calendar that day was also the **9th of Av**. If you count 360 days from the time that Nebuchadnezzar first arrived at Jerusalem's front door, that would bring you to (**605 B.C**.), on the **9th of Av** on the Hebrew calendar. It is no coincidence that this was the day that Nebuchadnezzar left Jerusalem with all of his loot also known as the **first deportation**.

In (**605 B.C.**) on the **9th of Av,** the first captives left Jerusalem and were taken to Babylon which includes Daniel, Shadrach, Meshach and Abednego. A thousand of the most important captives were taken which also included the priests, and all of the Temple gold and furnishings. Seventy Hebrew years later (**536B.C.**), on the **9th of Av,** God undoes what he did and construction of the 2nd Temple begins.

In (**586 B.C.**) on the **9th of Av**, the first Temple is burned. Seventy Hebrew years later (**515B.C.**), the second Temple is completed. Once again, for the third time, God undoes what he started 70 years earlier and this is all a part of the punishment prophesied by God against Israel.

Now that the 2483.78 (Roman) solar years of punishment are complete, what do we find if we count 70 Hebrew/lunar years from May 14,1948? (68.99 Roman years) Do we start counting on that day or do we start counting on the next Rosh Hashanah? When we start counting is an important issue. Another issue to consider is whether to use the Hebrew calendar or the Roman calendar. There is a possibility that the Roman calendar may be involved because we are using a Roman system and we are looking at a revival of the Roman Empire. We will just have to wait and see.

We have a pattern here. If you think about this, you may wonder if there is anything else that needs to be undone concerning these prophecies about Israel's punishment. Well, there is.

The next event on God's Prophetic Calendar is the removal of the Church at the end of Rosh Hashanah, just **before** the Seventieth Week of Daniel begins. That is the same moment that I expect Israel to begin construction of the Third Temple. It will take three days. The seventieth week of Daniel, also known as the Tribulation, should begin with a fully functioning third Temple and the Church gone for eight days.

The punishment concerning the land was completed May 13, 1948. The very next day Israel was recognized as a Nation. If you count 70 Hebrew years (69 Roman years) from May 14, 1948 that brings you to 2017. Count 70 Hebrew years (69 Roman years) from the end of the Blood Moon Tetrad that followed the rebirth of Israel in 1948. That would be the Blood Moon Tetrad that occurred from April 13, 1949 to September 26, 1950 that brings you to 2019.

But if it doesn't happen by 2019, then count 70 Hebrew years from a different starting point/event such as the period of the Blood Moon Tetrad

On May 14, 2018, Prime Minister B. Netanyahu dedicated the new U.S. Embassy. In his dedication speech, Prime Minister Netanyahu called it the **70th anniversary of Israel's Independence.** The Sanhedrin priests issued a half shekel Temple coin to commemorate what they called the 70th year anniversary. On one side of the coin was the Temple and on the other side was a profile of Trump in front of the Persian King credited with giving the order for the Jews to go and rebuild the second Temple.

If God is using the 9th of Av as a day to remind Israel of past sins and punishments and they are not yet acknowledging it, what is God going to do next? God says that Israel will seek Him under affliction in the last days and here we are. This 70 year interval between the beginning and ending of tasks concerning Israel may not be finished yet.

In his book "The Coming Prince" Sir Robert Anderson lists a series of prophetic cycles in Appendix 1 on page 227. There is a pattern of about 490 years between several associated events.

From the entrance into Canaan (B.C.1586) to the establishment of the kingdom under Saul (B.C. 1096) was about 490 years. From the Kingdom under Saul (B.C. 1096) to Nebuchadnezzar showing up to conquer Jerusalem (B.C, 606) was about 490 years. The time between the dedication of the first Temple (1005 B.C.) known as Solomon's Temple and the second Temple (515 B.C.) was also about 490 years.

There is also a pattern of 70 year intervals between associated events. Remember that 70 Hebrew years equals 68.99 Roman years. Rounded off, that equals 69 years.

However you count the years, the Prime Minister of Israel, a veteran of the six day war, and the Sanhedrin both publicly acknowledged May 14, 2018 as the 70th anniversary of Israel's Independence. They must by counting solar years <u>or</u> using a different starting date.

25.7 On May 14, 1948 The "Latter Days" Began.

"*For the children of Israel shall abide* ***<u>many days</u>*** *without a king, and without a prince, and without a sacrifice (No Temple), and without an image, (Idols) and without an ephod (Priesthood), and without teraphim"* (Household gods) (Hosea 3:4)

The term "***many days***" refers to the time when Israel would be without a priesthood; without a sacrifice, which means **<u>no Temple</u>**; no king or heir to the throne and no more household gods or idol worship either.

Throughout their long history, Israel has always had a sacrifice and God was continually reprimanding them for going after false gods and bowing down to images or idols, but all of that ended in 70AD just like the Bible said.

Israel has been without these things that God mentioned since the 9th of Av (Hebrew) 70 A.D. when Rome destroyed Jerusalem and the Temple. Since the destruction of the temple there has been no King; No heir to the throne; no priesthood; no temple sacrifice and no household gods. Since 70 A.D. you could not find a Jew bowing down to anything, anywhere. Israel has been without these things for ***"many days"*** now for about two thousand years.

*"**<u>Afterward</u>** shall the children of Israel **<u>return</u>**, and seek the LORD their God, and David their king; and shall fear the LORD and his goodness in the* ***<u>latter days</u>***". (Hosea 3:5)

It is important to remember that Hosea was written before Israel was taken off their land the first time. But the Bible talks about a **second time**!

"And it shall come to pass in that day, that the Lord shall set his hand again the ***second time*** *to recover the remnant of his people"*, (Isaiah 11:11)

This second return will happen in the ***"latter days"***. On May 14, 1948, that is exactly what happened. We have been living in the ***"latter days"*** ever since!

The Bible prophesied that Israel would return to their land (1st time); rebuild their Temple; and the Messiah would be presented to them 69 weeks of years (69 X 7 = 483 years) from the time that the commandment to rebuild and restore the walls of Jerusalem was given. In the twentieth year of his reign; on the 1st day of Nisan (Heb.) or March 14, 445 B.C. (Greg.), **Artaxerxes** gave the command! That was prophesied as well.

At that time the Messiah, who is God in flesh, would be rejected by his own people…

"And one shall say unto him, What are these wounds in thine hands? Then he shall answer, Those with which I was wounded in the house of my friends" (Zechariah 6:13)

Even the details of Christ's crucifixion were told in advance.

As a result, God would return to his place and Israel would be scattered until the last days. After they return in the ***"last days"*** God will afflict them until they seek him and acknowledge their offence.

"I will go and return to my place, till they acknowledge their offence, and seek my face: in their affliction they will seek me early". (Hosea 5:15)

This Old Testament passage is God speaking to Israel about their future rejection of himself and is similar to the words spoken later by Jesus (John 14:3) after he was rejected by Israel. So if this is God speaking, when did he visit Israel; get rejected and return to heaven? The answer to that question is obvious.

"And if I go and prepare a place for you, I will come again, and receive you unto myself; that where I am, there ye may be also". (John 14:3)

They killed their King and Messiah in 32A.D. and the Church age began. Shortly afterward, Israel was judged again. (Daniel 9:26) prophesied that the Temple would be destroyed after the Messiah was cut off, but not for himself. Before Jesus was crucified, he promised to return for the Church to take them back to his prepared place in Heaven. This is a reference to the Rapture which marks the end of the Church Age and the beginning of the 70th week of Daniel.

The Temple was destroyed, the Jews were scattered and now God has brought them back just like the Bible said!

Continuing with (Hosea 5:15)…

"Come, and let us return unto the LORD: for he hath torn, and he will heal us; he hath smitten, and he will bind us up" (Hosea 6:1).

"***After*** ***two days*** *will he revive us: in the* ***third day*** *he will raise us up, and* ***we shall live in his sight***" (Hosea 6:2)

Here the third day is tied to Christ living in Jerusalem. That doesn't happen until the Millennium which is defined in (Revelation 20: 4-7) as a thousand years. If the third day is defined as 1,000 years, then the first two days mentioned must also be a thousand years each. That is **2,000 Hebrew years**.

A Hebrew year is based on the lunar cycle, it only has 360 days. 2,000 Hebrew years can be converted to Roman years by multiplying by 360(days in a Hebrew year) and dividing by 365.25 (days in a Roman year) which equals 1,971.25 Roman years. (2,000 X 360 divided by 365.25)

(Hosea 5:15) is a reference to Christ being rejected. (Hosea 6:1) is the very next verse. Remember the chapter and verse divisions are not inspired by God, they were put there by man and usually break in the wrong places in such a way as to obscure the meaning. (Hosea 5:15) tells us when to start counting and (Hosea 6:1and2) tells us how many years to count. So if we start counting on the year of Christ's crucifixion and we count 2,000 (Hebrew) years.

If we consider the time between 32A.D. and the Rapture as the time where God would turn to the Gentiles to spread the gospel which is also known as the Church Age, then we would start counting that time beginning with Pentecost the year Christ was crucified, 32 A. D. If we start counting the years at 32 A. D. then that takes us to about 2,003. That is the time for the completion of this "***many days***" spoken of in (Hosea 3:4, 5) which says that the Jews shall return "***afterward***" and "in the latter days".

May 14, 1948 is considered Israel's rebirth. That was the day after their punishment away from their land was complete.

The bottom line of all of this is that we have been living in the "**latter days**" since May 14, 1948 and the rapture has been imminent since about 2003. Look at what has been happening since that time!

"For, lo, the days come, saith the LORD, that I will bring again the captivity of my people Israel and Judah, saith the LORD: and ***I will cause them to return to the land that I gave to their fathers, and they shall possess it".*** (Jeremiah 30:1-3)

Israel has returned right on time! God keeps his promises on time!

25.8 It Is Kind Of Like A Love Affair Gone Bad

God has made promises to Abraham about his descendants, but the children turned out naughty! God wants to bless them but they keep doing things that God has said not to do!

God treats Israel like a wayward Son and an adulterous wife!

God scatters the Jews and turns to the Gentiles for 2,000 years "***to provoke them to jealousy"*** (Romans 11:11)

*"They have moved me to jealousy with that which is not God; they have provoked me to anger with their vanities: and **I will move them to jealousy with those which are not a people**; I will provoke them to anger with a foolish nation."* (Deuteronomy 32:21)

*"But I say, Did not Israel know? First Moses saith, **I will provoke you to jealousy by them that are no people**, and by a foolish nation I will anger you"* (Romans 10:19)

*"I say then, Have they stumbled that they should fall? God forbid: but rather through their fall salvation is come unto the Gentiles, **for to provoke them to jealousy**"* (Romans 11:11)

For the last 2,000 years God has used Gentiles to dispense the Gospel throughout the world. This position of honor was intended to be Israel's! For the last 2,000 years this honor was taken from them and given to the Gentiles. This was…*"**to provoke them to jealousy**"* (Romans 11:11).

I believe the Church Age is about over. At the time of this writing, I believe the Church is almost ready for the Groom and the 70th week of Daniel is about to begin!

God is preparing Israel for the position that he promised Abraham. God keeps his promises!

(Hosea 3:5) (Jeremiah 30:1-3; 23:7-8)

25.9 God Said That Israel Would Be Scattered And Returned.

Ezekiel 36:19-36 (verse19) *"**<u>And I scattered them among the heathen</u>**, and they were dispersed through the countries: according to their way and according to their doings **I judged them**"* (verse20) *"And when they entered unto the heathen, whither they went, they profaned my holy name, when they said to them, These are the people of the LORD, and are gone forth out of his land"* (verse21) *"But I had pity for mine holy name, which the house of Israel* *had profaned among the heathen, whither they went"* (verse22) *"Therefore say unto the house of Israel, Thus saith the Lord GOD; I do not this for your sakes, O house of Israel, but for mine holy name's sake, which ye have profaned among the heathen, whither ye went"* (verse23) *"And I will sanctify my great name, which was profaned among the heathen, which ye have profaned in the midst of them; and the heathen shall know that I am the LORD, saith the Lord GOD, when I shall be sanctified in you before their eye"* (verse24) ***"For I will take you from among the heathen, and gather you out of all countries, and will bring you into your own land"***. (verse25) *"Then will I sprinkle clean water upon you, and ye shall be clean: from all your filthiness, and from all your idols, will I cleanse you"*. (verse26) *"A new heart also will I give you, and a new spirit will I put within you: and I will take away the stony heart out of your flesh, and I will give you an heart of flesh"*. (verse27) *"And I will put my spirit within you, and cause you to walk in my statutes, and ye shall keep my judgments, and do them"* (verse28) *"And ye shall dwell in the land that I gave to your fathers; and ye shall be my people, and I will be your God"* (verse29*)" I will also save you from all your uncleannesses: and I will call for the corn, and will increase it, and lay no famine upon you"* (verse30) *"And I will multiply the fruit of the tree, and the increase of the field, that ye shall receive no more reproach of famine among the heathen"* (verse31) *"Then shall ye remember your own evil ways, and your doings that were not good, and shall lothe yourselves in your own sight for your iniquities and for your abominations"*. (verse32) *"Not for your sakes do I this, saith the Lord GOD, be it known unto you: be ashamed and confounded for your own ways, O house of Israel"*. (verse33) *"Thus saith the Lord GOD; In the day that I shall have cleansed you from all your iniquities I will also cause you to dwell in the cities, and the wastes shall be builded"*. (verse34) *"And the desolate land shall be tilled, whereas it lay desolate in the sight of all that passed by"*. (verse35) *"And they shall say, This land that was desolate is become like the garden of Eden; and the waste and desolate and ruined cities are become fenced, and are inhabited"* (verse36) *"Then the heathen that are left round about you shall know that I the LORD build the ruined places, and plant that that was desolate:* ***I the LORD have spoken it, and I will do it"*** (Ezekiel 36: 36)

God speaks with certainty! He does what He says and He warns **everyone** in advance!

God said in no uncertain terms, that he would bring his people back to their land and they would keep it. God keeps his word.

"For, lo, the days come, saith the LORD, that I will bring again the captivity of my people Israel and Judah, saith the LORD: and I will cause them to return to the land that I gave to their fathers, and ***they shall possess it****"*(Jeramiah 30:3)

25.10 He'll Be Back!

"I will ***go*** *and return to my* ***place****,* ***till they acknowledge their offence****, and seek my face: in their affliction they will seek me early"*. (Hosea 5:15); 780BC

After he was rejected, Jesus Christ spoke those same words about 800 years later in 32 AD!

"In my Father's house are many mansions: if it were not so, I would have told you. ***I go to prepare a place for you.****"* (John 14:2)

"And if I go and prepare a ***place*** *for you,* ***I will come again, and receive you unto myself****; that where I am, there ye may be also"* (John 14:3).

God spent a considerable amount of verbiage on scolding Israel throughout the Old and New Testaments. He also said that he would ***"go"***, ***"till they acknowledge their offence"*** (Hosea 5:15) 780 BC

Well this is talking about Yeshua and it says that he isn't coming back for Israel until they admit their offence and **the offence of their fathers**. But that is not all. Israel must admit that they were punished by God and he did it justly! (Leviticus 26:40-42)

*"****If they shall confess*** *their iniquity, and the iniquity of their fathers, with their trespass which they trespassed against me, and that also they have walked contrary unto me;"* **(Leviticus 26: 40)**

"And that I also have walked contrary unto them, and have brought them into the land of their enemies; ***if*** *then their uncircumcised hearts be humbled,* ***and*** *they then accept of the punishment of their iniquity"* (Leviticus 26: 41)

*"****Then*** *will I remember my covenant with Jacob, and also my covenant with Isaac, and also my covenant with Abraham will I remember; and I will remember the land"*.(Leviticus 26: 42)

We are talking about Israel having a major attitude change toward God. God says that …"***in their affliction they will seek me early***"! (Hosea 5:15)

"I will go and return to my place, till they acknowledge their offence, and seek my face: ***: in their affliction they will seek me early****"*. (Hosea 5:15)

According to this passage God has already visited Israel. This was written hundreds of years before Christ. It is prophetic in nature in the sense that Jesus Christ spoke those very same words when he was here…

"And if I go and prepare a place for you, I will come again, and receive you unto myself; that where I am, there ye may be also" (John 14:3)

So ask yourself, when did God visit Israel?

"Jesus saith unto him, Have I been so long time with you, and yet hast thou not known me, Philip? He that hath seen me hath seen the Father; and how sayest thou then, Shew us the Father?" (John 14:9)

God visited Israel in the human form of Jesus Christ. He has said that he will not return until Israel admits their offence. (Hosea 5:15) Israel would seek God early when they are under affliction. The 70th week of Daniel will be a time of affliction like never before or ever will be again! The last half or three and a half years of the Tribulation will be the worst for Israel.

"For I say unto you, Ye shall not see me henceforth, till ye shall say, Blessed is he that cometh in the name of the Lord". (Matthew 23:27)

God is going to bring more affliction to Israel in particular and the whole world in general, to teach them to trust and obey Him.

This offence goes back quite a way, but I believe that it came to a climax with the crucifixion of Christ!

26

Seven Dispensations Mentioned In The Bible.

26.1 The 1st Dispensation; (Dispensation Of Innocence).

(Genesis 1:28-30 & 2:15-17).**This is the first and the shortest dispensation.**

(1) Dispensation of Innocence from the creation of Adam in (Genesis 2:7) to the expulsion from Eden (Genesis 3: 6) About 6,000 years ago or from 4,000 BC. Till Adam sinned.

"By C. I. Scofield

"This dispensation extends from the creation of Adam in Genesis 2:7 to the expulsion from Eden. Adam, created innocent and ignorant of good and evil, was placed in the Garden of Eden with his wife, Eve, and put under responsibility to abstain from the fruit of the tree of the knowledge of good and evil. The dispensation of innocence resulted in the first failure of man, and in its far-reaching effects, the most disastrous. It closed in judgment": "So he drove out the man." See (Genesis. 1:26; Genesis. 2:16, 17; Genesis. 3:6; Genesis. 3:22-24.)

26.2 The 2nd Dispensation; (Dispensation Of Conscience)

IT LASTED FROM THE TIME OF ADAM'S FALL UNTIL THE FLOOD (Genesis 3:8–8:22)

(2) Dispensation of Conscience from the time of Adam's fall until the flood (Genesis 3:8–8:22) from about ? BC till about 2400 BC (About ? years)

"By C. I. Scofield"

"By the fall, Adam and Eve acquired and transmitted to the human race the knowledge of good and evil. This gave conscience a basis for right moral judgment, and hence the race came under this measure of responsibility-to do well and eschew evil. The result of the dispensation of conscience, from Eden to the flood (while there was no institution of government and of law), was that "all flesh had corrupted his way on the earth," that "the wickedness of man was great in the earth, and that every imagination of the thoughts of his heart was only evil continually," and God closed the second testing of the natural man with judgment: the flood. See (Genesis. 3:7, 22; Genesis. 6:5, 11-12; Genesis. 7:11-12, 23.)"

26.3 The 3rd Dispensation; (Human Government).

It began in (Genesis 8). Right after the Flood and lasted till Abraham. (Genesis 17: 5)

"By C. I. Scofield"

"Out of the fearful judgment of the flood God saved eight persons, to whom, after the waters were assuaged, He gave the purified earth with ample power to govern it. This, Noah and his descendants were responsible to do. The dispensation of human government resulted, upon the plain of Shinar, in the impious attempt to become independent of God and closed in judgment: the confusion of tongues. (See Genesis. 9: 1, 2; Genesis. 11: 1-4; Genesis. 11:5-8.)"

26.4 The 4th Dispensation: (Dispensation Of Promise)

(Genesis 12:1) (Exodus 19:25).From Abraham till Moses; (From about 1,900 BC till about 1600 BC)

"By C. I. Scofield"

Out of the dispersed descendants of the builders of Babel, God called one man, Abram, with whom He enters into covenant. Some of the promises to Abram and his descendants were purely gracious and unconditional. These either have been or will yet be literally fulfilled. Other promises were conditional upon the faithfulness and obedience of the Israelites. Every one of these conditions was violated, and the dispensation of promise resulted in the failure of Israel and closed in the judgment of bondage in Egypt.

The book of Genesis, which opens with the sublime words, "In the beginning God created," closes with, "In a coffin in Egypt." (See Genesis. 12:1-3; Genesis. 13:14-17; Genesis. 15:5; Genesis. 26:3; Genesis. 28:12-13; Exodus 1: 13-14.)

26.5 The 5th Dispensation; (Dispensation Of Law)

From Moses till Pentecost 32 A.D. and the 70th week of Daniel which will commence on the year of the Rapture, at the Feast of Atonement.

"Man under Law" By C. I. Scofield

"Again the grace of God came to the help of helpless man and redeemed the chosen people out of the hand of the oppressor. In the wilderness of Sinai He proposed to them the covenant of law. Instead of humbly pleading for a continued relation of grace, they presumptuously answered: "*All that the Lord hath spoken we will do.*" The history of Israel in the wilderness and in the land is one long record of flagrant, persistent violation of the law, and at last, after multiplied warnings, God closed the testing of man by law in judgment: first Israel, and then Judah, were driven out of the land into a dispersion which still continues. A feeble remnant returned under Ezra and Nehemiah, of which, in due time, Christ came: "Born of a woman-made under the law." Both Jews and Gentiles conspired to crucify Him. (See Exodus 19:1-8; 2 Kings 17:1-18; 2 Kings 25: 1 -11; Acts 2:22-23; Acts 7:5152; Romans 3:19-20; Romans10:5; Galatians 3: 10.)"

26.6 The 6th Dispensation; (Dispensation Of Grace/ Church Age)

"*If ye have heard of the* ***dispensation of the grace of God*** *which is given me to you-ward*" (Ephesians 3:2)

The Dispensation of Grace is the Dispensation that we are currently in. It is mentioned by name in (Ephesians 3:2) and referred to as a "***mystery***" in (Ephesians 3: 3, 4) and (Colossians 1:26). Furthermore In (Ephesians 3: 5) and (Colossians 1:26)…it says that this "***Mystery***" Dispensation was not revealed in **previous Ages**, but now it is. So now we have more "Ages" or Dispensations mentioned in the past.

"*How that by revelation he made known unto me the* ***mystery****; (as I wrote afore in few words*", (Ephesians 3:3)

"*Whereby, when ye read, ye may understand my knowledge in the* ***mystery*** *of Christ"* (Ephesians 3:4)

"*Which in* ***other ages*** *was not made known unto the sons of men, as it is now revealed unto his holy apostles and prophets by the Spirit*"; (Ephesians 3:5)

"*Even the* ***mystery*** *which hath been hid from* ***ages*** *and from Generations, but now is made manifest to his saints*" (Colossians 1:26)

The Dispensation of Grace is sandwiched between the 69th and 70th weeks of Daniel. (Daniel 9:25, 27) …"*from the going forth of the commandment to restore and to build Jerusalem* ***<u>unto</u>*** *the Messiah the Prince shall be seven weeks, and threescore and two weeks*": The word used and highlighted here is "***<u>unto</u>***" which expresses an exact point in time.

As we have already discussed in chapter 24, From the day that **Artaxerxes Longitmanus** gave the command to rebuild the walls of Jerusalem on Nissan 10, 445 B.C., **unto** the Messiah was **exactly** 69 weeks of Hebrew years. That can be counted **exactly to the day** that Jesus rode into Jerusalem on a donkey on Palm Sunday. That was when Jesus was first publicly declared to be the King of Israel, the Messiah, just like the Bible Prophesied! The Romans had that title posted on his cross above his head.

(Sir Robert Anderson intelligence officer, Scotland Yard. He authored the book titled "The Coming Prince" that documents these dates and facts.)

That week Jesus was crucified, on the first Feast of Passover; The next Feast began at sundown and that is when Jesus was buried, on the Feast of Unleavened Bread;

Exactly 72 hours later, Jesus rose from the dead on the Feast of First Fruits; Exactly Fifty days after that, the Age of Law was put on hold at the completion of the 69th week of years which was marked by the Feast of Pentecost 32 A. D. and the Church Age began.

It is important to note that the Feast of Pentecost is considered by Jews to represent the day of the "giving of the Law". (See link to Judaism 101) http://www.jewfaq.org/holidayc.htm

It makes sense to measure the Hebrew years of law from the beginning at the Feast of Pentecost and therefore to mark the end of the 69th week of years and the beginning of the following Dispensation with the Feast of Pentecost also. And that is exactly what the Bible records! The Bible is mathematically accurate!

The Dispensation of Grace immediately follows the 69th week of Daniel. The 69th week of Daniel is part of the Age or Dispensation of Law. The Dispensation of Law was put on hold and the Dispensation of Grace began in 32 A.D. on the Feast of Pentecost which is the 4th Feast; or the last of the four Spring Feasts. The Age of Grace began on Pentecost, 32 A.D. and divides the dispensation of law between the 69th and the 70th weeks prophesied by Daniel 9:25 by **about** 2,000 Hebrew years or 1,971.25

Roman years. The Age of Law is not completed yet. Ten days after the Age of Grace ends on the 5th feast, the last 7 years of the Age of Law will begin to be counted again starting on the 6th Feast.

As we have already discussed, Hosea prophesied about this time when Israel would be dispersed and said that it would end **after** 2,000 years. The 2,000 Hebrew years expired about 2,003. Since that time the Rapture could happen any year.

"*If ye have heard of the* ***dispensation of the grace of God*** *which is given me to you-ward:*" (Ephesians 3:2)

The Church Age is unique from all other ages by the way God deals with Saints. All Church Age saints are sealed with the Holy Spirit.

"*But ye are not in the flesh, but in the Spirit, if so be that the Spirit of God dwell in you. Now if any man have not the Spirit of Christ, he is none of his*" (Romans 8:9)

During the Church Age, all believers whether they by Jews or Gentiles are baptized into one body of saints called the Church.

"For by one Spirit are we all baptized into one body, whether we be Jews or Gentiles, whether we be bond or free; and have been all made to drink into one Spirit" (1Corinthians 12:13)

The Church Age will end at the last trump of Rosh Hashanah some year **after** the 2,000 years are completed. The Church Age ends and the "Day of the Lord" begins at the Rapture. The Church **and** the Holy Spirit will **both** be taken from the earth (2 Thessalonians 2:1-8) and the Church will be judged in heaven. (Revelation 4:1- 4) describes this event.

Eight days after the Rapture, the 70th week of Daniel will begin on Yom Kippur which is the last 7 years of the age of Law. This is the 7 years of Tribulation described by Daniel and Revelation. After the 70th week of Daniel, follows the Millennial Kingdom of Christ.

"By C. I. Scofield"

"*Verily, verily, I say unto you, He that believeth on me hath everlasting life*" (John 6:47). "*Verily, verily, I say unto you, He that heareth my word, and believeth on him that sent me, hath everlasting life, and shall not come into condemnation; but is passed from death unto life.*" (John 5:24). "*My sheep hear my voice, and I know them, and they follow me: and I give unto them eternal life; and they shall never perish*" (John 10:27-28). "*For by grace are ye saved through faith; and that not of yourselves: it is the gift of God: Not of works, lest any man should boast*" (Eph. 2:8-9).

The predicted result of this testing of man under grace is judgment upon an unbelieving world and an apostate church. (See Luke 17:26-30; Luke 18:8; 2 Thess. 2:7-12; Revelation 3:15-16.)

The first event in the closing of this dispensation will be the descent of the Lord from heaven, when sleeping saints will be raised and, together with believers then living, caught up "*to meet the Lord in the air: and so shall we ever be with the Lord*" (I Thess. 4:16-17). Then follows the brief period called "the great tribulation." (See Jeremiah. 30:5-7; Daniel 12:1; Zephaniah 1:15-18; Matthew 24:21-22.)

After this the personal return of the Lord to the earth in power and great glory occurs, and the judgments which introduce the seventh, and last dispensation. (See Matthew 25:31-46 and Matthew 24:29- 30.)

The sacrificial death of the Lord Jesus Christ introduced the dispensation of pure grace, which means undeserved favor, or God giving righteousness, instead of God requiring righteousness, as under law. Salvation, perfect and eternal, is now freely offered to Jew and Gentile upon the acknowledgment of sin, or repentance, with faith in Christ.

26.7 The 7th Dispensation; (Christ's Millennial Kingdom)

The final dispensation is where the Heavens rule over the Earth. Jesus Christ himself will be sitting on the throne and ruling from Jerusalem!

*"That in the **dispensation of the fullness of times** he might gather together in one all things in Christ, both which are in heaven, and which are on earth; even in him"* (Ephesians 1:10)

"By C. I. Scofield"

After the purifying judgments which attend the personal return of Christ to the earth, He will reign over restored Israel and over the earth for one thousand years. This is the period commonly called the millennium. The seat of His power will be Jerusalem, and the saints, including the saved of the dispensation of grace, namely the church, will be associated with Him in His glory. (See Isaiah 2:1-4; Isaiah 11; Acts 15:14-17; Revelation 19:11-21; Revelation 20:1-6.

But when Satan is "loosed a little season," he finds the natural heart as prone to evil as ever, and easily gathers the nations to battle against the Lord and His saints, and this last dispensation closes, like all the others, in judgment. The great white throne is set, the wicked dead are raised and finally judged, and then come the "new heaven and a new earth." Eternity is begun. (See Revelation 20:3, 7-15; (Revelation chapters 21 & 22.)

26.8 Salvation In All Dispensations Is By Just Faith.

The Gospel of Grace through Just Faith in The Blood of The Lord Jesus Christ and His Death, Burial and resurrection has always been the message! (Acts 15:1-11)

*"But we believe that through the grace of the Lord Jesus Christ **we shall be saved, even as they**"*. (Acts 15:11)

"*For what saith the scripture? Abraham believed God, and it was counted unto him for righteousness"* (Romans 4:3)

Before Jesus died on the cross, everyone who believed that Jesus was coming to pay for sins was saved. Now we look back in time 2000 years ago and believe that Jesus paid for all human sin for all time and eternity on the cross of Calvary.

God Talks About Different <u>Times</u>…

*"God, **<u>who at sundry times</u>** and **<u>in divers manners</u>** spake **<u>in time past</u>** unto the fathers by the prophets,* (Hebrews 1:1)

*"**Who at sundry times**"* is translated from the Greek word associated with the Strong's # 4181, *"polymerōs"* and is used in the following ways; by many portions; by many times; and in many ways.

Sounds like **different *"Ages"*** to me.

"***<u>In divers manners</u>***" is used in the Bible to say *"**many different** ways (or manners)"*

Strong's G4187 - polytrópōs, pol-oot-rop'-oce; adverb from a compound of **<u>G4183 and G5158</u>**; in many ways, i.e. variously as to method or form:—in divers manners.

"<u>In time past</u>" is a phrase translated from the Greek word associated with the Strong's # G3819 – *palai* it is talking about the past times.

God is talking about dealing with the Human Race by using different methods or "*<u>manners</u>*" or in different ways. The past times that are being described are periods of time associated with the method that God used to deal with men at that time. An easier way of saying all of that is to just call it **"Dispensations"**.

26.9 A Discussion About Dispensations. (By C. I. Scofield)

"By C.I. Scofield" written about 1909

(Ephesians 3:2) (Ephesians 3:5) (Ephesians 1:10)

"The Scriptures divide time (by which is meant the entire period from the creation of Adam to the "new heaven and a new earth" of (Revelation 21:1) into seven unequal periods, usually called dispensations (Ephesians 3:2), although these periods are also called ages (Ephesians. 2:7) and days, as in "day of the Lord."

These periods are marked off in Scripture by some change in God's method of dealing with mankind, or a portion of mankind, in respect of the two questions: of sin, and of man's responsibility. Each of the dispensations may be regarded as a new test of the natural man, and each ends in judgment, marking his utter failure in every dispensation. Five of these dispensations, or periods of time, have been fulfilled; we are living in the sixth, probably toward its close, and have before us the seventh, and last: the millennium."

27

Distinguishing Characteristics Of The Church.

27.1 The Church Was First Described As Future By Jesus

The Church was first mentioned by Jesus in 32 A.D., just prior to Pentecost when he said...

"*And I say also unto thee, That thou art Peter, and upon this rock **I will build** my church; and the gates of hell shall not prevail against it*". (Matthew 16:18)

He did not say that he was currently in the process of building his Church, but he said that he would build it in the future. This is the expression of a plan to do something new, in the future.

Ten days before Pentecost, Jesus tells his disciples to wait for the Holy Spirit.

"*For John truly baptized with water; but ye shall be baptized with the Holy Ghost not many days hence*". (Acts 1:5)

This is the last future expression of the Church. Jesus said that in not many days they will be baptized by the Holy Spirit. Ten days later, on the Feast of Pentecost, that is exactly what happened as described in (Acts 2)

When Peter was telling the story about how Cornelius received the gospel and the Holy Spirit, he said…

"*And as I began to speak, the Holy Ghost fell on them, as on us **at the beginning***". (Acts11:15)

In the next verse, Peter indicates that this event along with the Pentecost event recorded in Acts 2: were the fulfillment of the prophetic words of Jesus in (Acts 1:5). Furthermore, Peter is describing the event recorded in Acts 2: as "***the beginning***" of something.

"*Then remembered I the word of the Lord, how that he said, John indeed baptized with water; but ye* ***<u>shall be</u>*** *baptized with the Holy Ghost*". (Acts 11:16)

Jesus spoke of something new that would begin to happen in the future. In what is now the Age that we live in called the Church Age.

This is the Holy Spirit Baptism that all believers receive during this age. The process of being baptized into one body made up of both Jews and Gentiles by the Holy Spirit was a new thing. Nothing like it has ever happened before. This is the body and bride of Christ, the Church.

"*For by* ***one Spirit*** *are we all baptized into one body, whether we be Jews or Gentiles, whether we be bond or free; and have been all made to drink into one Spirit*" (1 Corinthians 12:13)

Nowhere was anything like this described in the Old Testament. Never was anything like this associated with any promise to Israel.

27.2 The Church Is The Betrothed Bride Of Christ.

In scripture the Church is described as a betrothed Bride. That is like being engaged but not married. The marriage is future. This is in contrast to Israel that is referred to as the adulterous wife of God.

"*Ye yourselves bear me witness, that I said, I am not the Christ, but that I am sent before him*". (John 3: 28)

"*He that hath the bride is the bridegroom: but the friend of the bridegroom, which standeth and heareth him, rejoiceth greatly because of the bridegroom's voice: this my joy therefore is fulfilled*". (John 3: 29)

The Church will be married to Christ after the Rapture. (Romans 7:4)

The Church is described as a chaste, betrothed virgin to be presented to Christ as the husband to be married. (2 Corinthians 11:2)

The relationship between Christ and the Church is likened to a betrothed future bride. (Ephesians 5:23-32) This passage describes the scene in heaven after the Rapture. The marriage has taken place and it is time for the feast (Revelation 19: 7, 8) from this point on the Church is no longer the betrothed Bride, but the Wife (Revelation 21:1-22:7)

27.3 The Church Is Baptized And Sealed By The Holy Spirit!

<u>All</u> believers are sealed and baptized into one body by the spirit and the name of Jesus (Yeshua)! This includes both Jew and Gentile believers being grouped together into one group called the Church…

"*There is **one body**, and **one Spirit**, even as ye are called in **one hope** of your calling*"; (Ephesians 4:4)

"*For by **one Spirit** are we **<u>all</u>** baptized into **one body**, whether we be Jews or Gentiles, whether we be bond or free; and have been **<u>all</u>** made to drink into **one Spirit***". (1Corinthians 12:13)

"*In whom ye also trusted, after that ye heard the word of truth, the gospel of your salvation: in whom also after that ye **believed**, ye were **sealed with that Holy Spirit** of promise*", (Ephesians 1:13)

Before the Church Age, believers were not sealed with the Holy Spirit. The Holy Spirit would come and go at His discretion. That changed at the 4th and last Spring Feast, the Feast of Pentecost. That Feast occurred fifty days after Christ was resurrected and marked the Beginning of the Church Age.

The Church Age is also called the Age of Grace or Dispensation of Grace and was placed in between the 69th week of Daniel and the 70th week of Daniel. (Daniel 9:25-27) See chapter 24.

During the Church Age, **all** Believers are baptized into the body of Christ and sealed with the Holy Spirit. That has never happened before.

"*But ye are not in the flesh, but in the Spirit, if so be that the Spirit of God dwell in you. Now if any man have not the Spirit of Christ, he is none of his*". (Romans 8:9)

In the past the Holy Spirit would come and go in Believers. In the near future, the Tribulation will begin with all unbelievers and 7 years later the Millennium will begin with only believers.

"*Saints*" is a term for all believers of all time. The "*Church*" is a term used exclusively for those who become Saints and are permanently sealed with the Holy Spirit from the time beginning at Pentecost, 50 days after Christ was crucified and continuing until the Rapture at the end of the Church Age which is marked by the Feast of Trumpets.

At the Rapture, all believers and the Holy Spirit are both removed from the earth. As we have already said, this event marks the end of the Church Age but also the end of the restraining influence of the Holy Spirit against evil in this world through the Church. When the Church and Holy Spirit leave the earth at the Rapture, I believe that the False Prophet will immediately begin performing fake miracles *like Jannes and Jambres did.* (2Timothy 3:8)

The next Feast is the 6th feast, or the 2nd Fall Feast. The Feast of Atonement also known as Yom Kippur will mark the beginning of the 70th week of Daniel and will be the day that the False Prophet will publicly meet with the Antichrist and endorse him as the "savior" of the world and enforcer of Peace and security. This is when the Antichrist will promise to enforce the Oslo Accords, signed by Rabin, for 7 years.

This is the day they will… "*say peace and safety*". However at sundown… "*For when they shall say, Peace and safety;* ***then sudden destruction cometh upon them,***" (1 Thessalonians 5:3)

27.4 How The Church Began.

The Church began on Pentecost and is sandwiched between the 69th and 70th weeks of Daniel's prophecy. The 70 weeks of Daniel involves the Dispensation of Law. The 70th week of Daniel is the conclusion of the Dispensation of Law.

The Jews say that the law was given on Pentecost. I believe that it is no coincidence that the 69th week of Daniel ended and the Church Age began on the Feast of Pentecost as well.

The Age of Law and the 70th week of Daniel were put on hold on the 4th Feast (Pentecost) and the Church Age began.

Although the Church Age was not revealed in the Old Testament, it was prophesied. Before the dispersion occurred, Hosea said how much time would elapse between the 2nd dispersion of Israel and their return in the last days. Hosea gave other details about this time we call the Dispensation of Grace but enough of the details were left out so that it remained a mystery.

The Church Age will end and the last week (7 years) of the Age of Law (70th week)will begin to tick off again at the Rapture which will be marked by the 5th Feast.

The Church Age is also be called the Dispensation of Grace.

"*If ye have heard of the* ***dispensation of the grace of God*** *which is given me to you-ward*": (Ephesians 3:2)

As you can see by the following verses, (Ephesians 3:3-5) the term "Dispensation" is made synonymous with Ages. I will use them interchangeably to keep this fact in the forefront of the reader's attention.

27.5 The Church Was A Mystery, Not Revealed In Past Ages.

"*How that by revelation he made known unto me the **mystery**; as I wrote afore in few words* (Ephesians 3:3)

"*Whereby, when ye read, ye may understand my knowledge in the **mystery** of Christ"* (Ephesians 3:4)

*"Which **in other <u>ages</u>** was not made known unto the **sons of men**, as it is now revealed unto his holy apostles and prophets by the Spirit"* (Ephesians 3:5)

The Church Age was not revealed in past Ages or Dispensations. The term "sons of men" is a Biblical term used exclusively for unbelievers. (Ephesians 3:5)says that unbelievers were not aware of the dispensation of Grace.

"*Even the **mystery** which hath been hid from ages and from Generations, but now is made manifest to his saints:*" (Colossians1:26)

The term "*mystery*" is a Biblical term for a truth not previously known or being revealed for the first time.

The term "saints" is a Biblical term used exclusively for believers.

The "mystery" spoken of here is the Dispensation of Grace (Ephesians 3:2-10) where Jews and Gentiles (verse 6) are both baptized into one body by the Holy Spirit and are joint heirs in Christ. In other words, with respect to this current Age which is about to end, the Jews and Gentiles are equal in Christ Jesus. (1Corinthians 12: 12, 13)

27.6 God Prophesied That He Would Use Gentile Tongues

God prophesied about a future Age where God would use Gentiles to proclaim his message to the Jews. God would reverse the order to provoke them too jealously.

On the day of Pentecost the Dispensation of Grace began. (Acts 2:4 -11) The Jews heard about their Messiah in Gentile languages on that day and it has continued for the past 2,000 years.

"*In the law it is written, With men of other tongues and other lips will I speak unto this people; and yet for all that will they not hear me, saith the Lord*". (1 Corinthians 14:21)

"*For with stammering lips and another tongue will he speak to this people*" (Isaiah 28: 11)

"*To whom he said, This is the rest wherewith ye may cause the weary to rest; and this is the refreshing: yet they would not hear*". (Isaiah 28: 12)

When the Church Age ends, the Holy Spirit and his influence leaves the earth and takes all the Church Age believers with him. The Tribulation begins with no believers, but I believe that changes quickly because God turns back to the Jews and prepares them for their proper place. Many of them will respond and take the place that God has planned for them.

27.7 The Church Replaced Israel For "Many Days"

"*I say then, Have they stumbled that they should fall? God forbid: but rather through their fall salvation is come unto the Gentiles, for **to provoke them to jealousy***" (Romans 11:11)

The Messiah came on the declared day, but Israel was not watching.

"***I will go and return to my place**, till they acknowledge their offence, and seek my face: **in their affliction they will seek me** early*". (Hosea 5:15)

As predicted the Messiah was rejected. The words of (Hosea 5:15) are prophetic. (John 14: 2, 3) records Jesus saying those same words after he was rejected.

After the 69 weeks, the Messiah would come, be cut off, but not for himself. Then the Temple would be destroyed (Daniel 9:25, 26) and Israel would be scattered for ***"Many Days"*** (Hosea 3:4, 5)

The Temple was destroyed and Israel was scattered. According to Daniel (600BC), **after** the Messiah would be cut off. The Messiah was cut off in 32 A.D. and the Temple was destroyed in 70 A. D.

"*For the children of Israel shall abide* ***many days*** *without a king, and without a prince, and without a sacrifice, and without an image, and without an ephod, and without teraphim*": (Hosea 3:4).

"***Afterward*** *shall the children of Israel* ***return****, and seek the LORD their God, and David their king; and shall fear the LORD and his goodness in the* ***latter days***. (Hosea 3: 5).

After the ***"Many Days"*** Israel would return to their land ***"in the latter days"***. On May 14, 1948 Israel was recognized by the United Nations and the period of time defined in (Hosea 3:5) as the "last days" began.

"Many Days" is defined by comparing (Hosea 3: 4 and 5) with (Hosea 6:1, 2) and (Revelation 20: 3-7).

Historically we can see that it has been about 2,000 years since Israel was scattered.

"*Come, and let us* ***return*** *unto the LORD: for he hath torn, and he will heal us; he hath smitten, and he will bind us up*". (Hosea 6:1)

"***After two days*** *will he revive us: in the* ***third day*** *he will raise us up, and we shall live in his sight.* (Hosea 6:2)

This passage makes the **third Day equivalent to the Millennium.**

The word Millennium is defined 6 times in Revelation 20 as 1,000 years.

(Hosea 6:2) said that we shall live in his sight in the Third Day. If the third day is the Millennium, then the first two days must also be 1,000 years each.

Hosea says "*After two days will he revive us*:" Well if one "Day" is a thousand years, then two "Days" is two thousand Years. That is 2,000 Hebrew, Lunar years.

One lunar year is 360 days. To convert 2,000 lunar years into Solar years which consist of 365.25 days, we multiply (360 X 2,000) and divide by 365.25 which gives us 1,971.25 Roman years.

Back to Hosea 3:4; it says… "Israel shall abide many days without… a King, Priesthood, sacrifice or household gods. Then in verse 5 it goes on to say…"*after many days*" they will return in the "***Last Days***". Well they are back just like God said, and that is not all…

27.8 "Many Days" Is At Least 1,971.25 Roman Years

So the last week, the "70th week of Daniel (7 years) was put on hold for "***Many Days***" which Hosea and Revelation define as 2,000 lunar years which equals 1,971.25 Roman (Solar) years.

Start counting from Pentecost, 32 A.D. 1,971.25 years and that brings you to 2,003 (late August or early September).

Now let us read God's word carefully…

"*For the children of Israel shall abide* ***many days*** *without a king, and without a prince, and without a sacrifice, and without an image, and without an ephod, and without teraphim*:" (Hosea 3:4)

Historically Israel has always had teraphim which are household gods. They have also had images to worship. Since they left Egypt, Israel has always had a sacrifice, except for the Babylonian exile. Since the time of Moses, Israel has also had a Priesthood which is the teraphim spoken of. All of that ceased in 70 AD and has not restarted for the last 2,000 years. That is the "many days" spoken of in Hosea

"***<u>Afterward</u>*** *shall the children of Israel* ***<u>return</u>****, and seek the LORD their God, and David their king; and shall fear the LORD and his goodness* ***<u>in the latter days</u>****"*. (Hosea 3:4);

<u>After</u> the *"many days"* (Church Age) is like saying after 2,000 (lunar) years or after 1,971.25 Roman years. So the Church Age should end on the 5th Feast, Rosh Hashanah some year **after** 2003, and we are waiting. But there is a cap to how long we must wait.

"*Now learn a parable of the fig tree; When his branch is yet tender, and putteth forth leaves, ye know that summer is nigh*:" (Matthew 24:32)

"*So likewise ye, when ye shall see all these things, know that it is near, even at the doors.*" (Matthew 24:33)

"*Verily I say unto you, This Generation shall not pass, till all these things be fulfilled*". (Matthew 24: 34)

The fig tree represents Israel. The new leaves on a young and tender plant represents the first and only growth of a young and vulnerable Israel. Israel was still a "teenager" when it regained Jerusalem for the first time in nearly 2,000 years. They also gained five other new territories. There has been no new growth since the 1967 war.

If verse 34 is talking about the Generation that fought in the 1967 war, then they should still be alive to see Jesus! **Israeli Prime Minister Benjamin Netanyahu** fought in that war and has become a highly visible icon of that Generation that will live to see the return of Jesus Christ (Yeshua the Messiah)

27.9 God Uses Gentiles To Provoke Israel To Jealousy

The Church Age begins on Pentecost, 32 A.D. (Acts 2:1-21)

The Church Age was inserted between the 69th and 70th weeks of Daniel to provoke Israel to jealousy.

A time was prophesied where Gentiles would accept a Jewish Messiah.

"*And in that day there shall be a root of Jesse, which shall stand for an ensign of the people;* ***to it shall the Gentiles seek****: and his rest shall be glorious*". (Isaiah 11:10)

God would use Gentile believers to provoke Israel to jealousy because they provoked God to jealousy by turning to false gods…

"***They have moved me to jealousy*** *with that which is not God; they have provoked me to anger with their vanities: and I will move them to jealousy with those which are not a people; I will provoke them to anger with a foolish nation.*" (Deuteronomy 32: 21)

This was written over 3400 years ago and prophesied this current Dispensation of Grace. God has temporarily turned to the Gentiles to spread His word during this Dispensation of Grace.

"*But I say, Did not Israel know? First Moses saith,* ***I will provoke you to jealousy by them that are no people,*** *and by a foolish nation I will anger you*". (Romans 10:19)

That passage is talking about the Gentile nations replacing Israel as the message bearers of God to the world for this current period of time that we call the Church Age.

The Church Age, also known as the Dispensation of Grace began on a Sunday on the 4th Feast, the Feast of Pentecost 32 A.D. That began God's new program. It will also end some year in the future on the 5th Feast, the Feast of Trumpets which is also known as Rosh Hashanah. Maybe even this year!

28

The End Of The Church Age Is Near.

(Revelation 1:19)

28.1 Jesus Returned To Heaven And Said He Will Be Back!

When we compare (Hosea 3: 15) and (John 14: 2-4) we find that Jesus's words in (John 14: 2-4) were prophesied by Hosea!

Jesus returned to heaven and promised to come back to Israel when the Jews changed their attitude. So far they haven't.

God says they will seek him after they have been afflicted. Although Israel has already been afflicted since 606 B. C., the affliction spoken of in Hosea, I believe, is the affliction that Israel will receive during the last 3 ½ years of the 70th week of Daniel.

"I will go and return to my place, till they acknowledge their offence, and seek my face: ***in their affliction they will seek me early****".* (Hosea 5:15)

"In my Father's house are many mansions: if it were not so, I would have told you. I go to prepare a place for you". (John 14:2)

"And if I go and prepare a place for you, ***I will come again, and receive you unto myself****; that where I am, there ye may be also".* (John 14:3)

The Feast of Trumpets is God's New Year. It is also the 5th Feast and the next one in line to be fulfilled. Also known as Rosh Hashanah, the Feast Day ends with the final blast of the Shofar at sundown. That final blast marks the end of the former year and the beginning of the New Year and the end of the Old Year. The Feast of Trumpets is also when **epochs** begin and end.

On the year that Jesus comes back to receive the Church unto himself, it will be at the end of Rosh Hashanah, at the sound of the last "trump" (Shofar). That is the event that we call the "Rapture".

That year the final sound of the Shofar will not only mark the end of the former year and the beginning of the New Year, but it will also mark the end of the Church age and the beginning of the **Day of the Lord.** It will happen at the sound of... ***"the last trump".*** (1 Corinthians 15: 52)

"In a moment, in the twinkling of an eye, ***at the last trump: for the trumpet shall sound, and the dead shall be raised*** *incorruptible, and we shall be changed"* (1 Corinthians 15:52)

Every year, as believers who are faithfully watching for the Lord's return, we should be thinking about these things as we approach Rosh Hashanah (Feast of Trumpets). When the Feast of Trumpets ends in Jerusalem at sundown, about 6: 30 pm their time and about 11:30 pm Eastern Standard Time, if I am still here then I will know that I have another year left to serve the Lord.

I also believe that if we are watching, we will already know before it is blown! As we approach the time, signs are increasing with frequency and significance. I believe that before the day of the Rapture, that Christians who are obediently watching may be able to recognize the final significant signs and recognize the year when it arrives!

Although one of the common misconceptions about the Rapture is that you can't know anything, because Jesus comes as a "thief in the night". Read that passage again, (1Thessalonians 5:4) because it is saying that if you are watching you will not be caught off guard. Jesus will come as a thief to those who are lost or just not watching. If you are not watching, it just won't matter to you. You will miss the greatest witnessing opportunity in history!

If you are lost, you may not even know that it happened. Satan is already at work trying to cover up this event.

28.2 Revelation Begins With A Unique Promise Of Blessings.

Revelation is the only book in the Bible that promises a blessing to all those that Read, Hear and "*Keep the words of this prophecy*". (Revelation 1:3)

"*Blessed is he that **readeth**, and they that **hear** the words of this prophecy, and **keep** those things which are written therein: for the time is at hand*". (Revelation 1:3)

No other book in God's Word makes such a promise and that fact requires further consideration.

Revelation begins by talking about a "promise of blessings" and "***<u>things which must shortly come to pass</u>***". (Revelation 1:1)

"*The Revelation of Jesus Christ, which God gave unto him, to shew unto his servants things which must **<u>shortly come to pass</u>**; and he sent and signified it by his angel unto his servant John*": (Revelation 1:1)

The book of Revelation is in contrast to the book of Daniel with respect to the fact that Daniel 12:9 records God's instructions that the words were to be "***sealed** till the time of the end*". However in Revelation 22:10

the Bible says … "***Seal <u>not</u>** the sayings of the prophecy of this book: for the time is at hand*".

There has been much confusion about this. I'll attempt to clarify.

God divides up the Book of Revelation into 3 parts…

"*Write the things which thou **hast seen**, and the things **which are**, and the things which **shall be hereafter***;" (Revelation 1:19)

The book of Revelation included things which were **past** and **things which were just getting underway** and things that would happen ***"hereafter"***.

Revelation deals with this transition period from what is called "*the things which thou **hast seen***", (Revelation 1:1-18) to the **things that are.** (Revelation 1:20- 3:22) The Church is in view in (Revelation 1:20- 3:22). God starts out at the beginning of Revelation by talking about … "*things which must shortly come to pass*" in (Revelation 1:1); and he says … that ***"the time is at hand"*** (Revelation 1:3) These are references to the fact that the Church Age is already under way and the letters to the 7 Churches are for them. That was and at this time still is current material.

Preterists use these passages in an attempt to try to make a case that all of this is past. However, Revelation 4:1 is a dividing point between ***"the things which are, and the things which shall be hereafter;"***

(Revelation 4:1) is a description of the Rapture and it begins by saying…

"***After this*** *I looked, and, behold, a door was opened in heaven: and the first voice which I heard was as it were of a* ***trumpet*** *talking with me; which said, Come up hither,* ***and I will shew thee things which must be hereafter***." (Revelation 4:1)

Notice that the passage begins with "**after this**" and ends with **"I will shew thee things which must be hereafter"** (Revelation 4:1) This marks the end of the Church Age and the beginning of "*the things which must be hereafter*".

"And immediately I was in the spirit*: and, behold, a throne was set in heaven, and one sat on the throne*". (Revelation 4:2")

From this point on the Church is in heaven getting judged at the Judgment Seat of Christ. The Church is no longer mentioned in the book of Revelation. Verse two is the pretribulation Rapture of the Church. The phrase "***And immediately I was in the spirit***" (vs. 2) is equivalent to **…*"we shall all be changed… In a moment, in the twinkling of an eye, at the last trump"*** (1 Corinthians 15:1, 2)

The Saints mentioned in Revelation after Chapter 4 are Saints just like any other Age or Dispensation. The Church Age Saints are unique from all other previous Ages.

The worldwide scattering of the Jews was prophesied many times in the Old Testament; The Church Age was prophesied; the regathering of Israel in the last days was prophesied; the end of the Church Age with a pretribulation Rapture was prophesied; and so was the 70th week of Daniel also known as the Tribulation. (Revelation 4:1-22:21)

28.3 Conditions For Receiving The Blessings Of Revelation

If you want the blessings spoken of here, then you must <u>read</u>, <u>hear</u> and keep these things.

"*Blessed is he that **readeth**, and they that **hear** the words of this prophecy, and **keep** those things which are written therein*" (Revelation 1:3)

The message to the seven Churches contains messages for not only those 7 Churches it was written to at that time, but also churches throughout the Church Age. Those 7 Churches represent 7 types of Churches and the problems associated with them.

There is a lot to be said about the characteristics of these churches but for now I want to focus on the Church of Sardis.

The Church of Sardis was **not watching**. They are told twice to watch. In Revelation 3:2 they are told to **"*be watchful*"**.

"*Remember therefore how thou hast received and heard, and hold fast, and repent. **If therefore thou shalt not watch**, I will come on thee as a thief, and thou shalt not know what <u>**hour**</u> I will come upon thee*". (Revelation 3:3)

This passage is warning the Church of Sardis that if they don't watch then they won't know what **hour** the Lord will come. This implies that the whole point of watching is so that you will know the hour. They are warned to watch or else the Lord will come on them as a thief.

This implies that if you are watching you will not be caught off guard like the victim of a thief, but you will actually know the very **hour** that the Lord comes for his Church. This implies the opposite of what is commonly taught on the Rapture. Compare this with …

"But ye, brethren, are not in darkness, that that day should overtake you as a thief". (1 Thessalonians 5:4)

You can know the day and the hour!

29

Scoffers Deny The Promise Of Jesus Return

29.1 Satan Is At War Against God's People and the promises He made to them.

"*Knowing this first, that there shall come in the last days scoffers, walking after their own lusts*", (2Peter 3:3)

"*And saying, Where is the promise of his coming? for since the fathers fell asleep, all things continue as they were from the beginning of the creation*". (2Peter 3:4)

Preterism denies God's promises to Israel.

Preterism was introduced by the Roman Catholics.

According to Dr. Guinness in his book The Approaching End of the Age, Preterism originated with the Jesuit Alcazar toward the end of the sixteenth century. Dr. Guinness mentions that Preterism had few supporters in 1887.

OPENING MASS OF THE 11TH ORDINARY GENERAL ASSEMBLY OF THE SYNOD OF BISHOPS

Excerpt of Homily of Pope Benedict XVI - Vatican Basilica, Sunday, 2 October 2005.

Benedict XVI: "The judgment announced by the Lord Jesus refers above all to the destruction of Jerusalem in the year 70."

Dr. Guinness' statement taken from page 113 of Romanism and the Reformation:

That pretty much sums up what Preterists believe about Bible Prophecy. They don't believe it literally, they just don't believe it. They just say that Jesus already came in 70 A.D. and make up the rest as well.

Problem with this thinking is that it denies lots of clear scripture that says that God would bring Israel back to the Promised Land in the last days.

"After many days thou shalt be visited: in the latter years thou shalt come into the land that is brought back from the sword, and is gathered out of many people, against the mountains of Israel, which have been always waste: but it is brought forth out of the nations, and they shall dwell safely all of them". (Ezekiel 38:8)

"Afterward shall the children of Israel return, and seek the LORD their God, and David their king; and shall fear the LORD and his goodness ***<u>in the latter days</u>****".* (Hosea 3:5)

"And it shall come to pass in that day, that the Lord shall set his hand again the <u>second time</u> to recover the remnant of his people, which shall be left, from Assyria, and from Egypt, and from Pathros, and from Cush, and from Elam, and from Shinar, and from Hamath, and from the islands of the sea".(Isaiah 11:11)

300 years ago no one believed that Israel would ever return to their land except for the True Bible Believers. It seemed impossible, so the leadership of the Roman Catholic Church decided that God's Promises to Israel are just going to waist. The Vatican decided that there was no one around to claim those promises and it did not look like there ever would be. So some self-proclaimed, "infallible" Pope denies what the Bible has plainly said and makes up a story about the Church <u>permanently</u> replacing Israel and claiming those promises for the Church. Of course those promises include land, lots of it!

Apparently nobody in the Roman Catholic Church believed that God would or could bring the Jews back to their land "*in the last days*" as promised, but God did! Even though the Jews have returned home right on schedule, The Catholic Church and all of the other Preterists have not reconsidered their position. I see these denominations claiming the promises of Israel but not the Curse. It seems they overlooked that.

But now it is not just the Catholic Church promoting this. Long ago the reformed Churches brought some of that Catholic doctrine into their churches. Today there is an abundance of main-stream protestant denominations that are actively promoting in one form or another the idea that God is through with Israel.

They say that the Promises of God are somehow canceled or given to the Church. All of that is contrary to what the Bible has plainly and repeatedly declared throughout scripture. Contrary to history and contrary to what God says is happening in the world now and in the immediate future.

Generally Preterists believe in a works oriented, false gospel so when the Rapture occurs, they will still be here with all of the other unbelievers.

I know they say that they believe the Bible, but they admit that they don't believe it literally. That is a fancy way of saying they don't believe what it says. They just don't believe. They are wolves in sheep's clothing and they are misleading people!

Israel will endure as long as the sun and the moon! (Jeremiah 31:35-37) (Ezekiel 37:25)

"*Thus saith the LORD, which giveth the sun for a light by day, and the ordinances of the moon and of the stars for a light by night, which divideth the sea when the waves thereof roar; The LORD of hosts is his name*": (Jeremiah 31:35)

"*If those ordinances depart from before me, saith the LORD, then the seed of Israel also shall cease from being a nation before me forever.*" (Jeremiah 31:36)

God is saying that as long as the stars, moon and Sun are still in the sky and the seas still exist, there would be a nation of Israel.

"Thus saith the LORD; If heaven above can be measured, and the foundations of the earth searched out beneath, I will also cast off all the seed of Israel for all that they have done, saith the LORD." (Jeremiah 31:37)

Nobody has counted the stars yet and we still don't know what is going on down inside the earth.

God repeats his everlasting promise to Abraham again.

"*And they shall dwell in the land that I have given unto Jacob my servant, wherein your fathers have dwelt; and they shall dwell therein, even they, and their children, and their children's children for ever: and my servant David shall be their prince for ever*". (Ezekiel 37:25)

Anyone that questions God's intentions of literally fulfilling all of his promises is calling God a liar and that is exactly what Preterists do!

"God hath not cast away his people" (Romans 11:1-11)

"I say then, Hath God cast away his people? God forbid. For I also am an Israelite, of the seed of Abraham, of the tribe of Benjamin" (Romans 11:1)

"*God hath not cast away his people which he foreknew. Wot ye not what the scripture saith of Elias…*" (Romans 11:2)

… *"Say then, Have they stumbled that they should fall? God forbid: but rather through their fall salvation is come unto the Gentiles, for to provoke them to jealousy*". (Romans 11:11)

These are future promises that are in the process of being fulfilled right now!

"*And they also, if they abide not still in unbelief, shall be graffed in: for God is able to graff them in again*". (Romans 11:23)

"For I would not, brethren, that ye should be ignorant of this mystery, lest ye should be wise in your own conceits; that blindness in part is happened to Israel, until the fullness of the Gentiles be come in". (Rom 11:25)

The Feast of Trumpets will mark the end of the Church Age but the "fullness of the Gentiles" will end when the 70th week of Daniel ends. All of this begins at the Rapture, also known as the "blessed hope of the believer".

It is not good to question whether or not God really means what he says! This is the essence of unbelief! When God speaks literal, one should not say otherwise. That is exactly what Preterists do.

Anyone who suggests that they have a "code" to enable you to see another message other than the apparent, literal one is a Deceiver and a False Prophet. (Isaiah 8:20)

"*To the law and to the testimony: if they speak not according to this word, it is because there is no light in them."* (Isaiah 8:20)

God has given us literal prophecies, many of which have already been literally fulfilled. This fact is an indicator that all of the rest of God's prophecies will also be literally fulfilled.

29.2 Satan's Minions Oppose God's Promises.

"About the times of the End, a body of men will be raised up who will turn their attention to the prophecies, and insist upon their literal interpretation, in the midst of much clamor and opposition." (Sir Isaac Newton, 1642 – 1747)

Preterits attack God's Promises to Israel. Preterists deny end-time Biblical Prophetic Truths. Preterists deny the Rapture and deny God's Promises to Israel.

Preterists teach that the Promises from God to Abraham were not literal. (Not real) They deny that they were real.

Preterism teaches that End Time Prophecies are past and not literal.

The fact is that they have not happened yet but they literally are about to happen.

Not only do Preterists not recognize the promises of God to Israel but they are actively working against those promises. Preterists are taking sides with Israel's enemies on every issue!

Preterism promotes anti-Israel activity, Preterism is anti-Israel and anti-God!

If Satan could get people to believe that God doesn't keep his promises then he has won! People won't trust Christ as Savior because if God doesn't keep his promises, then how are you going to trust what he says about everything else?

Either God was lying to Israel because he never intended to keep his promises or things got out of hand and God was not able to keep his promises or they are yet future and God is about to keep his promises. His people are in their land looking for their Messiah and that is consistent with God keeping those promises in the near future.

Satan does not want people to trust or believe God's promises. Preterism does cause people to distrust God and creates hostility toward Israel and true believers!

God has not forsaken Israel! God keeps his promises!

Here is God's warning to Preterists...

"And I will bless them that bless thee, and curse him that curseth thee". (Genesis 12:3)

29.3 Preterists Teach That Christ Returned In 70 A.D.

"*Verily I say unto you, There be some standing here, which shall not taste of death, till they see the Son of man coming in his kingdom*"... (Matthew 16:28).

Preterits use this passage to prove that Jesus had to return before the death of the Apostles. They say that Jesus had to return before the Apostles died, however they fail to read the next verse.

..."*And **after six days** Jesus taketh Peter, James, and John his brother, and bringeth them up into an high mountain apart*", (Matthew 17:1)

"*And was transfigured before them: and his face did shine as the sun, and his raiment was white as the light*". (Matthew 17:2)

"*And, behold, there appeared unto them Moses and Elias talking with him*". (Matthew 17:3)

They got to see the second return as it is described in (Revelation 1:1-7) six days later.

29.4 Peter, James And John Saw 2,000 Years Into The Future.

They were actually there. According to the Bible, time travel is a fact, not a theory!

Peter and John would later say that they witnessed the return of Christ even though it still hasn't happened yet! (Revelation19, 20)

John talks about it in (Revelation 1:1-7)

"*Who bare record of the word of God, and of the testimony of Jesus Christ, and of all things that he **saw***". (Revelation 1:2)

"*Behold, he cometh with clouds; and every eye shall see him, and they also which pierced him: and all kindreds of the earth shall wail because of him. Even so, Amen*" (Revelation 1:7)

Peter talks about it.

"*For we have not followed cunningly devised fables, when we made known unto you the power and coming of our Lord Jesus Christ, but were* ***eyewitnesses*** *of his majesty*". (2Peter 1:16)

The Transfiguration of Jesus in (Matthew16:28-17:3) was a look into the future by Peter, James and John and they witnessed the second return of Christ.

29.5 Preterists Confuse The Parable Of The Fig Tree.

"*Now learn a parable of the fig tree; When his branch is yet tender, and putteth forth leaves, ye know that summer is nigh:*" (Matthew 24:32)

"*So likewise ye, when ye shall see all these things, know that it is near, even at the doors*". (Matthew 24:33)

"*Verily I say unto you, This Generation shall not pass, till all these things be fulfilled*". (Matthew 24:34)

Preterists teach that this is referring to the Generation living at the time. However the context says that this is a parable. Jesus has already said in other passages that parables are meant to hide truth from those who will not use it properly.

"*He answered and said unto them, because it is given unto you to know the mysteries of the kingdom of heaven, but to them it is not given*". (Mathew 13:11)

"For whosoever hath, to him shall be given, and he shall have more abundance: but whosoever hath not, from him shall be taken away even that he hath". (Mathew 13:12)

When men reject God's simple truth, God may stop giving more truth, and allows Satan's minions to fill you with lies.

The Generation this passage is talking about is the only Generation that expanded the borders of Israel since 1948. This is the Generation that fought in the 1967 war. Israeli Prime Minister Benjamin Netanyahu fought in that war and represents that generation.

29.6 Preterists Believe That Nero Was The Antichrist

Preterits believe that we are in the Millennium now.

"And he shall confirm the covenant with many for one week: and in the midst of the week he shall cause the sacrifice and the oblation to cease, and for the overspreading of abominations he shall make it desolate, even until the consummation, and that determined shall be poured upon the desolate" (Daniel 9:27)

When did Nero ever go to Jerusalem? When did he do any of the things that the Bible says that the Antichrist will do? It never happened.

Nero never went to Jerusalem;

Nero was not defeated by Christ returning in glory; The Bible says…"*every eye shall see him*", yet nobody did; The whole world will shake at God's presence, but that is not in the history books either. There is no evidence that any of the things spoken of by God, through the Prophets in the Bible about the return of Jesus has happened.

When did the 28 judgments spoken of by John take place? I don't see that in the history books either.

Preterists deny the rapture, the second return of Christ and everything that the Bible says is about to begin. They just don't believe what the Bible has plainly said.

Preterists are a major source of antisemitism.

Preterists consistently take sides with Israel's enemies. Preterists are false prophets that mislead people and preach a false gospel that can't save.

29.7 Preterists Teach That We Are In The Kingdom Now.

None of the things that the Bible describes about the Kingdom are happening today or have they happened over the last 2,000 years. Yet Preterists teach that we are in the Kingdom now.

"*The wolf also shall dwell with the lamb, and the leopard shall lie down with the kid; and the calf and the young lion and the fatling together; and a little child shall lead them*". (Isaiah 9:6)

"*And the cow and the bear shall feed; their young ones shall lie down together: and the lion shall eat straw like the ox.*" (Isaiah 9:7)

"*And the sucking child shall play on the hole of the asp, and the weaned child shall put his hand on the cockatrice' den*". (Isaiah 9:8)

"*The wolf and the lamb shall feed together, and the lion shall eat straw like the bullock: and dust shall be the serpent's meat. They shall not hurt nor destroy in all my holy mountain, saith the LORD*" (Isaiah 65: 25)

I don't see these kinds of changes in the animal kingdom. Carnivores still exist!

When was Satan bound for 1,000 years? It just hasn't happened yet.

29.8 Preterists Are Spreading Confusion!

Satan is the father of lies and the author of confusion. For every truth that God has revealed to us through his word, Satan has cooked up about a thousand lies. As a consequence, there is a lot of confusion about every doctrine in the Bible including the Pretribulation Rapture.

The doctrine of the Pretribulation Rapture is built on the foundation of several doctrines. These doctrines include the promises God made to Abraham, Isaac and Jacob which became the everlasting promises to Israel.

Preterism denies these promises and denies the second return as being literal and future. Furthermore, Preterism teaches that God is through with Israel and claims God's promises to Israel for themselves while ignoring the curses.

Denying the promises made by God to Abraham is a major step in undermining the credibility of the Bible and Bible prophecy. (ETBH chapter 23)

One of Satan's lies is that God is through with Israel and that all of the promises that God made to Abraham now belong to the Church. That is not true! If God would break a promise to Abraham, Isaac and Jacob, then how could he be trusted?

The idea that the Church has permanently replaced Israel is wrong. Those that teach such heresy are quick to claim for their denomination all of the promises that God made to Israel but ignore the punishment and curses that went along with those promises.

"For *God is not the author of confusion, but of peace, as in all churches of the saints"*.1Corinthians 14:33)

30

Three Fall Feasts Declare "The Day Of The Lord"

30.1 Prophecy And Science Confirms God's Word.

"*I have declared the former things from the beginning; and they went forth out of my mouth, and I shewed them; I did them suddenly, and they came to pass*" (Isaiah 48:3)

"*Because I knew that thou art obstinate, and thy neck is an iron sinew, and thy brow brass*" (Isaiah 48:4)

"*I have even from the beginning declared it to thee; before it came to pass I shewed it thee: lest thou shouldest say, Mine idol hath done them, and my graven image, and my molten image, hath commanded them*". (Isaiah 48:5)

"*Thou hast heard, see all this; and will not ye declare it? I have shewed thee new things from this time, even hidden things, and thou didst not know them.*" (Isaiah 48:6)

God confirms his word with Bible Prophecy and scientific knowledge that Mankind could not have had access to apart from God. The Author of the Bible consistently demonstrates knowledge of the Creation; the past; the present and the future.

God says that true science will always confirm his word. Whenever "science" disagreed with the Bible, the "science" was found out to be non-science and had to be changed.

30.2 The Bible Is Literal, Jesus Proves It

When Jesus debated with the Sadducees, Jesus proved there was a resurrection by the tense of the verb "am"…

"*I* ***am*** *the God of Abraham, and the God of Isaac, and the God of Jacob? God is not the God of the dead, but of the living*". (Matthew 22:32)

"*For verily I say unto you, Till heaven and earth pass, one jot or one tittle shall in no wise pass from the law, till all be fulfilled*" (Matthew 5:18)

God says what he means and means what he says **literally**; every sentence; every word, every letter and, every part of the letter!

Prophecy must be interpreted **literally** by comparing scripture with scripture. (2Peter 1:20)

All of the prophecies in the past were literally fulfilled, in their order and on their day and at their hour, just like the Bible said.

It is reasonable to expect all of God's promises and prophecies to continue to be literally fulfilled likewise, in their order; on their day; and where applicable, at the right "***moment***"!

"*In a* ***moment****, in the twinkling of an eye,* ***at the last trump****: for the trumpet shall sound, and the dead shall be raised incorruptible,* ***and we shall be changed***". (1 Corinthians 15:2)

30.3 Rosh Hashanah, Feast Of Trumpets, 1st of Tishri

Tishrei is the first month of the Jewish civil year and the seventh month of the ecclesiastical year, which begins on Nissan One. Tishrei is the month for counting the years and epochs. The count of the year is incremented on Tishrei 1 at the final sound of the Shofar blowing at sundown.

Rosh Hashanah, also known as **Yom Teruah (Day of Shofar Blowing),** begins at sunset at the beginning of Tishri 1 and ends at the next sunset. The Biblical Feast lasts for only one day.

Modern day Judaism extends the holiday to include an extra day. Although this is common practice, the observance of Tishri 2 is not Biblical and is therefore not a Sabbath day.

"*And the LORD spake unto Moses, saying,*" (Leviticus 23:23)

"*Speak unto the children of Israel, saying,* ***In the seventh month, in the first day of the month****, shall ye have a Sabbath, a memorial of blowing of trumpets, an holy convocation.*" (Leviticus 23:24)

"*Ye shall do no servile work therein: but ye shall offer an offering made by fire unto the LORD.*" (Leviticus 23:25)

Rosh Hashanah always occurs on the same Hebrew day every year and that is Tishrei One. "Tishrei" is the name of the seventh Hebrew month. Tishrei is the month for counting years and epochs. The year is increased by one at sundown and marked by the last sound of the Shofar at sundown.

Consider the following beginning and ending dates for the following years.

In **2017** Rosh Hashanah begins on Thursday, September 20, at sundown (9/21/**2017**). It will continue for two days until September 22, at night fall on Friday.

(2018: September 9-11,)

In **2018** Rosh Hashanah begins on Sunday, September 9, at sundown (9/9/2018). It will continue for two days until September 11, at night fall on Monday (9/11/2018).

Let me make a little comment about Rosh Hashanah in the year 2018 that makes this year a little more interesting than other years. When the sun sets in Jerusalem in 2018 on September 11th, this precise moment is when the Hebrew calendar and the Islamic calendars both start and stop their years at the same moment in time. At sundown, they are synchronized at that moment.

This is a rare event because there are 360 days in a Hebrew year and 354 or 355 days in the Islamic year. It happens on September 11, 2018 at sundown. **(9/11/2018) and it marks the 70th Hebrew year since Israel was reborn.**

It is important to note that two of these calendars are Pagan and only the Hebrew calendar is from God. Therefore, this unusual synchronizing of two pagan calendars with God's calendar is most likely the work of Satan. I suppose that Satan is preparing lies for his minions about what the Lord is about to do.

The Jewish years depicted in the following chart represent what the Jewish authorities have accepted as the correct Jewish year. However Christian scholars believe that those numbers are wrong and will be corrected on the day of the Rapture.

Below is a list of recent dates that this author considered while I was writing this book...

Tishrei One, 7th month; Rosh Hashanah, Yom Teruah (Day of Shofar Blowing), at Sundown		
Beginning/Ending **Roman** Dates		
Jewish	Begins	Ends
5777	Sunset, October 2, 2016	Nightfall October 4, 2016
5778	Sunset September 20, 2017	Nightfall September 22, 2017
5779	Sunset September 9, 2018	Nightfall September 11, 2018
5780	Sunset September 29, 2019	Nightfall October 1, 2019
5781	Sunset September 18, (Saturday) 2020	Nightfall September 20, 2020 (Sunday)
5782	Sunset September 6, (Tuesday) 2021	Nightfall September 8, 2021 (Wednesday)

According to (Revelation 3: 2,3) if we are watching we will know the hour of Jesus return. Those same passages indicate a connection between knowing the hour and watching. If you are watching, then you will probably know the time when it arrives.

Make special plans for where you will be when the Jewish New Year begins every year. This may be the greatest witnessing opportunity many believers will ever get. If the believers are still here when the shofar is blown, that means that we have another year left.

This is a good time to share with our Jewish friends because God says that they are looking for a sign. This could be it. One of these years, Jesus will return for the Church and when He does, He will prepare Israel to take it's promised place.

Sunset is defined as the moment that the sun disappears below the horizon. **Nightfall** is defined as when it is too dark to work outside. Sunset is when the shofar is blown for the last time for Yom Teruah (Rosh Hashanah). On the appointed year, that is when God will turn back to Israel forever. We know the day, the hour and the moment but not the year. I believe that the Shofar will be blown at Sundown, not nightfall.

<u>September 20, 2020</u> (Sunday), <u>Jerusalem Time</u>, will be the next Rosh Hashanah. Whatever year it happens, the Rapture will be based on Jerusalem Time. The time of sunset in Jerusalem on September 20, 2020 and the corresponding times in other selected time zones around the world are as follows…

Sunsets at 6:38 PM; That translates into **11:38AM (EST)**Eastern Standard Time; or 10:38AM Central Time ; 9: 38AM Mountain; 8: 38 AM Pacific; or 3: 38PM Greenwich Time…

<u>September 8, 2021</u> (Wednesday), <u>Jerusalem time</u>,

Sunsets at 6:54PM Jerusalem Time; That translates into 11:54AM (EST) Eastern Standard Time; or 10:54 AM Central Time; or 9:54AM Mountain Time; or 8:54 AM Pacific Time; or 3:54PM Greenwich Time…

At the 70th anniversary (Roman Time) of Israel's independence on May 14, 2018 the new U.S. Embassy was dedicated in Jerusalem. The official recognition of Jerusalem as the eternal capital of Israel by the United States and Donald Trump has finished paving the way for the third Temple. This day is a milestone according to the Sanhedrin and Benjamin Netanyahu.

The Half Shekel Temple coin is required by Old Testament Law for every male, rich or poor entering the Temple area. The Temple is prefabricated and reportedly can be erected in 48 hours. All of the furnishings and vestments are made. All personnel are trained and ready.

The Sanhedrin minted the half shekel Temple Coin commemorating the 70th anniversary of Israel's independence. The new Temple coin has a profile of king Cyrus with President Trump's profile in front.

King Cyrus is the Persian King that gave orders that allowed the Jews to rebuild the second temple. The Inscription "To Fulfill 70 Years" is on the Temple coin which draws attention to the prophecies about Israel.

In short, Jeramiah prophesied 70 years of punishment for Israel in Babylon. During that time, there were three judgements that each began on the 9th day of the Hebrew month called " Av" and ended 70 years later. The significance of the 70 year cycle is discussed in detail in...(ETBH chapters 25).

The Temple coin is the last material thing needed from Israel, that I know of, before building the 3rd Temple. The Temple half Shekel coin should be the last thing completed before the Temple is built because it will commemorate the event. I believe that a short countdown has begun to begin the construction of the third Temple soon!

My personal belief is that the Temple construction will begin at the Rapture. So born-again believers, in my opinion, will not get to see it. This is my speculation. We will just have to wait and see.

It appears to me that the only thing left is to Rapture out the Church and let the 70th Week of Daniel begin with a **New Third Temple**. Considering the message on the coin, I think that the next thing to do before building the 3rd Temple is going to be done by God himself. The Church may not see the construction of the third Temple because the Temple has nothing to do with the Church. We will just have to wait and see if I am right about that.

I don't believe that the Church will even see the Temple construction begin because we will be raptured first! The Church has nothing to do with the Temple. The Temple is for the Jews during the 70th Week of Daniel. The Dispensation of Law will continue for seven more years and the Church will be in Heaven at the Judgement Seat of Christ.

The Jews will not work until the last trumpet of Rosh Hashanah and that is the time to start the New Year or Epoch. I don't know for sure which day they will start building the Third Temple but I believe that it may be at the same moment as the Rapture..

The ten days beginning with Rosh Hashanah and ending with Yom Kippur are known as **"The Ten Days of Awe"** (Yamim Noraim) or "Days of Repentance". For religious Jews, this is a time for serious introspection; a time to consider the sins of the previous year; a time to change your mind about things and a time to repair broken relationships.

I believe that in the near future **"The Ten Days of Awe"** may actually be played out in real life. If you consider that the Antichrist and False Prophet are being "*restrained*" by a "*Restrainer*" until the Rapture, then it seems reasonable to expect the Antichrist and the False Prophet to suddenly come out of the closet at the Rapture so they can explain it away.

The best way to explain away the Rapture is to take credit for it. That is exactly what I expect for the Antichrist and the False Prophet to do for ten days. I expect for the traditional "Ten Days Of Awe" observed by the Jews to actually be the time that the Antichrist and the False Prophet will use to gain control of the United Nations and subsequently the United States.

The Antichrist and the False Prophet will begin performing their "miracles, signs and wonders" in the sight of men on Rosh Hashanah. You can be sure that CNN and all of the lame stream media will be there to say how wonderful the Antichrist is.

I believe this will begin on Rosh Hashanah and climax on Yom Kippur. I believe Yom Kippur is … "*when they shall say, Peace and safety*". The Antichrist promises peace but does the opposite. When Yom Kippur ends after sundown will be "*sudden destruction*". We just don't know the year. We can only speculate for now but it is close!

30.4 Atonement (Yom Kippur)

"*And the LORD spake unto Moses, saying,*" (Leviticus 23:26)

"*Also on **the tenth day of this seventh month** there shall be a day of atonement: it shall be an holy convocation unto you; and ye shall afflict your souls, and offer an offering made by fire unto the LORD.*" (Leviticus 23:27)

"*And ye shall do no work in that same day: for it is a day of atonement, to make an atonement for you before the LORD your God.*" (Leviticus 23:28)

"*For whatsoever soul it be that shall not be afflicted in that same day, he shall be cut off from among his people.*" (Leviticus 23:29)

"*And whatsoever soul it be that doeth any work in that same day, the same soul will I destroy from among his people.*" (Leviticus 23:30)

"*Ye shall do no manner of work: it shall be a statute for ever throughout your Generations in all your dwellings*". (Leviticus 23:31)

"*It shall be unto you a Sabbath of rest, and ye shall afflict your souls: in the ninth day of the month at even, from even unto even, shall ye celebrate your Sabbath*". (Leviticus 23:32)

The "**Ten Days of Awe**" ends at the same time Yom Kippur ends, at nightfall.

30.5 Tabernacles (Sukkot)

This year the Feast of Tabernacles marks the end of a "Blood Moon Tetrad" with a Super Blood Moon at the end on Sept. 28-29 2015.

A "Blood Moon Tetrad" is a combination of astronomical events that is very rare. This one that just passed is the eighth one since the time of Christ, and this one is unique.

The previous 7 are associated with major historical events involving Israel and the Church. A "Blood Moon Tetrad" is considered by some to be a sign from God, this author included. (ETBH Chapter 34)

*"And **the LORD spake unto Moses, saying.**"* (Leviticus 23:33)

"Speak unto the children of Israel, saying, The fifteenth day of this seventh month shall be the feast of tabernacles for seven days unto the LORD." (Leviticus 23:34)

"On the first day shall be an holy convocation: ye shall do no servile work therein." (Leviticus 23:35)

"Seven days ye shall offer an offering made by fire unto the LORD: on the eighth day shall be an holy convocation unto you; and ye shall offer an offering made by fire unto the LORD: it is a solemn assembly; and ye shall do no servile work therein." (Leviticus 23:36)

"These are the feasts of the LORD, which ye shall proclaim to be holy convocations, to offer an offering made by fire unto the LORD, a burnt offering, and a meat offering, a sacrifice, and drink offerings, everything upon his day:" (Leviticus 23:37)

"*Beside the Sabbaths of the LORD, and beside your gifts, and beside all your vows, and beside all your freewill offerings, which ye give unto the LORD."* (Leviticus 23:38)

*"Also in **the fifteenth day of the seventh month**, when ye have gathered in the fruit of the land, ye shall keep a feast unto the LORD seven days: on the first day shall be a Sabbath, and on the eighth day shall be a Sabbath."* (Leviticus 23:39)

"And ye shall take you on the first day the boughs of goodly trees, branches of palm trees, and the boughs of thick trees, and willows of the brook; and ye shall rejoice before the LORD your God seven days." (Leviticus 23:40)

"And ye shall keep it a feast unto the LORD seven days in the year. It shall be a statute forever in your Generations: ye shall celebrate it in the seventh month." (Leviticus 23:41)

"***Ye shall dwell in booths seven days****; all that are Israelites born shall dwell in booths:"* (Leviticus 23:42)

"*That your Generations may know that I made the children of Israel to dwell in booths, when I brought them out of the land of Egypt: I am the LORD your God.*" (Leviticus 23:43)

"And Moses declared unto the children of Israel the feasts of the LORD." (Leviticus 23:44)

This year there will be a "**Super Blood Moon**" on the Feast of Tabernacles, Sept. 28-29, 2015.

31

The Pretribulation Rapture Fact.

31.1 The Rapture Was First Revealed Around A.D. 59

"*Behold,* ***I shew you a mystery;*** *We shall not all sleep, but we shall all be changed*", (1Corinthians 15:51).

The Rapture was presented as a *"mystery"* in this passage in 59 A.D. That means that the information in this passage was not formerly known. Other passages that spoke of the Rapture later, no longer referred to it as a *"mystery"* because it was already revealed here.

Some preachers wrongly teach that (Matthew 24:36) is talking about the Rapture. The words spoken by Jesus Christ and recorded in Matthew the 24th chapter, were spoken in 32 A.D. That was 27 years before (1 Corinthians 15: 51) said..." *behold I shew you a mystery*".

The Rapture is mentioned nowhere in the book of Matthew. (Matthew 24) is talking to Israel, about the last days of the Age of Law, also known as the 70th week of Daniel or the Tribulation. The two in the field mentioned is talking about the unbelievers being taken by the Angels and killed. The believers being left to repopulate the earth and live into the Kingdom. This is the opposite of what happens at the Rapture. At the Rapture the believers are changed and taken to Heaven as a bride where Jesus has a place prepared for us. (John 14:2, 3)

31.2 ..." God is not the author of confusion,"...

"Even if I knew that tomorrow the world would go to pieces, I would still plant my apple tree." (Martin Luther)

Confusion about the pretribulation Rapture comes from several issues that have already been discussed. Those who deny the pretribulation Rapture often don't distinguish the Church from Israel. For that discussion refer back to (ETBH chapter 27).

Another thing that creates confusion is a lack of understanding of Bible Prophecy in General. Daniel's prophecy of 70 weeks (Daniel 9:25-27) tells us about the future of Israel and that there is a gap between the 69th week and the 70th week. Hosea tells us that it is a 2,000 year gap. Through careful study we find that gap is the Church Age. (See chapter 27) God prophesied Israel's disobedience from the beginning. (See chapter 22) God has prophesied the entire history of Israel, past, present and future! Israel's punishment was completed on May 13, 1948. For those discussions (ETBH chapters 25).

The Dispensation of Law began on the Feast of Pentecost, and was put on hold at the end of the 69th week of Daniel, also on the Feast of Pentecost. That same Feast day is when the Church Age began about 2,000 years ago. That is when the Holy Spirit came down and indwelt and sealed all believers. This is the period of time prophesied by God where he would turn from the Jews as his ambassadors to the world and give that honor to the Gentiles for a 2,000 year period of time to provoke Israel to jealousy. (See chapter 27.9) That time is about to end.

The next feast day to be fulfilled is the Feast of Trumpets. That is when the Church Age will end and God turns back to Israel to be His Ambassadors. The Holy Spirit works through believers as they yield themselves to Jesus Christ to withhold sin, evil and the evil one in particular. When the Church Age ends, the Holy Spirit leaves taking all of the believers with him. At that point the Restrainer has been removed and the Antichrist and False Prophet begin their campaign of deceiving miracles, signs and wonders.

Nine days after the Rapture is the Feast of Yom Kippur. Yom Kippur will mark the beginning of the 70th week of Daniel also known as the Tribulation. On that day the False Prophet will endorse the Antichrist with miracles, signs and wonders and declare him to be the guardian of world peace and security.

The Antichrist will guarantee the Oslo Accords for seven years. The Oslo Accords were officially signed on Yom Kippur, **September, 13 1993** on the White House lawn. They signed, shook hands and said "Peace and Security". Those are the same words that will be spoken by the False Prophet and the Antichrist on the day that marks the beginning of the 70th week of Daniel.

The Temple Mount will be divided with a wall separating the newly constructed third, Jewish temple from the Muslim Mosques.

At sundown when Yom Kippur ends on the day that precedes the 70th week of Daniel, Muslims will declare "holy war" or Jihad as a protest against building the 3rd Temple. At this point, the Tribulation has begun with violence and chaos and darkness!

This is a time of purification of Israel. This is a period of time where God turns back to Israel. God raises up 144,000 young, Israeli men as his Missionaries to the world and they preach the gospel to everyone.

There is no mention of the Church in the tribulation because we are not there! We miss it all! We are raptured **before** it begins. The Church is in Heaven getting judged and married to Jesus.

31.3 Darby Did Not Invent The Pre-Tribulation Rapture!

Because the opponents of the Biblical doctrine of the Pretribulation Rapture lack Biblical support for their position, they often resort to factoids. That is a polite way of saying that they often say things that are not true. Hitler once said that if you repeat a lie often enough, people will begin to believe it. That certainly has been the case with the assertion that Darby invented the doctrine of the Pretribulation Rapture in the 1830s.

John Darby did not invent the Rapture any more than Martin Luther invented the Gospel of Grace!

The doctrine of the Pre-Tribulation rapture has naturally become more talked about since the Reformation. Prior to the Reformation, if anyone was caught teaching contrary to Catholic doctrine they were tortured and burned alive. That is why there is very little written during that time that survived the tyranny of the Catholic Preterists. The authors and their works were both publicly burned!

As we approach the end times, people are becoming more concerned about end time prophecies. Daniel said that knowledge will be increased in the Last days and it certainly has. Darby has contributed to some of that along with plenty of others but he is not the first to write about it.

Dr. Thomas Ice of the Pretribulation Rapture Research Institute has been doing a bit of research on this issue because so many people are believing the cooked up story that John Nelson Darby invented the Pretribulation Rapture doctrine in the 1830s. Not so!

Dr. Ice has compiled a list of writings by men who taught about the pretribulation Rapture before Darby and has plenty of background research on this issue. Dr. Ice has traced the pretribulation Rapture doctrine all the way back to the first century. Dr. Ice has also revealed the consequences faced by those who taught about the Pretribulation Rapture.

Being burned alive or tortured by the Catholic's was the cost of teaching about the pretribulation Rapture prior to the 1700's.

Once again for a complete discussion of the matter, go to …

www.pre-trib.org or **icet@pre-trib.org**

For a direct link to the page with the list go to… http://www.pre-trib.org/articles/view/a-history-of-pre-darby-rapture-advocates

A Welsh Baptist named Morgan Edwards (1722–95) taught about the pretribulation Rapture before Darby did. Check it out in Wikipedia. Here is the link; https://en.wikipedia.org/wiki/Morgan_Edwards

There were others before that. Unfortunately, those who taught about the Pretribulation Rapture back then were either burned, tortured or jailed for life for teaching contrary to Catholic dogma.

Now Preterists did not have that problem did they? They are the ones who were in control and doing the killing.

Those who taught the truth were doing it under the threat of a horrible and vicious death at the hands of the Catholics, Preterists and sometimes Calvinists! True believers were forced into secrecy. Nevertheless some dared and some were caught. Foxes book of Martyrs is full of examples.

The Rapture was also taught in the first century as evidenced in Thessalonians, Corinthians and Revelation. It was referred to in the Old Testament but revealed in the New Testament.

Daniel did not understand what he wrote and when he asked he was told to shut up the book until the End Times.

"*But thou, O Daniel, shut up the words, and seal the book, even **to the time of the end**: many shall run to and fro, and knowledge shall be increased*". (Daniel 12:4)

"The time of the end "Is now and according to that passage, I would expect to see an increased knowledge on the subject! Well that is what Daniel was saying.

Daniel was also told that Knowledge will be increased in the last days. Knowledge certainly has increased about a lot of things including End

Times Bible Prophecy. Daniel's book is no longer sealed! Sir Isaac Newton wrote about it.

Knowledge about the end times should increase proportionately to the interest in the subject as we approach the end of the Church Age and it has. I don't think that the subject was as popular 300 years ago as it is today for obvious reasons!

The further back you go, you would expect to see less interest in End Times Bible Prophecy. Going back to about 70 AD is when Israel was so completely destroyed that nobody believed they would return to their land as prophesied in the Bible. A few centuries later, Preterism and replacement theology began.

Before 70 A.D. all kinds of false doctrine were going around about the rapture and the 2nd return of Christ. These false doctrines included forged letters claiming to be from Paul but were not! For that reason Paul defended the Pretribulation Rapture in (2 Thessalonians 2:1-11) against false teachers teaching exactly what false teachers are teaching about the Rapture today.

Satan's minions also tried to say that Martian Luther invented the doctrine of Grace! Satan's minions have always attacked the truth!

31.4 The Word "Rapture" Was Used In The 4th Century.

In the 4th century the Vulgate was the first to translate the Old and New Testaments from the original Hebrew and Greek into Latin.

The phrase "***caught up***" found in (1Thessalonians 4:17) is the source of the word "Rapture".

In (1 Thessalonians 4:17) the English phrase "caught up" was translated from the Greek word "*harpazo"* the Latin translation used in the Vulgate is *"rapere"*. It is found in the Latin Translation because it is Latin for "Harpazo".

Because the word "Rapture", which is Latin for *"caught up"* as found in (1Thessalonians 4:17) has been used to describe this event since the 4th century, everybody knows what it means and it has stuck with us to this day. Although the word "Rapture" is not found in the English translation, we will use the word in place of the English Phrase *"caught up"* as found in (1 Thessalonians 4:17) because the term has been used since the 4th century and its meaning is well recognized.

Those who protest the teaching of the Rapture on the basis that the word "Rapture" is not found in the Bible Generally ignore the above mentioned information. Furthermore, if they believe in the Trinity, then they are just being hypocrites because the word "Trinity" is not found in the Bible at all.

31.5 God Evacuates Believers Before He Judges The World!

The Church Age began with violence against Christians. Throughout the Church Age, Christians have been burned, tortured and murdered largely by the Catholic's. Today we see Muslims doing the same things but the Bible says that this will be worldwide during the Tribulation. This has not happened yet. Christians have seen some bad times and today Christian persecution is on the rise but not to the degree that Tribulation Saints will have to endure.

God has demonstrated that he warns or removes believers before judgment begins. In the case of the Tribulation, God indicates that they will be removed.

Enoch was raptured <u>before</u> the Flood and Methuselah died on the year of the flood. Methuselah was the oldest man recorded in the Bible and was the last believer on the earth other than Noah and his family.

"*The righteous **perisheth**, and no man layeth it to heart: and merciful men are taken away, none considering that the **righteous** is taken away from the evil to come*". (Isaiah 57:1)

The word ***"righteous"*** is a reference to believers. One of the definitions for the word *"**Perisheth**"* is to vanish and is translated in other places as "escape" (Job 11:20) and "to flee" (Jeremiah 25:35).

*"[To the chief Musician upon Sheminith, A Psalm of David.]] Help, LORD; for the godly man <u>ceaseth</u>; for the faithful **fail** from among the children of men"*. (Psalm 12:1)

The term "godly man" refers to believers. (Saved) The term *"Children of men"* refers to unbelievers. (Lost)

"Fail" translated from pacac; Strong's # H6461 – to **disappear**, **vanish**, cease, **fail** (Qal) **to vanish** : This word is only used <u>once</u> in the Bible, and this is it and it describes the Rapture perfectly.

Noah and his family were protected from the flood by being in the Ark. Enoch was raptured out first.

Lot and his wife had to be removed from Sodom **before** the judgment could begin.

God said he would not destroy Sodom if he could find 10 believers there. (Genesis 18:20-33) The angels sent to destroy Sodom had to take hold of Lot and his family and drag them out of the City. (Genesis 19: 16) They said… *"For I cannot do any thing till thou be come thither"* (Genesis 19: 22).

When the Holy Spirit is removed evil will run unchecked.

"Remember ye not, that, when I was yet with you, I told you these things"? (2 Thessalonians 2: 5)

"And now ye know what withholdeth that he might be revealed in his time." (2 Thessalonians 2: 6)

That could be translated better as… "And now you know that it is your presence that withholdeth the man of sin, so that he shall be revealed at the appointed time".

This is talking about the Holy Spirit restraining evil in the earth.

God has a time appointed for the man of sin to be revealed, and that time is 9 days after the Holy Spirit is removed from the earth. It is God's presence in his living, breathing and working Church on earth that is withholding the man of sin.

When we, the Church, get Raptured out at Rosh Hashanah, the False Prophet will begin his 10 day campaign of miracles, signs and wonders to deceive the world into accepting the big lie. The Antichrist and False Prophet will make their big move 10 days after the Rapture, on the 6th Feast, the Feast of Atonement.

"*For the mystery of iniquity doth already work: only he who now letteth will let, until he be taken out of the way*". (2 Thessalonians 2:7)

This is talking about the Holy Spirit holding back the evil until he is removed out of the way. That is what this chapter is talking about, the Holy Spirit being removed from the earth and the consequences that follow.

This Satanic, Globalist conspiracy was already at work 2000 years ago when Paul was writing this. Satan would have created his New World Order long ago except for the actions of believers under the influence of the Holy Spirit of God. At the Rapture the Holy Spirit leaves the earth along with all of the believers. Then there will be nothing to restrain evil.

"*And then shall that Wicked (one) be revealed, whom the Lord shall consume with the spirit of his mouth, and shall destroy with the brightness of his coming*": (2 Thessalonians 2:8)

Nine days after the Rapture, on Yom Kippur, the 6th Feast, is when the Man of Sin shall be revealed with the help of the False Prophet. This is when the "***treaty with many***" (Daniel 9:27) will come into play.

On the 6th Feast of Israel, the Antichrist will assume the responsibility of enforcing the covenant with Rabin for one "week": = (7 Years)

"*Even him, whose coming is after the working of Satan with all power and signs and lying wonders,*" (2 Thessalonians 2:9)

Because people would not receive the love of the truth, they will be sent a lie. The Antichrist and the False Prophet will use our latest, secret technology to fake "miracles" and deceive the world.

"*And with all deceivableness of unrighteousness in them that perish; because they received not the love of the truth, that they might be saved*". (2 Thessalonians 2:10)

"*And for this cause God shall send them strong delusion, that they should believe a lie:*" (2 Thessalonians 2:11)

When God gives you the truth, you better treat it with respect!

The lie is probably concocted up to explain the sudden, global epidemic of missing persons at the end of Rosh Hashanah, all at the same time of day, the same moment!

Some over the years have speculated that the Rapture will be blamed on Aliens. Who knows? It sure seems like we have been conditioned for it! God describes a time of Wrath…

"*For God hath not appointed us to **wrath**, but to obtain salvation by our Lord Jesus Christ*", (1 Thessalonians 5:9)

"*Behold, the day of the LORD cometh, cruel both with **wrath** and fierce anger*", (Isaiah 13:9)

"*The day of the LORD is great and very terrible; and who can abide it?*" (Joel 2:11)

So the Rapture happens at sundown, Jerusalem time, at the end of Rosh Hashanah in the appointed year. That is the believers hope!

31.6 Enoch, Was Raptured, Before The Flood.

Before God judged the earth with Noah's flood, he raptured Enoch.

"And Enoch walked with God: and he was not; for God ***took*** *him".* (Genesis 5:23, 24)

It is interesting to note that one of the definitions for the Hebrew word translated as "took" is to take in marriage. See Strong's # H3947 – *laqach* Strong's definition…" to take, get, fetch, lay hold of, seize, receive, acquire, buy, bring, **marry**, **take a wife**, snatch, take away"

Jesus is portrayed as the Bridegroom and the Church is the betrothed bride… (John 3: 28-30) and Jesus is going to "snatch" us away.

"*By faith Enoch was* ***translated*** *that he should not see death; and* ***was not found****, because God had* ***translated*** *him: for before his* ***translation*** *he had this testimony, that* ***he pleased God****".* (Hebrews 11:5)

(Hebrews 11: 6) "*But* ***without faith it is impossible to please him****: for he that cometh to God* ***must believe*** *that he is, and that he is a rewarder of them that diligently seek him".*

Enoch is a **believer** who had an unusually close walk with the Lord when he lived on the earth as a mortal man and he was well known for it too.

Translated is … "*to cause one thing to cease and another to take its place"* (Strongs)

Translated = (Strongs) G3346 – metatithēmi defined as… to transpose (two things, one of which is put in place of the other); to transfer; to **change**; to transfer one's self or suffer one's self to be transferred

"**Was not found**" is a phrase that implies that Enoch disappeared.

This passage also refers to Enoch's Rapture as being "***translated"*** three times in one verse. Being ***"translated***" is like being "***changed***" as mentioned in (1 Corinthians 15:51, 52)

In (Revelation 4:1) it says… ***"I was immediately in the spirit"***. That is the same as ***…"in a moment, in the twinkling of an eye"***. (1 Corinthians 15: 52)

The fact that Enoch was the first person recorded in the Bible as being Raptured leads me to believe that God has revealed something to us about the Rapture through Enoch.

Enoch was raptured **before** God judged the earth with the second flood.

Something else that I consider to be noteworthy about Enoch is the fact that he lived 365 years.

"And all the days of Enoch were ***three hundred sixty and five*** *years:"* (Genesis 5:23)

That is one year for every day in a solar year. Why did God do that? He did it for a reason and it is our honor to search out the answer.

"The heart of the righteous studieth to answer: but the mouth of the wicked poureth out evil things" (Proverbs 15:28)

"It is the glory of God to conceal a thing: but the honour of kings is to search out a matter" (Proverbs 25:2)

So ask yourself, what could a solar year possibly have to do with the Rapture? Will there be a sign in the sun, like Joel spoke of in (Joel 2:30)? Well if so, then that just happened in the second year of the 2014-2015 Blood Moon Tetrad.. We should consider what this may mean.

All doctrines of the Bible are attacked by satan and his minions and the doctrine of the Pretribulation Rapture is no exception. Some important facts which are first mentioned here about the Rapture are that **it is a**

translation; a marriage; and a disappearing without announcement to the unbelieving world before God's Judgment. The only thing the unbelieving world knew was that Enoch was gone and everything else continued on the same until Noah entered the Ark. See … (Matthew 24: 38-39)

31.7 2 Things Must Happen Before The 70th Week Of Daniel Begins

The Church Age began on the 4th Feast of Pentecost. That is when the Holy Spirit sealed every believer on earth.

"*In whom also after that ye* ***believed****, ye were* ***sealed*** *with that holy Spirit of promise,*" (Ephesians 1:13, 14)

Furthermore, God says that we are sealed until the day of redemption.

"*And grieve not the Holy Spirit of God, whereby ye are sealed unto the day of redemption*" (Ephesians 4:30)

At the Rapture we will be redeemed, the Holy Spirit that seals us will leave with us because he said that he will **never** forsake or leave us once we are sealed.

The Church Age will end at the Rapture, at the end of the 5th feast, which is when the Holy Spirit departs the earth with all of the believers.

After the first letter to the Church of Thessalonica was written, some joker forged a letter from Paul and even forged his signature on it.

Forged letters pretending to be from the Apostles happened more than once. Apparently this letter said that the rapture had already happened and they were all left behind and the day of the Lord had begun, or something like that. So Paul is straightening things out in his second letter to Thessalonica…

*"Now we beseech you, brethren, by the coming of our Lord Jesus Christ, and by **our gathering together unto him**"*... (2 Thessalonians 2:1)...

This is talking about the rapture. "***Our gathering together unto him***"

..."*That ye be not soon shaken in mind, or be troubled, neither by spirit, nor by word, nor by letter as from us, as that **the Day of Christ** is at hand"* (2 Thessalonians 2:2)...

In this passage, *"The Day of Christ"* should be translated as "*The Day of the Lord"*. In other translations, it is.

This is talking about the forged letter pretending to be from Paul. Notice that this verse is telling them to not be troubled or shaken in mind. They were in distress because false teachers sent letters pretending to be from Paul. Apparently the letter stated that the tribulation had begun and they were left behind! False teachers were already attacking the doctrine of the Pretribulation Rapture.

From the beginning of the Church Age False teachers were at work deceiving and corrupting Church doctrine! Most of the letters written to the early Churches were dealing with false teachers attacking and deceiving the early Church with false doctrine. In this passage, (2 Thessalonians 2:1-12), Paul is defending the doctrine of the Pretribulation Rapture.

*"Let no man deceive you by any means: for **that day** shall not come, except there come a **falling away <u>first</u>**, and **that man of sin be revealed**, the son of perdition;"* (2 Thessalonians 2:3) ...

This passage says that **two things** must happen **<u>before</u>** "***that day***". **The first thing** mentioned here that must happen <u>before</u> "***that day***" is... "*the falling away*".

Paul says... "***That Day shall not come, except there come a <u>falling away</u> first***". In other English translations prior to the King James Translation, the words "***falling away***" were translated as "**Departure**". The King

James Translators changed "departure" to "Falling Away" without any explanation for why.

The Greek word "apostasia" was translated as "Departure" in the first seven English translations.

The first 7 English translations are: the Wycliffe Bible (1384); the Tyndale Bible (1526); the Coverdale Bible (1535); the Cranmer Bible (1539); the Breeches Bible (1576); Beza Bible (1583); Geneva Bible (1608).

The Latin Vulgate also known as Jerome's translation is dated about 400 A.D. The Vulgate is the first translation into Latin to include a direct translation from the original Hebrew instead of the Septuagint. That translation is also noted for being the original source for the word "Rapture" that we use today. Now that we know which translation we are talking about, that early translation translated *"apostasia"* into the word *discessio*, which means departure in Latin.

A more recent translation that still uses "departure" is the Hebrew Names Version (HNV)

In (2 Thessalonians 2: 3) *"apostasia"* is preceded by the direct article "the" indicating two things: The "Apostasia" is an event separate from the man of sin being revealed and the Church of Thessalonica was familiar with the event being spoken of (vs. 5). **This means that the Greek word "apostasia" cannot mean a decline in doctrinal purity because that would be a process over time and not a single event.**

The following context defines "The Departure" as the departure of the Holy Spirit which is the Restrainer.

So in verse 3 the statement "*for **that day** shall not come, except there come a **falling away first**, and **that man of sin be revealed**"*, It is talking about the Holy Spirit being removed **before** the Antichrist is revealed. Continuing on in verse 4...

"Who opposeth and exalteth himself above all that is called God, or that is worshipped; so that he as God sitteth in the temple of God, shewing himself that he is God". (2 Thessalonians 2:4)

This is talking about the Antichrist.

"Remember ye not, that, when I was yet with you, I told you these things?" (2 Thessalonians 2:5)

This passage indicates that Paul had already discussed this matter with them and he is now reminding them of that. The point being that **the direct article "the" before "*apostasia*" indicates a single event that they had discussed previously**. Paul is referring to that event with a direct article and those discussions about that event.

*"And now ye know what **withholdeth** that he might be revealed in his time".* (2 Thessalonians 2:6)

This is talking about the Holy Spirit.

*"For the mystery of iniquity doth already work: only he who now **letteth** will **let**, until he be taken out of the way".* (2 Thessalonians 2:7)

The words *"withholdeth"* in verse 6 and *"Letteth"* in verse 7 are both translated from the same Greek word. A better translation of this word would be "Restrainer" or "Hinderer". The Restrainer or Hinderer spoken of here is the Holy Spirit which is Hindering Satan's plan to reveal the Antichrist to the lost world. The Muslims are looking for their 12th Imam, this will probably be it.

*"**And then shall that Wicked be revealed**, whom the Lord shall consume with the spirit of his mouth, and shall destroy with the brightness of his coming":* (2 Thessalonians 2:8)

The second thing that must be done before "*that day*" begins is that the "Wicked" (one) shall be revealed. This is talking about the Antichrist and calling him the wicked (one). I believe that this event will occur on

Yom Kippur day. Yom Kippur ends at sundown and I believe that is the beginning of the first day of the seventieth week of Daniel. In this context, the term "that day" refers to the 70th week of Daniel which will begin about nine days after the rapture of the Church.

"*Even him, whose coming is after the working of Satan with all power and signs and lying wonders,"* (2 Thessalonians 2:9)

"And with all deceivableness of unrighteousness in them that perish; because they received not the love of the truth, that they might be saved". (2 Thessalonians 2:10)

Paul is defending the Doctrine of the pretribulation Rapture because it is important! The context mentions comfort as one reason in this case but I believe there is more to it.

The Bible says …"*For the Jews require a sign*" (1 Corinthians 1:22). I believe that the Rapture may be the sign that they are looking for. After all, the Rapture marks the end of the Church Age and the beginning of the preparing of Jacob's unbelieving descendants for the position God has promised.

If an unbelieving Jew is in the company of believers at the end of Rosh Hashanah when the shofar is blown for the last time to mark the end of the Feast; and it is the year of the Rapture; and they see the believers disappear; I think that would be the perfect sign at the perfect time for them to believe! This is a good reason to start an annual tradition of believers uniting with Jews at the end of Rosh Hashanah to pray for a third Temple! You could call it the annual day of Judeo-Christian Unity.

In conclusion, the two things that must happen before the 70th week of Daniel begins is first the Rapture on Rosh Hashanah, at sundown Jerusalem time. The second thing is the revealing of the Antichrist nine days later on the Feast of Yom Kippur. Remember that the end of Yom Kippur, sundown Jerusalem time, marks the beginning of the 70th week of Daniel which is all about Israel. The Church is not included in that. They will be in Heaven.

And that is why we call it the Pretribulation Rapture Fact.

Dr. Thomas Ice offers a complete discussion on this matter on his website. For a thorough understanding of this subject visit…

www.pre-trib.org or ICET@PRE-TRIB.ORG

31.8 "Changed" Is A Term Associated With The Rapture.

"*Behold, I shew you a mystery; We shall not all sleep, but we shall all be **changed***", (1Corinthians 15:51)

This is where God reveals for the first time that there will be a future Generation that will never die. I believe that we may be that Generation!

"***Changed"*** is a term associated with the Rapture not the 2nd return. At the 2nd return unbelievers are taken **first** and killed by Jesus Christ. There is no mention of anyone being "changed" or "translated" at the 2nd return of Christ.

"***Changed***", was translated from Strong's G236 – allassō, and means to **transform**, **change** or exchange one thing for something else,

"***In a moment**, in the twinkling of an eye, **at the last trump**: for the **trumpet shall sound**, and the dead shall be raised incorruptible, and we shall be **changed***". (1Corinthians 15:52)

At the Rapture, people are going to disappear instantly. Remember that (Revelation 4:1, 2) starts out with a voice from Heaven like a trumpet saying "*come up hither*" and then John says "***immediately I was in the spirit***". From that point on the Church is in Heaven and no longer mentioned in the Book of Revelation. That is why it is called the Pretribulation Rapture Fact.

Once again, I cannot over emphasize that the Rapture of the Church is nothing like the second return of Christ. The Rapture is instantaneous the second return is not! Trumpets are associated with the Rapture, but not the second return.

"*For this corruptible must put on incorruption, and this mortal must put on immortality*". (1Corinthians 15:53)

"*So when this corruptible shall have put on incorruption, and this mortal shall have put on immortality, then shall be brought to pass the saying that is written, Death is swallowed up in victory*". (1Corinthians 15:54)

"*O death, where is thy sting? O grave, where is thy victory*"? (1Corinthians 15:55)

"*The sting of death is sin; and the strength of sin is the law*". (1Corinthians 15:56)

"*But thanks be to God, which giveth us the victory through our Lord Jesus Christ*". (1Corinthians 15:57)

The point of this passage about the Rapture is to motivate and encourage believers in the work of the LORD. This teaching should not be neglected.

"*Therefore, my beloved brethren, be ye steadfast, unmoveable, always abounding in the work of the Lord, forasmuch as ye know that your labour is not in vain in the Lord*". (1Corinthians 15:58)

Remember, Enoch was "*translated*" or "*changed*" **before** the flood.

31.9 "We Which Are Alive And Remain Shall Be Caught Up"

"*But I would not have you to be ignorant, brethren, concerning them which are asleep, that ye sorrow not, even as others which have no hope*". (1Thessalonians 4:13)

"*For **if we believe** that Jesus died and rose again, even so them also which sleep in Jesus will God bring with him*". (1Thessalonians 4:14)

God does not want us to be ignorant about the hope of the resurrection for all those who **believe** on Christ. When Christ comes for the living Church, he will bring the dead with him.

"*For this we say unto you by the word of the Lord, that **we which are alive and remain unto the coming of the Lord** shall not prevent them which are asleep*". (1Thessalonians 4:15)

"*For the Lord himself shall descend from heaven with a **shout**, with the voice of the archangel, and with the **trump** of God: and the dead in Christ shall rise first*": (1Thessalonians 4:16)

"*Then we which are alive and remain shall be **caught up** together with them in the clouds, to meet the Lord in the air: **and so shall we ever be with the Lord***" (1Thessalonians 4:17)

"caught up" is translated from Strong's G726 – *harpazō* . This is the Greek word that was translated into Latin and gave us the word "Rapture". (Rapere) is the Latin

Also notice that believers are caught up to the Lord in the air and it says***..."and so shall we ever be with the Lord."*** Compare that phrase with... "*I go to prepare a place for you*"..." *And if I go and prepare a place for you, I will come again, and receive you unto myself; **that where I am, there ye may be also***". (John 14:2, 3)

Both of these passages are talking about the Rapture and both indicate that **believers are going to be removed from the earth and taken to their new home in Heaven which has been prepared for them.**

The Rapture differs from the second return in this respect also. In contrast, the second return leaves the believers who survive the Tribulation on the earth to live into the Kingdom and repopulate it. However, unbelievers are taken first and killed by Christ at the second return.

Several other things are important to make note of here. First is that the Lord himself shall descend from Heaven with a **shout**. In (Revelation 4:1) John describes the Rapture as a **voice** from Heaven. Notice also that…"*the first voice which I heard was as it were of a **trumpet** talking with me*". So we have a voice and trumpets again in (Revelation 4:1). It is the same event.

Also notice that the LORD meets us in the air. It does not say that the LORD comes all the way to the earth like he does at his 2nd return. (Zechariah 14:4)

Another important thing to remember is the statement…"

"*Wherefore comfort one another with these words*". (1Thessalonians 4:18)

It does **<u>not</u>** say for the Church to prepare for the Antichrist and the Tribulation. It does say for the Church to look for Jesus to come and take us to our home that he has prepared for us in Heaven. Living in anticipation of the Tribulation is **<u>not</u>** a comforting thought! Living in anticipation of seeing Jesus is!

"*But of the **times and the seasons**, brethren, ye have no need that I write unto you.*" (1Thessalonians 5:1)

"***The times and seasons***" is what the Pharisees were not paying attention to and Jesus scolded them for it…

"*The Pharisees also with the Sadducees came, and tempting desired him that he would shew them a sign from heaven*". (Matthew 16:1)

"*He answered and said unto them, When it is evening, ye say, It will be fair weather: for the sky is red*". (Matthew 16:2)

"*And in the morning, It will be foul weather to day: for the sky is red and lowring. O ye hypocrites, ye can discern the face of the sky; but **can ye not discern the signs of the times?***" (Matthew 16:3)

Now the subject switches from the Rapture to the "***Day of the LORD,***" for two reasons. The first reason is that the Rapture marks the end of the Church Age and the beginning of the Day of the LORD. The second reason is that there are signs for the Day of the Lord. Those signs are compared to weather signs. Because there are signs for the "Day of the LORD", as we see it approaching we should know the times and seasons. "*The Day of the LORD comes as a thief in the night*" for those who are not watching for it.

The words translated into "day" as it is used in the phrase "*Day of the Lord*" or "*That Day*" means "time" in a General sense or period of time. This is true in both the Old and New Testaments. The Day of the LORD can be used as a technical term referring to the 1,000 year Millennial reign of Christ which will begin on the Feast of Tabernacles 13 days after the Tribulation ends. The term "*Day of the LORD*" is most often used in a broader sense which begins at the Rapture and includes the Millennium.

"*For yourselves know perfectly that **the day of the Lord** so "cometh as a thief in the night*". (1Thessalonians 5:2)

I have had so many people quote this verse and say that you can't know when Jesus will come because he comes as a thief in the night. Well this just is not true. (See vs. 4) This passage is talking about unbelievers and any believers who are **<u>not</u>** watching.

"*For when they shall say, Peace and safety (security); then **sudden destruction** cometh upon them, as **travail upon a woman with child**; and they shall not escape*". (1Thessalonians 5:3)

Compare this passage with (Isaiah 13:6)…

"*Howl ye; for **the day of the LORD** is at hand; it shall come as a **destruction** from the Almighty*". (Isaiah 13:6)

Both of these passages are talking about the "***day of the LORD***" and both passages describe it as **destruction**.

"*And they shall be afraid: pangs and sorrows shall take hold of them; they shall be* ***in pain as a woman that travaileth***" (Isaiah 13:8)

Birth pain is used in both passages to describe the Tribulation.

"*But ye,* ***brethren****, are* *__not__* *in darkness, that that day should overtake you as a thief*". (1Thessalonians 5:4)

It says here that **that day** should not overtake believers as a thief **if** they are watching.

"*Ye are all the children of light, and the children of the day: we are not of the night, nor of darkness*". (1Thessalonians 5:5)

"*Therefore let us not sleep, as do others; but let us* ***__watch__*** *and be sober*". (1Thessalonians 5:6)

Here it is again. Believers are commanded to "***watch***". If you are watching, Jesus will __not__ come as a thief in the night to you. You will see it coming and God has a special reward for you if you are watching! You are either watching or sleeping. Which one are you doing?

"*For they that sleep sleep in the night; and they that be drunken are drunken in the night*". (1Thessalonians 5:7)

"*But let us, who are of the day, be sober, putting on the breastplate of faith and love; and for an helmet, the hope of salvation*". (1Thessalonians 5:8)

"***For God hath not appointed us to __wrath__****, but to obtain salvation by our Lord Jesus Christ*", (1Thessalonians 5:9)

Compare this passage with (Isaiah 13:9)…

"*Behold,* ***the day of the LORD*** *cometh, cruel both with* ***__wrath__*** *and fierce anger, to lay the land desolate: and he shall destroy the sinners thereof out of it*". (Isaiah 13:9)

"***That day is a day of wrath***, *a day of trouble and distress, a day of wasteness and desolation, a day of* ***darkness*** *and gloominess, a day of clouds and* ***thick darkness,***" (Zephaniah 1:15)

"Neither their silver nor their gold shall be able to deliver them in the day of the LORD'S ***wrath****; but the whole land shall be devoured by the fire of his jealousy: for he shall make even a speedy riddance of all them that dwell in the land*". (Zephaniah 1:15)

It says in (1Thessalonians 5:9) that **we are not appointed to wrath**. A time of "wrath" is how God describes ***"the Day of the LORD"*** which is another name for the tribulation or the 70th week of Daniel. God says that **believers are not appointed to wrath!**

And that is why we call it the Pretribulation Rapture Fact.

"*Who died for us, that, whether we wake or sleep, we should live together with him*". (1Thessalonians 5:10)

"*Wherefore* ***comfort*** *yourselves together, and edify one another, even as also ye do*". (1Thessalonians 5:11)

It is **comforting** to know that believers will not go through the Tribulation!

32

The Rapture Compared To The 2nd Return Of Christ

32.1 Some Distinctions Between The Church And Israel

One area of confusion about end time's prophecies involves a failure to distinguish between Israel and the Church. When a passage is talking to Israel about the last days, it is talking about the 70th week of Daniel and the Kingdom to follow. The last seven years of the Dispensation of Law is the seven years of Tribulation or the 70th week of Daniel.

When a passage is talking to the Church about the last days, it is talking about false teachers growing worse and worse and the Church Age ending with the Rapture.

"*Now the Spirit speaketh expressly, that in the latter times some shall depart from the faith, giving heed to seducing spirits, and doctrines of devils*" (1 Timothy 4:1)

An important point made by John Walvoord in his book titled "The Rapture Question" is about the use of the word "***ecclesia***" in the New Testament. Walvoord discuses four different ways that Ecclesia is used in the New Testament.

The meaning of the **first type** is used in the most General sense of a group of people;

The meaning of the **second type** is used of an assembly of professing Christians in a local Church or in the plural form when speaking of a group of Churches. This method is used when speaking the messages to the seven Churches of Asia (Revelation 2:-3 :). From that use we can determine that these groups are not necessarily made up exclusively of true believers. They may be professors that associate themselves with other professors and outwardly appear to be so, but some are not! It is a mixed group.

The meaning of the **third type...**" is also used of the total of professing Christians without reference to locality and is practically parallel in this sense to Christendom (Acts 12:1); (Romans. 16:16); (1Corinthians 15:59); (Galatians 1:13); (Revelation 2:1-3:22).

The meaning of the **fourth type** is the most specific because it isolates true believers from all others by their being baptized into the Church by the Holy Spirit. (1Corinthians 12:13) This meaning of "ecclesia" when it is used in this way becomes a technical term for the true believers of this Dispensation of Grace which is sandwiched between the 69th and 70th weeks of Daniel.

The meaning of the **first type** of use of the word is in the most general since. In this application, ecclesia is used to describe any group of people. In this meaning, "ecclesia" is used to refer to Israel as a gathered people out of the wilderness, (Acts 7:38) It is also used to refer to a regular assembly of citizens (Acts 19:39); or a group gathered for religious worship (Hebrews 2:12).

It is important to note however, that every time that ecclesia is used to describe Old Testament groups, it is always in the most General sense and **is in <u>contrast</u> to the way it is used in the fourth type of meaning.**

John Walvoord says..."The issue is whether ecclesia is **<u>ever</u>** used of Israel in the sense of the second, third, and fourth meanings. A study of every use of ecclesia in the New Testament shows that all references where ecclesia is used in the New Testament in reference to people in the Old Testament can be classified under the first meaning. **Of particular importance is the fact that ecclesia is <u>never</u> used of an assembly or body of saints except in reference to saints of the Present age.**" (Chapter 2, pg. 22)

The Dispensation of Grace was a mystery not revealed until Pentecost (Ephesians 3:2-5) (Colossians 1:26)

The Rapture of the Church was not revealed until (1Corinthians 15:51, 52) and that was 59 A.D. However, the 2nd return was taught throughout the Old Testament. In the Old Testament the 2nd return was expounded in detail but the Rapture was concealed.

32.2 The Rapture Is About The Groom Coming For His Bride.

The Church is called the Bride of Christ. The term "Bride" is not the same as a wife. A bride in Jewish custom represents a virgin that is engaged to be married. After the wedding and consummation, then she becomes the wife.

The Church is represented as the betrothed bride of Christ throughout the New Testament. In a sense the Dispensation of Grace is the betrothal period.

*"He that hath the **bride is the bridegroom**: but the friend of the bridegroom, which standeth and heareth him, rejoiceth greatly because of the bridegroom's voice: this my joy therefore is fulfilled".* (John 3:29)

"*Wherefore, my brethren, ye also are become dead to the law by **the body of Christ**; that ye should be to another, even to him who is raised from the dead, that we should bring forth fruit unto God".* (Romans 7:4)

This passage indicates a future marriage and refers to the Church as ***"the body of Christ".*** Nowhere in the Old Testament are Old Testament Saints referred to as the body of Christ." That terminology is exclusive to Church Age Saints.

"*For I am jealous over you with godly jealousy: for I have espoused you to one **husband**, that I may present you as a **chaste virgin** to Christ".* (2 Corinthians 11:2)

The marriage takes place between the translation of the Church and the 2nd coming.

"*Let us be glad and rejoice, and give honour to him: for the marriage of the Lamb* ***is come****, and his wife hath made herself ready*". (Revelation 19: 7)

In the Greek, the words "*is come*" indicates a completed act. The marriage took place after the translation and before the 2nd return.

"*And to her was granted that she should be arrayed in fine linen, clean and white: for the fine linen is the righteousness of saints*" (Revelation 19: 8)

The future wife; (Revelation 21:1- 22: 8)

32.3 At The Rapture, Jesus Takes The Church To Heaven.

At the Rapture the Lord comes to take the Church back home to Heaven. First believers are changed and then "*caught up*" in a moment, in the twinkling of an eye. Jesus said…

"*In my Father's house are many mansions: if it were not so, I would have told you. I go to prepare a* ***place*** *for you*". (John 14:2)

"*And if I go and prepare a* ***place*** *for you, I will come again, and receive you unto myself; that where I am, there ye may be also*". (John 14:3)

Notice that the purpose of the Rapture is to take us from where we are, to where he has prepared a place for us.

Notice also that Jesus catches us out and meets the believers in the air at the Rapture. At the Rapture, only believers are involved and Angels are not mentioned. Also, unbelievers are left behind. They don't understand what happened and some may not believe that it happened at all. The Antichrist and False Prophet will cover it up with a lie.

However at the 2nd return, Jesus comes with his armies which are made up of **Cherubim, Seraphim, Angels and returning saints**. The **Angels gather up the unbelievers <u>first</u> and kill them**. The believers are left behind to live on into the Kingdom.

32.4 Distinctions Between The Rapture And Christ's 2nd Return

The Rapture involves believers only. The Believers are changed and taken. The unbelievers, (unsaved) are left behind to face the Antichrist.

"*But I would not have you to be ignorant, brethren, concerning them which are asleep, that ye sorrow not, even as others which have no hope*". (1 Thessalonians 4:13)

"For if we ***believe*** *that Jesus died and rose again, even so them also which sleep in Jesus will God bring with him*". (1 Thessalonians 4:14)

"*Then we which are alive and remain shall be caught up together with them in the clouds, to meet the Lord in the air: and so shall we ever be with the Lord*". (1Thessalonians 4:17)

At the Rapture, only the believers are taken before the time of wrath begins. (1 Thessalonians 5:9)

In his book, "The Rapture Question", page 37, John Walvoord concludes his section titled "The Meaning of the Church" with the following paragraph;

"It is significant that none of the truths discussed as distinctive of the church are found in the description of saints in the Tribulation. Never are tribulation saints referred to as a church or as the body of Christ or as indwelt by Christ or as subject to translation or as the bride. As the church is a distinct body with special promises and privileges, it may be expected that God will fulfill His program for the church by translating the church out of the earth before resuming His program for dealing with Israel and with the Gentiles in the period of the Tribulation".

32.5 At The 2nd Return, Unbelievers Are Taken First.

The 2nd return of Christ is also called the harvest. In the Parable of the wheat and tares, Notice that the unbelievers are gathered first and the believers are left behind to live into the Kingdom.

"*Another parable put he forth unto them, saying, The kingdom of heaven is likened unto a man which sowed good seed in his field*": (Matthew 13:24)

"*But while men slept, his enemy came and sowed tares among the wheat, and went his way*". (Matthew 13:25)

"*But when the blade was sprung up, and brought forth fruit, then appeared the tares also*" (Matthew 13:26)

"*So the servants of the householder came and said unto him, Sir, didst not thou sow good seed in thy field? from whence then hath it tares?*" (Matthew 13:27)

"*He said unto them, An enemy hath done this. The servants said unto him, Wilt thou then that we go and gather them up?*" (Matthew 13:28)

The reason God doesn't destroy unbelievers is because of the undesired collateral damage…

…"*But he said, Nay; **lest while ye gather up the tares, ye root up also the wheat with them***". (Matthew 13:29)

"*Let both grow together until the harvest: and in the time of harvest I will say to the reapers, **Gather ye together <u>first</u> the tares, and bind them in bundles to burn them**: but gather the wheat into my barn.*" (Matthew 13:30)

"*Then Jesus sent the multitude away, and went into the house: and his disciples came unto him, saying, **Declare unto us the parable of the tares of the field***". (Matthew 13:36)

"*He answered and said unto them, He that soweth the good seed is the Son of man*"; (Matthew 13:37)

"The field is the world; the good seed are the ***children of the kingdom****; but the tares are the children of the wicked one;"* (Matthew 13:38)

"The enemy that sowed them is the devil; the harvest is the end of the world; and ***the reapers are the angels****".* (Matthew 13:39)

"As therefore the tares are gathered and burned in the fire; so shall it be in the end of this world". (Matthew 13:40)

According to this parable, the tares represent the unbelievers and they are gathered into bundles **first** and burned.

It is important to remember that **the unbelievers are gathered up first by the Angels and killed at the second return**. The believers are left behind to live into the Kingdom and repopulate the earth. The whole purpose of the 2nd return is to destroy the wicked (unbelievers) (Isaiah 13: 9); set up the Kingdom and fulfill the remainder of unfulfilled promises that God made to Abraham, Isaac and Jacob. Those promises were made to Israel and this parable is about Christ returning to set up his Kingdom.

The Rapture has nothing to do with any of that. That is why the Church should look for the Rapture at the end of the Church Age, which is seven years before the 2nd return.

Nobody sees the Rapture except those who are involved and perhaps those who are in the right place at the right time to see it. Of those who do see it, if they have not already been told about the event, they simply won't understand. If they have been told, it will be a sign for them to believe.

Satan has already prepared a lie to explain away this event and use it to further deceive the blinded and lost. But to those who have been informed by witnessing Christians, they may remember and understand that the Rapture is a sign from God and believe!

32.6 The Rapture Is Not Mentioned In The Book Of Matthew

In (Matthew 24) the Rapture was still a mystery. The Rapture was first revealed, A.D. 59" The book of Matthew begins in verse one with a reference to the Genealogy that connects Jesus with the promise to Abraham and the promise to David.

"*The book of the Generation of Jesus Christ, the son of David, the son of Abraham*". (Matthew 1:1)

This sets the tone for the rest of the book as being written to Israel about the Abrahamic promise and the Davidic Kingdom Covenant.

Likewise, (Matthew 24) is talking to Israel about the Kingdom and the seven years of tribulation that shall proceed it. The Tribulation is also called "the 70th week of Daniel". (Daniel 9:25-27)

Remember that the Rapture was first revealed in (1Corinthians 15:51, 52). That was about 27 years after this conversation recorded in (Matthew 24:) between Jesus and his disciples took place. Nowhere in the book of Matthew is the Rapture mentioned, not even once!

In verse 3 the disciples asked Jesus when the age would end and what would be the sign of his coming. The Age that they were in at the time was the Age of Law, more specifically the 69th week of Daniel. ("Week" means 7, it is used that way again in (Genesis. 29: 27, 28) He answered in verse 4 with a warning about deception.

In verse 5 he said that "*many*" would come in his name agreeing that Jesus is the Christ and deceive **many**. This is saying that many or the majority that preach in the name of Jesus and admit that Jesus is the Christ are actually false prophets and will deceive many. In the Last Days, the **Majority** of those that claim to be preaching in the name of Jesus are false teachers or false prophets!

It is important to understand that these warnings are all concerning the 70th week of Daniel. The end of the 70th week of Daniel will be the end of the Age of Law. Thirteen days after the Dispensation of Law ends on Yom Kippur, the Millennial Kingdom of Jesus Christ will begin. When the disciples ask about the end, they are talking about the end of the Dispensation of Law.

In verses 6, Jesus warns about wars and rumors of wars. In verse 7 he continues with a description of racial wars; national wars; and a description of the first four seals. There shall be **famines, pestilences** and **earthquakes in divers places**. (Revelation 6:1-8)

In (Revelation 6:1-8) it is stated that these first **four seals kill one fourth of the world's population**! "*And power was given unto them over the **fourth part** of the earth, to kill with sword, and with hunger, and with death*" (Revelation 6:8)

"*All these are the beginning of sorrows*" (Matthew 24:8)

What we have read so far sounds bad, but verse 8 says that this is only the beginning of sorrows. This is the first 3 ½ years of the 70th week of Daniel, which is also known as the Tribulation.

Verse nine, I believe is the 5th seal judgment of Revelation 6 which is in the middle of the Tribulation. This is when things really get bad! Up to verse eight was the beginning of sorrows, but now is the beginning of great tribulation. This beginning of Great Tribulation starts in verse nine and continues through verse 21. Verse nine marks the middle of the Tribulation.

"*For then shall be **great tribulation**, such as was not since the beginning of the world to this time, no, nor ever shall be*" (Matthew 24:21)

From verse nine to verse 21 is a description of what happens in the middle of the tribulation. Death, violence, deception betrayal and hatred are intensified characteristics of this second half of the 70th week of Daniel. This is the midpoint of the Tribulation and the point when things go from bad to worse. Twenty five percent of the worlds population will die during the first half of the Tribulation. The second half of the Tribulation will be much worse, but in verse thirteen it says…

"*But he that shall endure unto the end, the same shall be saved"* (Matthew 24:13)

This verse is saying that if a believer can endure all of this and survive for another three and a half years, then they will live on into the Kingdom.

There are plenty of false teachers today that ignore the context of what is being said here and use this passage out of context to imply works for salvation. As you can see from the context and the next verse in particular, this verse is talking about **flesh** being saved, **not souls**. We are talking about survival here!

*"And except those days should be shortened, there should **no flesh be saved**: but for the elect's sake those days shall be shortened"*. (Matthew 24:22)

In verses 23 through 36 we are warned of false prophets and false Christ's doing miracles, signs and wonders to deceive. We are also warned not to believe claims about Jesus returning before we can see and hear the things described in God's word that relate to his return. God says, every eye shall see him; as lightening shinning from the east unto the west; the sun and moon darkened; the whole world will shake while the Angels are gathering the unbelievers first.

"But of that day and hour knoweth no man, no, not the angels of heaven, but my Father only". (Matthew 24:36)

This passage tells Israel that no man can know the **day or the hour** of the Lord's return. But in contrast to that, in (Revelation 3:2, 3) the warning to the Church of Sardis is to **watch** for the Lord's return or else they would not know the **hour** of his return. Therefore, the whole point of watching for the Lord's return is to know the hour of the Rapture.

If you are watching, you should know the hour of the Rapture according to this verse (Revelation 3:2, 3). The difference between these two passages is due to the fact that one passage is talking to Israel about the second return of Christ to set up his Kingdom and the other passage is talking to the Church about the Rapture which occurs seven years earlier. These are different events separated by 7 years with different purposes.

The 2nd return of Christ will occur between Yom Kippur and Tabernacles. There are thirteen days between those two feasts. Which one of those thirteen days is the day of Christ's return is not known.

"But as the days of Noe were, so shall also the coming of the Son of man be" (Matthew 24:37)

"*For as in the days that were before the flood **they** were eating and drinking, marrying and giving in marriage, until the day that Noe entered into the ark"*, (Matthew 24:38)

*"And knew not until the flood came, and **took them** all away; so shall also the coming of the Son of man be"* (Matthew 24:39)

Verse 37 compares the days of Noe with the 2nd return of Christ. It continues on in verse 38 to say that **they** were eating drinking, marrying etc. until Noe entered into the Ark…

…*"And knew not until the flood came, and **took them** all away; so shall also the coming of the Son of man be.* (Matthew 24:39)

In the days of Noe, the unbelievers were **taken** by the flood and killed. Just like the Parable of the Wheat and Tares, the unbelievers were **taken** first and killed and the believers were left to live on into the Kingdom and repopulate the Earth.

"*Then shall two be in the field; the one shall be **taken**, and the other left"*. (Matthew 24:40)

"Two women shall be grinding at the mill; the one shall be ***taken****, and the other left".* (Matthew 24:41)

At the 2nd return of Jesus, the unbelievers are **taken** by the Angels and killed by Jesus. (Matthew 13:24-43) The believers are left to live on into the Millennial Kingdom and repopulate the earth.

In contrast to the 2nd return, at the Rapture believers are "translated" or "changed" "in a moment, in the twinkling of an eye" and taken to heaven as the bride of Christ.

32.7 Comparisons, Contrasts And Conclusions

The Rapture, when it occurs, will be at the end of the Feast of Trumpets, "***at the last trump***", at **sundown Jerusalem time**, which is approximately about 6:37 pm (or about 11:37 am Eastern Standard Time) That is an approximate time because the exact time changes from year to year. The exact time for the sunset in Jerusalem that marks the end of the old year and the beginning of the new one should be checked every year. Every believer should know when the sun sets in Jerusalem on that day. Compare your time zone to Jerusalem and know when their sundown occurs in your time zone.

Perhaps someone you have been witnessing to about Christ has been stubborn and remained in unbelief. It would be good to bring in the Jewish New Year with someone like that. If you get raptured in front of their eyes, they may just change their mind.

Remember (Revelation 3:3)…" ***If therefore thou shalt <u>not</u> watch****, I will come on thee as a thief, and thou shalt not know what hour I will come upon thee"*… This implies that **<u>if</u>** you **<u>are</u> watching**, you will know the hour.

This could be an annual tradition until the year that the Lord comes for the Church. Let this be the year of the first annual Day of Judeo-Christian Unity on Rosh Hashanah. The Bible says that the Jews are looking for a sign. What perfect timing, this could be it.

At the Rapture, Christ meets believers in the air and they are translated or changed. There is no mention of Christ's feet touching the ground. They receive their new bodies and Christ takes his bride home to heaven. The Church is judged in Heaven first and then the marriage celebration begins. Meanwhile back on earth, No one else is involved.

The world continues on uninterrupted as the 70th week of Daniel begins and the Antichrist and False Prophet cook up lies to explain away the missing people who disappeared. They probably will even take credit for the event.

Seven years later, when the 70th week of Daniel ends on the Feast of Yom Kippur, Christ will return between Yom Kippur and the Feast of Tabernacles to stop the battle of Armageddon.

At the second return ***"every eye will see him"***, even as lightening shines from east to west, everything shakes at God's presence. Christ comes down to earth, stands on the Mount of Olives, it cracks open (Zechariah 14:4) and creates a big valley. The Angels **first** gather up all of the unbelievers (Matthew 13:30) and cast them into the newly created valley and they are killed by Jesus. (Joel 3:12 & 13) The believers are not changed or translated like described at the Rapture. They are not taken to heaven, but are left on the earth in their mortal bodies to rebuild and repopulate the earth. (Isaiah 4:3)

There is a lot of differences between the Rapture and the second return. Nevertheless, false teachers are multiplied in the last days and so is their false teachings to deceive about the Rapture.

33

God Commands Believers To "Watch"

(1Thessalonians 5:6) (Matthew 24:48)

33.1 The Sign Of An Evil Servant.

*"For yet a little while, and he that shall come will come, and **will <u>not</u> tarry**."* (Hebrews 10:37)

If you say or even think that Jesus will delay his return, then God says that is the sign of an evil servant…

*"But and if that **evil servant** shall say in his heart, My lord delayeth his coming"* (Matthew 24:48)

*"The lord of that servant shall come in a **day** when he looketh not for him, and in an **hour** that he is not aware of"* (Matthew 24:50)

The Pharisees are called wicked for not watching and not knowing the **<u>day</u>** and the **<u>hour</u>** of our Lord's first visit while other individuals did recognize the signs

"The Pharisees also with the Sadducees came, and tempting desired him that he would shew them a sign from heaven". (Matthew 16:1)

"He answered and said unto them, When it is evening, ye say, It will be fair weather: for the sky is red". (Matthew 16:2)

*"And in the morning, It will be foul weather to day: for the sky is red and lowring. O ye **hypocrites**, ye can discern the face of the sky; **but can ye not discern the signs of the times?**"* (Matthew 16:3)

*"**A wicked and adulterous Generation** seeketh after a sign; and there shall no sign be given unto it, but the sign of the prophet Jonas. And he left them, and departed"*. (Matthew 16:4)

It is important to note that Jesus is comparing weather signs to signs of his first visit. Furthermore, Jesus is scolding the Pharisees for not paying attention to the signs of his first visit. God has also made the same comparison of weather signs to the signs of his return for the Church at the Rapture. (1Thessalonians 5:1-6) In verse 6 we are commanded again to watch.

Every important detail about God's incarnation and subsequent sacrifice for the sins of humanity was prophesied well in advance. On the morning known as Palm Sunday, Jesus rode into Jerusalem on a donkey. The time of that event was prophesied to that exact day! (Daniel 9:25) The details of that day (Matthew 21:5) were written about five centuries before it happened. (Zechariah 9:9) All of the details of that week were declared by God long before it happened so there would be no question about who it was! Yet the religious leaders of Israel were clueless!

The Pharisees did not believe, therefore they did not watch for the LORDS return and did not recognize him.

33.2 Believers Are Commanded To Watch For Christ's Return.

(1Thessalonians 5: 6) says that if we are watching, we **will** know the **day** and (Revelation 3: 2, 3) tells us if we are watching, we will know the **hour** of our Lord's return. The only way that I know of to be able to know the day and the hour of the Rapture of the Church is if it has its own Feast day and it does.

Trumpets are often mentioned with the Rapture in the Bible. The Feast of Trumpets is next in line to be fulfilled. The Feast of Trumpets is the Hebrew New Year for counting years and epochs. The next prophetic event is the Rapture of the Church. The Rapture of the Church will mark the end of one epoch that we call "the Church Age" and the Beginning of a New Epoch called "*the Day of the Lord*".

The Feast of Trumpets is intended by design to mark the end of one epoch and the beginning of another. I think that we can conclude that the Feast of Trumpets will end, some year in the future with the Rapture ..." *at the last trump*". The moment the shofar marks the end of the year and epoch with its final toot. That is the **moment** that... "*we shall be changed*" (1Corinthians 15: 51, 52).

This Greek word translated as "*watch*"(Strong's #G1127 *grēgoreō*) has a Hebrew equivalent which is (Strong's #H8104 *shamar*). This Hebrew word is used in (Leviticus 26: 2) and translated as "***keep***".

*"Ye shall **keep** my sabbaths, and reverence my sanctuary: I am the LORD".* (Leviticus 26: 2)

This draws our attention to the seven Feasts or Sabbaths of the Lord spoken of in (Leviticus 26). The first four Feasts, which are observed in the Spring. They were fulfilled on their corresponding days, hours and in their numerical order by Jesus Christ (Yeshua).

The remaining three Feasts are all observed in the Fall and will also be fulfilled on their corresponding days, hours and in their numerical order by Jesus Christ (Yeshua). If you believe that simple fact, then you are ready to start "*watching*" because the next Feast in line is the fifth Feast.

The fifth Feast is called the Feast of Trumpets. Every year it is to be observed on the **first day of the seventh Hebrew month called Tishrei.** Tishrei one is the day for counting years and epochs. The Rapture of the Church will mark the end of the Church Age and the beginning of the "*Day of the Lord*". Those are both epochs.

Don't confuse Nisan one with Tishrei one. Nisan one is for counting months, not years or epochs. The Bible says that the Jews are looking for a sign. I believe that the Rapture of the Church may be that sign.

God will be sending two witnesses to Israel <u>before</u> The Day of the Lord. Those two witnesses will have the power to do the same miracles that Moses and Elijah did. God identifies Elijah by name as being a witness to Israel in the last days. The presence of Moses and Elijah at the Transfiguration of Jesus Christ (Matthew 17:1-3) further identifies them as the two witnesses. Michael the archangel contended and disputed with the devil over the body of Moses(Jude 1:9) because God had plans for it.

(1Corinthians 1:22) *"For the Jews require a sign, and the Greeks seek after wisdom:"*

If a Jewish person sees Moses and Elijah doing miracles, they may be inclined to listen and believe. If a Jewish person witnesses the Rapture of a Christian friend, he will likely begin the 70th Week of Daniel, as a new believer and there will be no stopping him.

The Church is commanded to "*watch*" for Jesus to come for them at the Rapture. At the Rapture, the Church will be changed and taken to the place that God has prepared for them in Heaven.

This will happen at the end of the "***dispensation of the grace of God***" (Ephesians 3:2) also known as the "**Church Age**" or the "**Age of Grace**". Those are all terms used to describe the same thing.

The first term ***"the dispensation of the grace of God"*** is a direct quote from the Bible. (Ephesians 3:2) The other terms are descriptive terms used to make the concept easier to understand for more people.

The Rapture marks the end of the Church Age and the beginning of the **Day of the LORD.**

It is clearly stated here that Jesus will <u>not</u> come as a thief in the night to the believers who are faithfully watching!

"*But ye, brethren, are not in darkness, that **that day** should overtake you as a thief*" (1Thessalonians 5:4).

"*Ye are all the **children of light**, and the **children of the day**: we are not of the night, nor of darkness.*" (1Thessalonians 5:5)

"*Therefore let us not sleep, as do others; **but let us watch** and be sober*" (1 Thessalonians 5:6)

"*For they that sleep, sleep in the night; and they that be drunken are drunken in the night*" (1Thessalonians 5:7)

The Church of Sardis is told to "*watch*"!

"***Be watchful**, and strengthen the things which remain, that are ready to die: for I have not found thy works perfect before God*". (Revelation 3:2)

In the next verse, God repeats the command to watch. This passage is warning the Church about the dangers of not watching. It warns that you will "*not know what **hour**"* our Lord will return **if** you are **not watching** for it.

"***If** therefore thou shalt **not watch**, I will come on thee as a thief, and thou shalt not know what **hour** I will come upon thee*". (Revelation 3: 3)

The reason implied here is for believers to watch **so that they will know what hour the LORD will return**!

In verse 3, they are warned that if they don't watch they won't know the **hour** of Christ's return. The implication being that **if they are watching** they **will** know the **hour**!

If you are not watching then you are not obeying God's command and you won't know the hour of Christ's return! Do you know the hour? The Bible tells us (Revelation 3: 3) to watch for it.

This passage commands believers to watch, it implies that there are things to watch for and furthermore it implies that you should know the hour of our Lords return if you are watching for it!

In the following passage, God is talking to believers that are watching for his return and instructing what they should be doing as they ***"see the day approaching"***.

"Not forsaking the assembling of ourselves together, as the manner of some is; but exhorting one another: and so much the more, ***as ye see the day approaching****"*. (Hebrews 10:25)

If God says that you can ***"see the day approaching"*** then there must be something to see! Are you watching for it?

These passages are talking about the Lord's Return for the Church. They are talking to the Church about preparing for our future reunion in the air when we will be changed and meet Jesus in person at the Rapture.

The Rapture is where believers are "***translated***" or "***changed***" ..."***in a moment, in the twinkling of an eye***" (1Corinthians 15: 52)

We will receive our new bodies and be taken to Heaven where Christ has prepared a place for us. (John 14:2, 3) The first thing that believers will see at the Rapture is the Judgment seat of Christ. (Revelation 4:2) That is what we should be thinking about and preparing for!

The Rapture has been considered to be eminent because God did not give us the year, but he did give us the Feast day, hour, moment and plenty of signs to watch for.

The Feast day is Rosh Hashanah or the Feast of Trumpets. The Feast of Trumpets is the Biblical New Year for counting years and epochs. It is the first Fall Feast or the fifth Feast or next in line to be fulfilled. The first four spring Feasts were fulfilled in their order and on their day and hour. The Church Age began on the fourth feast which was the last spring Feast called Pentecost. The Church Age will end on the fifth Feast which is next in line and is the first of the three remaining fall Feasts.

I believe that (1Corinthians 15: 52) Identifies the hour and the ***moment*** as the final trumpet blast of the Shofar that ends the Feast of Trumpets.

"***In a moment****, in the twinkling of an eye,* ***at the last trump****: for the trumpet shall sound, and the dead shall be raised incorruptible, and* ***we shall be changed***". (1Corinthians 15: 52)

33.3 Can't Know The Hour Of The Rapture? Not So!

God would not command us to watch unless there was plenty to watch for.

"*Remember therefore how thou hast received and heard, and hold fast, and repent.* ***If*** *therefore thou shalt* ***not watch****, I will come on thee as a thief, and thou shalt not know what* ***hour*** *I will come upon thee.*(Revelation 3: 3).

Twenty three times, God commands believers to watch. In this passage we are given a good reason for watching. It says that if you are not watching, Jesus will come as a thief to you and you will not know the "**hour**" that he is coming. Therefore, **the whole point of watching is so that you will know the hour** and you will not be surprised as by a thief. God is saying here that if you are watching for the Lords return for the Church at the Rapture, then you should know the hour. Knowing the hour is an inseparable part of watching.

Most people say that you cannot know the day or the hour of the Rapture of the Church. They come to this wrong conclusion because they don't understand that (Matthew 24:36) is not talking about the Rapture of the Church. It is talking to Israel about God's promises to Israel and about the 70th week of Daniel.

the Bible says here that you will know the hour if you are watching. Knowing the hour of our Lord's return for the Church (John 14: 3) (1Corinthians 15: 53) is an indication that someone is watching. God is connecting the act of watching with knowing the hour of his return for the Church at the Rapture. Believers should be watching for the **hour**. (Revelation 3: 3)

Compare this passage with … (1Thessalonians 5: 4-6) which talks about knowing the day of the rapture

(1Thessalonians 5: 4-6) *"But ye, brethren, are not in darkness, that* ***that day*** *should overtake you as a thief".*

If you are watching for it, then you will see the day coming and you will know the day and the hour.

I have often heard preachers quoting Matthew 24:36 … They use this passage to say that you cannot know the day and the hour of the Rapture. But this passage is not talking about the rapture.

"But of that day and hour knoweth no man, no, not the angels of heaven, but my Father only". (Matthew 24:36)…

Many use that passage to say that no one can know the day and the hour of the Rapture. But this passage is not talking about the Rapture, it is talking about the second return of Christ, approximately 7 years after the Rapture. A completely different event.

The Rapture is not mentioned in the book of Matthew at all and certainly not in the 24th chapter.

The Church is told to watch for the Rapture in the Last Days, but Israel is told to watch for the 2nd return in the Last Days. The Last Days for the Church is the Last Days of the Church Age. The Last Days for Israel is the Last Days of the Dispensation of Law or the 70th week of Daniel which is followed by the Return of Jesus to set up his Kingdom! Matthew 24: is talking to Israel about the 70th week of Daniel.

Some preachers don't read carefully and willfully do not rightly divide God's word.

*"But ye, **brethren**, are **not in darkness**, that that day should overtake you as a thief."* (1Thessalonians 5:4)

This passage is saying that *"that **day** should overtake you as a thief"*... only if you are not watching for it. Revelation 3:3) is saying the same thing about knowing the hour.

*"Therefore let us not sleep, as do others; but let us **watch** and be sober".* (1 Thessalonians 5:6)

God says that if his people are watching as ordered they will not be surprised or caught off guard when the Rapture happens.

*"In a moment, in the twinkling of an eye, **at the last trump**"*: (1Corinthians 15:52)

I believe that this passage reveals that we will all be changed at the same **moment** as the last trump of Rosh Hashanah is blown at sundown in Jerusalem. Therefore we know the day, the hour and moment, but not the year.

The Jews have two days set aside to observe Rosh Hashanah even though the Feast of Trumpets is only one day in the Bible. That is because this feast is based on the new moon and the moon is not visible. It crosses the sky with the Sun at this time of year.

Traditionally the Priest would consult his lunar records and look at the barley crops to determine which day is correct. This was done every year. You could expect for Rosh Hashanah to fall on one of those two days. But you had to wait for the Priest to decide which of those two days would be the correct day to observe.

On the first day, the priest would either say that this is the correct day, therefore let the Trumpets begin their proclamation at sundown; or he would say wait until tomorrow to start. So at that moment, you would either know that you had 24 hours left in the year or 48 hours. On the year of the rapture, I believe that those two days will be full of signs for those who are watching.

I became a believer in the summer of nineteen sixty-seven, in the month of June, Just after the six day war in Israel. Since that time I was taught by what I believe to have been one of the best Bible Teachers of our time, Dr. Hank Lindstrom. His specialty was Bible Prophecy and difficult passages.

Since before I was in High School I was made aware of current events as they pertain to the Bible. Hank's church was taught how to *"watch"*. Dr. Hank Lindstrom went home to be with the Lord on Oct. 13, 2008. I have continued Hank's tradition and I can tell you that the signs are picking up.

Trump has just recently recognized Jerusalem as the "eternal capital of Israel" and promised to help defend her. The half shekel silver Temple coin that is required by Old Testament law was just minted by the Sanhedrin to commemorate Trump's recognition of Jerusalem as Israel's capitol and **The <u>Seventieth</u> Anniversary Of Israel's Independence**. The Old Testament law required every male to pay to enter the Temple area with one of these coins.

On one side of the coin is an image of the **Third** Temple. On the other side is the profile of Trump over the profile of King Cyrus. This is a major sign that the Jews are getting ready to build the third Temple and Trump is involved! The Temple coin is the last step before construction on the third Temple begins. The Jews are getting ready to build the Temple and when they do, there will be "sudden destruction" following promises of "***peace and safety***"!

The Church has nothing to do with the Dispensation of Law, but the Third Temple does. That is one reason why I believe that the rapture of the Church will happen at the same moment that Temple construction begins. That moment marks the beginning of the New Year and the Day of the Lord when it is legal to go back to work after the high holy day of Rosh Hashanah ends at sundown. Someday the Rapture of the Church will mark the end of one epoch and the beginning of another at the same final sound of the shofar. may also mark the end of the Church Age and the beginning of the Day of the Lord. that moment that Temple construction becomes the first line of business of the evening.

The exact day that Rosh Hashanah begins was uncertain until the Priest checked his lunar records and checks his barley harvest on the appointed day to determine which day is the correct one on any given year.

But today NASA has Lunar calculations that take away the uncertainty of knowing which day it is but the Jews are still following a two day tradition.

The Bible does not give us the year, therefore the Rapture has always been considered to be imminent. Nevertheless, I believe that believers who are faithfully watching probably will be able to know when the year arrives.

The Bible says that we won't be caught off guard and compares our watching for the Rapture to watching for signs of the weather and a woman's labor pains. Although you may not be able to predict these kinds of events far in advance with precise accuracy, but when they do get near, you can tell when they are about to happen. That seems to be the situation now.

(Matthew 16:3) "... *O ye hypocrites, ye can discern the face of the sky; but can ye not* ***discern the signs of the times****?*"

Considering the quantity of signs and verbiage on this matter, it is reasonable to consider that believers will probably be able to recognize the year when it arrives by recognizing current events described in scripture.

For example, I believe that the Church will be raptured before the Third Temple construction begins. The construction must be done in secret otherwise the Muslims will attack. A corona virus lockdown should do the trick. If everyone in locked in their homes they will not know that the Temple is being built. I believe the Temple will be built in three days. Of course this is speculation, but that is what we are told to do. If you are watching, you have to speculate about how current events might match scripture.

Temple construction should begin while Trump is in office and willing to defend Israel. The Bible mentions a "*covenant with death*" that Israel makes to avoid becoming a victim of Islamic violence. God says that the treaty will fail. It is used to set up Israel for a betrayal in the middle of the Tribulation.

If that Treaty is depending on Trump to defend Israel and something happens to Trump at the same time that Temple construction begins, Israel will be in big trouble. If Trump gets raptured that would fit the timing perfectly.

If the rapture happens and Trump or Pence are left behind; the Christians who were watching their backs will be gone and the Globalists will take them out and replace them with the Antichrist. That is my speculation, now let's watch and see what happens.

We are only speculating and not making predictions, there is a difference. That is part of watching.

33.4 A Crown Of Righteousness… If You Are Watching.

(II Timothy 4:8)…

God has prepared a special reward for believers who are watching for his return.

"*Henceforth there is laid up for me a **crown of righteousness**, which the Lord, the righteous judge, shall give me at that day: and not to me only, but unto all them also that love his appearing*". (II Timothy 4:8)

THE CROWN OF RIGHTEOUSNESS, is a crown that will be given to those who love His appearing because they live their lives anticipating Christ's soon return and are watching for the signs. This is talking about people who are "watching" for our Lords return. All of this is connected together.

Jesus Scolded The Pharisees For Not Watching The sign of an evil servant is someone who says that Jesus may delay his coming.

God would not tell us to "Watch" unless there was something to watch for.

Jesus scolded the Pharisees because they could *"not discern the signs of the times?"* (Matthew. 16:3) Do you know the **signs** of the times?

Daniel told the very year and day the Messiah would be announced on Palm Sunday. (See chapter 23 ETBH)(Daniel 9:25) Why didn't the religious leaders recognize him? Jesus was recognized as a child by those who knew the scriptures, but the official religious leadership were clueless. (Luke 2:21-38). This is a good reason for home schooling!

"It is the glory of God to conceal a matter, But the glory of kings is to search out a matter" (Proverbs 25:2)

Roger Bacon, a famous twelfth Century English mathematician, scientist and proponent of scientific experimentation, once said… "Experimental science is the queen of sciences and **the goal of all speculation.**"

The Bible says…

"*I wisdom dwell with prudence, and **find out knowledge of witty inventions***". (Proverbs 8:12)

(Sir Isaac Newton), once said… "No great discovery was ever made without a bold guess." Newton is a believer!

Speculation is a natural function for an inquisitive, thinking person. When speculation is expressed verbally to communicate thoughts and ideas of common interest to others there is the possibility of gaining new ideas and understanding through sharing. One person may have additional knowledge, hence a different possible view. The other person gives his opinion based on different knowledge. That opinion is speculation. It does not necessarily mean that it is correct, but it could be if the speculation is consistent with the supposed missing facts. This process of elimination is called the scientific method.

Newton speculated, Galileo speculated, Thomas Edison speculated, Einstein is famous for speculating. Dr. Hank Lindstrom speculated and I speculate.

We are commanded 23 times to "watch" it behooves us to find out what we are supposed to be watching for. The process of "watching" entails paying attention to God's Word and heeding Gods commands. If we know what to watch for, we shouldn't have a problem but you can't watch if you are not speculating about new information. Speculating out loud among other believers is the fast track to sharing and gaining new information and understanding.

Satan hates it when Christians are gathered together and speculating about God's word out loud. Satan's minions will attack and try to shut down any such attempt by criticism, ridicule or by any other means.

"Speculation" is a scientific term used in science. "Meditating means the same thing but the word is not used in science as often as speculate. Because Meditate is used in the Bible, some may add a "religious" or spiritual meaning. Without speculation within one's own mind, discovery will cease. Without speculation no new ground is covered.

God commands us to speculate when he tells us to meditate on his word. Think about it. If you are thinking about something you have learned and are trying to figure out something about it that is new to you, then you have to go through the process of gathering information, speculation and finally experimentation.

Here are some definitions for "speculate"…

Dictionary.com … *Speculate* definition, to engage in thought or reflection; meditate

Merriam Webster Dictionary… 1a: to meditate on or ponder a subject

Cambridge Dictionary… *speculate* meaning: 1. to guess possible answers to a question when you do not have enough information to be certain.

Vocabulary.com … "When you *speculate*, you use what you know to make a prediction about an outcome, like when you *speculate* about what someone will do; to reflect deeply on a subject; **Synonyms**; chew over; contemplate; excogitate; meditate; mull; mull over; muse; ponder; reflect; ruminate; think over; talk over conjecturally, or review in an idle or casual way and with an element of doubt or without sufficient reason to reach a conclusion"

The only ones who don't want Christians to speculate about God's word is Satan and his minions. They don't want Christians to talk out loud amongst ourselves and share thoughts, ideas and possibilities with each other because that will direct the search for answers in the right direction. Satan hides under a cloche of lies. If Christians freely speculate out loud without fear of being criticized for it, Satan's cloche of lies will be exposed.

34

What Should We Watch For?

34.1 The "Last Days" Began When Israel Was Reborn.

God said he would bring Israel back to their land in the "*last days*". Since May 14, 1948 Israel has been back and according to God, this time it will be to stay! God prophesied the very day Israel was reborn on May 14, 1948! (ETBH chapter 25)

..."*O LORD, save thy people, the remnant of Israel ... Behold, I will bring them from the **north country**, and gather them from the coasts of the earth ... a great company shall return thither ... They shall come with **weeping**, and with supplications will I lead them: ... hear the word of the LORD, O ye nations, and declare it in the isles afar off, and say, He that scattered Israel will gather him**, and keep him**, as a shepherd doth his flock*". (Jeremiah 31:7-10)

The North Country mentioned here is Russia and the weeping spoken of, I believe, is from the Holocaust.

"*For the children of Israel shall abide **many days** without a king, and without a prince, and without a sacrifice, and without an image, and without an ephod, and without teraphim*": (Hosea 3:4)

That passage describes the last 2,000 years (Hebrew years) since 70 A.D. For 2,000 years, the Jews have been without a King or a Temple. Throughout their history, Jews have always had household gods except for the last 2,000 years.

"***Afterward** shall the children of Israel **return**, and seek the LORD their God, and David their king; and shall fear the LORD and his goodness in the **latter days***" (Hosea 3:5)

The "***many days***" mentioned in verse 4 is defined in (Hosea 6:1) as the 2,000 years known as the Dispensation of Grace. (ETBH chapters 26)

"*And it shall come to pass in that day, that the Lord shall set his hand again the **second time** to recover the remnant of his people, which shall be left, from Assyria, and from Egypt, and from Pathros, and from Cush, and from Elam, and from Shinar, and from Hamath, and from the islands of the sea*". (Isaiah 11:11)

God had taken Israel off their land twice and brought them back twice. This second time will be the last!

But remember that the Catholic church said that God was through with Israel. Nobody in the Catholic church believed that the Jews would ever regain their land. But some, like Sir Isaac Newton, believed the Biblical record that said that Israel would return.

Newton believed in the literal interpretation of the Bible. In other words, he believed that God said what he meant and meant what he said! It is all quite simple. Satan complicates things.

"*For, lo, the days come, saith the LORD, that I will bring again the captivity of my people Israel and Judah, saith the LORD: and **I will cause them to return to the land that I gave to their fathers, and they shall possess it***". (Jeremiah 30:3)

Israel has been back in their land since 1948 and that means that we are living in the "***latter days***" (Hosea 3:5) ever since. God will not remove them from their land again and He will not allow anyone else to remove them either! However, they will be invaded by the Antichrist in the middle of the Tribulation.

34.2 God Promised Signs On Earth… (Matthew 24)

"For nation shall rise against nation, and kingdom against kingdom: and there shall be famines, and pestilences, and ***earthquakes****, in divers places".* (Matthew 24:7)

"Nation shall rise against nation" is talking about race wars. The word "nation" here was translated from the Greek word ***ethnos*** **and is designated in the Strong's Greek dictionary as G1484. It means race. So we are talking about Race Wars.**

"**Kingdom against kingdom**" is a reference to ***"wars and rumors of wars"***. (Matthew 24: 6) That certainly describes the time we are living in.

Famines and pestilences are naturally the result of wars. War causes a lack of food. A lack of food causes disease. Because of the Corona virus, farmers destroyed crops, livestock and milk in quantities that are likely to cause food shortages next year (2020). I think that we may begin seeing some of that this fall. Is this just the beginning? If so, it is not going to get better. We should be warning the world about "*sudden destruction*" before it is too late.

(1 Thessalonians 5: 3)*" For when they shall say, Peace and safety; then* ***sudden destruction*** *cometh upon them, as travail upon a woman with child; and they shall not escape"*

Food control has been used in the past to weaken and eliminate opposition. Violent Revolutions typically exterminate any potential future enemies. The more wrong thinkers that are eliminated up front, the less chance of opposition springing up later. The French Revolution or the First Masonic Revolution is called a bloody and violent revolution. Heads were being chopped off daily. Christians were being killed because of their faith and Freemasons controlled the whole thing.

Memoirs Illustrating the History of Jacobinism by Augustin Barruel, a French Jesuit priest. It was written and published in French in 1797 and translated into English in 1799

Stalin planed a food shortage as his weapon against his own people. Stalin committed genocide by starving millions! Many Russians had to choose between starvation and cannibalism. It was like the Zombie Apocalypse! There were stories about the government sending people out into the countryside with survival instructions. Instructions like digging up dead animals to boil the remains for soup. They called it "bone soup". The streets in the cities were littered with the dead and dying.

Epidemics of diseases like typhus, typhoid fever, smallpox, influenza, dysentery, cholera, even bubonic plague were the result of starvation in Russia during the early1920's, just after the Bolshevik Revolution began. Disease spread all over the country. Can you imagine something like that going on in America and Europe?

If you don't think that it can happen here, it can! Our food reserves are low and dependent on our power grid which is vulnerable to terrorist attacks and we can't trust our leaders! Our leaders are bringing terrorists into our western countries and protecting them in the name of being politically correct. The Bible talks about heads being chopped off during the tribulation. That sounds like Islam to me. These are all signs that God's judgement is near!

Earthquakes have been increasing in frequency and strength since the middle of the last century. Volcanic activity has also increased more than linearly since the middle of last century. The Ring of fire in the pacific has become active with volcanic eruptions, increasing in number and intensity.

The Super Volcano underneath Yellowstone National Park has been showing signs of change. The ground has lifted and temperatures have increased. The ground has become so hot that the asphalt on some roads has melted rendering them unusable. Parts of the park have been closed down. The animals have been leaving the park.

The frequency and magnitude of earthquakes have also increased in the area. The super volcano in Yellowstone National Park has a history of erupting about every 600,000 years and the last eruption was over 640,000 years ago. According to geological data, it is past due for another eruption!

I would not be surprised if Yellow Stone turned out to be involved with one of the "***Thunder Judgements***" spoken of in Revelation. There are Scientists who say that if Yellow Stone erupts in our lifetime, it will devastate most of our country! It may never be the same again! Yellow Stone could completely destroy everything in the North West to the Midwest. Most of what is left will be covered in volcanic dust and uninhabitable.

Volcanic dust is super fine splinters of glass. That stuff causes lung disease, bone disease and a painful death. Much of this dust will fall back to the earth and cover the Midwest of the U. S. and reach to the East Coast. Perhaps millions of tons of this dust will be pushed up beyond our atmosphere and surround the planet. The sun would be blocked, creating a "nuclear winter". That term was coined to express the idea that so much dust in the atmosphere would block out the sun for at least a year. If you lose one season of planting, a lot of people are going to starve!

During the tribulation, the sun will actually become brighter and then darker for a while. (Revelation 16:7-11) Men will be scorched and fires will be set… *"And men were scorched with great heat, and blasphemed the name of God, which hath power over these plagues: and they repented not to give him glory*" (Revelation 16:9) *"And the fifth angel poured out his vial upon the seat of the beast; and his kingdom was full of darkness; and they gnawed their tongues for pain,"* (Revelation 16:10)

The Bible talks about signs in the sun. I suppose that there will be plenty of smoke from volcanic eruptions that could make the sun and moon appear red when it can be seen. One third of the vegetation will be burned up. (Revelation 8:7) We are seeing some of that in California and Australia already.

Just think, if God destroyed Sodom because of Sodomy and Pompeii was doing the same things when Mount Vesuvius erupted, what do you think God will do to the Nation of the Antichrist? Mr. Obama was promoting sodomy, perversion and abortion while making fun of Christians and the Bible. I can see the U.S. getting it from all directions by God during the

Tribulation! A Yellow Stone eruption would seem to be consistent with end times descriptions in the Bible.

The *Nile River* would dry up in the *"latter days"* and it is… (Isaiah 19: 6-10)

The **Euphrates River** would dry up to prepare the way for a 200,000,000 man army from the East. (Revelation 9:14, 16:12) This happens at the **end** of the Tribulation. That is about seven years after the rapture.

China has had the ability to field a 200, 000, 000 man army since the nineteen sixties. That is the same exact, unbelievably enormous number that the Bible uses about two thousand years ago to describe the army of the Kings of the East. At the time, there wasn't that many people on the whole earth.

There are dams along the **Euphrates River** that are often controlled by terrorists. With just the push of a button the whole river could be dried up. Until recently, none of this seemed possible.

Israel would be hated above all Nations… (Matthew 24:9) Nobody believed that Israel would ever exist after 70 A.D. or ever be significant again, but they have and it is making the Muslims angry. The religion of Islam did not exist until the seventh century, but the Bible describes it thousands of years ago.

There are too many signs to talk about them all.

34.3 God Established The Lights In The Heavens For Signs.

"And God said, Let there be lights in the firmament of the heaven to divide the day from the night; and ***let them be for signs****, and for* ***seasons****, and for days, and years:"* (Genesis 1:14)

The word "***signs***" used in (Genesis 1:14) is the same Hebrew word used in (Genesis 9:13) to describe a rainbow as a sign. In (Genesis 17:11) that word is used to call circumcision a sign. Again in…(Exodus 4: 8, 9, 17, 28, 30) the same word is used to describe the miracles that Moses did in the sight of Pharaoh and the people.

The Hebrew word used here for "***seasons***" is the same Hebrew word used in (Leviticus 23:2, 37) to describe the Feasts of the LORD. The Feasts of the LORD are based on the Sun and the Moon.

The *"lights in the firmament of the heaven to divide the day from the night"*(Genesis 1:14) is a reference to the Hebrew calendar.

The Hebrew calendar is lunar/solar. The months are based on a lunar cycle which is 360 days per year. God gave Israel a calendar of events based on heavenly signs particularly the moon on feast days.

The seven Feasts of Israel are based on the Hebrew calendar. Each of those seven feasts were intended to be signs of events that would happen in the future. The first four feasts declared the Crucifixion of Christ. Christ was crucified on the Passover. Christ was buried on unleavened Bread. Christ rose from the dead on First Fruits. The Church Age began on Pentecost Sunday.

Israel's Feast days are Devine Appointments! The first four Feasts, also called the "Spring Feasts" were fulfilled on their day, in their order and according to their Biblical traditions.

God has used signs in the heavens in the past: The rainbow; the Star of Bethlehem; and when the Sun was darkened at the Crucifixion of Christ… are some examples of how God used heavenly signs in the past. God said that he would use heavenly signs again in the future, maybe now.

34.4 God Uses The Heavens To Communicate With Man.

Many references to God's use of the sun, moon and stars. After all who can duplicate what God does in the sky? **Satan can't copy what God does with the planets so he lies about what it means.**

Satan spreads lies about the constellations and perverts their use so that believers will not understand God's intended use. Nevertheless, God still uses the sun, moon and stars to confirm his word to believers who study his word and watch for the promised signs.

"[To the chief Musician, A Psalm of David.]] The heavens declare the glory of God; and the firmament sheweth his handywork" (Psalm19:1)

"Day unto day uttereth speech, and night unto night sheweth knowledge". (Psalm19:2)

"There is no speech nor language, where their voice is not heard". (Psalm19:3)

"Their line is gone out through all the earth, and their words to the end of the world. In them hath he set a tabernacle for the sun" (Psalm19:4)

"Seek him that maketh ***the seven stars*** *and* ***Orion****, and turneth the shadow of death into the morning, and maketh the day dark with night: that calleth for the waters of the sea, and poureth them out upon the face of the earth: The LORD is his name"* (Amos 5:8)

"By the word of the LORD were the heavens made; and all the host of them by the breath of his mouth" (Psalm33:6)

"Which maketh ***Arcturus, Orion, and Pleiades****, and the chambers of the south".* (Job9:9)

"Canst thou bind the sweet influences of ***Pleiades****, or loose the bands of* ***Orion****?"* (Job38:31)

Pleiades is the constellation consisting of a cluster of 7 stars. Also known in mythology as "the 7 sisters".

"Canst thou bring forth Mazzaroth in his season? or canst thou guide ***Arcturus*** *with his sons"?* (Job38:32)

Here Job is naming off the constellations and stating that no one can duplicate what God has done! Not only does God know, but he shares his knowledge with us. He lets us know about his Devine appointments through the stars that only he controls.

"Knowest thou the ordinances of heaven? canst thou set the dominion thereof in the earth"? (Job38:33)

34.5 God Confirmed His Last Visit With Signs In The Heavens

God announced the birth of Jesus with a star at the Feast of Tabernacles.

God announced his Death by darkening the sun for three hours! This happened at the Feast of Passover.

The Bible indicates that the Sun was darkened much longer than any normal Solar Eclipse. In other words, it would require something much larger than the moon passing between the earth and the sun. It is believed by some that such a large heavenly body would produce earthquakes and tsunamis worldwide.

I don't know how the sun could be darkened for the length of time that the Bible records, all I know is that it happened.

According to NASA's website there were four solar eclipses in the year of 32A.D. March 29, April 28, September 23 and October 23. None of those solar eclipses lasted more than a couple of minutes and none of them occurred at the right time in 32A.D. see link… (http://eclipses.gsfc.nasa.gov/SEcat5/SE0001-0100.html)

Solar eclipses Generally only last a couple of minutes or so, sometimes a little more. The longest lasting one that I could find was in 96A.D. and lasted just over 11 minutes. No normal solar eclipse can cause the three hours of darkness described during the crucifixion of Christ...

...*"Now from the sixth hour there was darkness over all the land unto the ninth hour"*. (Matthew 27:45)

That was **three hours of darkness, beginning at noon, lasting till three in the afternoon and an earthquake**. Such an event could have been caused by a very large planet passing between the earth and sun during the time mentioned. The additional gravitational pull on the earth would explain the big earthquakes.

More specifically the planet would have to be much larger than the Earth; it must pass between the orbits of Venus and Earth. The planet would also have to be traveling much faster than the Earth, in the same General direction and in a very large, elongated, elliptical orbit so that the shadow lasts no longer than 3 hours. The 3 hours begins with noon and ends at 3 pm

Such an orbit would give it the velocity to pass by quickly and carry it back out to the edge of the solar system. Such an elliptical orbit would carry it far out into space way beyond Pluto, where it would be hard to detect. Such a planet has been theorized.

Astronomers are looking for such a planet. A large, dark, non-reflective planet traveling far out of the solar system can explain a lot of things. It can partially explain why such a planet has not been discovered yet. If it only comes around in cycles of about every 2,000 years or so.

Such a planet passing between the Sun and Earth could exert additional gravitational forces on the Earth which could affect earth's tectonic plates. Upon its return, the planet could cause extreme weather, earth quakes, activation of otherwise dormant volcanoes and create environmental issues like the book of Revelation describes.

A large planet traveling close to earth's orbit could bring with it plenty of space debris which could create meteor showers (Revelation 16:21)and asteroid impacts (Revelation 8:8) like the ones described in the book of Revelation..

If such a planet was discovered with an orbit that put it in the right place at the Crucifixion of Christ, then it would be just one more incredible piece of undeniable evidence to present to someone like Richard Dawkins! God could have accomplished what is described in the Bible in many ways, but **I observe in scripture a consistent pattern of God proving and confirming his word with discoverable science and verifiable prophecy.** So I believe that we should be looking for such a planet, just like we should be looking for Noah's Ark. Some people have been looking for these types of things.

Sir Robert Anderson confirmed the year of Christ's crucifixion in his book titled "The Coming Prince". After much research, the retired assistant commissioner of Scotland Yard concluded it to be 32A.D. Sir Robert Anderson used the Royal Astronomical Society and other prestigious sources to confirm many dates including the date of Christ's crucifixion.

It seems like we should be able to be certain of the year of Christ's crucifixion with all of this data. Some people say they still are not certain. However I accept Anderson's date of 32 A.D.

34.6 God Will Use Signs In The Heavens Again, Maybe Now

Mark Biltz discovered the Blood Moon Tetrad phenomenon around the year 2008, during Obama's Presidential campaign. Mark taught that unusual astronomical events occurring during Hebrew Feast days should be considered potential signs and deserve some attention. Mark Biltz demonstrated a historical pattern of significant events that involved God's people and this blood moon tetrad phenomenon.

"*The sun shall be turned into darkness, and the moon into blood, **before** the great and the terrible day of the LORD come.*" (Joel 2:31)

"*The sun shall be turned into darkness*" sounds to me like a solar eclipse and "*the moon into blood*" sounds like a lunar eclipse.

God says that there will be a solar eclipse and a lunar eclipse **before** the Day of the Lord. (Joel 2:31)

Solar and lunar eclipses occur all the time, but when they occur on a Feast Day then it is something to take note of. If you have an unusual pattern of some sort that ties astronomical events with feast days and scripture then you may want to consider what the sign could mean.

My God is a reasonable God, **not** a god of confusion! God gave us a calendar based on the moon and the sun for a reason, why not pay attention to God's calendar and his Feast Days?

"*And God said, Let there be lights in the firmament of the heaven to divide the day from the night; and let them be for **signs**, and for **seasons**, and for days, and years*": (Genesis 1:14)

Signs and **seasons** are the **reasons** God says that he gave us the heavens. God has also given Israel 7 Feasts to mark past and very near-future events or Divine Appointments.

Since A.D. 1, "Biblical Blood Moon Tetrads" have occurred a total of 8 times. Blood moons that occur on the holy days for two consecutive years' are called a "Biblical Tetrad". In 2014-2015, it happened for the 8th time since Christ was here.

This event is unlike any that has ever happened before, and will not happen again because it also involves **a full solar-eclipse on Nissan 1** and **a partial solar-eclipse on Rosh Hashanah**, and ends with a **super blood moon at the Feast of Tabernacles.** Altogether, there was seven significant "signs in the heavens" from 2013 – 2015. God was calling man's attention to events transpiring on earth.

There will not be another tetrad for nearly 600 years. But there will never be another one like this last one nor has there ever been one like it in the past.

That year we also had a total solar eclipse on Nisan one which is the Hebrew New Year for counting months. A partial solar eclipse occurred on September 13, 2015 on Rosh Hashanah which is the Biblical New Year for counting years, epochs or ages; also known as the Feast of Trumpets; The Fifth Feast or the First Fall Feast; or the next Feast in line to be fulfilled which I believe will some year soon, mark the Rapture.

God Promises More Heavenly Signs.

(Joel 2:30, 31) (Leviticus 23:1, 2) (Exodus 12:1-14)

"*And I will shew **wonders** in the heavens and in the earth, blood, and fire, and pillars of smoke*" (Joel 2:30)

The word "***wonder***" as it is used here is talking about heavenly signs. The word "*blood*" implies the color "red". It does not necessarily mean blood unless the context requires it. The same word is used for wine. A lunar eclipse is called a "blood moon" because of its red color. Or it could be the color of the sky. I suppose the sky could also have a red color if a super volcano erupted and spewed the right kind of dust into the air. Or this could be the result of something man made.

"*The sun shall be turned into darkness, and the moon into blood, **before** the great and the terrible day of the LORD come.*" (Joel 2:31)

The phrase *"The sun shall be turned into darkness"*, I believe is a reference to a Solar ---eclipse on a Jewish Holy Day, otherwise it would not be a sign.

The phrase "and the moon into blood", I believe, is a reference to the color of the moon during a lunar eclipse. I believe that it is talking about an unusual lunar eclipse on a Feast day **just before The Day of the LORD**.

The Hebrew word translated as "wonder" is; Strong's # H4159 – (mowpheth): wonder, sign, miracle, portent, wonder (as a special display of God's power) sign, token (of future event)

"The sun shall be turned into darkness" sounds like a solar eclipse to me.

"*The moon into blood"* is another way of saying the moon will turn red. It is talking about a lunar eclipse.

This sounds like God is saying that solar and lunar eclipses will occur **before** the *"Day of the Lord"* and it will be a sign. Therefore it will be unusual and associated with the Jewish Feast and the Hebrew Calendar.

The "*Day of the Lord*" begins with the rapture of the Church. The Rapture of the Church is the next event on God's prophetic calendar. The next Feast in line to be fulfilled is the Fifth Feast of Israel or the Feast of Trumpets, also called **Rosh Hashanah**.

Eclipses occur all the time, but when they occur on Feast days then they are not so common and are considered potential signs. Furthermore there is God's calendar of events which consists of 7 Feasts to mark events <u>before</u> they happened.

In the past, God has used heavenly signs such as a rainbow or a solar or lunar eclipse to confirm that the event or message was of God, before, during or after an event occurred.

34.7 Biblical Blood Moon Tetrads In History

Since the first Century there has been eight blood moon tetrads. Here is the list in order.

<u>162-163 A.D</u>. A Biblical tetrad occurred that involved Passover and Tabernacles. **Marcus Aurelius was Emperor of Rome from 161-180 A.D**. during **which time Christian persecution intensified**. It is said that Christian blood flowed more profusely during the reign of **Marcus Aurelius** than any other time!

<u>**795-796 A.D.**</u> A Biblical tetrad occurred that involved **Passover** and **Atonement** (Yom Kippur) Feasts two years in a row. This tetrad is associated with an **Islamic invasion**.

<u>**In 842 A.D. and 843 A.D.**</u>**,**A Biblical Tetrad occurred**.** This tetrad involved lunar eclipses on **Passover** and Feast of **Atonement (Yom_Kippur**). Those dates are: **Passover**; March 30, 842 A.D. **Yom_Kippur**; September 23, 842 A.D. **Passover**; March 19, 843 A.D. **Yom Kippur** (Atonement); September 12, 843 A.D. This tetrad is associated with another **Islamic invasion**.

<u>**860 A.D. and 861 A.D**</u>**.** A Biblical Tetrad occurred. This tetrad involved lunar eclipses on **Passover** and Feast of **Atonement** (Yom Kippur). This tetrad is sandwiched between two events involving Islam's attempt to conquer Christendom. From 851A.D. to 859A.D. Muslim conquerors beheaded Christians in Southern Spain for speaking about their faith. On one occasion, they beheaded forty eight Christians in a mass execution.

The first lunar eclipse in this series occurred on **Passover, April 6, 860 A.D.** which was followed by another on <u>**Yom Kippur**</u> **October 3; in 861A.D**. there was another lunar eclipse on **Passover March 30**; which was followed by another on <u>**Yom Kippur**</u>**, September 22, 861A.D.**

In 863A.D. the Byzantine Empire successfully defended itself from an Arab Muslim invasion. After the Byzantine Empire's success of defending itself from the Muslims, they followed through by crossing the border into enemy territory and completely defeating the enemy, thus ending Muslim persecution of Christians in that area and beginning the freedom to be Christian.

Something about the Biblical Tetrads that involve a <u>**Blood moon on Yom Kippur**</u> that I think is interesting is the fact that they involve Islam attacking Christians.

I believe that the future 7 year Tribulation will begin on **Yom Kippur**, at the end of the Feast, after sundown with a surprise Jihad attack against the west from within. The Believers would be "Raptured" nine days earlier, so we will miss the whole thing.

<u>In 1493-1494</u>, A Biblical Tetrad occurred that involved lunar eclipses on the first Spring Feasts and the first Fall Feasts, two years in a row. That is **Passover** and the Feast of **Trumpets** (**Rosh Hashanah**). This is probably related to the expulsion of the Jews from Spain. The Edict of Expulsion was in 1492. Not much is said about Christians receiving the same persecution as the Jews, but the Catholic "Church" burned and tortured Christians as well as Jews. The Jews were easier to identify and more profitable targets for evil people looking for profit without working for it.

<u>In 1949-1950</u>, A Biblical Tetrad occurred which many relate to Israel becoming a nation in 1948. Again this tetrad also involves a lunar eclipse on the first and last feasts two years in a row. On 11 May 1949, Israel became the 59th member of the United Nations.

<u>In 1967-1968</u> there was another Biblical blood Moon Tetrad which is associated with the recapture of Jerusalem by Israel. This tetrad involved a lunar eclipse on the first and last feasts two years in a row.

In June of 1967 two important things happened about that time. First of all that is when I got saved. Secondly that is when Israel defeated her enemies, regained Jerusalem and tripled the size of its territories. On July 23, 1968 was the El Al Flight 426 hijacking.

In **<u>2014 and 2015</u>** was the last Blood Moon tetrad which occurred on **Passover** and **Tabernacles**. It was called a super blood moon because the moon was at its closest point to the earth and appeared 20 percent larger.

There was no blood moon tetrad during Jesus Christ's time on earth. Since 1 AD, a "Biblical tetrad" has occurred a total of 8 times. This last blood moon tetrad (2014-2015) is the eighth one since the time of Christ. This last tetrad is unique and will never be repeated.

Another Biblical blood moon tetrad won't occur again for almost 600 years. It makes good sense to consider what transpired on earth while God is getting our attention In the sky.

34.8 Seven Pre-Tribulation Heavenly Signs!

Just prior to the 8th Biblical Blood Moon Tetrad, the **Comet ISON** appeared on Hanukah, November 28, 2013. I believe this comet is the first of **Seven Pre-Tribulation Heavenly Signs**. These seven signs fit the description that Joel the prophet gave when he wrote... *"The sun shall be turned into darkness, and the moon into blood, before the great and the terrible day of the LORD come"*. (Joel 2:31)

February 28, 2013 Pope Benedict XVI (2005-2013) resigned and his Papacy ended: It has been over 600 years since a Pope resigned.

Thirteen days after Pope Benedict XVI resigned, Pope Frances began his papacy on **March 13, 2013**. Look at all of those thirteens.

Nine months after Pope Frances began his papacy, the Comet ISON appeared. The nine months may be symbolic for human gestation.

(1)November 28, 2013, **Comet ISON appeared in the night sky as Hanukah began.. Both Thanksgiving and Hanukah occurred on the same day.** This is rare and never to be repeated again! That night the Comet ISON passed perihelion which is when it is closest to the sun. This is when it swings around the sun and back out into deep space.

Thirteen days later, on December 11, 2013 Pope Frances becomes Time magazine's "Person of the Year". That is a lot of thirteen's. I'm calling the Comet ISON the Sign of the False Prophet. The Comet ISON is the first of seven heavenly signs that fulfill Joel's pretribulation prophecy.

*"The sun shall be turned into darkness, and the moon into blood, **before** the great and the terrible day of the LORD come"*. (Joel 2:31)

There were noteworthy events that occurred about the time of the last Blood Moon Tetrad that we should consider.

March 27, 2014, Mr. Obama made a surprising visit to the Vatican. Relations between the Vatican and the Whitehouse made a reversal for the better. The Pope began traveling to promote the **United Nations Sustainable Development Agenda 2030.** Nineteen days after the meeting with Obama, the first Blood Moon is seen on Passover.

(2)April 15, 2014, Passover, is the first Feast Day in the spring and the first of a series of 4 total lunar eclipses that occurred on Feast days in 2014 and 2015. Four weeks later, a 500 day warning was given...

May 13, 2014, The **Environmental Chaos Warning was given.** On the eve of Israel's Independence Day, a 500 day warning was given. The French Foreign Minister Laurent Fabius did the talking with John Kerry standing at his side. A warning was given three times that the world had 500 days to avoid "climate chaos".

Why did they chose to give this warning on the eve of Israel's Independence Day? What is the connection between Israel and the **UN's Sustainable Growth agenda 2030**? About five months later is the second blood moon…

(3)October 8, 2014, Tabernacles (Succoth, Last Feast Day in the fall) is the second total lunar eclipse that occurred in this series of four.

Remember that these blood moons are also accompanied by two solar eclipses, one on Nisan 1 and another on Rosh Hashanah 2015.

(4)March 20, 2015, Nisan 1 solar eclipse, The Hebrew religious New Year for the counting of months begins with a total solar eclipse on Nisan 1 which is the first day of the ecclesiastical New Year. This day is marked by a full solar eclipse inside of this series of four blood moons.

(5)April 4, 2015, Passover, the first Feast Day in the spring; the third total lunar eclipse!

(6)<u>**September 13, 2015**</u>, **Rosh Hashanah,** is marked by a partial solar eclipse on the Feast of Trumpets. This marks the beginning of the Hebrew New Year for counting years or Epoch years. It will be followed by another blood moon <u>two weeks later</u>. . Whatever year it happens, the Rapture of the Church will occur at the end of Rosh Hashanah, "*at the last trump*".

<u>**September 22-23, 2015**</u>, **Yom Kippur** (Feast of Atonement) 500 days since the warning was given on the eve of Israel's Independence Day 2014, an important meeting takes place. **Barak Obama meets with the 266th Pope on the 266th day of the year.** What are the chances of that happening by accident? Why did they pick Yom Kippur for this event? Why was the announcement made 500 days in advance, on Israel's Independence Day? What does Israel have to do with this?

Five hundred days after the warning was given, Obama met with the **266th Pope on the 266th day of the year** on Yom Kippur. The Pope gave a sermon and declared the environment to be a moral issue. The following two days were filled with sermons, seminars and discussions about the environment and agenda 2030.

Agenda 2030 is all about total control of the world's population and its resources. It is all based on lies about key issues like the climate to justify total control. Population control and reduction are at the center of Agenda 2030.

<u>**On September 25, 2015 the UN adopted Agenda 2030**</u> for sustainable Development titled "Transforming our world: the 2030 Agenda for Sustainable Development"

Top headline of "**The UN News**" 1/22/2020 reads …

"UN chief outlines solutions to defeat <u>'four horsemen'</u> threatening our global future

(7)Tabernacles (Succoth), September 28, 2015. This final blood moon of the sequence is called a <u>**super**</u> **blood moon** because the earth and moon are 20% closer together, therefore the moon appears 20% larger.

The unique characteristics about this tetrad includes a full solar eclipse on Nissan 1, which is the Jewish secular New Year for counting months and a partial solar eclipse on Rosh Hashanah, which is the Biblical New Year for counting years and epoch years. It ends with a **Super Blood Moon** on the Feast of Tabernacles.

34.9 What Did The 8th Blood Moon Tetrad Signify?

Just prior to the beginning of the eighth Blood Moon tetrad, the <u>ISON</u> comet appeared in the sky on **Hanukah.** This day was the day that Thanksgiving and Hanukah both fell on the same day. Astronomers tell us that this will never happen again. Traditionally comets are considered warnings of coming judgment.

I associate the Hanukah comet with Pope Frances The First. The comet was observed 261 days after Pope Frances the First began his Papacy on 03/<u>13</u>/20<u>13</u>, <u>thirteen</u> days after his predecessor resigned. That is an awful lot of 13's. Considering what I know about Pope Frances, I'm calling the comet **<u>ISON</u> … "The Sign of the False Prophet".**

If you are wondering about what God's 8th Blood Moon Tetrad signifies, consider the significance of the events that occurred during that time. I am not the only one to make these connections. Furthermore, consider the disagreement and borderline hostility between key players such as the Prime Minister of Israel, the Pope and Mr. Obama prior to Yom Kippur 2015. What changed all of that?

On **May-13, 2014**, an announcement was made by the French Foreign Minister, Laurent Fabius, with John Kerry standing at his side. The occasion was the eve of Israel's Independence Day (May 14, 1948). That coming sundown marked 66 years since the Nation was reborn.

A speech about the environment was made citing overpopulation as the cause of the problem. During that speech, the French Foreign Minister (Laurent Fabius) announced 3 times that… **"We Have 500 Days to Avoid Climate Chaos".** At what appeared to be the end of the speech, Mr. Fabius began to walk away from the microphone while John Kerry stayed put and looked around like something had gone wrong.

Someone off camera directed Mr. Fabius back to the microphone. Mr. Fabius walked back to the microphone and stuttered a moment and then said… "Uh… oh yea, **the world has 500 days to avoid environmental chaos**." The point being that just saying it 2 times is not enough. It had to be said 3 times.

Someone wanted to make a point about that and planned how the third warning would be given in such a way as to emphasize the 500 days in a conspicuous way.

Throughout the first administration of Mr. Obama, there was verbal conflict between Obama, the Pope and Prime Minister Netanyahu.

The Pope had made quite a few negative statements about Mr. Obama's antichristian rhetoric throughout his first term. Pope Francis responded to Obama's remarks with public rebukes to Mr. Obama. During his second term, on **March 27, 2014**, Mr. Obama made a surprising visit to the Vatican. Pictures of Obama smiling and bringing gifts to the Pope were shown by CNN which called the meeting…"a step toward smoothing tensions with Catholics". CNN also called the meeting…"a fresh start of sorts between the administration and Catholic leadership after years of strained relations."

Since that first meeting they seem to be getting along just fine. The negative rhetoric stopped and Pope Francis began traveling around the world promoting the United Nations environmental agenda 2030 and meeting with world leaders and he has not stopped.

Pope Francis is the first Pope to address the U.S. Congress. He talked about the environment, climate change and the United Nations new environmental agenda called "**Sustainable Growth 2030**". Pope Francis "the first" is **the 266th Pope**. He is also the first Jesuit Pope and the first of a list of other things as well. Robert Spencer, a Greek Orthodox Catholic and most highly respected expert on Islam, has called Pope Frances "**the first Islamic Pope**".

First published by Benedictine monk Arnold Wion in 1595, were a number of prophecies he attributed to the Irish monk Malachy of Armagh. Malachy who purportedly wrote the prophecies about all of the Popes since 1139, said that there would only be 266 Popes. There are 266 days in the human gestation period.

According to this 876 year old prophecy, the **266**th Pope will be the last. So now the Vatican can call Pope Frances the "first and the last." Can you see where this may be going? I don't believe in this prophecy. The Catholic Church has an interest in getting people to believe in the prophecy or it would be forgotten long ago. So what are they up to?

If we know our Bibles, it is a good idea to know what the enemy is saying. Then we can compare it to scripture and warn people. The only time that Satan will give truth is to gain your trust so he can sell a bigger lie. Some scholars believe that these "prophecies" were forgeries that were written just before they were published. So why would anyone cook up a story like this?

Remember that 500 day announcement made by the French Foreign Minister on May 13, 2014 that announced <u>the</u> meeting between Pope Francis and Barack Obama. The meeting was announced publicly and globally 500 days in advance and it turns out that the meeting would be held on **the 266th day** of that following year (2015) and was also on **Yom Kippur 2015**. There are 266 days in the human gestation period. What are the chances of that?

On that peculiarly announced public meeting, the **266th Pope** announced on the **266th** day of the year that the environment was **a <u>moral issue</u>**. That 2nd meeting between Pope Francis and Barack Obama was on **Yom Kippur, September 23, 2015**. Morality was equated to "saving the environment" which to them is equal to human population reduction! To achieve the population reduction levels talked about by Ted Turner (Club of Rome), you would have to murder billions of people. This is stated in indirect terms so that most people don't notice.

Two days later, Pope Francis addressed the United Nations and again said that the environment was a moral issue. Later that day, **September 25, 2015, the 2030 Agenda for Sustainable Development was adopted at the United Nations Sustainable Development Summit.**

UN Agenda 21/Sustainable Development was the action plan to inventory and control all land, all water, all minerals, all plants, all animals, all construction, all means of production, all information, all energy, and all human beings in the world. INVENTORY AND CONTROL EVERYTHING! Agenda 21 expired in 2015. This vote was to replace it with agenda 2030, a more aggressive plan I suppose.

Seven heavenly signs showed up in our skies beginning on Hanukah, **November 28, 2013** with the comet ISON and continued through to **September 28, 2015,** ending with a super blood moon over Jerusalem.

During this time the 266th pope began his papacy on 03/13/2013 just thirteen days after the former pope resigned. I call the comet ISON "**The Sign Of The False Prophet**". Maybe I am wrong, we will just have to wait and see.

The 8th Blood Moon covered the span of time that the new **United Nations Sustainable Development Agenda 2030** was announced, promoted and accepted by the United Nations.

34.10 The Sign Of The Four Horseman

"And I looked, and behold ***a pale horse****: and his name that sat on him was Death, and Hell followed with him. And power was given unto them over* ***the fourth part of the earth****, to kill with* ***sword****, and with hunger, and with death, and with the beasts of the earth"* (Revelation 6:8)

This is talking about the "Four Horseman of the Apocalypse" (Revelation 6:1-8). They are the first four judgements or "*seals*" of the Tribulation. They begin on Yom Kippur and they span three and a half years. According to this passage, **One fourth of humanity...** (7.6 billion divided by 4 =**1.95**

billion) … will be destroyed in those first 3 ½ years of the Tribulation by those first four judgements alone! This may be the greatest Genocide in human history up to that point! According to the Bible, the following 3 ½ years will be even greater Genocide! The last 3 ½ years of the Tribulation will be the worst time in all of human history, future included!

Notice at the beginning of (Revelation 6:8) it says… *"And I looked, and behold* ***a pale horse****":* The word ***"pale"*** is translated from the Greek word …"chlōros"… found in the Strong's Greek dictionary under the number G5515.

The meaning of that word is… "verdant, dun-colored:—green, pale". (Strong's) "tender green grass or grain color". The point being that it is not a pale skin color or anything like that. It is a pale green color that they are talking about, **just like the color of the Saudi Arabian flag**. That is the color that this passage is talking about. This rider of the four horses is the Antichrist and his religion is Islam. (Revelation 6:8) I don't know of a Muslim country flag that does not incorporate this same color into some part of their flag, but the Saudi Arabian flag is all that color, solid.

The form of Islam followed in Saudi Arabia is called Wahhabism. It is the "purest", most violent and aggressive form of Sunni Islam.

The Saudi Arabian flag is **a pale green flag** with **a white sword** over the Shahada or Muslim creed also written in white. **The Muslim creed** is as follows…"There is no god but god; Muhammad is the Messenger of god." The Saudi Arabian flag makes it clear that Allah is a god of force by putting a sword below the Shahada.

Clearly this passage is talking about Islam taking over the world **by force**. The Bible says that the Antichrist will "*honour the* ***god of forces"…*** The Koran teaches that it is the responsibility of all Muslims to fight against unbelievers until they either convert or are subdued. This forced conversion under the threat of death or enslavement Identifies "alah" as "*the god of forces*" that Danial was talking about.

"But in his estate shall he honour the ***god of forces****: and a god whom his fathers knew not shall he honour with gold, and silver, and with precious stones, and pleasant things."* (Danial 11:38)

Furthermore, the Bible identifies this god and says that he will take over the whole earth during the Tribulation. The first three and a half years will kill twenty five percent of the world's population. The last 3 ½ years of the Tribulation will be much worse because Muslims will gain control of the world by then.

"Therefore hath the ***curse devoured the earth****, and they that dwell therein are desolate: therefore the inhabitants of the earth are burned, and* ***few men left****.* (Isaiah 24:6)

The word translated as "***curse***" here is "**alah**", which is pronounced like "Allah". The Hebrew and Arabic words are similar in sound, spelling and origin, which is Arabic. There is no doubt that the Hebrew word "alah" existed before the name "Allah" did. Around 550 BC when Daniel was written, there was no "Allah" except for in the Saudi Arabian peninsula there was a moon god named Allah. The Koran did not exist until the 7th Century AD.

When Alexander the Great conquered Egypt he established a city and a library with his name on both. His first order was to translate the ancient Hebrew scriptures into his language which was Greek.

Notice also that in (Revelation 6: 4) the Bible says*..."****and there was given unto him a great sword"****.* Again in (Revelation 6: 8) ... "***to kill with sword***" is mentioned. I believe that **about 1.95 billion people will be slaughtered by Muslims during the first 3 ½ years of the Tribulation.**

I also believe that a disproportionately large portion of that number is from the West in general, but from the United States of America in particular. These will not all be quick deaths. Islam requires "painful chastisement" to those who believe in Christian doctrine and that means torture and humiliation. Psychological, spiritual torture!

(Daniel 7:25) *"And he shall speak great words against the most High, and shall* ***wear out*** *the saints of the most High, and think to change times and laws: and they shall be given into his hand until a time and times and the dividing of time"*.

This is talking about the Tribulation Saints being Psychologically worn down, abused, humiliated. That is what the Quran teaches them to do to non-Muslims. (Quran 4:18)

According to (Isaiah 24: 6) and (Zechariah 5: 3), Islam will take over the whole world during the 7 years of Tribulation.

(Isaiah 24: 6) "***Therefore hath the curse devoured the earth****…*".

(Zechariah 5: 3) "*Then said he unto me, This is the* ***curse*** *that goeth forth* ***over the face of the whole earth****:*"

The first four seals represent Islam taking over the world under the Antichrist. This final Jihad will cause a lot of death and destruction during those first three and a half years. but there are 24 more judgements that covers the last 3 ½ years of the Tribulation that will follow! As you can see, the stage is set for this kind of population reduction. God calls it "***sudden destruction***" (1Thessalonians 5:3)!

The Globalists want complete and total control of everything and everyone. They know that it is easier to control one million needy and dependent persons than three million well fed; well rested and well educated individuals with lots of information, time and resources on their hands.

The writers of UN policy are asserting their New Age lies that human population is the major cause of "**environmental chaos**". The Masonic doctrine "**ordo ab chao**" means "**order out of chaos**". The plan is to destroy the old system and establish a one world government with one religion, Islam. The Globalists have many minions to do their bidding such as the Free Masons.

With the help of Pope Francis, the United Nations is claiming the moral high ground for whatever is required to reduce the human population to what they consider to be acceptable levels. In essence they are attempting to "legitimize Genocide against Christians and Jews" through their New Age Morality and Islam.

The Georgia Guide Stones, erected in 1980 in Elbert County, Georgia, USA begins with an arrogant requirement to reduce the human population down to 500,000,000. That is Genocide on Biblical levels! Seriously!

Here is the Guide Post list…

1. Maintain humanity under 500,000,000 in perpetual balance with nature.
2. Guide reproduction wisely— improving fitness and diversity.
3. Unite humanity with a living new language.
4. Rule passion— faith— tradition— and all things with tempered reason.
5. Protect people and nations with fair laws and just courts.
6. Let all nations rule internally resolving external disputes in a world court.
7. Avoid petty laws and useless officials.
8. Balance personal rights with social duties.
9. Prize truth— beauty— love— seeking harmony with the infinite.
10. Be not a cancer on the earth— Leave room for nature— Leave room for nature.

The World Population clock says that the world's population today is about 7.8 billion. To reduce the world's population down to 0.5 billion would require eliminating 7. 3 billion people!

The Bible says that the first four judgements will kill one fourth of humanity. That is four out of twenty eight judgements killing about 1.95 billion in the first half of the Tribulation with 24 more judgements following over the next 3 ½ years. Dr. Hank Lindstrom said that he did not think that more than 100,000,000 would survive to live on into the Millennium of Christ. I think that the number is closer to about 30,000,000. (See Ch. 47 ETBH)

Cultural Purification is standard communist revolutionary policy, or so it seems. The French Revolution is an example of secret Masonic deception and subversion to lead a rebellion against the French government. It was all based on lies and deception. Almost all of the key players of the French Revolution were Masons. A very influential Mason was Erasmus Darwin, an Englishman.

Erasmus Darwin's influence comes from his writings about evolution and his New Age religion. His books include: "**The Botanic Garden: The Loves of the Plants**"; "**Zoonomia**"; "**The Temple of Nature**" and more. His grandson, Charles Darwin had a complete set of his grandfather's books with notes written throughout. Every bit of Charles Darwin's work came from his grandfather, but there is no mention about Erasmus Darwin's contribution to "The Origin of Species" or anything else Charles did. Erasmus concealed his involvement in the American and French revolutions. Erasmus Darwins son and famous grandson Charles concealed the family secret about where Charles got his information from.

Erasmus Darwin's influence was all over the French Revolution, as well as the American Revolution. After his death, his legacy lives on. The Bolshevik Revolution and the Third Reich and every communist revolution since was influenced by Erasmus Darwin.

These ideological revolutions use force to enforce a new morality based on lies, deception, terror and Genocide!

After gaining power, purging wrong thinkers is the first priority of ideological Rebels. The Rebels must eliminate all opposition first. The success of a communist revolution is directly proportionate to the number of "wrong thinkers" who are eliminated at the beginning. That means Genocide! That happened in the French Revolution, Communist revolutions and also Muslim Conquests.

New Age Pantheists promote the idea that the human race is an infestation like a cancer that must be controlled by a powerful central government run by them. I call these people "Satan's minions". They are motivated by lust for power and pleasure.

Satan hates the whole human race but he especially hates born-again believers because they interfere with his activities. After the believers are taken away at the Rapture, there will be no one to resist evil.

Even though God changed Lucifer's name to Satan, he doesn't admit his fate to those he wishes to deceive so his followers call him by his pre-fallen name which is Lucifer. Satan also goes by the names of men he has possessed or the names of false god's that men worship.

Established by Alice Bailey in 1922, the **Lucifer Publishing Company** changed its name to Lucis Trust in 1925 for obvious reasons. The name "Lucifer" means "light bearer". God changed Lucifer's name to Satan after Satan led a rebellion against God. Satan means "adversary" and he doesn't like to be called that, so his followers call him by his pre-fallen name. Free Masons insist on calling Satan "Lucifer" and persuade others to do the same. (Morals and Dogma, chapter 19, "Grand Pontiff" Page 321, first paragraph);

Lucis Trust was based at 120 Wall Street, they have expanded all over the world and moved to the UN. Lucis Trust is the official publisher of the United Nations. Lucis Trust manages the U.N. "Meditation Room and oversees their spiritual matters." There are plenty of books written to expose the satanic secrets and agendas of Free Masonry and the U.N.

On the Lucis Trust website is this statement… "In order to place a closer focus on the work of the UN, the Lucis Trust has set up a new blog, World Goodwill at the UN, which will focus on **defining new Sustainable Development Goals** for humanity after 2015 when the Millennium Development Goals expire".

According to that statement, **Lucis Trust is defining goals for the UN and for humanity!** These people are behind the scenes planning the UN agenda for all of humanity! These are Satan worshipers, but they just call him Lucifer.

They moved their headquarters to the United Nations just to prepare for this September 25, 2015 vote during the Summit on Sustainable

Development. Their main interest seems to be population reduction and control of everything.

During the time of preparation leading up to that vote there were six unusual, astronomical events that occurred on Feast days. Then three days after the vote was taken was the seventh in this series of **seven rare astronomical events on Feast Days**.

The Super Blood Moon of 2015 was in full view from Jerusalem. It appeared to be 20 percent larger in the sky than normal. This was a special event, God was talking to us through the sky, just like he said he would do… "*Before the great and the terrible day of the LORD come.*" (Joel 2:31)

The point being, the Day of the Lord is at Hand! In other words, time is running out fast! People need to get saved now to avoid the Tribulation! Yeshua is trying to get the attention of anyone who cares.

Yeshua is about to bring judgement to planet earth!

March 13, 2013 Pope Frances The First begins his Papacy. Eight and a half months later we see the Hanukkah Comet.

(1) The **Hanukkah Comet of November 28, 2013** seems to be the first in a series of seven total astronomical events on Biblical Feast days, that appear to be associated with events on earth that transpired during that same time. A comet on a Jewish Feast day is considered by many to be a warning from God.

About four and a half months later…

(2)Passover evening, (First Blood Moon) April15, 2014. One month later…

(3)May 13, 2014; The 500 day warning to avoid climate chaos marked the beginning of preparations for the United Nations year of light and Light Based Technologies. This announcement was made on the eve of Israel's rebirth day and the eve of a full moon.

(4)May 24-26, 2014 Pilgrimage of Pope Francis to the Holy Land…

Six weeks from the first blood moon, Pope Frances began his Pilgrimage. It is important to note that 2014 and 2015 was filled with meetings, seminars, sermons and visits from the Pope to different heads of state throughout the world. He was promoting Agenda 2030 for Obama.

The preparations for the ***"year of light"*** began "officially" in 2014, but preparations have actually been going on for quite some time and accelerating. The promotions and preparations climaxed on September 25, 2015 with the world vote to accept **Sustainable Development Goals 2030**. All of these events occurred during the time of the 2014-2015 Blood Moon Tetrad. **I am <u>not</u> the only one who noticed this.**

On the LucisTrust.org website is a Blog called "World Goodwill at the UN Blog". On the Home page Archives is an article that mentions these events that occurred during the time of the four Blood Moons. Here is a sampling of that article from lucistrust.org …

The article was titled**…"Sustainable Development Goals, the United Nations and the Libra Full Moon"** The article went on to talk about the unusual events going on in the sky **<u>during</u> the promotion of the summit.** The article said…

…"A significant event in the heavens occurred <u>during</u> the period of the Sustainable Development Summit at UN Headquarters in New York"…

…"It should not surprise us that <u>a significant alignment in the heavens</u> was <u>accompanied</u> by such a <u>significant</u> gathering at the UN."

The United Nations and Lucis Trust were officially working on preparations for the "Year of Light" from May 13, 2014 when the "Five Hundred Day Warning of Climate Chaos" was given. Five hundred days later, on September 23, 2015, Obama met with the 266th Pope on the 266th day of the year. (Remember; there is 266 days in the human gestation period).

After the Pope's sermon about the environment, he announced to the world that **Climate Change was a "moral issue"**. Two days later the U.N. Agenda 2030 for Sustainable Development was adopted. This marked the climax of the United Nations Sustainable Development Summit and the U.N. "Year of Light" on September 25, 2015. Three days later, on **September 28, 2015 was a super blood moon over Jerusalem**.

The abrupt change in rhetoric between Pope Francis and Mr. Obama after their first meeting leads me to believe that they may have made an agreement at that first meeting. **Robert Spencer**, who is a well-respected authority and author of about 17 books on Islam …"that can be found in any self-respecting book store", and is the director of **JihadWatch.com**, has called Pope Frances 1st … "**The First Islamic Pope**".

David Hunt's book "A Woman Rides the Beast" pretty much tells the whole story. Let me just say that the agreement that allows the Roman Catholic Church to escape Islamic persecution from the Antichrist for the first 3 ½ years of the Tribulation, will be nullified in the middle of the 7 years..

The Bible talks about another agreement. An agreement between the leaders of Israel and the Antichrist. The Bible calls it the ***"covenant with death and hell"*** (sheh-ole) (Isaiah 28:15) … Basically, the Bible is saying that the leadership of Israel will be depending on an agreement that is made in exchange for protection.

"*Because ye have said, We have made a covenant with death, and with hell are we at agreement; when the overflowing scourge shall pass through, it shall not come unto us: for we have made lies our refuge, and under falsehood have we hid ourselves*": (Isaiah 28:15)

I believe that Israeli leadership has made that deal spoken of in (Isaiah 28:15) soon if not already. The agreement of Israel will not hold up. Something unexpected happens, maybe more than is mentioned here.

"And your covenant with death shall be disannulled, and your agreement with hell shall not stand; when the overflowing scourge shall pass through, then ye shall be trodden down by it". (Isaiah 28:18)

Is this talking about the seven year guarantee of protection given by the Antichrist on Yom Kippur at the beginning of the Tribulation or is this talking about some other side deal made earlier?

This is a good time for a little speculation. Let's suppose that Israel made a deal with Trump. What would happen if both Trump and Pence got raptured? Nancy Pelosi would become President! I believe that the Temple construction will begin at the same moment as the Rapture.

Temple construction will lead to Muslim violence worldwide. The Bible says that a kingdom will be given to the Antichrist. Can you imagine what Pelosi would do if an all-out Jihad broke loose in the U.S. just after she was sworn in?

Obama to the rescue with his fake credentials like his Nobel Peace Prize that he was nominated for just after becoming President.

He will have the news media in his hand. He will control the U.N. He will have the support of the Pope, the far left. A.N.T.I.F.A. will do some of the "dirty work" but most of the muscle will come from Muslim immigrants and the U.N.

These Jihad soldiers in waiting are well paid with the $1.8 billion in cash from the Iranian deal. Iran is the largest sponsor of world Jihad Terror. The Bible describes the Nation of the Antichrist. The U.S. is it.

God says that Israel's plan will fail. (Daniel 9:27)describes how the treaty will be broken in the middle of the 70th week of Daniel. (seven years Tribulation)

KOL NIDRE, also known as "all vows" is a legal formula written in Aramaic, not Hebrew and is recited in the synagogue at the beginning of the evening service on the Day of Atonement (Yom Kippur) that proactively cancels all oaths made the following year, so as to preemptively avoid the sin of breaking vows made to God which cannot be or are not upheld.

In essence, any oath or promise made the following year is made void. This tradition was used by the Catholics during the dark ages as an excuse to persecute Jews and accuse them of being untrustworthy.

It is important to note that the Rabin Treaty which is referred to in the Bible as "*the covenant with many*" (Danial 9:27) was made on Yom Kippur. I believe that the Antichrist will also make his promise to enforce that covenant on Yom Kippur. I believe that is the day that they will say "*peace and safety*" (security), but at sundown will come sudden destruction. **KOL NIDRE**, or "all vows" will be his justification to break all treaties.

You could say the "***Dispensation" of Law**" will start back up 8 days, after the Rapture and after the Feast of Atonement, at sundown.

The "Ten Days of Awe" begins with Rosh Hashanah and ends at sundown with Yom Kippur.

With a full **solar eclipse** on Nissan 1 (March 20, 2015) which is the Jewish secular New Year for counting months; and a **partial solar eclipse** on the Feast of Trumpets which is the Biblical New Year for counting years. God is directing our attention to the Hebrew calendar.

This Biblical Blood Moon Tetrad meets the description of (Joel 2: 31)

"*The sun shall be turned into darkness, and the moon into blood, **<u>before</u>** the great and the terrible day of the LORD come*". (Joel 3: 31)

When a blood moon occurs on God's Feast days, it meets the Biblical definition as a sign. (ETBH chapters 24, 25 and 26)

The Blood Moon Tetrad of 2014/2015 was clearly an unusual astronomical event that occurred during the time that the UN was promoting the **"Year of Light"** and the **Sustainable Development Summit,** which climaxed with the signing of the U.N. Sustainable Development Agenda 2030 on September 25, 2015.

Pope Francis endorsed and promoted the new UN plan for Sustainable Development ever since his first meeting with Obama.

The previous plan, called "Agenda 21" expired in 2015. Agenda 21 names overpopulation as environmental enemy #1. Depopulation is a tool Globalists are using to gain control of the population and justify genocide in the name of saving the planet. https://www.youtube.com/watch?v=BkRirCcmrX0

The new program for sustainable development was put together by LUCIS Trust, formerly known as Lucifer Publishing Company. They moved their headquarters into the UN building for this project that was publicly formalized on Yom Kippur 2015. The moral issue about Genocide has been taken care of by the Pope.

The Church Age ends and the Day of the Lord will begin one of these years and it will happen at the sound of the last trump of the Feast of Trumpets also known as Rosh Hashanah. Sundown generally occurs around 6:30 PM Jerusalem time or about 11:30 AM Eastern Standard Time which is my time zone. The actual time varies from year to year so this coming Rosh Hashanah Sep 19-20, 2020 will end at sundown in Jerusalem on September 20, at 6:38 AM. Next year, (2021) Rosh Hashanah ends on Wednesday, September 8th at sunset in Jerusalem at 6: 54 P.M.

Find out where you will be when Rosh Hashanah ends at sunset in Jerusalem, when the Shofar is blown to mark the end of the old year and the beginning of the new one. That is the moment we should be watching for!

Where will you be and who will you be with when Rosh Hashanah ends? If you are with an unbeliever, will they understand the significance of the holiday? Did you tell them?

34.11 The Solar Eclipse Across The U.S. On August 21, 2017.

In addition to the seven signs on Jewish Holidays, there was also a solar eclipse that crossed the U.S. on August 21, 2017. Yom Kippur followed this solar eclipse by forty days and ended at sundown on Saturday September 30, 2017. That was the night that the "Harvest Festival was celebrating just after sundown, that Saturday night in Las Vegas, Nevada

Astronomers call it **"The Grand Coincidence".** The only way that an eclipse can occur is if the size, proportions and distances all fit into a nice neat little equation. Whereas the distance from the center of the earth to the center of the moon is represented by "x"; the distance from the center of the earth to the center of the sun is 400x.

The sun is **400** times larger than the moon, and is **400** times farther away from the earth. These exact proportions and distances, make a "blood moon" eclipse possible. Likewise with the solar eclipse. The exact sizes, distances and proportions of the sun, moon and the earth make an eclipse possible. Nowhere else in our solar system can such an eclipse be found. Our situation with the planets, sun and moon is unique to earth so far! Earth is also unique because it has life. No thinking astrophysicist would expect to find that unusual eclipse characteristic anywhere else in a life time.

When God restored the earth for man, He said in (Genesis 1:14), that the Sun, Moon, Stars and planets would be used… "*for **signs**, and for seasons, and for days, and years*". God said that he would be using these things because that is what He Created them for "*in the beginning*". However, He may have made a few adjustments to the alignment of this planet about 6,000 years ago.

It is quite a miracle the way God uses the sizes, distances, proportions and motions of the planets, stars and moon for signs. This can only happen by design and is in itself another proof that there is a God!

The Hebrew letter "**tav**" means **"sign"** and has a gematria value of **400**. I don't believe that this is a coincidence. This sort of solar/lunar/planet arrangement is not known to be anything that an Astrophysicist would expect to find any place else. The chances of an arrangement similar to our solar system that could produce such eclipses are quite remote. An eclipse is like God's signature on his Creation just to remind us that He is there.

Every time there is a solar or a lunar eclipse, it is a sign from God telling us that He is the Creator and final Judge.

A solar eclipse crossed the United States on August 21, 2017 beginning at the **33rd state** of Oregon and the **town of Salem** at the 45th parallel. Salem is the ancient name for Jerusalem.

As the sun sets in Jerusalem half way around the world, the moon is also beginning to block the sun's light in the US. The eclipse in the US begins at the same time as the sun sets in Jerusalem half way around the world. Is this another coincidence or is (Genesis 1:14) talking about this sort of thing?

Traveling at about 1500 miles per hour, the 70 mile wide shadow will cross the nation in 1 hour and 33 minutes and end at the 33rd parallel. This event occurs on the 233rd day of the year with 133 days left till 2018. The eclipse completes its 1 hour and 33 minute trip across the US, 33 days before the "last trump" of Rosh Hashanah. September 23, 2017 This eclipse is only seen in the U.S.

This eclipse divides the U.S. in half and runs from the Yellowstone calderas to the New Madrid Fault Zone!

About 7 years later, another eclipse crosses the path of its predecessor at the New Madrid Fault Zone making an "x marks the spot". The paths also looks like the Hebrew letter "**tav**" in the ancient script. As we have already mentioned, "tav" is the Hebrew letter that means "sign" and has a gematria value of "400".

I don't know for sure what all of this means, but it seems like God is warning us about a judgment that is coming to our once Godly Nation. According to Jewish tradition, God uses the sun for signs of judgement to the nations or Gentiles. The moon is used for signs to Israel.

Four days after the solar eclipse, a category four hurricane called "Harvey" hits the Texas town of Corpus Christi on August 25, 2017. The name "Corpus Christi" literally means "Body of Christ". Could this be a sign?

The Yellow Stone calderas are showing signs of activity and this super volcano is past due on its eruption schedule. Just think about what Mount Vesuvius did to Pompeii! The people were engaged in every perversion they could think of and the volcano exploded! It happened so suddenly that there was no time to escape. Many were buried alive in volcanic ash. Yellow Stone is big enough to do that to much of the U.S. and spread a dark dust cloud around the globe for a year or more.

One thing that I have always believed is that the Rapture will happen at the end of the 5th feast of Israel known as Rosh Hashanah or The Feast of Trumpets. The 70th week of Daniel begins nine days later at the end of Yom Kippur. The day begins with promises of "***Peace and Safety***" (or "security"), but just after sundown comes ***"Sudden destruction"***. That will mark the beginning of the ***"Seventieth Week of Daniel"***. America will be blind-sided and become the Nation of the Antichrist! We know the Feast Days but we just don't know the year yet.

34.12 "Tav" Signs In The Sun

"Tav" is the Hebrew letter that represents a sign and has a numerical value of 400.

"And there shall be signs in the sun, and in the moon, and in the stars; and upon the earth distress of nations, with perplexity; the sea and the waves roaring;" (Luke 21:25)

If God says that he will give us signs in the sun before the *"Day of the Lord"*, then we should look for them…

"The sun shall be turned into darkness, and the moon into blood, ***before*** *the great and the terrible day of the LORD come."*(Joel 2:31)

The Old Testament made it clear that this sign would occur **before** the *"Day of the Lord."* The New Testament also makes it clear that there will be a sign in the sun **before** the ***"Day of the Lord."***

"The sun shall be turned into darkness, and the moon into blood, ***before*** *that great and notable day of the Lord come:"* (Acts 2: 20)

Usher has set a date of 862 BC for the writing of the book of Jonah. I don't know if Usher got the date right or not but it just so happens that on June 15, 763 B.C. there was a total Solar Eclipse around Nineveh. That Eclipse occurred exactly 40 days **before** the Feast of Atonement. This may have happened just before Jonah went to Nineveh. I believe that this passage is talking about the resurrection of Christ Jesus and that generation for sure.

"A wicked and adulterous Generation seeketh after a sign; and there shall no sign be given unto it, but the sign of the prophet Jonas. And he left them, and departed". (Matthew 16:4)

Apparently God had already been working on the evil Ninevites. Secular sources indicate that they were already suffering from war, food shortages and plagues resulting from years of war and bad weather. When Jonah showed up, the work was already done to prepare their stubborn hearts.

Ancient Nineveh is modern day Mosul and I.S.I.S. headquarters in Iraq. The ancient city of Nineveh changed their minds toward God and believed and were spared. Many got saved! Now look at their descendants today. What happened?

Yom Kippur will, some year in the future, mark the beginning of the 70th week of Daniel, which is also called the "Tribulation". Whatever the year,

I believe that it will occur at the end of the 6th Feast called Yom Kippur or Feast of Atonement.

I believe that the Antichrist will probably be revealed to the world on the day that precedes the Tribulation. The Tribulation will begin after sundown following Yom Kippur because Jewish nights precedes the day.

On that Yom Kippur, the Antichrist will promise peace and security to the World for seven years, but at the end of the Feast, at sundown, two things will happen.

First of all, the dispensation of Law will pick up where it left off 2,000 years ago on the Feast of Pentecost. God's clock will begin to count off the last 7 years of the Dispensation of Law, which will be the worst time the world has ever seen or ever will see.

The Second thing to happen is the Tribulation will begin with *"sudden destruction"* and darkness. From the Biblical descriptions of the Tribulation, I believe that this ***"sudden destruction"*** spoken of in (1Thessalonians 5:3) is Islamic Jihad. The darkness spoken of may be connected to the power grid being taken down by terrorists.

But also the year 2018 is interesting because Rosh Hashanah falls on September 11. But that is not all, it is also the 70th year since 1948. The Muslim New Year also falls on that same day. Imagine that, the Muslim new year and the Jewish new year both fall on the same day and it is 9/11 2018 and on the 70th anniversary!

34.13 "As A Woman In Travail"

"*For when they shall say, Peace and safety; then* ***sudden destruction*** *cometh upon them, as* ***travail*** *upon a woman with child; and* ***they shall not escape***". (1 Thessalonians 5:3)

God uses birthing pangs to describe what Israel and the world must endure during the 70th week of Daniel, also called the Tribulation.

*"And they shall be afraid: pangs and sorrows shall take hold of them; **they shall be in pain as a woman that travaileth**: they shall be amazed one at another; their faces shall be as flames"*. (Isaiah 13:8)

*"And she being with child cried, **travailing in birth**, and pained to be delivered"*. (Revelation 12:2)

It is interesting to note that God makes repeated references to human gestation in connection with the 70th week of Daniel. There are **266 days in the human gestation period**. As the time of birth approaches, there are a number of signs that labor is near. These signs increase in frequency and magnitude as the day approaches. When labor begins, the pain becomes more intense with each contraction. Contractions increase in frequency and magnitude until birth.

I believe this may be a hint about the spacing of the judgments throughout the tribulation or some other aspect of timing associated with the second return of Christ. It is of particular interest to me that the current Pope is the **266th Pope**. The French Foreign Minister announced a meeting between the Pope and Mr. Obama in a very mysterious and unusual manner, as a riddle. The announcement was made three times which said **"the world has 500 days to avoid environmental chaos"**. That announcement was made on May 13, 2014. It turns out that **Mr. Obama and the 266th Pope were scheduled to meet on the 266th day of the year** which just happened to be Yom Kippur. The 70th week of Daniel, also known as the Tribulation will begin on Yom Kippur on the appointed year.

What does all of this mean? I don't believe that these are coincidences that should be ignored. I believe that **The Conspiracy of the Four Horsemen of the Apocalypse** was completed and agreed to by all parties on Yom Kippur 2015. The successful completion of agreements to depopulate the earth was marked in the sky by God as completed on the Feast of Tabernacles 2015!

During the time that our leaders were betraying our trust in them God was warning us in the sky. Since that time, terrorists are flooding into western countries. I believe the "Four Horsemen are mounting up and getting ready to ride! We are that close!

34.14 There Will Be A Famine For Truth.

"Behold, the days come, saith the Lord GOD, that I will send a famine in the land, not a famine of bread, nor a thirst for water, but of hearing the words of the LORD": (Amos 8:11)

"*Now the Spirit speaketh expressly, that in the* ***latter times*** *some shall depart from the faith, giving heed to seducing spirits, and doctrines of devils*". (1 Timothy 4:1)

"*This know also, that in the* ***last days*** *perilous times shall come*" (2Timmothy 3:1)

"For men shall be lovers of their own selves, covetous, boasters, proud, blasphemers, disobedient to parents, unthankful, unholy" (2Timmothy 3:2)

"Without natural affection, trucebreakers, false accusers, incontinent, fierce, despisers of those that are good", (2Timmothy 3:3)

"*Traitors, heady, highminded, lovers of pleasures more than lovers of God*" (2Timmothy 3:4)

"*Having a form of godliness, but denying the power thereof: from such turn away*". (2Timmothy 3:5)

"For of this sort are they which creep into houses, and lead captive silly women laden with sins, led away with divers lusts," (2Timmothy 3:6)

"*Ever learning, and never able to come to the knowledge of the truth*" (2Timmothy 3:7)

"Now as Jannes and Jambres withstood Moses, so do these also resist the truth: men of corrupt minds, ***reprobate concerning the faith****"* (2Timmothy 3:8)

The Bible says that in the last days we will have deceivers of the caliber of Jannes and Jambres, who were two of Pharaoh's magicians who were able to duplicate most of the miracles done by Moses. Today we are already seeing trickery in the form of the Tongues movement and faith healers.

34.15 Jesus Warned About Deception In The "Latter Days"

"And as he sat upon the Mount of Olives, the disciples came unto him privately, saying, Tell us, when shall these things be? And what shall be the sign of thy coming, and of the end of the world?" (Mathew 24:3)

This passage is talking about the end of the Age, not the end of the world. The word "world" is a poor translation and should be translated as "Age". The Age that they are talking about is the Age or Dispensation of Law. More specifically, the answer is about the last days that lead up to and include the 70th week of Daniel.

"And Jesus answered and said unto them, ***Take heed that no man deceive you****"*. (Mathew 24:4)

When the disciples asked Jesus about the second coming the next words he said were... "***Take heed that no man deceive you***". The Bible describes the last days as a time of deception.

"For many shall come in my name, saying, I am Christ; and ***shall deceive*** ***<u>many</u>****"*. (Mathew 24:5)

Jesus said "many" shall come in his name. This is describing a broad, widespread deception because it says *"many"* shall come in his name and deceive "*many*".

Most people think this passage is talking about False Christs, but I don't think so. Jesus is doing the talking and he is talking about False Prophets and False brethren. False Christs are not mentioned until verse 24.

Jesus is doing the talking and he is saying that these deceivers will come in his name and acknowledge that he (Jesus) is the Christ. In other words these people will claim to be Christian. Either Christian Preachers (False Prophets) or Christian Politian's or Christian business men. They are not the real deal. They intend to deceive and use people to satisfy their own lusts and purpose.

The Catholics claim to be Christian, so does the Watch Tower Society, the 7th day Adventists, the Mormons the Church of Christ" and many more, but they preach a false gospel that cannot save.

*"And **many** false prophets shall rise, and shall deceive **many**"* (Matthew 24:11)

"For there shall arise false Christs, and false prophets, and shall shew great signs and wonders; insomuch that, if it were possible, they shall deceive the very elect". (Matthew 24:24)

34.16 False Prophets Will Be Multiplied In The "Last Days".

*"This know also, that in the **last days** perilous times shall come"* (2Timothy 3:1)

"But evil men and seducers shall wax worse and worse, deceiving, and being deceived". (2Timothy 3:13)

"But there were false prophets also among the people, even as there shall be false teachers among you, who privily shall bring in damnable heresies, even denying the Lord that bought them, and bring upon themselves swift destruction" (2 Peter 2:1)

False teachers speak against the gospel of Grace because they lose control of the money. False preachers claim to be believers but they are not because they have believed on a false Gospel of human merit.

*"And many shall follow their pernicious ways; by reason of **whom the way of truth shall be evil spoken of**"*. (2 Peter 2:2)

*"And through covetousness shall they with feigned words **make merchandise of you**: whose judgment now of a long time lingereth not, and their damnation slumbereth not"*. (2 Peter 2:3)

False teachers are in business for the money. Their followers are their merchandise.

"Beware of false prophets, which come to you in sheep's clothing, but inwardly they are ravening wolves". (Matthew 7:15)

They put on an act to look good for appearance sake so they can gain your trust and mislead. The Pharisees of Jesus' time are examples of what will be multiplied many times in the last days. Jesus explained it…

…*"Woe unto you, scribes and Pharisees, hypocrites! for ye are like unto whited sepulchres, which indeed appear beautiful outward, but are within full of dead men's bones, and of all uncleanness"*. (Matthew 23: 27)

"Even so ye also outwardly appear righteous unto men, but within ye are full of hypocrisy and iniquity". (Matthew 23: 28)

Many preachers claim to be Christian (come in the name of Jesus) and acknowledge that Jesus is the Christ but are preaching a false gospel of human righteousness that cannot save and are deceiving many people about many things. God calls them wolves in sheep's clothing.

Pretending to be something that they are not is the nature of false teachers. To make an emotional connection with people, they use voice inflection to sound like they really care when they really don't. Wearing big, fake smiles some act extremely sweet to appear "Godly".. They appeal to your feelings to gain peoples trust and support so that they can take advantage of your trust. There are many examples of this

After deception comes betrayal. "The Bible says…

"And no marvel; for Satan himself is transformed into an angel of light. Therefore it is no great thing if his ministers also be transformed as the ministers of righteousness; whose end shall be according to their works". (2Corinthians 11:14, 15)

Pray for the truth and God will send it. If you are faithful with the truth that God gives you, he will send more. If you ignore God's truth he may just let Satan have you and deceive you with lies.

34.17 Watching For The Antichrist And The False Prophet.

Many people believe that you can't know who the Antichrist is before he is "*revealed*" as described in (2Thessalonians 2:3). My view on that passage is that the Antichrist will be revealed to the children of darkness, not to the children of light. The children of darkness are unbelievers who will be left behind. If the believers who get raptured out were watching, they should recognize the antichrist from scripture, because the Bible describes him. God the Holy Spirit revealed all of this in his word thousands of years ago.

The False Prophet and the Antichrist will reveal themselves to the world as the leaders of all faiths. This is where the Satanic, one world religion emerges from Free Masonry; Islam; The Vatican; Lucis Trust; The occult; The U.N.; the globalists and every secret, satanic organization working together as one to establish a New World Order. It is going to be an Islamic New World Order for seven years.

God reveals the identity of the Antichrist and those who are associated with him through God's Holy Word. The Antichrist is described by his religion, his words, his deeds, his associations; and the country that brings him to power. (ETBH chapters 44, 45 and 46)

"Neither shall he regard the God of his fathers, nor the desire of women, nor regard any god: for he shall magnify himself above all" (Daniel 11:37)

Obama quoted the following passage in one of his speeches. The speech caught the attention of many believers.

(Psalms46:10) *"Be still, and know that I am God: I will be exalted among the heathen, I will be exalted in the earth"*

Although there are no signs specific to the Rapture, other than the Feast of Trumpets and the prophecy of Joel 2:31, there are plenty of signs for the 70th week of Daniel.

The Tribulation begins on the 10th day of Tishri, at the end of the 6th Feast which is called the Feast of Atonement or Yom Kippur.

I believe that is the day the False Prophet and the Antichrist will meet to make their big announcement to the world. That is when the Antichrist will be revealed to the children of darkness. (1Thessalonians 5: 4, 5)

The announcement may be that Mr. Obama will be the enforcer of the Oslo Accords for 7 years. That is also the day that they will say *"peace and security"*. It may also be the day that the False Prophet tells the world to follow the Antichrist. The Antichrist may be revealed to Muslims as the 12th Imam. Acceptance of the Antichrist as the 12th Imam will mean that about 1.8 billion Muslims will believe that their only guarantee to escape hell and enter into paradise and get 70 virgins is to fight and die for him.

"*For when they shall say,* ***Peace and safety****; then sudden destruction cometh upon them"* (1Thessalonians 5:3)

Peace and safety is the same as "Peace and security". The Strong's definition for the Greek word translated as safety here is Strong's # G803 – *asphaleia* ; it is defined as: firmness, stability, certainty, undoubted truth, **security from enemies** and dangers, s**afety**

On the U.N. website, under the heading of "U.N. Charter; Chapter one; Article one; The first six words written are "To maintain international **peace and security**". The central mission for the U.N. since its creation in 1945 is International **"Peace and Security"**.

This is clear identification that the U.N. will provide the Antichrist with a path to controlling the world.

Follow this link to the U.N. website … http://www.un.org/en/sections/what-we-do/maintain-international-peace-and-security/index.html

"The United Nations came into being in 1945, following the devastation of the Second World War, with one central mission: the maintenance of international **peace and security".**

"Peace and Security" are the same words spoken on live T.V. when the Oslo Accords was signed September 13, 1993. Those are the same words spoken of repeatedly by Mr. Obama.

https://www.youtube.com/watch?v=JQoyFcaZOC8

https://www.youtube.com/watch?v=jmU1eDYFn6w

https://www.youtube.com/watch?v=-hl3jcNKies

https://www.youtube.com/watch?v=7so2yCmaio0

Twelve days after being in office as President, the nominations for the 2009 Nobel Peace Prize was closed and Obama's name was included. How did that happen? Imagine that, after only 12 days in office he is nominated for the Nobel Peace Prize! He has not even had the time to finish moving into the White House yet.

At some point in time someone must ask themselves, is this conspiracy so big and pervasive that it covers the whole world? The Bible says yes and that "Peace Prize" confirms that!

There were 205 nominations for the 2009 award and Mr. Obama's name was one of them. How could this be? What did he do in those first 12 days to earn such an award? He did not even finish his first term as senator before he ran for president and won. Talking about a fast track!

Mr. Obama received a Nobel Peace Prize in Oslo Norway, on October 9, 2009 less than a year after he was elected for no apparent reason. It is interesting to note that Oslo Norway is where the Rabin peace treaty was negotiated. That is why it is called the Oslo Accords. It was later officially signed in the U.S. Whitehouse on Yom Kippur. (ETBH chapter 42.8)

I think that it is significant that Mr. Obama was the senator from the 13th district of Illinois before he became President. I also think that it is noteworthy to mention that the morning that Mr. Obama was declared to be the 44th president on November 5, 2008, his home state, Illinois, had a winning pick three lottery number of 666. That evening, the winning four was 7779.

According to TRUTHSTREAMMEDIA.COM mass shootings increased by six times during Obama's administration compared to Bush. One headline reads "Why have there been more mass shootings under Obama than the four previous presidents combined"? Another headline reads... "Five of the twelve deadliest gun massacres in U.S. History took place during Obama's first term." Read their articles in their entirety on their website; TRUTHSTREAMMEDIA.COM There is much more to be said about Mr. Obama, but I will save that for later.

Many say that believers can't know who the Antichrist is because he will not be revealed until after the Rapture. Consider who the Antichrist will be revealed to. He will be revealed to a lost, deceived and unbelieving world that will be left behind at the rapture. He will be revealed to the children of darkness as they're Messiah or the 12th Imam.

Although it is true that the Man of Sin will be revealed to a lost and blinded world after the Rapture, that does not mean that believers who were taken at the Rapture did not recognize him! We are children of the light and we are supposed to be watching. God has given us plenty to watch for, including many details about the Antichrist that we are supposed to recognize.

Just because the Antichrist will be revealed nine days after the believers are raptured out of this world, is no reason to think that believers should not be diligently watching for him and recognize him. If believers are obediently watching for the Antichrist and False prophet and we are acting like the children of light, then I believe that we will be able to recognize the signs, warn people and receive the reward that God has promised to those who are watching.

The Bible tells us which country the Antichrist will come from. We have a time frame to be looking. We have his number, 666. We have descriptions of him; we know what he will do in the near future and seven years after that. We also may have his name! (ETBH chapter 45 and 46)

I believe that there are believers and unbelievers who know who he is. There is no reason for believers to not recognize who he is now and recognize from all of the other signs that our time is at hand!

The Antichrist and the False Prophet are here! In plain view of anyone who is watching!

I believe that we are looking at the players of the Tribulation in the news on a daily basis. The Bible has given us plenty of details about the main players and what they are about to do. When you see two men matching all of the descriptions of the Antichrist and False Prophet then that is a big sign that we are very close!

God has also provided us with signs in the earth, in the sea, in the sun and in the moon.

34.18 Don't Be Left Behind!

The Bible seems to imply that the deception that occurs during the Tribulation will be very convincing! People will be amazed, even frightened. People will wonder, "HOW CAN THIS BE?!

The fake miracles will be so great that the tricks will seem like magic or super power. But it will only be the miracles of Satan's Demons working through men. Just like Pharos's magicians could copy the miracles of

Moses, Satan's miracles will pretend to be miracles from God. In the case of Pharaoh, God's miracle through Moses literally ate the fake miracles of the magicians.

Satan will use men's technology to persuade people to follow him because of those faked miracles and demonstrations of powers.

I believe that some of that technology has already been tested and used. But that technology has advanced a lot since September 11, 2001.

I believe that the Globalists and their Antichrist has already planned what they are going to do and they are waiting for the day that the Restrainer is taken away at the Rapture.

Those who are taken at the Rapture are those who have believed on the name of Jesus (Yeshua in Hebrew). A person is Born Again when they understand and believe the only Gospel that saves. It is the Simple Gospel that Saves. Just Faith in Jesus alone!

The Belief that Jesus is God who loves us and who came to the Earth as a Man, born of a virgin, and lived among us for about 33 years. Jesus lived a perfect life and offered himself and his blood on the cross as a perfect sacrifice for the sins of the whole world, past, present and future.

"The wages of sin is death, but ***the Gift of God is Eternal Life*** *through Jesus Christ our Lord"* (God Almighty!)(Romans 6:23).

Eternal Life is a Gift, it is Free, and it has already been paid for. God is not an Indian giver. You don't have to promise or give anything to God or Jesus to receive Eternal Life. God does the giving and the believer does the receiving. It is Just Faith. If you believe that, tell God now that you believe and thank him for paying for your sins and giving you Eternal Life and a free citizenship and home in Heaven!!! Amen! Amen! Amen! You are saved!

34.19 When God's Clock Starts!

God's clock starts ticking at the sound of the last trump, at sundown ending Rosh Hashanah on the year of the Rapture. There is a Jewish tradition about the ten days between the beginning of Rosh Hashanah and the end of Yom Kippur. It is called the "**Ten Days of Awe**".

On Yom Kippur, I believe that the Antichrist will have gained control of the U.N. and the U.S.

(Revelation 6:2) *"And I saw, and behold a white horse: and he that sat on him had a bow; and* ***a crown was <u>given</u> unto him****: and he went forth conquering, and to conquer"*

That "crown" represents a kingdom that is given to the antichrist at that time.

(Danial 11:21) *"And in his estate shall stand up a vile person, to whom they shall not give the honour of the kingdom: but* ***he shall come in peaceably, and obtain the kingdom by flatteries****"*.

The antichrist is going to lie his way into power. He will take over the U.S. and the U.N. and next Europe, probably starting with France, England and Germany.

(Revelation 17:12) *"And the ten horns which thou sawest are ten kings, which have received no kingdom as yet; but receive power as kings one hour with the beast"*.

(Revelation 17:13) *"These have one mind, and* ***shall <u>give</u> their power and strength unto the beast****"*.

(Revelation 17: 17)*"For God hath put in their hearts to fulfil his will, and to agree, and* ***give their kingdom unto the beast****, until the words of God shall be fulfilled"*

A Kingdom will be given to him. I believe that Kingdom is the United States. If the President and vice president are both taken at the rapture, then the speaker of the house will be in the position to do that. Of course this is speculation, but that is what watching is all about.

The 70th Week of Daniel begins at sundown ending Yom Kippur. That is when God is going to start counting out the days for the 70th week of Daniel.

The next thing that happens is ***"sudden destruction"*** and God's two witnesses begin their ministries. That is how the Tribulation will begin.

"I will give power unto my two witnesses, and they shall prophesy a thousand two hundred and threescore days" (Revelation 11:3)

God tells us how many days his two Witnesses will preach. If you divide 1,260 by 30 (days per month) you get 42 months. Their ministry will end in the middle of the Tribulation. (5th seal) That is when the Antichrist is indwelt by Satan and he kills the two Witnesses!

"And from the time that the daily sacrifice shall be taken away, and the abomination that maketh desolate set up, there shall be a thousand two hundred and ninety days" (Daniel 12:11)

"Bessed is he that waiteth, and cometh to the thousand three hundred and five and thirty day" (Daniel 12:12)

The second part of the Tribulation is 75 days longer than the first half (1260). 1290+45=1335

34.20 Rosh Hashanah Is Observed On The 1st Day Of Tishrei.

Rosh Hashanah begins on the first day of Tishrei and ends at sunset or at the sound of the last trump of the Feast of Trumpets. But something unusual also happens at the same **moment** that the Hebrew New Year changes in 2018. At the same moment the Jewish Feast of Trumpets ends, the Islamic New Year celebration also begins.

The Islamic New Year is also called The Arabic New Year or also known as "**Al Hijra**" meaning ("flight") and the celebration begins when Rosh Hashanah celebration ends at sundown in 2018. "Al Hijra" is the flight of Muhammad from Mecca to Medina with the plan to immigrate, and conquer . The Islamic New Year begins on the first day of the first month of Muharram in the Islamic lunar calendar.

Muharram is known as the time that Muhammad "**immigrated in the way of Allah**". That is to say, immigration with the intent to blend in, betray and conquer. Muharram is one of four "sacred months" when war is not permitted. Muharram is second to Ramadan in importance.

The Islamic year has 354 days and the Hebrew year has 360 days which makes this alignment rare and it happens on September 11, 2018 which really makes this event unusual. What we have here is the converging of significant dates on two **pagan calendars** with the Hebrew calendar. The Roman and the Muslim calendars converge on the same day as a Biblical Feast day as Rosh Hashanah in 2018.

Over 2,600 years ago we have a connection between "alah" and 9:11 (Daniel 9: 11). In Danial 9:11, the word translated as "curse" is "alah". In the book of (Isaiah 24:6) God says that "alah" will take over the world during the 70th week of Daniel.

At the same instant, at the same moment in time that the final trump of the day will sound off at sundown in Jerusalem marking the end of the former year; the end of The Feast of Trumpets; and the beginning of the New Year; That is the moment when every believer will find out if this is going to be it or do we have another year left. That happened on September 11, 2018, at sundown, Jerusalem time and that was the 70th Roman year since Israel's Independence.

In 2018 America has been flooded with Muslim immigrants. It appears like America has been invaded.

At this point in time, Rosh Hashanah 2018 has already come and gone. It is February 2019, Trump has recognized Jerusalem as the capital of Israel and moved our embassy there. The embassy was dedicated on the 70th anniversary on the Roman calendar along with a half-shekel Temple coin that was minted by the Sanhedrin.

The Temple coin has Trump's profile with the profile of King Cyrus in the background. King Cyrus is the Persian King that helped the Jews to rebuild the second Temple. On the same side of the coin is the inscription "To fulfill 70 years". Now that sounds like they are following the Roman calendar to make 2018 the 70th anniversary of Israel's independence in May 14, 1948.

They are still talking about impeaching Trump. Trump appears to be fighting the radical left; the Muslim invasion of America and the Main Stream Media all at the same time. I expect, at this point that these impeachment proceedings are in some way, a part of laying the groundwork for the Antichrist to take over America.

It just appears like Trump is fighting against this globalist agenda by himself.

34.21 Yom Kippur Is Observed On The10th Day Of Tishrei.

In the year 2017 it began on September 29 just after sundown and ended just before nightfall **Saturday night September 30, 2017**. At midnight the date changes to October 1.

Since writing the previous lines and before publishing this book, sometime has elapsed and things are happening pretty fast that should be mentioned here.

On Saturday night, September 30, 2017, just a couple of hours after Yom Kippur ended at sundown, a mass murder occurred in Las Vegas. It happened about 10pm, Right in front of the pyramid, Sphinx and Obelisk at the Luxor Hotel. Three Masonic symbols of the 32 degree Egyptian Shriner variety are lined up with and overlook **the harvest festival**. The FBI claims that a lone shooter fired from the 32nd floor.

Who picked that name, **the harvest festival**? Who picked that place in front of the Luxor Hotel with all of the secret Masonic symbols? Who picked that time for the festival, just after Yom Kippur ended?

That is quite a name for a place for a mass murder, **the harvest festival;** in front of Masonic symbols; just after sundown on Yom Kippur of all nights. It seems more like an occult human sacrifice of some sort, well planned and covered up by the F.B.I.

The shooting started just after the crowd finished singing God Bless America. The lights were turned off with all attention on the flag. They all held up their Bic lighters as they all sang God Bless America together. Soon, the shooting began!

I know that this was not the work of one man. But I do believe it is the work of many government agents working together with other nongovernment agencies or individuals to pull this off and make it look like the work of one man. But it doesn't seem like the perps of this evil deed consider public awareness to be much of a threat.

The F.B.I. changed their story more than once, but stuck with the lone shooter story. Plenty of witnesses are challenging the F.B.I. story and some of those witnesses are now reportedly dead.

There are reports about three helicopters flying and hovering within range of the festival and tracked on radar. One helicopter was reported to be using a false transponder I.D. There are reports that it was masquerading as a Southwest airlines passenger plane, Southwest Airlines Flight number 4119. That is impossible. Southwest airlines can't hover, fly slow or change directions like a helicopter. I don't want to go into this in detail here but I will say that it smells like deception, betrayal, treason and a sloppy cover-up from within our own government!

Check out The Vegas shooting conspiracy… https://www.youtube.com/watch?v=6bxyR3B3jow

Las Vegas has "Aerial Shooting Adventure Tours". Check out "the gunship experience" for a description of the weapons and their capabilities… https://gunshiphelicopters.com/ The airport and facilities that house these helicopters is about a mile from where the shooting took place.

I.S.I.S has claimed credit for this deed. In the past, I.S.I.S. has denied responsibility for terror acts and claimed credit for terror acts and to this day, investigators have confirmed them to be accurate in this respect. Islam is a religious, military and political system built around warfare against non-Muslims.

Maintaining credibility among Muslims in warfare against non-Muslims is a most important requirement to establish a caliphate. As I understand it, I.S.I.S is Obama's creation and that $1.7 billion dollars in cash from the Iran deal will pay his army of terrorists that he brought into the U.S. I believe that they will bring "*sudden destruction*" again, just after sundown some year in the near future just after sundown, on Yom Kippur. I believe this event that took place on Yom Kippur 2017 is sort of like a dry run or sampling of the real thing coming in the near future.

Yom Kippur 2018 began at sundown on Tuesday, September 18 and ended at nightfall on Wednesday, **September 19**.

Trump is president now and it is 2018. Trump has recognized Jerusalem as the capital of Israel and is determined to move our embassy there! This is the first time that Israel has had this kind of support from the U.S.

The **Sanhedrin** issued a Temple coin to commemorate Trump's recognition and the building of the 3rd Temple. On one side of the coin is a profile of Persian King Cyrus who gave the order that allowed the Jews to rebuild the 2nd Temple. Super imposed over King Cyrus is the profile of Trump.

Trump said in a speech recently, that **this year marks the 70th anniversary since the rebirth of Israel**. I was under the impression that the Temple would be rebuilt just after the 70th year, but I thought that it was 70 Hebrew years not 70 Roman years. But Trump made a reference to the 70th year anniversary being this year and so did Prime Minister Benjamin "Bibi" Netanyahu and the Sanhedrin.

I believe that the construction of the 3rd Temple will begin right after the rapture. I believe that the 3rd Temple will be built soon without any warning. It will take three days to build. I also believe that the Church will not be here to see the Temple built. Of course I am speculating. We will just have to watch, wait and see.

34.22 Satan Will Empower Two Men

The Antichrist and False prophet will use our technology to do "miracles" to deceive. It has been known that the military has been studying holographic technology for quite some time now. Perhaps to direct the enemies attention away from real targets such as troops, tanks and aircraft.

We have seen what can be done with lasers in the civilian realm. Just think about what the military can do with them!

"*Now as Jannes and Jambres withstood Moses, so do these also resist the truth: men of corrupt minds, reprobate concerning the faith*". (2Timothy 3:8)

Satan will have his own show of power. He will empower two powerful men of his own to do his bidding on Earth during the Tribulation.

Those two men are known as the Antichrist and the False Prophet. They come up with their own sort of miracles, signs and wonders to deceive mankind.

Satan's "miracles" will not be of the caliber used by God's witnesses but many will be deceived anyway.

Satan will use feelings, lust, terror and deception to control men.

At the end of the "Great Tribulation", is when the Battle of Armageddon takes place. That is when Satan and his demons make their last stand. It will be violent and destructive but they will finally be subdued.

The upcoming 7 years of terror and Genocide known as the "Great Tribulation" is time set aside by God to persuade men and fulfill his promises to Israel. It is the time for winning as many unbelievers as possible just before the great harvest.

144,000 Jewish witnesses will be warning the world on God's behalf, persuading them to believe. These are God's witnesses proclaiming that there is only one way to get saved and Jesus is the way.

"*Jesus saith unto him, **I am** the way, the truth, and the life: no man cometh unto the Father, but by me*". (John 14:6)

Faith Only Or, Just Faith!

35

The False Prophet.

(Revelation 13:7, 8, 16, 17) (Revelation 17) (II Thessalonians 2:4)

35.1 The Roman Catholic False Prophet.

"The False Prophet will definitely come out of the Roman Catholic Church and I think that he will probably be the Pope himself".

That is a quote from one of Dr. Lindstrom's audio tapes and what Dr. Hank Lindstrom always taught, for as far back as I can remember.

Joel said … (Joel 2:31) …there would be signs in the heavens… "***<u>before</u>* the great and the terrible day of the LORD come**". In my opinion, for reasons that have already been stated in (chapter 34), the signs that we have recently seen in the Sun and the Moon and the stars meet that description. Furthermore, we are seeing and hearing about events that are occurring and are about to occur which relate to the events which will begin the Tribulation. A significant event was scheduled to climax with a Yom Kippur meeting between Pope Frances and Barack Obama on September 23, 2015, followed by a U.N. vote to replace "Agenda 21" with a new U.N. plan for sustainable development called Agenda 2030.

That is the day that Obama was scheduled to meet with Pope Francis the first. This meeting was publicly promoted by the French Foreign Minister and John Kerry. On May 13, 2014, the eve of Israel's' independence day, they made a speech together and said that the world had 500 days to avoid "climate chaos".

If you count 500 days from **May 13, 2014** that brings you to the end of Yom Kippur, **September 23, 2015**. What we have here is a link between Israel's anniversary and the **United Nations vote on September 25, 2015 to accept the new 2030 goals for sustainable development**. The actual vote was preceded by days of sermons from the Pope and other dignitaries

Coincidentally or maybe not, September 23, 2015 was the 266th day of the year and Pope Francis the First is also the 266th Pope.

And to make it even more interesting there are 266 days in a human gestation period, from conception to birth. This number 266 is also the number of days from the beginning of the first of a series of astronomical events associated with the eighth blood moon tetrad of 2014-2015 to a midpoint in that event. Someone is paying attention to these things.

Who do you think is coming up with all these numbers? It seems to me that any reasonable person must realize that somehow, someone is putting significance on numbers and dates and overseeing their application over long periods of time. That person knew where the planets would be thousands of years in advance!

Pope Benedict XVI (2005-2013)was the 265th Pope he resigned on February 28, 2013. **Thirteen days later**, Pope Frances the First was ordained and became the 266th Pope on 3/13/2013

See link; http://www.newadvent.org/cathen/12272b.htm

Pope Francis the first began his service on March 13, 2013. He is traveling around the world persuading governments to treat climate change as a moral issue. Their meaning of treating climate change as a moral issue is what concerns me!

Pope Francis the first is the first pope to take the name of Francis; the first South American pope; the first Jesuit Pope; the first Pope to have one lung and according to Robert Spencer…" **Pope Francis is the first Islamic Pope**". Not only is Pope Francis the First, first in a list of some interesting things but the number one just keeps popping up in relation to him as does the numbers 13 and 266.

A unique feature about this last Blood Moon Tetrad event is the fact that it is the only Biblical tetrad that is said to be symmetrical about a midpoint. The first event of this last Biblical Blood Moon Tetrad was on Passover 2014 and was 266 days from the midpoint. The last event was on Tabernacles and it was also 266 days from the midpoint. This number 266 is clearly associated with Pope Frances the First and associated with an astronomical event that seemed to tie together events leading up to the meeting of Obama and Pope Frances the First on September 21, 2015.

Pope Francis was prophesied to be the last Pope by St. Malachy (1094-1148) or so they say. If Pope Francis turns out to be the False Prophet, and I believe that he probably is, then he could be called "*The First and The Last*". Now where have I heard that title before? That is the title that Jesus claims in the book of Revelation.

The Muslims are looking for their messiah and their version of Jesus to break all of the crosses and tell the world to listen to the Antichrist.

March 13, 2013 - Pope Francis the 1st was elected after the previous Pope resigned. This is very rare. Interestingly enough is the fact that the predecessor to Pope Francis 1st was Pope Benedict XVI (2005-2013)

Dr. Lindstrom once said that he often wondered if the frequency that the number 13 was associated with Satan, the Antichrist and False Prophet was more than a coincidence. He went on further to say that the chapter and verse divisions are not inspired. They were put there by man and often divide things up wrong and hide the meaning of the context.

The combination of facts surrounding this Pope and President and this upcoming meeting happening at a time when we are seeing signs in the Sun and Moon as Joel described, made me think that this is a sign. A sign to get men's attention and warn believers and unbelievers about the situation that is at the door.

Dr. "Hank" Lindstrom has written papers on Bible prophecy that name the United States as the first Nation to be controlled by the Antichrist. He even starts out one video on YouTube saying that the United States would most definitely be the nation the Antichrist would come from. All of this was written long before 2008. The fact is that Hank died long before he could have been influenced by biases or prejudices concerning anyone involved.

Hank died from a botched heart surgery on October 13, 2008, twenty one days before the election.

Revelation 13, first mentions the False Prophet and the Antichrist. The False Prophet will use Miracles signs and wonders to deceive the world.

The beast and the False Prophet are the first to be cast into the Lake of Fire and are the only ones mentioned to be cast in alive!

"And the beast was taken, and with him the false prophet that wrought miracles before him, with which he deceived them that had received the mark of the beast, and them that worshipped his image. These both were cast alive into a lake of fire burning with brimstone". (Revelation 19:20)

Another distinction of the false prophet is that he will come out of the Roman Catholic Church. (Revelation 17)

"And upon her forehead was a name written, MYSTERY, BABYLON THE GREAT, THE MOTHER OF HARLOTS AND ABOMINATIONS OF THE EARTH".

"*The seven heads are seven mountains, on which the woman sitteth".* (Revelation 17:9)

And the woman which thou sawest is that great city, which reigneth over the kings of the earth. (Revelation 17:9)

This city that sits on seven hills has traditionally been accepted as Rome. The pagan/Babylonian origin of the Roman Catholic Church is well established.

35.2 The False Prophet Will Use Miracles To Deceive.

The False Prophet will not be able to do the miracles until after the rapture because that is when the Holy Spirit or the Restrainer is removed. However, I believe that the False Prophet and the Antichrist may get started early promoting their Agenda. After the Rapture, they will only have 9 days to persuade the U.N. to accept the Anti-Christ.

(Revelation 19: 20) *"And the beast was taken, and with him the **false prophet that wrought miracles before him**, with which he deceived them that had received the mark of the beast, and them that worshipped his image. These both were cast alive into a lake of fire burning with brimstone"*.

The ten days beginning with Rosh Hashanah and ending with Yom Kippur is known as **"The Ten Days of Awe".** That will most probably be the days of the miracles, signs and wonders used by the Antichrist and the False Prophet to deceive the world into following him. The Feast day of Yom Kippur marks the beginning of the seven years of Tribulation and the completion of **"The Ten Days of Awe"**.

Yom Kippur is most likely to be the day the Antichrist, the False Prophet and the United Nations will say *"peace and safety"*. (1 Thessalonians 5:3)

*"For **when they shall say, Peace and safety; then sudden destruction cometh** upon them, as travail upon a woman with child; and they shall not escape."* (1 Thessalonians 5:3)

The first day of the 7 year Tribulation begins at sundown ending Yom Kippur. That is the day the Antichrist will assume responsibility for enforcing the Rabin Treaty for 7 years. I believe that is when they will say *"peace and safety"*.

If the Rapture were to happen on this coming Rosh Hashanah, people would want an "official explanation" from the government. I suppose that the Pope would begin his display of fake miracles right after the Rapture. After nine days of fake miracles, on Yom Kippur the Antichrist and the Pope meet and give the "official version" of the story which will be supported by the N.W.O. controlled news media.

I believe the Pope will endorse the Antichrist as one sent and empowered by Allah to be the Guardian of World Peace. But he is actually sent and empowered by satan to take away World Peace.

The Antichrist probably will claim responsibility for those missing at the Rapture. He may say that they were useless eaters, undesirables, either abducted by aliens or something else.

Nine days after the Rapture, on Yom Kippur day, the Antichrist will promise to enforce the Treaty of Rabin for 7 years. That is when they will say *"Peace and Security"*, but at sundown, suddenly a global Jihad begins. We still don't know what year it will be.

Three and a half years later, in the middle of the Tribulation the false prophet and the Antichrist will betray those who trusted in them, including Israel and the Catholic "Church".

It is interesting to know that the Catholic "Church" has its own giant telescope. The Vatican's telescope is called L.U.C.I.F.E.R. which is an acronym. After people became aware of its curious name and began to talk and write about it, the Vatican was forced to do damage control. They changed the name to "L.U.C.I.", changed their website and set up damage control websites to misdirect people doing a superficial search. But their original articles were copied and are available on line. Do a careful search.

I believe that their telescope will be used to spot the New Jerusalem coming from heaven to orbit the earth. If the Catholic Church spots it first, they will call it an attack by aliens. That may be why the armies of the world unite to fight against Christ and His armies coming to the earth along with the New Jerusalem. The New Jerusalem is a cube measuring 1500 miles on each side which makes it about the size of the moon.

35.3 The Origins Of The Roman Catholic Church

H. A. Ironside, in his commentary of the Book of Revelation, written in 1920, gives us the following insights into the origins of the Roman Catholic Church.

"The wife of Nimroud-bar-Cush was the infamous Semiramis the First. She is reputed to have been the foundress of the Babylonian mysteries and the first high-priestess of idolatry. Thus, Babylon became the fountainhead mystery-religion that was originated there and spread in various forms throughout the whole earth; it is with us today.

Building on the primeval promise of the woman's Seed who was to come, Semiramis bore a son whom she declared was miraculously conceived! And when she presented him to the people, he was hailed as the promised deliverer. This was Tammuz, whose worship Ezekiel protested against in the days of the captivity. Thus was introduced the mystery of the mother and the child, a form of idolatry that is older than any other known to man. **The rites of this worship were secret. Only the initiated were permitted to know its mysteries.** It was Satan's effort to delude mankind with an imitation so like the truth of God that they would not know the true Seed of the woman when He (Jesus Christ) came in the fullness of time.

From Babylon this mystery-religion spread to all the surrounding nations. Everywhere the symbols were the same, and everywhere the cult of the mother and the child became the popular system; their worship was celebrated with the most disgusting and immoral practices. The image of the queen of heaven with the babe in her arms was seen everywhere, though the names might differ as languages differed. It became the mystery-religion of Phoenicia, and by the Phoenicians was carried to the end of the earth. Ashtoreth and Tammuz, the mother and child of these hardy adventurers, became Isis and Horus in Egypt; Aphrodite and Eros in Greece; Venus and Cupid in Italy; and bore many other names in more distant places. Within 1,000 years Babylonism had become the religion of the world, which had rejected the Divine revelation. In 316 AD, it became Mary and Jesus.

Linked with the central mystery were countless lesser mysteries, the hidden meaning of which was known only to the initiates, but the outward forms were practiced by all the people. Among these were the doctrines of purgatorial purification after death; salvation by countless sacraments, such as priestly absolution; sprinkling with holy water; the offering of round cakes to the queen of heaven as mentioned in the book of Jeremiah; dedication of virgins to the gods, which was literally sanctified prostitution; weeping for Tammuz for a period of 40 days, prior to the great festival of Ishtar, who was said to have received her son back from the dead; for it was taught that Tammuz was slain by a wild boar and afterwards brought back to life. To him the egg was sacred, as depicting the mystery of his resurrection even as the evergreen was his chosen symbol and was set up in honor of his birth at the winter solstice, when a boar's head was eaten in memory of his conflict and a yule-log burned with many mysterious observances. The sign of the cross was sacred to Tammuz, as symbolizing the life-giving principle and as the first letter of his name. It is represented upon vast numbers of the most ancient altars and temples, and did not, as many have supposed, originate with Christianity.

From this mystery-religion, the patriarch Abraham was separated by the divine call; and with this same evil cult, the nation that sprang from him had constant conflict, until under Jezebel, a Phoenician princess, it was grafted onto what was left of the religion of Israel in the northern kingdom in the day of Ahab, and was the cause of their captivity at the last. Judah was polluted by it, for Baal-worship was but the canaanitish form of the Babylonian mysteries, and only by being sent into captivity to Babylon itself did Judah become cured of her fondness for idolatry. Baal was the Sun-god, the life-giving one, identical with Tammuz.

When Christ came into this world the mystery of iniquity was holding sway everywhere, save where the truth of God as revealed in the Old

Testament was known. Thus, when the early Christians set out upon the great task of carrying the gospel to the end of the earth, they found themselves everywhere confronted by this system in one form or another; for though Babylon as a city had long been but a memory, her mysteries had not died with her. When the city and temples were destroyed, the high-priest fled with a company of initiates and their sacred vessels and images to Pergamos, where the symbol of the serpent was set up as the emblem of the hidden wisdom.

From there, they afterwards crossed the sea and immigrated to Italy, where they settled in the Etruscan plain. There the ancient cult was propagated under the name of the Etruscan Mysteries, and eventually Rome became the headquarters of Babylonism. The chief priests wore miters shaped like the head of a fish, in honor of Dagon, the fish-god, the lord of life-another form of the Tammuz mystery, as developed among Israel's old enemies, the Philistines. The chief priest when established in Rome took the title Pontifex Maximus, and this title was held henceforth by all the Roman emperors down to Constantine the Great, who was at one and the same time, head of the church and high priest of the heathen! The title was afterwards conferred upon the bishops of Rome, and is borne by the pope today, who is thus declared to be, not the successor of the fisherman-apostle Peter, but the direct successor of the high priest of Babylonian mysteries, and the servant of the fish-god Dagon, for whom he wears, like his idolatrous predecessors, the fisherman's ring."

"***Verily, verily, I say unto you, He that entereth not by the door into the sheepfold, but climbeth up some other way, the same is a thief and a robber***". (John 10:1)

A false religion is man's made-up way to heaven, a fantasy that doesn't work.

36

The Rise Of The "Little Horn".

36.1 Babylonian Origins Of Our Countries Symbols.

Babylon is the originator of all of the pagan mystery religions through the ages. Most can be traced back to Babylon.

In his book "Things to Come" J Dwight Pentecost wrote about the Babylonian mystery-religion. On page 366 he wrote…

"From Babylon this mystery-religion spread to all the surrounding nations… Everywhere the symbols were the same, and everywhere the cult of the mother and the child became the popular system; their worship was celebrated with the most disgusting and immoral practices. The Image of the queen of heaven with the babe in her arms was seen everywhere, though the names might differ as languages differed, It became the mystery religion of Phoenicia, and by the Phoenicians was carried to the ends of the earth. Ashtoreth and Tammuz, the mother and child of these robust adventurers, became Isis and Horus in Egypt, Aphrodite and Eros in Greece, Venus and Cupid in Italy, and bore many other names in more distant places. Within 1000 years Babylonian religion had spread throughout the world and became the religion of those who rejected the Divine revelation."

America is the "*Daughter of Babylon*" in Bible prophecy. Peter uses "Babylon" as a type of code word for Rome when talking to The Church in Rome. Christians were being persecuted in Rome, so **they used code words to hide their activity**. At the time that this was written, there was no "Babylon". Babylon was long gone.

"*The church that is at **Babylon**, elected together with you, saluteth you; and so doth Marcus my son*". (1Peter 5:13)

36.2 The Layout of Washington D.C.

The Capitol Building mirrors Rome's Pantheon

The design of the streets, monuments and government buildings were all done by high ranking Free Masons and reflect occult signs and symbols.

The Masonic Temple in Washington D.C. is exactly 13 blocks from the capitol building.

God refers to Jerusalem as Sodom and Egypt in the last days because of all the false religions and wickedness. Instead of calling Jerusalem by its name, he called it "*Sodom and Egypt*".

Peter does the same thing when talking about the Nation of the Antichrist. Peter used the name "Babylon" as a code word to refer to the Church of Rome…

"***The church that is at Babylon**, elected together with you, saluteth you; and so doth Marcus my son"* (1Peter 5:13)

This was written long after Babylon was gone. There was no Babylon. Christians were being hunted down and killed in Rome, so Peter did not want to bring attention to the Church in Rome so he called it "***The church that is at Babylon***". Babylon was well known as the source of Pagan Roman religion. Babylonian religion was in competition with Christianity. The Roman leadership wanted to keep Paganism alive and destroy Christianity. At that time, Rome was the present representation of the Babylonian religion passed down to them.

When the Bible talks about Babylon in the last days and its leader the Antichrist, I believe that it is talking about the United States; the10 European nations and the Eastern countries which were once called

"Assyria" and made up the Eastern part of the Roman Empire. The term "Babylon" when used in reference to The Nation of The Antichrist is talking about this revival of the Ancient Roman Empire with the United States as its head.

36.3 The Washington Monument

Excavation for the foundation of the Washington Monument began in spring 1848. Its cornerstone was laid on July 4, 1848, as part of an elaborate Fourth of July ceremony hosted by the Freemasons,

While the world's largest obelisk, *The Washington Monument*, reflects Egypt's Heliopolis (the City of the Sun) with its temples and monuments to the sun god, Ra or Atum (the single creator god at the helm of the nine gods of the Ennead). That its height is **6,660** inches and its width **666.0** inches (555' x 55') hardly seems coincidental.

Click on the link below… http://www.jesus-is-savior.com/

36.4 Statue of Liberty

Before beginning the statue of liberty project, its creator, Bertholdi, was seeking a commission to construct a giant statue of the goddess "Isis," the Egyptian Queen of Heaven, to overlook the Suez Canal. The statue of Isis was to be of "a robed woman holding aloft a torch" (Statue of Liberty: 1st Hundred Years, Bernard Weisberger, p.30, quoted in Beyond Babylon, James Lloyd, p.103).

The Statue of Liberty is yet another example, and is actually a replica of another Babylonian goddess "Ishtar" the Mother of Harlots and the goddess of Liberties. A European Freemason who wanted to honor a Masonic doctrine that dates back to the time of Nimrod and Babylon created this "artwork"!

The Statue of Liberty is simply another Brotherhood symbol, highlighting the lighted torch of the Illuminati. It is in actual fact the **Statue of Liberties** more like the "liberties" perpetrated on the American people by the Freemasonry Brotherhood. There she stands on her island in New York Harbor holding her torch of freedom and Americans believe she is the symbol of their liberty in the Land of the Free. Nothing could be further from the truth. The Statue of Liberty was given to New York by French Freemasons and her mirror image stands on an island in the River Seine in Paris. These statues of "liberties" are in reality representations of **Queen Semiramis and Isis**, with the rays of the Sun around her head, similar to that of the venerated saints of Catholicism. The ancients typically symbolized the Sun in this way. They are not holding the torch of liberty, but the torch of the illuminated ones, the Masonic, Luciferian Initiated Elites!

36.5 The United Nations

Dr. Robert Muller served as the assistant Secretary General to the United Nations for 40 years. During that time he promoted the New Age movement in and through the U.N. He was considered the U.N. philosopher.

Dr. Robert Muller's Thoughts on Global Religion…

"Do not worry if not all religions will join the United religions organization. Many nations did not join the UN at its beginning, but later regretted it and made every effort to join. It was the same with the European community and it will be the case with the world's religions because whoever stays out or aloof will sooner or later regret it".

"Peace will be impossible without the **taming of fundamentalism** through a United Religion that professes faithfulness only to the global spirituality and to the health of this planet".

"My great personal dream is to get a tremendous alliance between all the major religions and the UN".

Lucifer Publishing Company, changed their name to Lucis trust and set up their headquarters in the U.N. building so that they could oversee the development of new goals for Sustainable Development which revolves around population control. https://en.wikipedia.org/wiki/Lucis_Trust

Wikipedia reports… "In order to place a closer focus on the work of the UN, the Lucis Trust has set up a new blog, World Goodwill at the UN, which will focus on defining new Sustainable Development Goals for humanity after 2015 when the Millennium Development Goals expire"

So what are the new goals? On May 13, 2014, the eve of the anniversary of Israel's recognition as a Nation by the United Nations, a speech was given by the French Foreign Minister. John Kerry was standing by his side during the entire speech with his hands folded, head slightly bowed and looking very reverential, silently nodding his head in agreement from time to time.

In that speech, the French Foreign Minister said three times…"The world has 500 days to avoid "climate chaos". The French Foreign Minister had repeated that statement twice and began to walk away from the microphone when someone off camera sent him back to repeat it a third time, which he did. Two times was not enough. It was preplanned for him to make that statement 3 times.

So what happened in 500 days that was so important? In 500 days the 266th Pope met with Barack Obama on the 266th day of the year (2015)

which also just happens to be Yom Kippur. It is significant to note that there are 266 days in a human gestation period. From conception to birth. More about that connection later.

The new Goals for Humanity that Lucis Trust is promoting with the help of the Pope is Population control. The Pope and the U.N. have linked saving the environment with population control and has defined them both as moral issues.

The Club of Rome is an International Think Tank that developed Agenda 21 for the U.N. Agenda 21 is all about population control.

Ted Turner, who was also a member of the Club of Rome, was quoted as saying in 1996: "A total population of 250 to 300 million people is ideal. That means a reduction of 95 -98% from present levels would be even more ideal? Anyone who abhors the China One Child Policy Is simply a Dumb Dumb."

To achieve those numbers would require Genocide on a Biblical scale, unlike anything ever seen!

So what is the new, more aggressive Goals for Humanity that Lucis Trust is promoting with the help of the Pope? It is **Population control by "vaccination!**.

36.6 God Declares The Antichrist!

Some truths are slightly encrypted to hide them from those that really don't care about knowing the truth.

It seems like the number 13 is associated with Satan and the Antichrist more often than mere coincidence can explain, and in some of the most peculiar and surprising ways.

It is important to note that there were no chapter and verse divisions when the Bible was first written. The chapter divisions of the Bible were added by man for convenience and referencing purposes in the 13th century A.D. For the full story follow this link to the Blue Letter Bible website; https://www.blueletterbible.org/faq/don_stewart/don_stewart_273.cfm

Dr. Lindstrom said that most of the time the chapter divisions obscure the meaning because they are placed incorrectly. With that said, Dr. Lindstrom made a comment about that. He expressed curiosity about how often the number 13 is associated with Satan, Lucifer, the Antichrist and the False Prophet in the Chapter and verse divisions. He said that he wondered if it was more than just a coincidence. He warned that they were not inspired by God but were added by man much later. Furthermore, the chapter and verse divisions often seem to break chapters and verses in the wrong places. As a result, many times the meaning is obscured by the location of the chapter and verse divisions.

I think that it is important to note that the Number of 666 being associated with the Antichrist is **inspired by God**, but the number 13 maybe is not. Now with that said, did Jesus give us the name of the man of sin in Luke's Gospel? Let's consider the two words "***lightning***" and "***heaven***" in (Luke10:18).

The name Baraq (Hebrew for lightning and name of Mohamed's horse) Pronounced just like the name used by the illegal alien that occupied the White House for eight years with a stolen social security number.

The way Jesus used the word lightning in (Luke 10: 18) is interesting. The disciples had just returned from their mission of proclaiming the Kingdom message. They had been casting out demons. They were excited about the power that the name of Jesus had over demons. Jesus responded with an unexpected statement in verse 18.

"And he said unto them, I beheld Satan as ***lightning*** *fall from* ***heaven****".* (Luke 10: 18)

This passage is talking about a future event, a kind of "de Ja vu"; of a similar event that I believe occurred about 65 million years ago. The future event occurs in the "midst" of the 70th week of Daniel, which is as of this writing, still future. I am talking about the 7 years of Tribulation discussed in (ETBH chapters 38 through 44).

The middle of the Tribulation is when the ***"Prince of the Power of the Air"*** (Ephesians 2:2) loses that title.

Satan is cast back down to the earth. He is in a rage and indwells the Antichrist. He is getting closer to being just plain old satan. Until the return of Jesus, Satan will still be called "the god of this world". He losses that title about 3 ½ years later, at the Battle of Armageddon. That is when Jesus returns and takes away that title for ever!

(Verse 18) was spoken in Aramaic, the most ancient form of Hebrew. It was written in Greek and our King James Bible is an English translation of that. To understand the original words spoken of by Jesus in this passage we consider the meaning of the Hebrew words associated with the Aramaic that it was spoken in.

To get the correct Hebrew translation I first looked at passages in the Old Testament which uses the word ***lightning.*** When I did the search in the Blue letter Bible, the message popped up which said that the word "lightning" "occurs 13 times in 13 verses in the KJV". When I got that message I thought to myself that I must be on the right track.

Then I noticed that the first time the word Lightning is used in the Bible is in (2Samuel 22:15). I also noticed that there were two other words used in other verses later on. But I noticed something interesting about that first word.

The Strong's number for that first word is **# H<u>13</u>00 (baraq)** and is pronounced as ***bä·räk*** (בָּרָק) and it is associated with violence by definition. I also noticed that this word is pronounced the same as the name of Mohamed's horse and I began to feel a little creepy.

Also related to it is Strong's definition for; בָּרָק bârâq, baw-rawk'; from H1299; is… lightning; by analogy, a gleam; concretely, **a flashing sword**:—bright, glitter(-ing **sword**), lightning.

Baraq is also associated with violence by the way it is used. In the first verse in the Bible where it is used (2Samuel 22:15) it is associated with arrows. Likewise that first mention rule holds up by the way it is used elsewhere.

Satan is negative and is associated with violence also. In (Ezekiel 28: 13-19) Lucifer's fall is described and is associated with sin and violence.

However in contrast to that is the way the other two Hebrew words for lightening are defined and used. They are: H2385 which is associated with weather only and H216 'ôwr, ore; from H215; illumination or (concrete) luminary (in every sense, including lightning, happiness, etc.):—bright, clear, day, light (-ning), morning, sun. The last two words for lightning are positive. The first one used is **H1300**, it **is negative** and is associated with weapons, war, violence, hatred, and vengeance as well as lightening. **It has a violent and negative connotation** and it is pronounced the same way as Mr. Obama's first name.

Continuing on in verse 8…

The word "*<u>heaven</u>*" comes from the Greek word associated with Strong's #G3772 – ouranos Strong's Greek Dictionary Defines "ouranos" in the following ways…

…"the vaulted expanse of the sky with all things visible in it; the universe; the world; the aerial heavens or sky; the region where the clouds and the tempests gather; and where thunder and lightning are produced; the sidereal or starry heavens; the region above the sidereal heavens; the seat of order of things eternal and consummately perfect where God dwells and other heavenly beings";

In (Luke 8:5) the same word is used again to refer to Satan's abode.

"A sower went out to sow his seed: and as he sowed, some fell by the way side; and it was trodden down, and the fowls of the ***air*** *devoured it".* (Luke 8:5)

The word ***"air"*** used here was also translated from ***ouranos***.

Although the word can mean Heaven as in the abode of heavenly beings and God, it is most often used to refer to the air surrounding the earth, the atmosphere, the clouds, the abode of Satan and the stars or heavens.

Satan is also called the "***Prince of the power of the air***".

"*Wherein in time past ye walked according to the course of this world, according to the* ***prince of the power of the air****, the spirit that now worketh in the children of disobedience*:" (Ephesians 2:2)

Satan's original, pre-fallen name was Lucifer. The Hebrew equivalent to the Greek word "ouranos" is found by referencing the only Old Testament passage that talks about Lucifer. (Isaiah 14: 12, 13) In verse (13) of this passage, a distinction is made between the abode of God (Strong's H8064 – shamayim) and the abode of Satan as found in the next verse (14).

"*For thou hast said in thine heart, I will ascend into* ***heaven****, I will exalt my throne above the stars of God: I will sit also upon the mount of the congregation, in the sides of the north*:" (Isaiah 14: 13)

"***Heaven***" is translated from "**shamayim**" in verse (13) and that is where Satan wanted to go.

"*I will ascend above the* ***heights*** *of the clouds; I will be like the most High*". (Isaiah 14: 14)

Verse 13 says where Satan wanted to go and distinguishes that from the place he must pass through to get there. It is important to know that there are three "*heavens*". The first heaven is the atmosphere around the earth where birds fly. The second "heaven" is where the stars are. The third "Heaven" is where God is. No sin can enter God's perfect Heaven, therefore Satan never made it that far since his fall. When the Bible talks about Satan being cast out of "heaven" it is **<u>not</u>** talking about God's residence.

The passage in Job where the Sons of God came together and Satan was among them, is not describing a meeting in Heaven. It is talking about believers gathering themselves together at the only place on earth designated by God for that purpose. Believers are the Sons of God. Angels are never called Sons!

Verse 13 is talking about where God is and verse 14 is talking about Satan's territory. This passage uses the word ***"heights"*** to describe the abode of Satan.

The word ***"heights"*** is translated from…

Strong's H1116 – bamah, with a prefix of either "U" or "O" to connect two words or ideas.

Strong's definitions are…

…high place, ridge, height, bamah (technical name for cultic platform)

…high place, mountain

…high places, battlefields

…high places (as places of worship)

…funeral mound?

It is important to note that Satan, A.K.A. Prince of the power of the air will be cast down to earth in the middle of the Tribulation. That is when he indwells the Antichrist.

The context of (Luke 10: 18) makes me believe that Jesus is making reference to those events. Remember that in verse 17 the disciples cast out demons in the name of Jesus. They were amazed at how the demons respond to the name of Jesus. And Jesus replies in the next verse (18) that he saw Satan as Barak – (O or U) - O - Bama.

In the middle of the Tribulation, about 9 days and 3 ½ years from the Rapture, the Antichrist will be indwelt by Satan himself and take over most of the Mideast and Jerusalem. He will rule the West and what used to be the Assyrian Empire from the Temple in Jerusalem. The Antichrist will be called "the Assyrian" by right of conquest.

I believe that in verse (18) Jesus is referring to the fact that the Antichrist will be indwelt by satan in the middle of the tribulation and Jesus will cast him out of the Antichrist at the end of the Tribulation.

At that time, satan will be bound for a thousand years in the heart of the earth and the Antichrist and the False Prophet will both be cast alive into the lake of fire. I believe that the Antichrist will become the last Islamic caliphate, the world leader of Islam and demand to be worshiped as god.

The Strong's Hebrew dictionary has listed Hebrew word number 1300 which is translated into the English word "lightening".

That Hebrew word translated as lightening is "Baraq". Interestingly a Greek word for lightening was assigned the number 913. The similarity for the Greek word for lightening is surprising. "Barak" also has a Strong's number containing 13 in both the Hebrew **and** Greek! Both words are pronounced the same. What could cause something like that to happen so often? Could this be more than Just a coincidence? If it is more than just a coincidence, consider what it may mean.

Consider also the word "***ascend***" in (Isaiah 14: 14). It is translated from the Hebrew word **"*alah*"**.

Strong's H5927 - **`*alah***

"*I will **ascend** above the **heights** of the clouds* (Isaiah 14: 14)

The Hebrew word "alah" is pronounced the same as the Arabic word which contains two "L"s, except it has negative connotations in the Hebrew culture.

What the god of the Koran calls evil, The God of the Bible calls good! What the Koran calls good, the God of the Bible calls evil. They are opposites!

Go to the address below... http://worldtruth.tv/did-jesus-reveal-the-name-of-the-anti-christ-2/

"Change" is associated with the Antichrist and is translated from Strong's H8133 - *shĕna'* (Aramaic)—alter, **change, (be) diverse.** It is used in Daniel...

..."*and think to **change** **times and laws***": (Daniel 7:25)

"Times" is a reference to Holidays or the calendar.

... **Sharia Law** is the "law" of Islam. It is the responsibility of every Muslim to impose **Sharia Law** by force on the whole world or die trying. The only guarantee of a place in their "heaven" is to kill or die in that fight. Then 70 virgins are promised.

The Antichrist is associated with the word ***"change"***. Does that sound familiar?

He will be Jewish, but not respect the God of his fathers...

"Neither shall he regard the God of his fathers*, nor the desire of women, nor regard any god: for he shall magnify himself above all"* (Dan. 11:37)

This term ***"God of his fathers"*** is used in scripture to indicate Jewish ancestry.

If your mother is Jewish then the Jews consider you to be Jewish. It is also interesting to note that if your father is Muslim, then the Muslims consider you to be Muslim. So If his father is openly Muslim and he is raised as a Muslim and he can quote the Koran like a Muslim, the Muslim world will consider him Muslim and even keep it a secret.

It would be necessary to keep his mother's Jewish ancestry a secret, so that he will be accepted by both sides. Once in power, he can betray whosoever he desires. He will betray Israel, the Catholic Church and more.

Deception, betrayal and Terror are core elements in the strategy for the world domination of Islam. It is a "religion" of deception, betrayal and force.

The Antichrist will not regard the desire of women (Daniel 11:37). This may be a reference to the fact that women will be required to serve in the military like men, with no special consideration given for raising children and making families.

He will not regard ***"the desire of women"*** (Daniel 11:37). Some people believe that this is a reference to the fact that the **Antichrist is a Sodomite**, a pervert and a **reprobate**. Others believe that the **Antichrist will not regard family values, <u>at all</u>**. Both may be true.

*"Neither shall he regard the **God of his fathers**, nor **the desire of women**, nor regard any god: for he shall magnify himself above all"*. (Daniel 11:37)

He will conquer all of what used to be Ancient Assyria and will be called *"**the Assyrian**" by right of conquest, not birth.*

The Antichrist will conquer Saudi Arabia, Syria, Lebanon and Egypt during the first half of the tribulation. This will be the result of an attack by Egypt and most of the Mideast against Israel. Most agree that this short war will occur during the first half of the Tribulation. I believe that this war may be part of the "*sudden destruction*" mentioned In (1Thessalonians 5:3)

I said that this will be a short war because the Antichrist and his country which I believe can only be the United States, will astonish the world by how fast he is able to destroy the attacking armies.

"And the Egyptians will I give over into the hand of a cruel lord; and a fierce king shall rule over them, saith the Lord, the LORD of hosts." (Isaiah 19:4)

The Antichrist is called "*cruel*". Egypt will suffer greatly under the hand of the Antichrist. When the Antichrist takes Egypt, he will be warned by **Russia** and **China** not to touch Ethiopia and Libya. So because of that warning he stops. (Daniel 11:40)

"but these shall escape out of his hand, even Edom, and Moab, and the chief of the children of Ammon". (Daniel 11:41)

God says that Jordan will escape the Antichrist! It is <u>the only place on earth</u> that I know of that God calls by name and says that <u>they will escape the Antichrist</u>! That makes Jordan real-estate quite valuable!

36.7 Political/Religious Considerations… The Antichrist

The Muslim connection.

https://www.youtube.com/watch?v=tCAffMSWSzY

"*He will claim to be god and will oppose the worship of any other*". (II Thessalonians 2:4)

In a speech which made fun of the Christian faith, Mr. "O" referred to Abraham talking to God. Mr. "O" said that not everyone can hear God's voice. In his own words Mr. "O" said that we need a voice that everyone can hear. This was said when he was running for President and the implied voice that everyone could hear should be his.

"And the Egyptians will I give over into the hand of a ***cruel*** *lord; and* ***<u>a fierce</u> king*** *shall rule over them, saith the Lord, the LORD of hosts*" (Isaiah 19:4)

The Antichrist will be cruel! When God calls something **cruel**, then you can bet on it being **cruel beyond a doubt!**

The Antichrist is called the "***man of sin***". He is also called "the **lawless one**". Therefore he will be a liar, murderer, pervert and everything else evil that you can think of and more. I would expect for the Man of Sin to surround himself with every kind of demon-possessed, lying, deceiving, betraying, cruel reprobate pervert there is. The man of sin, along with the false prophet will earn the distinction of being the **only ones said to be cast alive into the lake of fire** and the **first two in**! Currently, the Lake of Fire is empty as far as I know.

I believe, as do others that **the Antichrist will be a Sodomite**. Sodomites who claimed to be homosexual "lovers" with Mr. "O" were found dead from mysterious circumstances. These Sodomites openly boasted about their activities with Mr. O. The names of the dead Sodomites are:

November 17, 2007; Larry Bland was murdered execution-style

December 24, 2007, Donald Young ; 5 weeks later, was murdered execution-style;

December 26, 2007; Nate Spencer 2 days later, reportedly died of septicemia, pneumonia, and HIV. That was 2 days after **Donald Young** was murdered execution-style. **Nate Spencer's death certificate is not available**. Septicemia pneumonia can be suddenly induced or faked on anyone who is already hospitalized for something else. Remember that **his death certificate is missing!** Copies of the other two death certificates are available online by copying this link… http://www.rense.com/General82/cdet.htm Into the address bar of your browser. Notice the **"multiple gunshot wounds"** listed as the cause of death for both victims. They were both shot in the head; multiple times; from close range; execution style; within six weeks of each other. They were both bragging about their homosexual relationship with Mr. "O" at the beginning of his campaign.

What does **Donald Young**, **Larry Bland** and **Nate Spencer** have in common? They are all openly homosexual black men who attended "**Reverend" Jeremiah Wright's "Trinity United Church of Christ**" where Mr. "O" attended; All bragged about their Homosexual relationship with Mr. "O" the candidate; all died of suspicious circumstances within six weeks.

, Another admitted Sodomite who boasted about his relationship with Mr. "O" survived till now by going into hiding periodically. Initially he went into hiding when members of Rev. Wright's church began showing up dead in unusual fashions. He found a sodomite attorney to protect him. His name was **Larry Sinclair**. He went public in hopes that his murder would create too much public awareness and suspicion to be a benefit to anyone to cover up things. Author of "**Barack Obama & Larry Sinclair: Cocaine, Sex, Lies & Murder", By Lawrence W. Sinclair.**

Another piece of information that we have about the Antichrist is that **he will persecute the saints and prevail**. (Revelation 13:7-17) If God says that he will prevail then those who are left behind at the rapture should be thinking first about the gospel and trusting Christ alone for their salvation and not their own works and secondly about getting as far away from the Antichrist as soon as possible. (Within about a week after the rapture)

The number *"666"* is associated with the Antichrist. (Revelation 13:18)

On the morning of November 5, 2008 the announcement was made that the winner of the 2008 presidential election was the U.S. Senator from the **13th district of Illinois**, Barack Obama**.** That same morning the Illinois state lottery's winning pick 3 numbers on the morning of **November 5, 2,008 was "666"** and can be viewed in the archives online by clicking on the link… http://www.illinoislottery.com/en-us/winning-number-search-year.html#loadingImg2

The Antichrist will be a Muslim, the Mahdi. (Daniel 9:11) (Isaiah 24:6) (Revelation 20:4) The Koran commands its followers, the Muslims, **to** "kill **them** wherever **you find them**" three times (2:191, 4:89, 9:5)"

He will be trusted by the Jews (John 5:43) (Isaiah 28:14-18)

He will betray the Jews (Daniel 9:27)

The Antichrist will deceive most of the world.

The False Prophet is definitely Roman Catholic and probably The Pope himself. (Revelation 17 :)

The Pope was scheduled to meet with Mr. Obama on September 23, 2015 at the Whitehouse. This day is Yom Kippur (https://www.youtube.com/watch?v=TdOJUGPaLvA)

The Antichrist & The False Prophet will be the **first** to be cast into the lake of fire and **the only ones** said to be cast in **alive!** (Revelation 19:20)

Dr. Lindstrom wrote a paper about the Antichrist more than a decade ago. It is interesting what was said back then. Here is a link to that paper...

http://www.biblelineministries.org/articles/basearch.php3?action=full&mainkey=THE+ANTICHRIST

https://www.youtube.com/watch?v=EFCQvvXSwx4

36.8 The "Board Of Governors"

During the Obama Administration, the White House website posted an emergency executive order that divides up the US into 10 territories. Our constitution which established our freedom is an agreement between 50 states, not 10 territories; this may be an intermediary step toward setting up new boundaries that could become 10 states shortly after it is up and running.

Under this particular executive order called "the Board of Governors", the President has total control of the 10 governor's indefinitely and no means to remove him. I know it may be hard for some to believe, so here is a link to that page on the White House website... http://www.gpo.gov/fdsys/pkg/FR-2010-01-14/pdf/2010-705.pdf

It seems like the number 10 keeps popping up in relation to this new global order that the Bible talks about.

Could it be that there may also be a similar emergency plan to divide Europe into 10 regions and is just waiting for the right time to be activated? Or could this be the universal plan to be used when Europe goes into chaos 9 days after the Rapture, at sundown? I would not be surprised. A little speculation to consider. (Speculation is the first step on the road to discovery)

We have already entered Orwell's police state world.

I know that you are probably thinking that the US would never accept a totalitarian dictator, but President Bush has already made up Executive orders that take away our constitutional rights which includes a right to privacy.

Currently if anyone is declared to be a terrorist by the President, that person may be detained indefinitely without formal charges, hearing or legal counsel.

Drone patrols are currently being set up all over America and the technology to increase their capabilities is being updated continually.

https://www.youtube.com/watch?v=WyBJScMDeU8

https://www.youtube.com/watch?v=2RGtNX4_9pA

At the May 1 White House Correspondents Dinner. Mr. Obama made a joke May 3, 2010 5:43 PM ET about Predator Drones. He said…

"The Jonas Brothers are here; they're out there somewhere. Sasha and Malia are huge fans. But boys, don't get any ideas. I have two words for you, 'predator drones.' You will never see it coming...You think I'm joking?" The audience breaks out in laughter and clapping.

Operating for years in Afghanistan and Pakistan as an officially secret counterterrorism program, the drones have drawn controversy for their high civilian casualty rate.

The Drone technology has been rapidly advancing. They are adding artificial intelligence, they are increasing in numbers and flying high over our heads!

https://www.youtube.com/watch?v=ehS1UaIZPco

www.lulu.com/spotlight/islamic_antichrist

36.9 The 7 Year Treaty

The Oslo Accords were signed at a public ceremony in Washington, D.C., on September 13, 1993 in the presence of PLO chairman Yasser Arafat, the then Israeli Prime Minister Yitzhak Rabin and U.S. President Bill Clinton. At the signing ceremony, they shook hands and said "**Peace and Security**". This happened on Yom Kippur.

The documents themselves were signed by Mahmoud Abbas for the PLO, foreign Minister Shimon Peres for Israel, U.S. Secretary of State Warren Christopher for the United States and foreign minister Andrei Kozyrev for Russia.

US adopted an even-handed approach and pledged to promote a settlement that would provide security for Israel and justice to the Palestinians. Negotiations were to be based on UN resolution 242 of November 1967 and the principle of land for peace that it incorporated. God warned about trading land for peace.

The 11th Horn Revealed

https://www.youtube.com/watch?v=PlT9SYR6jq4

https://www.youtube.com/watch?v=4Yro63c7B7A

37

The 10 Nation European Union (Daniel)

(Deuteronomy 28:49)

37.1 Nebuchadnezzar Had A Dream, Daniel Interprets.

(Daniel 2:31-34; 2:41, 42)

"*Thou, O king, sawest, and behold a great image. This great image, whose brightness was excellent, stood before thee; and the form thereof was terrible*". (Daniel 2:31)

"*This image's head was of fine gold, his breast and his arms of silver, his belly and his thighs of brass*", (Daniel 2:32)

*"His legs of **iron**, his feet part of **iron** and part of clay"*. (Daniel 2:33)

Nebuchadnezzar is identified as the head of gold. (Daniel 2:37, 38)

The Meads and Persians are the next ruling power; *"arms of silver"* (vs 32) and after them is the Greeks. *"belly and his thighs of brass"* (Daniel 2:39)

The fourth world ruling power was Rome, it is represented by ***"legs of iron".*** The fifth world ruling empire is (also represented by **Iron**. (Daniel 2:33)

I believe that iron is representative of Roman military strength and economic power. Rome was the military super power of its time so likewise the 5th world ruling power which is also represented by iron will be the superpower of its time and of Roman descent in many ways.

Some other important details revealed in this dream are that this image will stand in Iraq (Babylon) in the last days. We know that it is the last days because in verses 34, 35 and verses 44, 45 is a reference to Christ's return in the days of these kings. (Christ being referenced as a stone cut out without hands)

"And whereas thou sawest the feet and toes, part of potters' clay, and part of ***iron****, the kingdom shall be* ***divided****; but there shall be in it of the strength of the* ***iron****, forasmuch as thou sawest the* ***iron*** *mixed with miry clay"*. (Daniel 2:41)

"And as the toes of the feet were part of ***iron,*** *and part of clay, so the kingdom shall be partly strong, and partly broken."* (Daniel 2:42)

The Nation of the Antichrist will be the undisputed military superpower of the world but it will be divided internally. The internal division probably will be a conflict between Sharia law and the current system. ("***Suddenly"*** coming to a neighborhood near you, **sooner than you think**.)

The Babylonian Empire is the first Empire and is represented by the Head. Historically we can see that Babylonian pagan religion spread to all of the following Empires represented in the image and more. The names of their pagan deities changed somewhat due to language differences, but the emblems; stories; monuments and practices remained similar or the same to this day.

37.2 The 5th World Power Will Have Internal Conflict.

The fifth power is represented by iron plus clay. Clay is just something else that doesn't stick to iron. Iron and clay are not compatible and neither is Islam and western culture. (Daniel 2:33; 2:41, 42) Western culture is Bible based and Islam is based on the Koran and they are opposites in every way! The Koran is specifically anti-Bible! Islam is specifically at war with Christians. The Koran specifically says that believing the Christian doctrine of salvation is the greatest sin that can be committed and is punishable by beheading or torture! The Koran condemns "the people of the Book", that means Christians and Jews.

The other "Infidels" they will just enslave or behead later.

"People of the Book" or Christians will be special targets during the Tribulation. During the first 3 ½ years, everyone that professes Christ will get it first. Next target will be non-Muslim's except the Jews and Catholics will be exempt from persecution for the first 3 ½ years.

The fact that iron represents the fourth world ruling power and the fifth world power is represented by iron plus something else, leads me to conclude that the old Roman system is used in these nations that are represented in chapter 7 as 11 horns.

The descendants of the Roman Empire (Europeans) and Americans will arise in the last days as a superpower under the control of the Antichrist.

37.3 Daniel Had A Dream

God reveals to Daniel the four future world ruling powers in a dream. (Daniel 7:1)

"*behold, the four winds of the heaven strove upon the great sea";* (Daniel 7:2)

"*four great beasts came up from the sea, diverse one from another"* (Daniel 7:3)

The first three beasts are described… verses 4-6.

The first is Babylon (vs. 4),

The second is the Medes and the Persians (vs.5),

The third is Alexander the Great, (vs.6)

37.4 The Fourth Beast Is Described With 10 Horns.

*"After this I saw in the night visions, and behold a fourth beast, dreadful and terrible, and strong exceedingly; and it had "great **iron** teeth": it devoured and brake in pieces, and stamped the residue with the feet of it: and it was diverse from all the beasts that were before it; and **it had ten horns."*** (Daniel 7:7)

The Roman Empire conquered the Grecian Empire and replaced them. The fourth beast was the Ancient Roman Empire which has become modern day Europe. In verse 7 it goes on to say that the fourth beast is diverse from the others and ***..."it had 10 horns".***

Notice the **iron** teeth and 10 horns (Daniel 7:7). Remember that the image of King Nebuchadnezzar's dream had toes of iron and clay.

This is the fourth world ruling power. *"It had great **iron** teeth"* (Rome) *"and it had ten horns".*

The Bible indicates that Europe will also be divided into "10 Kingdoms" again. That happened in Ancient Rome under Constantine for administrative purposes.

The Western half of the Roman Empire was fully divided into 10 Kingdoms by the end of the 4th century. As the central government weakened the 10 regions turned into separate countries.

*"And the ten horns out of this kingdom are ten kings that shall arise: and **another shall rise after them**; and he shall be diverse from the first, and he shall subdue three kings".* (Daniel 7:24)

37.5 Sir Isaac Newton Wrote About The 10 Horns...

In his book titled "Observations Upon the Prophecies of Daniel and the Apocalypse of St. John", in the 6th chapter, Newton, wrote about the ten Kingdoms represented by the ten horns of (Daniel 7:7). He wrote that the 10 horns represented...

"1. the kingdom of the Vandals and Alans in Spain and Africa; 2. of the Suevians in Spain; 3. of the Visigoths; 4. of the Alans in Gallia; 5. of the Burgundians; 6. of the Franks; 7. of the Britons; 8. of the Hunns; 9. of the Lombards; 10. of Ravenna; who gives an account of the various kings of these kingdoms; and these, as the same learned writer says (n), whatever was their number afterwards, they are still called the ten kings from their first number; and though they have not always been in the same form and order, yet they have been Generally about, if not exactly, the same number; as they are now near the same; and may be thus reckoned, as the kingdoms of France, Spain, Portugal, Germany, Great Britain, Sardinia, Denmark, the two Sicilies, Swedeland, Prussia, and Poland;"

Newton believed that in the future Europe will be divided into ten Nations or regions with boundaries similar to those established by the end of the 4th century A.D. He compares the divisions of his time with those of the Ancient Roman Empire. In some respects this is descriptive of a revival of the Roman Empire. The future time that he was talking about is our present!

Sir Isaac Newton Wrote About An 11th Horn

In his book titled; "Observations Upon the Prophecies of Daniel and the Apocalypse of St. John", Newton titled the 7th chapter… "The Eleventh Horn of Daniel's Fourth Beast"

CHAPTER 7 Page 54.

Newton dedicated the entire 7th chapter to speculating about the 11th horn. Newton was born in 1642 and lived 84 years. He died 49 years before the American Revolution. When he was born the 13 American colonies of England were still being formed.

In his time there was no United States of America to consider, so he speculated about the Roman Catholic Church being the 11th horn because of the Roman connection.

Almost 300 years ago there were too many pieces of the puzzle missing. Nevertheless, he made observations that are missed by modern day Bible teachers. Dr. Lindstrom was an exception.

Dr. Lindstrom is the only other Bible teacher that I know of that has recognized an 11th horn.

Lindstrom used to joke about other teachers missing the 11th horn. He would say… "I learned in grade school that if I have 10 apples and then I am given **another** apple, I now have 11 apples. So I figure that if there is 10 horns and then **another** horn appears, there must be 11 horns altogether."

37.6 An 11th Horn Is Identified.

"<u>Another</u> little horn" means one more, an 11th horn or Kingdom. I don't understand why Bible teachers don't pick up on the fact that there is an 11th Kingdom, the *"little horn"* which came out of the 10. It is younger than the 10; it came after the 10 and is more powerful than the 10.

When we compare the previous passages already discussed from (Daniel 2:31-34; 2:41, 42) with chapter 7: verses 8, 20 & 24 we find that the 10 horns are a confederation of 10 European nations. They are the descendants of the western half of the ancient Roman Empire that will rise up in the last days which is now!

The Roman Catholic Church was traditionally called the Holly Roman Empire and has been the common link between these Kingdoms since the 4th century.

From this description, I would expect for the Nation of the Antichrist to be on the same calendar system as the western half of the Ancient Roman Empire. That would be a solar calendar with the names of the days and months to be names of Roman, pagan gods and emperors.

I would also expect to see Roman Pagan religious influence in all 11 nations. Customs such as the worship of Ishtar (Easter in English) the sex goddess have replaced Christian customs celebrating the resurrection of Jesus at The Feast of First fruits. The Roman pagan custom. Likewise, the worship of Saturn on December 25 is called Christmas with Santa and Christmas trees and much of the associated activities all come from Roman Pagan origin. The early Christian church had no such customs! Christ was born on the Feast of Tabernacles when it was nearest to the last day of October. Because the solar calendar changes with respect to the Hebrew calendar, those two dates come closer at times.

I would also expect for the countries laws and customs to have developed from the Roman system like ours has and use Latin words, and concepts like we do.

The countries that descended from the western half of the Ancient Roman Empire are all on that old Roman system and so is the U.S. Culturally we are Roman.

Ancient Roman Empire is modern day Europe which will be reunited and then taken over by the 11th horn or the *"little horn"*.

"*I considered the horns, and, behold, there came up among them* ***<u>another</u> little horn****, before whom there were three of the first horns plucked up by the roots: and, behold, in this horn were eyes like the eyes of man, and a mouth speaking great things"*. (Daniel 7:8)

"There came up ***among*** *them"* (vs. 8) is a phrase that indicates the origin of the 11th horn or Kingdom. It came out of the 10 which is a reunited Europe.

"*<u>Another</u>*" means another Kingdom. The "little horn" is an 11th Nation. A younger nation, a more powerful nation which came up among the ten and later than the 10. The United States was settled by European immigrants. We came out of Europe and nobody would argue that we are more powerful than all of Europe combined.

"And of the ten horns that were in his head, ***and of the <u>other</u>*** *which came up, and before whom three fell; even of that horn that had eyes, and a mouth*

that spake very great things, whose look was ***more stout than his fellows.*"** (Daniel 7:20)

***"More stout than his fellows*"** (vs. 20) means that he is the stronger nation. The 11th horn or Kingdom takes over three of the 10. The 11th Nation is a Super Power. (More about that later)

There is no distinction made between the man running the nation and the nation itself because he has total control. He is a dictator.

The 11th horn is the nation of the Antichrist and he will be its dictator. The Nation of the Antichrist takes over 3 of the 10 European countries by military force and the other 7 give up without a fight. You can probably guess which countries will be taken over.

The EU-3, also known as the G-3 nations are the three most powerful and influential nations in Europe. They are France, England and Germany and they are said to control the rest of Europe and we already have our military bases there.

"And the ten horns out of this kingdom are ten kings that shall arise: and ***another*** *shall rise* ***after*** *them; and he shall be diverse from the first, and he shall subdue three kings."* (Daniel 7:24)

The phrase ***"shall rise after them"*** is synonymous with "younger than". The size of the horn is an indication of age not strength.

The Hebrew were herdsmen. Herdsmen knew how to judge the age of an animal by the size of its horns. This message is written to their descendants. Its message is only slightly obscured by metaphor.

You judge the age of an animal by the size of its horns, therefore a "little horn" is a younger nation.

The Historical Origin, Relative Age and Military Strength of The Nation of the Antichrist was revealed in… (Daniel 2) & (Daniel 7)

37.7 The Location Of The Nation Of The Antichrist

In the middle of the Tribulation, the Nation of the Antichrist comes from the other end of the earth and attacks Israel. God describes it …*"[as swift] as the **eagle** flieth"*. (Deuteronomy 28:49) says…

"*The LORD shall bring a nation against thee **from far, from the end of the earth**, [as swift] as the eagle flieth*" (Deuteronomy 28:49);

The Location of the Nation of the Antichrist relative to Jerusalem is the opposite side of the earth.

Notice that the Bible says a nation "*from far, from **the end of the earth***". Just saying far away is not enough to describe the Location of the Nation of the Antichrist; the Bible goes on further and says, "*from the end of the earth*". That sounds to me like the other side of the Planet.

If you took a world globe and put one finger on Jerusalem and with the other hand put a finger at the opposite end of the globe, your other finger would be off the southern coast of California, between California and Hawaii in the territorial waters of the United States of America. The United States would be the closest country to your finger on that globe and the only superpower that matches the description of the Nation of the Antichrist in the world!

37.8 National Emblems Of The 11th Horn

(Deuteronomy 28:49) (Daniel 7:7)

The phrase, ***"as swift as the Eagle flieth"*** In, (Deuteronomy 28:49), may be a reference to Americas National emblem as well as our air superiority.

The Eagle was also the emblem of the Roman Empire! I don't believe that is a coincidence.

Both the 4th and 5th world powers are both represented by iron in the scriptures. (Daniel 2:33) (Daniel 7: 7, 8) However the fifth world ruling power is represented by iron plus something else. (Daniel 2:33)

They both are associated with eagles and use the eagle as a national emblem.

We know beyond any doubt that Rome is the fourth world ruling power mentioned in (Daniel 2:33) & (Daniel 7: 7, 8)

This implies that the fifth world empire must be the revival of the Roman system plus something else. Modern day Europe is what used to be Ancient Rome.

Roman influence in American and European culture is pervasive because we are culturally of Roman decent.

37.9 Lindstrom Taught That The U.S. Was The 11th Horn.

Dr. Hank Lindstrom taught that the United States of America was the 11th horn spoken of by Daniel. See link to… www.biblelineministries.org . Hank's Paper about the Antichrist begins by stating that he will come out of the United States… wehttp://www.biblelineministries.org/articles/basearch.php3?action=full&mainkey=THE+ANTICHRIST © 2004-2006,

Also… "The End time Controversy" tape-4, 09.00 min www.Biblelineministries.org

Link to youtube video of Hank talking about the **"little horn"** being the United States and its totalitarian leader the Antichrist. This was said during the Bush administration and he made no claims about who the Antichrist was. https://www.youtube.com/watch?v=EqRHNYRzpCo

37.10 Dr. Lindstrom Was An Expert In Bible Prophecy

Dr. Lindstrom (Hank) used

(Deuteronomy 28:49) and (Daniel 7:24) to identify The Nation of the Antichrist by…

Location (The US is on the most opposite side of the earth from Jerusalem);

National emblem (eagle); Ancient Rome and modern day U.S. both used the eagle as their national emblem.

Historical Origin (Roman/European); until recently, The United States of America was colonized and dominated primarily by West Europeans.

Relative age (Young) The United States of America has only been around since July 1776. That is just about 238 years ago.

Strength, we are a Superpower Nation compared to those European nations from which our American ancestors immigrated.

Only the United States meets those criteria. Although Hank would continually offer up further evidence such as current events, our global positioning and political statements made by our leaders and leaders around the world, he felt those two passages alone pretty much settled it.

Global position; Hank would also talk about What the Antichrist Does during the Tribulation and how The US is already prepared to do exactly those same things right now!

Dr. Lindstrom (Hank), graduated from the University of Florida in Gainesville in a five year Electrical Engineering program.

Dr. Lindstrom then entered into his Biblical Studies at Florida Bible College in Miami.

In 1965. Dr. Lindstrom began directing the Tampa Youth Ranch and he established Bible Study clubs in most every high school in Hillsborough County.

Hulk Hogan believed on Jesus Christ at the age of thirteen in one of Dr. Lindstrom's meetings.

In 1966 Dr. Lindstrom became the Pastor of Calvary Community Church.

Dr. Lindstrom was on the Moody Bible Institute faculty for eleven years.

Dr. Lindstrom had the longest continuously running weekly television program on the Time/Warner Access channel in Hillsborough County [over 20 years]. Also he was heard on Christian radio in the Tampa Bay area for many years.

Dr. Hank Lindstrom went home to be with the Lord on Oct. 13, 2008.

Twenty one days after Hank died of an apparent heart attack, (November 5, 2008) Mr. Obama, the U.S. Senator from the 13th district of Illinois, claimed the office of the presidency of the United States.

That same morning, the lottery from Illinois, Obama's home state produced a winning morning lottery number of 666. That evening, the winning four was 7779.

I know you are going to have to see that for yourself before you believe it, so here is a link to the Illinois state lottery website. The entire past winning lottery numbers are posted and archived for public access. Just click on this page which contains all of the winning numbers for the month of November 2008… and find day 5.

Here is a direct link to that page… http://www.illinoislottery.com/en-us/winning-number-search-year.html#loadingImg2 … and scroll down to November 5, 2008 and see it for yourself; if you want their home page first, you can go to… http://www.illinoislottery.com/ which is the home page. Click word Numbers at the top of the page and a drop-down menu will appear. The last item at the bottom of the drop-down menu is Past Winning Numbers. Click on that to take you the month of November 2008, and scroll down to November 5.

You may want to believe that this 666 thing is just a coincidence and so is the 7779 in the evening marking the end of what someone wanted others to believe was the completion of a perfect day for agenda "666". But to me, there seems to be an unholy connection going on there. Who and what is Agenda 666? The Bible has already told us.

Hank would regularly refer to the Nation of the Antichrist as being the US and often pointed out Biblical references with descriptions that could only fit the US. Hank was always looking and speculating about who might be the Antichrist. I wonder if he had just figured it out when he had his heart attack?

Hank often used passages like (Deuteronomy 28:49) and (Daniel 7:24) to identify the Nation of the Antichrist by location (most opposite side of the globe), National Emblem (eagle); Historical Origin (Roman Empire); Unusual Strength (a super power nation); and our global, political and military posture.

37.11 The Little Horn Is The Antichrist And His Host Nation.

The Antichrist will be a totalitarian dictator similar to Hitler but much worse. In scripture it may be sometimes confusing whether you are talking about a person or a nation when talking about the little horn because of his totalitarian control. The Bible talks about the Antichrist just like you would talk about Hitler doing this or Hitler doing that because he totally controlled Germany and likewise the Antichrist will totally control the United States and Europe and most of the Mideast for three and a half years.

"And he shall speak great words against the most High, and shall wear out the saints of the most High, and think to change times and laws: and they shall be given into his hand until a time and times and the dividing of time." (Daniel 7:25) (3 1/2 years)

"And the ten horns out of this kingdom are ten kings that shall arise: and ***another*** ***shall rise after them****; and he shall be diverse from the first, and he shall subdue three kings"*. (Daniel 7:24)

"He shall speak great words against the most High", (Daniel 7:25)

This countries current administration has mocked Christianity and promoted sodomy and Islam.

Our current leadership has said that "one of the sweetest sounds is the Muslim call to evening prayer".

Furthermore he told the U.N. that "the future must not belong to those who slander the prophet of Islam". According to Islam, unbelief **is** slander and so is unwelcomed truth. If you say that Islam has doctrines that mandate killing Infidels, which it does, that can be considered slander if it is being used to warn the unbelievers. If you warn people that Islam mandates the killing and subjugating of "infidels" you are guilty of slander according to Islamic law books. Under Shira law, "slander" can carry a death penalty. Muslims don't want their intended victims to be warned.

If you are watching, then you should be seeing "the handwriting on the wall" just about everywhere you look.

The Antichrist will brainwash/torture/murder believers throughout his empire.

"and shall ***wear out*** *the saints of the most High"*... (Daniel 7:25) *"Wear out"* is meant in a mental way such as brain-washing and torture.

"***wear out***" is translated from Strong's H1080 - bĕla' (Aramaic) and means to… to wear away, wear out; to harass constantly (used only in a mental sense)

This is a constant, wearing out. Kind of like the Catholics did to Christians during the Dark Ages and the time of the Spanish Inquisition. Or like Muslims are doing to Christians in the Mideast and all over Africa right now!

"*And he shall speak great words against the most High, and shall* ***wear out*** *the saints of the most High, and think to* ***<u>change</u> times and laws****: and they shall be given into his hand until a time and times and the dividing of time".* (Daniel 7:24-25)

For 3 ½ years the Antichrist will be gaining power and tightening his grip on the west. That process requires changing laws starting with our constitution.

Religion is great for controlling the masses, Constantine, the Emperor of Rome, knew that. That is why he started Catholicism in the 3rd century A.D. Constantine repackaged Roman paganism in a way to attract ignorant Christians. He called his new repackaged religion the Roman Catholic Church. He made himself the first Pope.

The Antichrist will be Muslim and with the help of the False Prophet, he will repackage Catholicism by combining it with Islam. That has already been done and it is called Chrislam. Pope Frances the First has been called the first Islamic Pope by Robert Spencer director of Jihad Watch . The antichrist will have a working relationship with the Roman Catholics and Israel for the first 3 ½ years of the Tribulation. In the middle of the Tribulation he will betray them both.

37.12 The Little Horn Is Referred To As Babylon.

Peter refers to Rome as Babylon in his letter. At the time it was written, there was no Babylon. Babylon as it is used here is a code word for the Church in Rome.

"*The church that is at Babylon, elected together with you, saluteth you; and so doth Marcus my son*" (1Peter 5:13).

This statement is prophetic in nature. Rome is called "Babylon" because that is where Roman religion came from. The "little horn" is a revival of the Roman Empire and it too is referred to as "Babylon". (1Peter 5:13)

This is not the only time that a country is identified by attributes of another country. Jerusalem in the last days is referred to as Sodom and Egypt.

"*And their dead bodies shall lie in the street of the great city, which spiritually is called Sodom and Egypt, where also our Lord was crucified*". (Revelation 11:8)

The point is that Babylon is being used here to describe Rome. The 11th horn is the head of the revived Roman Empire in the last days.

Remember that the dream of King Nebuchadnezzar described in Daniel 2: had a head of Gold which represented Babylon and stood on its iron and clay feet in what is now modern day Iraq and used to be Babylon. The U.S. already has its footprints in Iraq! **The fact that Peter referred to Rome as Babylon is significant considering how the term Babylon is used in Bible Prophecy of the last days.**

38

Feast Of Trumpets Declares "The Day Of The Lord"

38.1 Trumpets Are Associated With The Rapture

When the Rapture was first revealed in (1 Corinthians 15:50-58) trumpets are conspicuously mentioned twice in the verse that describes the moment that we are all changed. (Verse 52)

*"In a moment, in the twinkling of an eye, at the **last trump**: for the **trumpet** shall sound, and the dead shall be raised incorruptible, and we shall be changed"*. (1 Corinthians 15:52)

Again trumpets are mentioned in (1Thessalonians 4:16)

*"For the Lord himself shall descend from heaven with a shout, with the voice of the archangel, and with the **trump of God**: and the dead in Christ shall rise first"* (1Thessalonians 4:16)

Trumpets are mentioned again in (Revelation 4:1) where the Rapture is also described.

*"After this I looked, and, behold, a door was opened in heaven: **and the first voice which I heard was as it were of a <u>trumpet</u> talking with me**; which said, Come up hither, and I will shew thee things which must be hereafter"*. (Revelation 4:1)

*"**And immediately I was in the spirit**: and, behold, a throne was set in heaven, and one sat on the throne"*. (Revelation 4:2)

After we hear the trumpet and the voice of Christ we will changed, translated and immediately be in the spirit. The first thing we will see will be the judgment seat of Christ. The Church will be judged at that time. We will be judged separately from all other Saints of all other Dispensations.

After (Revelation 4:1) the Church is no longer mentioned in the Book of Revelation. From that point on we are in Heaven.

38.2 The Feast Of Trumpets Is The Biblical New Year

The Feast of Trumpets or Rosh Hashanah occurs on the first and second days of Tishri. (Leviticus 23:24-25) Rosh Hashanah begins and ends at sundown. This is considered the Feast for counting years and epoch years such as Sabbatical and Jubilee years or Ages.

The Feast of Trumpets is observed by the blowing of trumpets in predetermined sets of high, low, short and long blasts throughout the 24 hours of the Feast day. During that 24 hours the trumpets are blown over 100 times.

This Biblical Holy Day marks the end of the current year and the beginning of the following year at the sound of the last trump at about 6:30 PM sundown Jerusalem time.

Every year the priest must determine which day is the proper day to observe the Feast of Trumpets. The Jews set aside two days to observe The Feast of Trumpets and it is up to the Priest to determine which day is the correct day by checking lunar records and comparing with the barley harvest on the first day at sundown.

If the Priest determines it to be the correct day, then the trumpets will begin their sounding. If not, then the trumpets must wait for another day.

The Feast of Trumpets is the 5th Feast and **the next Feast in line to be fulfilled;** The Church Age began on the 4th Feast and it will end on the

5th. The Holy Spirit came down and sealed every believer on the 4th Feast and he will leave on the 5th Feast taking all believers with him.

When the appropriate year arrives the Rapture will be the fulfillment of the 5th Feast, also known at the Feast of Trumpets (1Corinthians 15: 50-58)

"*In a moment, in the twinkling of an eye,* ***at the last trump****: for the* ***trumpet shall sound****, and the dead shall be raised incorruptible, and we shall be changed*". (1Corinthians 15:52)

Trumpets are associated with the Rapture in other passages as well. (1Thessalonians 4:16) (Revelation 4:1)

38.3 The Feast Of Trumpets Will Mark The Rapture!

Although the Rapture was spoken of in the Old Testament it was not actually revealed until (1Corinthians 15: 51-58)

"*Behold, I shew you a mystery; We shall not all sleep, but we shall all be* ***changed***", (1Corinthians 15: 51)

"***In a moment****, in the twinkling of an eye,* ***at the last trump****: for the* ***trumpet shall sound****, and the dead shall be raised incorruptible, and we shall be* ***changed***". (1Corinthians 15: 52)

"*For this corruptible must put on incorruption, and this mortal must put on immortality*". (1Corinthians 15: 53)

This passage is where the Rapture is revealed to the Church for the first time. This passage was dated by Usher to be 59 A.D.

The Rapture is as unique to the Church as the Church Age Itself.

The Church Age began on the fourth Feast and it will end on, the fifth feast. The Feast of Trumpets is next in line to be fulfilled and is also known today as Rosh Hashanah.

The 5th Feast will mark the end of the Church Age and the Beginning of the "Day of The Lord"

"*For the children of Israel shall abide* ***many days*** *without a king, and without a prince, and without a sacrifice, and without an image, and without an ephod, and without teraphim*:" (Hosea 3:4)

This Old Testament passage is talking about the Church Age. It is prophetic in nature but it does not reveal all of the facts concerning the Church Age, only enough to indicate that God foretold this age.

For the last 2,000 years the Jews have been scattered throughout the world just like this passage describes. They have been without a King or heir to the throne. There has been no Temple, Priesthood or sacrifice. They have been without household gods and images, just like the Bible prophesied!

The term *"many days"* is defined later in (Hosea 6: 1, 2) as 2000 Hebrew years.

"***Afterward shall the children of Israel return****, and seek the LORD their God, and David their king; and shall fear the LORD and his goodness in the* ***latter days***". (Hosea 3:5)

God turned to the Gentiles after Israel rejected their Messiah.

For 2,000 Hebrew years God has used Gentiles to carry God's message. Israel was scattered just after the Church Age began just like the Bible Prophesied. Now Israel is back in the land as Prophesied. Their punishment mentioned by Ezekiel, Moses and Jeremiah was satisfied on May 14, 1948.

We have just passed the 2,000 Hebrew years discussed in (Hosea 6:1, 2) and now we are waiting and "watching" for the Lord's return for his Bride, the Church. This event is called the Rapture.

"*Therefore let us not sleep, as do others; but let us* ***watch*** *and be sober*". (1 Thessalonians 5:6)

38.4 The Church Age Ends And The Day Of The Lord Begins

Rosh Hashanah is the modern Jewish name for the "Feast of Trumpets" and one of these years it will mark the end of the **Church age** and the beginning of the **Day of the Lord**.

The first Fall Feast marks the next prophetic event to be fulfilled and is the 5th Feast of the total 7 Feasts given. The first four Feasts are called the "Spring Feasts". The Spring Feasts were all literally fulfilled on their day and in their order. Likewise, the 3 remaining feasts, known as the Fall Feasts will also literally be fulfilled, on their day and in their order as well.

The Church Age began on the 4th Feast (Pentecost) and whatever year it ends, it will end on the 5th Feast (Trumpets) which is the next in line to be fulfilled.

It is interesting to note that the Jews believe that the Law was given on the Feast of Pentecost. If so, then the Dispensation of Law began on Pentecost. Well it gets even more interesting when we consider that the Dispensation of Law was put on hold and the Dispensation of Grace began at the end of the 69th week of Daniel, on the Feast of Pentecost. This Dispensation of Grace will end on the next Feast which is Trumpets, on the appointed year.

The Holy Spirit came to earth and entered into believers on Pentecost (4th Feast) and it will leave the earth taking sealed believers with him on the 5th Feast of Trumpets.

More specifically, the Church Age will end, and the "Day of the Lord" will begin "*At the last trump*". (1Corinthians 15:52) The Church Age will end and 8 days later, the 70th week of Daniel will begin.

38.5 The Day Of The Lord Begins With The Rapture

According to (1Thessalonians 5) & (2 Thessalonians 2) The Day of the Lord begins at the Rapture.

When the last trumpet of Rosh Hashanah sounds, it marks the end of the former year and the beginning of the New Year.

The previous year changes to the New Year when the final Shofar is blown at Sunset at about 6:30 pm Jerusalem time at the end of Rosh Hashanah. That time of Sunset changes from year to year. Sunset would be the point in time where the year changes over to the New Year. This next Rosh Hashanah, September 20, 2020 sunset is at 6:38 pm Jerusalem Time or 11:38 am EST. In 2021 Rosh Hashanah ends at sundown on September 8, at 6:54 pm. The time of sunset changes every year.

Rosh Hashanah (Feast of Trumpets) is a Devine Appointment that will mark the end of the Church Age and the beginning of the Day of the Lord on the heretofore unknown, appointed year!

That is also the point in time, the "***moment***" identified in scripture (1Corinthians 15: 52) that the Rapture will occur on the appointed year. But which year will it be? We don't know yet. In that sense, it is imminent.

God commands every Believer to ***Watch***! (1Thessalonians 5:6) Every Year Believers Should be ***"Watching"*** for that Day, for that **Moment** when the Shofar is blown in Jerusalem halfway around the world at about 6:30 pm sundown, Jerusalem Time, at the end of the Feast of Trumpets! If I am still here when that trumpet is blown halfway around the world, then that means that I have another year left to serve the Lord.

The Rapture of the Church will mark the end of the Age of Grace and the beginning of the Day of the LORD. The Age of Grace was an interim period of time that God sandwiched between the 69th and 70th weeks of Daniel. (Daniel 9:24-27) Hosea identifies it as two thousand years and other passages identify it as a time where God will turn to the Gentiles to bear his message to the world.

During this time God will provoke Israel to jealousy by using Gentiles to spread the Gospel instead of Israel. (Deuteronomy 32: 21) At the end of the Church Age, the 70th week of Daniel will commence. God will turn back to the Jews and prepare them to take their promised place.

The Rapture is the next Prophetic event on God's list of events; The Rapture represents the end of the Church Age and the beginning of The Day of the Lord. Eight days later, on the Feast of Atonement, (Yom Kippur) begins the final week of the Age of Law, also known as the 70th week of Daniel. The Church will be gone and God will turn back to Israel and use them to reach the world.

39

After The Rapture, Only The Unsaved Are Left Behind

39.1 Many will be left behind who call themselves "Christian".

Christendom is the subject of the letters to the seven Churches. Christendom is a term that loosely refers to all of those who profess to believe in Christ and identifies themselves as being followers of Christ. The term "Christian" was first recorded In (Acts 26:28)and used by the pagan Agrippa.

The term "Christendom does not indicate that all of these individuals have ever understood and believed the gospel. It just means that they call themselves "Christian". That is how they identify themselves. The Bible indicates that in the *"last days"* the majority have not. These people may be religious and appear to be pious, but they have not understood the Gospel because their minds were blinded by satan through false teachers!

The Body of Christ is an exact term that refers to all of those during this age or Dispensation who have understood and believed the Gospel. This group has been born again and they may or may not have anything in their life noticeable to the average person to indicate that they are believers. When the Rapture happens, Christ will take all of those who have been born again regardless of their lifestyle.

There will be "Many" or the majority of "Christendom" who will be left behind at the Rapture because they did not understand and believe the Gospel. You only get Saved by trusting in the Lord Jesus Christ apart from works! They received another-Jesus and another-spirit and another-gospel. (2Corinthians 11:4)They were trusting in what they considered good works (Matthew 7:21,22), not Jesus alone. (Romans 4: 4,5)

The Rapture of the Church marks the beginning of the "Day of the Lord". When asked about this time, Jesus said…

"And Jesus answered and said unto them, Take heed that no man ***deceive*** *you"* (Matthew 24:4)

"For ***many*** *shall* ***come in my name****, saying, I am Christ; and shall deceive many"*(Matthew 24:5)

Jesus is doing the talking and He said that many will come in his name and acknowledge that "He"(Jesus)is the Christ spoken of by God's Prophets. These false teachers are not claiming to be Christ, they are claiming to be followers of Christ or in modern day terms, they are claiming to be "Christian".

You must be born again by Just Faith in Jesus Christ's shed blood <u>apart</u> from works to take part in the Rapture. The Rapture is only for those who have once tasted of the **<u>Grace</u>** and Spirit of our Lord Jesus Christ by faith in his **<u>finished</u> work** on the cross of Calvary (Hebrews 6:4-6). The false brethren majority will be left behind because they trusted in their works. Their false teachers who misled them will also be left behind and they will still be misleading people.

I believe that the main stream news media will be under the control of The Antichrist during those seven years. At first many will not be aware that a rapture has occurred but word will get around about the Rapture and the Third Temple being built in three days. The Antichrist and the False Prophet will

USE MIRACLES, SIGNS AND WONDERS DURING THE "TEN DAYS OF AW" TO CREATE CONFUSION ABOUT THE RAPTURE AND DISTRACT ATTENTION FROM IT.

THEY WILL USE THE CONFUSION ABOUT THE RAPTURE TO TELL A LIE SO THE ANTICHRIST CAN GAIN CONTROL OF THE U.N. FIRST AND THEN THE U.S.

(2 THESSALONIANS 2: 11) *"AND FOR THIS CAUSE GOD SHALL SEND THEM STRONG DELUSION, THAT THEY SHOULD BELIEVE A LIE:"*

THE ANTICHRIST MAY EVEN TAKE CREDIT FOR THE RAPTURE. MAYBE BLAME IT ON ALIENS OR SOMETHING ELSE JUST AS CRAZY. OF COURSE THIS IS SPECULATION, BUT WE ARE SUPPOSED TO SPECULATE ABOUT POSSIBILITIES AND PROBABILITIES. DISTINGUISHING THE DIFFERENCE IS CALLED CRITICAL THINKING.

I EXPECT THAT THERE WILL BE A PROPAGANDA BLITZ, FULL OF MIRACLES, SIGNS AND WONDERS TO PROMOTE THE FALSE PROPHET AND THE ANTICHRIST DURING THE JEWISH HOLLIDAY CALLED "THE **10 DAYS OF AWE**". THE FAKE "MIRACLES" WILL BEGIN AFTER THE RAPTURE. THE 10 DAYS OF AWE BEGINS ON ROSH HASHANAH AND WILL CLIMAX AT SUNDOWN ENDING YOM KIPPUR, NINE DAYS AFTER THE RAPTURE.

THE FALSE PROPHET AND THE MAIN STREAM NEWS MEDIA WILL PROMOTE THE ANTICHRIST AS A WORLD LEADER FOR PEACE. I BELIEVE THE U.N. WILL TAKE A VOTE TO MAKE THE ANTICHRIST THE HEAD OF THE U.N. THAT WILL BE THE DAY THAT THE ANTICHRIST IS "*REVEALED*" TO THE FOLLOWERS OF ISLAM AS THE TWELFTH IMAM.

THE ANTICHRIST WILL PROMISE TO GUARANTEE THE OSLO ACCORDS SIGNED BY ISRAELI PRIME MINISTER RABIN FOR SEVEN YEARS. THIS TREATY IS SPOKEN OF BY NAME IN (DANIEL 9:27) THE U.N. WILL SAY "***PEACE AND SAFETY***" AFTER THE ANTICHRIST PROMISES TO ENFORCE THE TREATY FOR SEVEN YEARS.

The first goal mentioned in the U.N. charter for the United Nations to "Maintain International Peace and Security". That phrase identifies The United Nations involvement in the Antichrist being revealed on that day.

… "*For when they shall say, Peace and safety; then* ***sudden destruction*** *cometh upon them*"… (1Thessalonians 5:3)

After sundown, Jihad will begin. Whoever is running the United States after the Rapture will be involved in the Antichrist taking over the Country. The influence of the Holy Spirit will be gone and evil will take over. The Chaos created from the Jihad will be the excuse to use the martial law plan from Obama's Presidency. The United States will be given to the Antichrist in hopes of protection from Jihad. The United Nations will rule over the United States.

No believers will be left behind, only false teachers, false brethren and other unbelievers. There will be plenty of professing "Christians" left behind because they never believed God's Gospel of Free Grace they were trusting in their works. They will stand out as the first targets of a brutal, merciless Jihad!

There will be a famine for the truth, but God will supply one hundred and forty four thousand Jewish Missionaries to preach the gospel to everyone on earth.

39.2 False Brethren Trusting In Dead Works Will Be Left Behind

There will be plenty of people who outwardly appeared like a Christian, but were not saved.

"*And he said unto them, Ye are they which justify yourselves before men; but God knoweth your hearts: for that which is highly esteemed among men is abomination in the sight of God*". (Luke 16:15)

There are plenty of people that consider themselves to be devout Christians but don't understand the Gospel of Salvation. You must understand the Gospel before you can believe it.

"*When any one heareth the word of the kingdom, and* ***understandeth it not****, then cometh the wicked one, and catcheth away that which was sown in his heart. This is he which received seed by the way side*". (Matthew 13:19)

Satan is the father of lies. He uses lies to keep people from understanding the Gospel. People are blinded from understanding God's simple truth because they have already heard and believed Satan's lies. This passage says that *"the wicked one"* *"catcheth away"* the truth from those who hear it. If a person hears the truth but does not understand it because he is already confused by Satan's lies, Satan will attempt to take away the truth with more lies.

"*In whom the god of this world hath* ***blinded the minds*** *of them which believe not, lest the light of the glorious gospel of Christ, who is the image of God, should shine unto them*". (2Corinthians 4:4)

Satan blinds the minds of the unbeliever with lies so that they can't understand the simple Gospel when they hear it because satan redefines the terms.

There are many false gospels that can't save, but there is only one Gospel that does save!

"*Enter ye in at the strait gate: for wide is the gate, and broad is the way, that leadeth to destruction, and* ***many*** *there be which go in thereat:"* (Matthew 7:13)

"*Because strait is the gate, and narrow is the way, which leadeth unto life, and* ***few*** *there be that find it*". (Matthew 7:14)

This passage uses the word ***many*** to describe those who choose the wrong way that leads to destruction and ***few*** to describe those who find the only way to Salvation. This passage identifies the word ***"many"*** as meaning **the majority**. It also identifies ***"few"*** as the minority.

*"**Many** will say to me in that day, Lord, Lord, have we not prophesied in thy name? and in thy name have cast out devils? and in thy name done many wonderful works"?* (Matthew 7:22)

"*And then will I profess unto them, I **never** knew you: depart from me, ye that work iniquity*". (Matthew 7:23)

These people were proud of their good works. All done in the name of Jesus! They were deceived by Satan into believing that their works could contribute to their salvation. They did not believe the simple gospel of grace so they did not get grace. (Romans 4:4,5) They got judgment! Furthermore, God calls their *"many wonderful works" Iniquity"*. God says that trying to earn your salvation is sin!

"But we are all as an unclean thing, and all our righteousnesses are as filthy rags; and we all do fade as a leaf; and our iniquities, like the wind, have taken us away". (Isaiah 64:6)

"*For **many** shall come in my name, saying, I am Christ; and shall deceive **many***". (Matthew 24:5)

I used to think of this passage as saying that many would claim to be Christ, but now I realize that is the wrong interpretation.

This is describing a widespread deception such as people claiming to be Christian brothers when they are not. The term "*many*" as it is used here means the majority. An individual claiming to be Christ just doesn't fit what is being described here. The majority doesn't claim to be Christ. Only a very tiny few have done that. After all, how can someone claim to be sent by Christ and also be Christ at the same time?

Jesus is doing the talking. He is saying that ***"many"*** will come in his name. They would acknowledge that he is the Christ and they would deceive ***"many"***. These people are saying that they are Christian, to gain your trust and deceive you!

False brethren may be careful to maintain outward appearances, but inwardly are lost. They may act "holier than thou", but are lost.

"*For we commend not ourselves again unto you, but give you occasion to glory on our behalf, that ye may have somewhat to answer **them which glory in appearance, and not in heart***". (2Corinthians 5:12)

"*But the LORD said unto Samuel, Look not on his countenance, or on the height of his stature; because I have refused him: for the LORD seeth not as man seeth; for man looketh on the outward appearance, but the LORD looketh on the heart.*" (1Samuel 16:7)

False Prophets and false Teachers pretend to be brothers in Christ, but they are not! They put on a good show and appeal to your feelings, but it is only a show. If you don't know your doctrine, you will be deceived.

"*I marvel that ye are so soon removed from him that called you into the grace of Christ unto **another gospel***" (Galatians 1:6)

"*Which is not another; but there be some that trouble you, and would **pervert the gospel of Christ***". (Galatians 1:7)

"*But though we, or an angel from heaven, preach **any other gospel** unto you than that which we have preached unto you, let him be accursed*". (Galatians 1:8)

"*As we said before, so say I now again, If any man preach **any other gospel** unto you than that ye have received, let him be accursed.*" (Galatians 1:9)

Satan offers a heavenly way to go to hell. Satan offers many counterfeit gospels to choose from but the real Gospel stands out from all the rest. The real Gospel is where God did all the work and he gets all of the credit.

God will not share credit with anyone. God offers Eternal Life for free to all those who believe.

39.3 God Does Not Save Those Who Are Trying To Be Saved!

God only Saves those who are **trusting Jesus** to be Saved! Those who are **trying** to be saved **are <u>not</u> trusting Jesus** to be saved and **they will be left behind**. They are not trusting to be saved because they have believed a false gospel, a lie from Satan that says that they must do something of themselves to be saved. Calling yourself a Christian, going to Church and trying to live the "Christian life" does not make you a born-again child of God! Asking Jesus into your heart doesn't save you either.

You become a "born again", "child of God" the moment that you speak to God in your thoughts and tell him that you are trusting in his Son Jesus as the only way to Eternal Life; and acknowledge to God that you have Eternal Life right now; Acknowledge to God that he is now your Father because you have just been born into his family and are his child forever! Amen! Amen! You become a believer by believing God's record that He gave of his Son. (1John 5: 10,11)

The term "Christian" was first used in (Acts 26:28) by Agrippa, a pagan. The term means "follower of Christ" and is a reference to **outward appearances**. However God uses different terms in the Bible to describe those that are his. Terms like "Child of God"; "*Sons of God*"; "*Children of God*" are some terms used throughout the Bible that refer to "believers" and the new birth. Jesus said "*Ye must be born again.*" You can't see a new birth on the outside. The Bible also uses terms such as "Believers", the "Righteous" and "Brethren".

Hell is a real place that God has prepared to contain all the sin of the world. Anyone who dies without the new birth will end up in Hell for all of eternity. There is no escape after death. Belief in the Gospel is obedience to the Gospel. Unbelief is disobedience to the Gospel. The Gospel tells you to believe on Jesus and you will be saved. That is what you are told to do. Obey the gospel while you still can!

Jesus said that you must be born again. Everyone receives a New Birth when they believe God's Truth. The Truth is that Jesus paid it all! **The Gift of God is Eternal Life to all those who believe!** Just Faith, nothing more! It's Free! You have to admit that you have nothing to offer God to merit salvation. You must admit, like the thief on the cross that you are helpless, that Jesus is God in flesh who paid it all and put your trust in **him alone**!

If you don't believe that what Jesus did on the cross was sufficient, then you are not believing the Gospel and you need to change your mind. The term "***Gospel***" simply means **"good news**" and nothing more.

The message that God loves you, has paid for all of your sins and will give you Eternal Life as a free gift is **Good News!** The moment that you ***believe*** God's ***record*** about Jesus, God gives you *Eternal Life* and he will never take it back no matter what! Now that is good news! Salvation can't be earned and it can't be lost! It is free! ***It is the Gift of God*** that is dispensed by just ***faith***. That means **believing God's *record* about Jesus**. You must believe the list of facts recorded in the Bible about Jesus.

God's record says that Jesus Christ is our Creator who came to the earth by being born of a virgin (Isaiah 7: 14; & 9:6) and offered himself in our place as a perfect and infinite sacrifice for our sins. (John 3:16)

When you believe that message, then you are believing God's record (1John 5:10-13); and you are trusting in Jesus and not yourself; and you **have** eternal life (John 6: 47); and you are born again (1John 5:1); and God will never cast you out or lose you no matter what (John 6: 37&39); and your citizenship is in heaven (Philippians 3:20) and your name is announced in Heaven (Luke10:20); you are sealed with the Holy Spirit until the day of redemption (Ephesians 1:13 & 4:30); you are no longer condemned (John 3:18); and you are passed from death unto life. (John 5:24)

God declared his record long before it happened so there would be no doubt about who the real Jesus is.

The real Gospel says that the work has already been done and Jesus did it! A false gospel denies that and says that you must also do something of yourself to be saved. If you believe that, then you need to **change your mind** about what you are trusting in and believe the Gospel. There is only one Gospel that saves and this is it!

"***Believe** on the Lord Jesus Christ, and thou shalt be saved,*" (Acts 16:31)

Jesus said…

"*Verily, verily, I say unto you, He that **believeth** on me **hath everlasting** life*". (John 6:47)

Just Faith! Eternal life is free, you get it when you believe it. It is called "eternal" because you can't lose it! It is the Gift of God so you can't earn it. (Romans 6:23) & (Ephesians 2:8 & 9)

The Bible says … *"for all have sinned and come short of the glory of God"* (Romans 3:23) God also said… *"There is none righteous, no not one"*. (Romans 3:10)

We are all guilty and in need of a Savior. Jesus paid for all the sins of all mankind, past, present and future. If you are a religious person trying to work your way to heaven by doing good deeds, then you need to change your mind and believe the Gospel because you are lost and you can't save yourself!

If you are someone who is chopping of heads and killing people in the name of some other god, you also need to change your mind and believe the Gospel. Either way, the wages of sin is still the same. When you believe God's record of his Son, you receive Eternal life and it cannot be undone. If you do not confess your faith in Christ to others, you are still saved, but you will lose rewards. If you deny Jesus after you are saved, you may lose rewards but you are still saved because Eternal Life is the ***Gift*** of God.

No sin can enter Heaven or exist in God's presence. When you change your mind about what you believe and **believe the Gospel**, you become born again; God becomes your father; God forgives you; cleanses you from **all** sin and gives you eternal life as a **free gift**.

"*For the wages of sin is death; but* ***the gift of God*** *is eternal life through Jesus Christ our Lord*". (Romans 6:23)

Notice that this passage says that eternal life is the **gift of God**. That means that God does the giving and the believer does the receiving. The Bible always describes eternal life or eternal salvation in that order. God does the giving and the believer does the receiving. God never tells you to give anything of yourself to be saved. It is the **Gift of God**! It is free! God does the working and giving and the believer does the receiving. God never reverses that order, but satan does.

"*But as many as* ***received him****, to them* ***gave he*** *power to become the sons of God, even* ***to them that believe*** *on his name:*" (John 1:12)

"*For God so loved the world, that* ***he gave*** *his only begotten Son, that whosoever* ***believeth*** *in him should not perish, but have everlasting life.*" (John 3:16)

"*To him give all the prophets witness, that through his name* ***whosoever believeth in him shall <u>receive</u>*** *remission of sins*" (Acts 10:43)

It is always God who does the giving and the believer is the one that does the receiving! Satan always reverses that order when his minions present a false Gospel. So be on guard, and don't be fooled.

"*For by grace are ye saved through faith; and that* ***<u>not of yourselves</u>****: it is the* ***gift of God***"...

"***Not of works****, lest any man should boast*". (Ephesians 2: 8 & 9)

God says that Eternal Life is a Gift by just faith or believing. It is not of yourselves so you can't work for it.

If you are working to be saved, then you are not trusting to be saved. If you are not trusting Jesus alone to be saved, then you are not obeying the

Gospel which says to trust Jesus alone to be saved. You need to change your mind and believe! Eternal Life is the Gift of God, not a reward!

"*Now to him that **worketh** is the reward not reckoned of grace, but of debt*". (Romans 4: 4)

"*But to him that **worketh not**, but **believeth** on him that justifieth the ungodly, his faith is counted for righteousness*". (Romans 4: 5)

If you are working to be saved, then you are not trusting to be saved and therefore you are lost!

You can't mix Grace and works, they are opposites. You either have one or the other. Satan knows that if he can sneak in some human works into a Gospel presentation then he has neutralized the message. Those who believe such a message are still lost!

"*And if by grace, then is it no more of works: otherwise grace is no more grace. But if it be of works, then is it no more grace: otherwise work is no more work*" (Romans 11:6)

Any human effort or works added to a Gospel presentation will neutralize the message!

"*A little leaven leaveneth the whole lump*". (Galatians 5:9)

39.4 Wolves In Sheep's Clothing Will Be Left Behind.

"*Beware of false prophets, which come to you in sheep's clothing, but inwardly they are ravening wolves*". (Matthew 7:15)

Wolves in sheep's' clothing indicates an intent to deceive and prey upon the deceived. The Bible describes false teachers or false prophets as wolves in sheep's clothing because they want to gain your trust by pretending to be a Christian Brother so that they can use you.

The term "False Prophet" is equivalent to saying "false teacher", "false preacher" or perhaps "false leader". Outwardly false prophets may appear to be "good" or "Christian". They sing the same hymns in church. They use the same terminology. False preachers often put on an act to gain people's trust.

If it seems too good to be true, it probably isn't true. False preachers pretend to be Godly by talking so sweetly as if they care. But when it comes to really caring to the point of doing something that they will not profit from, don't count on it. That is because the Bible says that they are motivated by their own lusts.

I believe that Joel Osteen is a classic example of exaggerated smiles and exaggerated "sweetness". He avoids teaching Bible doctrine and tells people what they want to hear. He acts so sugary sweet that anyone with an ounce of brains knows that it is just an act. As a result, he is known as one of the wealthiest preachers.

"*Their throat is an open sepulchre; with their tongues they have used deceit; the poison of asps is under their lips*": (Romans 3:13)

Before his election, I remember when George Bush Jr looked into the camera and said with a very straight face that he had a personal relationship with Jesus Christ. I remember saying... "Hey wait a minute, he is a member of the satanic skull and bones society, just like his dad". But my Christian friends just wanted to keep on believing that things were going to get better. They just weren't paying attention.

"*This people draweth nigh unto me with their mouth, and honoureth me with their lips; but their heart is far from me*". (Matthew 15:8)

"*Now the Spirit speaketh expressly, that in the **latter times** some shall depart from the faith, giving heed to seducing spirits, and doctrines of devils*"; "*Speaking lies in hypocrisy; having their conscience seared with a hot iron*"; (I Timothy 4:1-2);

"*For the time will come when they will not endure sound doctrine; but after their own lusts shall they heap to themselves teachers, having itching ears*"; (II Timothy 4:3);

"*And they shall turn away their ears from the truth, and shall be turned unto fables*". (II Timothy 4:4);

"*And **many** false prophets shall rise, and shall deceive many*". (Matthew 24:11)

"*He answered and said unto them, Well hath Esaias prophesied of you hypocrites, as it is written, This people honoureth me with their lips, but their heart is far from me*". (Mark 7:6)

False teachers look good on the outside. They use scripture out of context like Satan did when he tempted Jesus on the pinnacle of the Temple. Their doctrine cannot withstand scrutiny in the light of scripture.

False teachers preach works for salvation because they want you to give them your money. They say that you must turn from your sin to be saved. This implies that they have already done that and you need to follow their good example. Well they are lying if they say that they have turned from all of their sin. So how much do you have to turn from? How do you know if you have turned from enough? What if you backslide?

"*If we say that we have no sin, we deceive ourselves, and the truth is not in us.* (1John 1:8)

The fact is that the phrase "turn from sin" is equivalent to "keep the commandments". The commandments can't save, they only condemn! The commandments tells us what is sin and what is not and that we are all guilty. "Turn from sin" and "keep the commandments" mean the same thing.

"*Therefore by the deeds of the law there shall no flesh be justified in his sight: for by **the law is the knowledge of sin**.* (Romans 3:20)

"But now the righteousness of God without the law is manifested, being witnessed by the law and the prophets" (Romans 3:21);

"Even the righteousness of God which is by faith of Jesus Christ unto all and upon all them that ***believe****"*: (Romans 3:22)

"Therefore we conclude that a man is justified by faith ***without the deeds of the law****"*. (Romans 3:28)

If Satan can get people to work for their salvation he has won! Everyone who works for their salvation will be lost!

39.5 Some Born Again Believers Will Be Missed

Some genuine, born-again believers will be missed by those that knew them. The following passage describes what it will be like for those who were left behind and knew believers who were raptured. Believers will be translated before the 70th week of Daniel begins. Here is an example of the pretribulation Rapture and the Tribulation that follows.

"*Woe is me! for I am as when they have gathered the summer fruits, as the grapegleanings of the vintage: there is no cluster to eat: my soul desired the firstripe fruit*". (Mica 7:1)

He misses his Christian friends because they have all been raptured out or "harvested". His Christian friends are "*the first ripe fruit*" spoken of.

Those that heard about the Rapture and actually get to see it happen will have an advantage to sorting things out. To them, the Rapture will be a sign from God. Maybe it will be the sign that the Jews are looking for. They will realize that the Tribulation has begun and their Christian friends were raptured.

"For ***the Jews require a <u>sign</u>****, and the Greeks seek after wisdom:"* (1Corinthians 1:22)

The unbelievers who are left behind will miss their Christian friends. When they hear the lie cooked up by the False Prophet and the Antichrist they may not be so easily deceived if they heard enough truth from their Christian friends.

Over the years, born again believers who were watching for this event have speculated about how the Antichrist will explain or maybe even take credit for the Rapture.

"*The good man is perished out of the earth: and there is none upright among men: they all lie in wait for blood; they hunt every man his brother with a net*". (Mica 7:2)

"*The good man*" refers to believers. *"Perished"* is referring to the "***Departure***". The believers will be taken away and removed before God's wrath and destruction begins. Now the Holy Spirit or the "*restrainer*" is gone and things really get bad. There is no one left to oppose evil.

"*That they may do evil with both hands earnestly, the prince asketh, and the judge asketh for a reward; and the great man, he uttereth his mischievous desire: so they wrap it up*". (Mica 7:3)

"*The best of them is as a brier: the most upright is sharper than a thorn hedge: the day of thy watchmen and thy visitation cometh; now shall be their perplexity*". (Mica 7:4)

"Trust ye not in a friend, put ye not confidence in a guide: keep the doors of thy mouth from her that lieth in thy bosom" (Mica 7:5)

The Bible says that a house divided against itself cannot stand. (Matthew 12:25)Can you imagine what it is like to have your wife working against you? The Bible says …

"*Every wise woman buildeth her house: but the foolish plucketh it down with her hands*". (Proverbs 14:1)

"For rebellion is as the sin of witchcraft, and stubbornness is as iniquity and idolatry"... (1Samuel 15:23)

During the Tribulation, you won't be able to trust anyone, not even your own wife! We are seeing that already. It is witchcraft spreading.

"For the son dishonoureth the father, the daughter riseth up against her mother, the daughter in law against her mother in law; a man's enemies are the men of his own house." (Mica 7:6)

We are already seeing families divided from within. During the Tribulation the Bible says that *"a man's enemies are the men of his own house"*.

40

Feast Of Atonement Declares Daniel's 70th Week

40.1 End Times Prophecies Were Not Possible, Until Now

Ancient Middle Eastern knowledge did not extend as far as it does today. They did not travel far from home like we do. They did not have maps of the World, or of nearby territories and land masses. Yet the Prophets spoke accurately of future events, their geographical locations, dates and details!

The Prophets described the Geography; the time; the Technology; and the nature of this time that we are living in! Until recently, Biblical descriptions of events during the Tribulation just did not seem possible and were challenged, even ridiculed. But now, as the time approaches, the details are all in place and not many are laughing .

In the past, these judgements against the earth during the Tribulation has seemed too fantastic to be true. Men questioned how one man called the Antichrist could number everyone and control all buying and selling. Until recently, it was not humanly possible.

The idea of a two hundred, million man army crossing a dried up river Euphrates from the East was just too fantastic for many to believe at the time that it was written. Today however, we know that China can produce a two hundred million man army just like the Bible said 2,000 years ago. And now, the river Euphrates has five dams with five plugs that can shut off the water and dry up the river.

That was written about 2,000 years ago, when there wasn't two hundred million people on the entire planet! The idea of the Euphrates River drying up sounded impossible until recently. That river has never dried up in all of human history!

China has boasted the largest army in the world for quite some time now! Their boasting number of 200,000,000 identifies them as the Eastern country and this is the time! But yet the prophets knew that a two hundred million man army would cross over the dried-up riverbed of the Euphrates from the East to fight in Armageddon. That can only mean **"Red China"**, and it says **when** it will happen! Armageddon will occur at the end of the 70th week of Daniel. Between the Feast of Yom Kippur and the Feast of Tabernacles.

These are End Times Prophecies and we are living in the End Times. "End Times" and "Last Days" or "Latter Days" are terms that mean the same thing. They are talking about the end of the Age called the "Times of the Gentiles". The "Last Days" began on May 14, 1948. It is defined in Hosea 3:5. (ETBH chapters 20-32)

I recently read that terrorists have taken over one of the dams on the river Euphrates and are threatening to close it off. The technology required to do the things mentioned during the Tribulation did not exist until now.

Over the past centuries, nonbelievers have considered the descriptions of events that occur during the Tribulation as nonsense. They just could not believe that such technology could ever exist, but now it does.

Many things mentioned in the Bible about this time that we are living in have seemed too fantastic for many to believe, until now.

For example, no one believed that the Jews would ever return back to Israel after the Romans scattered them throughout the world back in 70A.D. It seemed impossible! But God said he would do it; he said when he would do it; he said why he would do it; and how he would do it. Not only did God perform a miracle to bring Israel back, but he did it when he said he would.

(ETBH chapters 22- 28,) … in chapter 21 we discover unconditional promises that God made to Abraham that he intends to keep in the near future! Later on there were other, promises made to Abraham's descendants through Moses that were conditional. There were promises of blessings if Israel obeyed God and curses promised if Israel chose not to obey God.

Israel chose not to obey God so they got the Curses that God promised. This is discussed in (ETBH chapters 23 & 24). The Curse ended on May 13, 1948, but there is still the 70th week of Daniel left! God has already kept some of those promises and the time has come for him to finish what he promised! (ETBH chapters 28 & 29)All of this was prophesied, every detail even the time.

The Bible talks about the whole world seeing God's two witnesses killed and left lying in the streets of Jerusalem; the two witnesses resurrected after 3 ½ days and ascending up into heaven; the image of the Antichrist visible for all of the world to see; and to see Christ return in the clouds. That was Impossible until recently, when satellite T.V. became operational. But now we can see how that can happen. **Remember, the Bible said it first!**

The Bible describes Russia before it existed. The Bible identifies Russia by name, by Geographical location relative to Israel and by description of its surrounding Geography.

The Bible describes an arms race throughout the Tribulation. The Bible also describes wars and rumors of wars leading up to the final battle. When Christ returns in the sky, every eye will see him because CNN will be there to see them turn their weapons against Christ!

Today the N.A.S.A. is developing the ability to track and destroy asteroids by using nuclear weapons. The plan is called the **Asteroid Redirect Mission** (**ARM**). The plan is to either divert their path or destroy them all together to protect the earth. Recently Trump created the U.S. Space Force. I believe that technology will be ready for Armageddon and used against Christ returning with his armies.

The Bible describes nuclear, biological and air warfare during the Tribulation. The Bible said that God would bring the whole world against Israel for a battle that would last for one hour and destroy a third of all mankind. That was not ever possible until now! **Remember, the Bible said it first!**

40.2 God Will Empower Two Men.

(Revelation 11) is about the two witnesses and the building of the third Temple. Apparently the third temple will share the Temple Mount with what is called profane by God. However, there will be a wall of separation.

"And I will give power unto my two witnesses, and they shall prophesy a thousand two hundred and threescore days, clothed in sackcloth" (Revelation 11:3)

God's two Witnesses will teach Israel for exactly 3 ½ Hebrew years which are 360 days each. During that time they will teach 144,000 Jewish male virgins to become missionaries throughout the world, (Revelation 7:4 & 14:1-5) 12,000 from each of the 12 tribes of Israel.

These two witnesses are identified as Moses and Elijah by the types of miracles that they perform. Elijah is also identified by name.

"*Behold, I will send you Elijah the prophet* ***before*** *the coming of the great and dreadful day of the LORD*": (Malachi 4:5)

Moses is also identified by the fact that Satan tried to steal his body, apparently so that he could not come back like Elijah.

"Yet Michael the archangel, when contending with the devil he disputed about the body of Moses, durst not bring against him a railing accusation, but said, The Lord rebuke thee." (Jude 1:9)

"These are the two olive trees, and the two candlesticks standing before the God of the earth" (Revelation 11:4)

"And if any man will hurt them, fire proceedeth out of their mouth, and devoureth their enemies: and if any man will hurt them, he must in this manner be killed". (Revelation 11:5)

*"**These have power to shut heaven, that it rain not** in the days of their prophecy: and have **power over waters to turn them to blood**, and to smite the earth with all **plagues**, as often as they will"*. (Revelation 11:6)

It is important to understand that the word "blood" as used in the Old Testament Hebrew does not necessarily mean literal blood. Most often it does, however it can also mean anything red in color like red wine or a red moon or sky etc. Likewise the same thing applies to the New Testament Greek.

Those are the same miracles that Moses and Elijah used during the Dispensation of Law. Remember that the 70th week of Daniel is the remainder of the Dispensation of Law.

40.3 Big Water Problems! Forests Will Dry Up

Moses and Elijah will come **before** the Day of the Lord.

"*Behold, I will send you Elijah the prophet* ***before*** *the coming of the great and dreadful day of the LORD***"**: (Malachi 4:5)

When Elijah was here last time, he wanted to get Israel's attention. So he prayed to God and asked for God to stop the rain to get Israel's attention. So God stopped the rain for 3 ½ years. They are called the two witnesses because they will operate as a team of two. Moses and Elijah both were present at the "Transfiguration" and they will also be involved on the earth for the first 3 ½ years of the Tribulation. After they are killed in the middle of the Tribulation, they will be in Heaven and return with us as part of Christ's great army.

40.4 The Euphrates And Nile Rivers Will Dry Up

The Nile River has been drying up for decades now. During the Tribulation the Nile River will dry up. (Isaiah 19:5-8)

The Euphrates River will also dry up to prepare the way of the Chinese army. There are five dams built on the Euphrates River that could be closed at any time and dry up the river.

That river will be completely dried up (Revelation 16:12)to allow a 200 million man army (Revelation 9:13-16), to cross over into Armageddon.

No one ever believed that either of these two largest rivers in the world would ever dry up until now… But **Remember, that the Bible said it first!**

40.5 Bodies Of Water All Over The World Are Turning Red.

I can't help but notice the similarity between (Joel 2:31) and (Malachi 4:5). Both passages speak of signs associated with the Day of the LORD and both say that these signs will commence **before** the Day of the LORD. It sounds like God will announce the event **before** it begins.

We are told to look for signs in the Heavens, the earth and the seas. We just had a unique astronomical event associated with the Biblical Feasts called a Blood Moon Tetrad. This occurred on 2014 and 2015. We are experiencing "*waves roaring*" such as Tsunamis and extreme weather like Hurricane Katrina.

"*And there shall be signs in the sun, and in the moon, and in the stars; and upon the earth distress of nations, with perplexity; the sea and the waves roaring*" (Luke 21:25)

I think that it is important to mention that Katrina came through Louisiana a week after Bushes "Road Map to Peace" was being implemented in the West Bank of Israel. One week after Israelis were being forced from their homes on the West Bank of Gaza by George Bush's Road Map to Peace plan, Americans were being forced out of their homes on the West Bank of Louisiana because of Katrina.

Now we are seeing something else that is very unusual. The waters are turning red all over the world, from the South Pole to the North Pole and in-between.

https://www.youtube.com/watch?v=ouc-0HDvTwQ

https://www.youtube.com/watch?v=8A34mJKVATE

I think that we have gotten all of the signs that we are going to get before the Rapture! We are running out of time! It is at the doors!

40.6 Bottomless Pit?

"And when they shall have finished their testimony, the beast that ascendeth out of the bottomless pit shall make war against them, and shall overcome them, and kill them". (Revelation 11:7)

"***The Beast that ascendeth out of the bottomless pit***" is a phrase that makes me curious. During the Tribulation the bottomless pit is mentioned more than once. During the Tribulation there seems to be interaction between the surface and the center of the earth that is unprecedented in human history.

Experiments done in the weightlessness of space on the international space station, has shed new light on the behavior of matter in space. The forces that control the gathering of particles together to form larger bodies of matter, even planets behave differently than previously imagined. Click on following link to view a short video of this experiment…

https://www.youtube.com/watch?v=JeMzOhoJpfw

I believe, as already discussed, that the bottomless pit leads to the center of the earth. I say that for several reasons. The main reason being that the Bible teaches that the heart of the earth is hollow. An opening that connects the surface to the center has no bottom if the opening connects the north and south poles through the center.

Planets are not formed in the manner previously assumed. For example, experiments in the International Space Station have demonstrated that as matter is attracted toward a common center of mass a spin will naturally occur. This spin is caused by the offset trajectories of the different masses relative to a common center of gravity. The faster the spin, the bigger the hollow center will be.

H They said that they had seismic data to prove it.

Traditionally men have believed that the earth was hollow because the Bible said so. But modern "science" said that the heart of the earth was made of molten Iron because of our planets unique magnetic field. Modern "science" said the belief in a hollow earth was superstitious nonsense and the Bible was wrong.

Men have dedicated a great deal of time investigating this issue and have published their works. Jan Lamprecht is one of them…

Jan Lamprecht; Hollow Earth Theory; Inner Earth Science; Evidence of Hollow Planets; is what you search for on line. He has plenty of videos and here is one of them… https://www.facebook.com/128488287345987/videos/440337489494397/

Now we have new interpretations of that same data and new data to add to it which indicates that planets are Generally hollow with narrow openings connecting the center to the surface at both polar ends. Our north pole is covered with water, but the South Pole is not. Furthermore, there seems to be a lot of unusual attention directed toward the South Pole lately. More about this later.

40.7 The 70th Week Of Daniel Begins at Sundown, Yom Kippur!

(Daniel 9)(Matthew 24: 21)(Isaiah 13:6 – 13)

Review chapters 26-28 to better understand how the Feasts of Israel represented and marked future events. The time and the nature of the event was marked by its corresponding Feast. The rituals of each Feast depicted the event so that it would be recognized by anyone who was watching for it. They were told back then to "watch" for signs. God gives Signs to his believers who are "watching".

The ninth chapter of Daniel starts out with Daniel discovering the number of years pronounced by God for the first exile in Babylon. It says in verse two that he discovered this by *"books"* (plural). So I believe that he was reading at least Jeremiah, Ezekiel and the Torah. (Daniel 9: 13)

By this we know that Daniel understood that the 70 years of exile in Babylon were almost up. (Verse 2) From the Torah he understood that the **curse** for disobedience came in two parts.

The **First part** was defined by Jeremiah as the removal of Israel from their land for 70 years. During this time God would let the land rest while Israel served their enemies in Babylon for those 70 years.

This was to be the first punishment for failure to observe the Sabbatical year for 490 years. Divide 490 by 7 and that gives you 70 which is how many Sabbatical years were owed to God by Israel.

The **second part** of the curse is conditioned upon Israel's continued disobedience. **The second part of the curse would be "seven times more**". (Leviticus 26:18) Daniel knew that Israel did not have a change of heart and was worried about what will happen next. (Daniel 9:6)

Daniel is confessing his sins and the sins of Israel to approach God with his questions. God hears his prayers and sends him the answer. While Daniel was confessing his sins and the sins of Israel (verses 20-23) God sends the Angel Gabriel to answer his questions. The answer is called the 70 weeks of Daniel (verses 24-27).

"Seventy weeks are determined upon thy people and upon thy holy city, to ***finish the transgression****, and to* ***make an end of sins****, and to* ***make reconciliation for iniquity****, and to* ***bring in everlasting righteousness****, and* ***to seal up the vision and prophecy****, and to* ***anoint the most Holy****".* (Daniel 9: 24)

As you can see, the answer has six parts. These six items include the completion of the two parts of punishment prophesied against Israel. Also included is God offering himself as the sacrifice for the sins of mankind.

This "seventy weeks" spoken of here takes us all the way to the second return of Christ and the setup of his millennial reign. The 70th week of Daniel is the last 7 years of the Dispensation of Law.

In verse 25 we have the first 69 weeks identified as beginning with the commandment to rebuild Jerusalem and ending with the Messiah identifying himself to Israel on Palm Sunday. (ETBH chapter 23) That was accurate to the very day!

"*from the going forth of the commandment to restore and to build Jerusalem unto the Messiah the Prince shall be seven weeks, and threescore and two weeks*" (verse 25) And that is 69 weeks. (483 Hebrew years) Start counting the weeks when the commandment was given to rebuild the streets of Jerusalem.

In verse 26 God says that the Messiah will be cut off but not for himself. Furthermore it says that the people of the Prince that shall come, shall destroy the city and the sanctuary. Rome invaded and destroyed the Temple in 70A.D. (the European Union is a revival of the Roman Empire and Babylon, the U.S. will be its head)

"He shall ***confirm*** *the covenant with* ***many****"* (Daniel 9:27) ***"Many"*** as it is used here, is translated from the Hebrew word **Rab.**– רב, "rab" Strong's

#H7227; Any Hebrew word can be changed into a person's name by adding a masculine or feminine ending to the word. The name of the Israeli Prime minister who negotiated and signed the Oslo Accord on September 13, 1993 was Yitzak Rabin. His name is the Hebrew word for ***"many"*** used in verse 27.

His father's name was Nehemiah Rubitzov. Nehemiah Rubitzov moved to the U.S. from the Ukraine and changed his name to Nehemiah Rabin. And that is how Yitzak Rabin got his name which has been in the Bible for over 2,500 years!

The last week of years prophesied by Daniel is also known as the Tribulation. Remember that this prophecy is about Israel, not the Church. The Church was a mystery not revealed until 32 A.D. by Jesus himself. (See ETBH chapters 26-28) The Church Age began on the 4th Feast and it will end on the 5th Feast.

The seven years of Tribulation or the 70th week of Daniel begins on the 6th Feast of Israel known as the Feast of Atonement or Yom Kippur. The Oslo Accords 1 was signed by Rabin in 1993 on Yom Kippur. At the signing ceremony Rabin shook hands with Yasser Arafat and they all said "**Peace and security**".

That treaty is identified in the Bible in three ways. First is the name of the Israeli Prime Minister who signed it. Second is the Feast day on which it was signed. Third is the pronouncement of "Peace and security" when it was signed. There you have it, three witnesses that the Oslo Accords is the ***"covenant with many"*** spoken of in (Daniel 9:27).

Therefore I believe that the 70th week of Daniel will begin on Yom Kippur with the False Prophet endorsing the Antichrist and the Antichrist making an oath to enforce the Oslo Accords for 7 years. That day they will say ***"peace and safety"*** but at sundown the Tribulation will begin with ***"sudden destruction"***. The destruction will be **sudden** and it will be worldwide.

Anyone who was warned about the Rapture, the Tribulation and the Gospel of Salvation and remained unbelieving till the time of the Rapture may recognize the event and believe. If so, the next thing to do is get out of the United States and Europe **before** sundown on Yom Kippur nine days after the Rapture, because that is when the sudden destruction begins***..."and they shall not escape"***. (1Thessalonians 5:3)

I expect for the borders to be closed after the Jihad begins after sundown. If you are in the wrong place you will probably be stuck there. The only country in the world that the Bible says will escape the Antichrist is **Jordan**. That sounds like a good place to be.

Consider the likelihood of the borders of the United States and Europe being closed to people trying to get into or out of any country in the world. Consider the likelihood of check points and random searches, any time, any place and for any reason!

Instant I.D. technology will be used to identify anyone, anywhere any time! Also consider that Americans have been betrayed by our secretly Luciferian / Marxist leaders who have been plotting against American citizens and Christian's in particular. They have been intentionally importing Muslim terrorists to create chaos and destruction for their purpose.

Importing terrorists is part of the "Ordo Ab Chao" doctrine which is Latin for Order out of Chaos. Which refers to the plans to build the New World Order on the ruins of the Old World Order. To 33rd degree Masons, the old world order is based on Christian principles. However, the New World Order will be based on Masonic principles and their Babylonian religion. There are a number of secret societies all working together on this common goal. John F. Kennedy made a speech about this threat and what he was going to do about it just days before he was assonated. Here is a link to that speech.

https://www.youtube.com/watch?v=BHCPj7VDX6o

The Marxists would import and supply Muslim terrorists to do the dirty work for them. The Marxists will control the U.N. and disarm the public while pretending to be on their side. The Muslims would exterminate the defenseless Christians while the Marxists in control would act like there is nothing that they can do to stop it. They will just set up the situation, instigate the response and then just sit back and let it happen! This has been happening in Africa and the Middle East for decades. Those were all beautiful cities that were safe, clean, hospitable, educated and prosperous in every way. But now all of that has changed!

The Christians allowed Muslims to immigrate to their countries and the Muslims rose up against their host country and took them by surprise, with the help of the Marxists.

During the Bush administration, photos of Muslims being tortured in the Guantanamo Bay Naval **Base** circulated around the world. The pictures were in the news in Muslim countries where their religion and law requires them to indiscriminately punish non-Muslims for such behavior. The Muslim's from Eastern countries want to have vengeance against the citizens of the United States, Europe and their allies who participated in the Middle East conflicts that have killed so many Muslims. Our leaders have brought into our country the very people who we have been "fighting" against. Many believe that these people are angry and are here with plans to kill us.

No doubt that Satan has been working on hiding the greatest sign from God of all time with misinformation! I fully expect for the Antichrist to take credit for the Rapture and say that he was just getting rid of the most undesirables. Perhaps there will be a U.F.O. connection explanation. We certainly have been conditioned for it.

"Howl ye; for the day of the LORD is at hand; it shall come as a destruction from the Almighty". (Isaiah 13:6 – 13)

41

Something For World Leaders To Think About

41.1 God's "Last Days" Message To The Rulers Of Israel!

Before Tribulation, God Uses Gentiles To Warn Israel.

"For with stammering lips and another tongue will he speak to this people." (Isaiah 28:11)

This is not a flattering way to speak about Gentiles, but God is not flattering the Jews either…(Isaiah 28:14-18)

"Wherefore hear the word of the LORD, ye scornful men, that rule this people which is in Jerusalem" (Isaiah 28:14)

"In the law it is written, With men of other tongues and other lips will I speak unto this people; and yet for all that will they not hear me, saith the Lord" (1Corinthians 14:21)

God said that he would turn to the Gentiles to spread his word for a period of time and he did. Now that time is almost up and he is about to deal directly with Israel again.

The Bible says that the Jews are looking for a sign. (1Corinthians1:22) Perhaps the Rapture will be that sign for many who see and recognize it. Christians and Jews should come together, join hands and sing praises to the Lord every New Year

God Uses Affliction To Prepare His Remnant_

"*Behold, I have refined thee, but not with silver; I have chosen thee in the furnace of affliction*". (Isaiah 48:10)

The 70th week of Daniel will be a time of purifying a remnant from Israel and also a remnant of Gentiles.

Conspiracy!

"*Wherefore hear the word of the LORD, ye scornful men that rule this people which is in Jerusalem*". (Isaiah 28:14)

"*Because ye have said, We have made a covenant with death, and with hell are we at agreement; when the overflowing scourge shall pass through, it shall not come unto us: for we have made lies our refuge, and under falsehood have we hid ourselves*": (Isaiah 28:15)

Satan has just about everyone deceived! Smart people, dumb people, powerful people, weak people, sophisticated people, crude people, timid people, aggressive people, Jews, Gentiles and people just like you and me.

Only God's word can discern between good and evil.

Betrayal!

God says that Israel has conspired with the wrong people! God says their Conspiracy will fail terribly because the Antichrist will be indwelt by Satan after 3 ½ years. The Antichrist will betray Israel in the middle of the Tribulation!

One third of the Jews living in Israel will be miraculously protected by God in the city of Petra Located in the Kingdom of Jordan. The other two thirds will fall into the cruel hands of the Antichrist. Worldwide, only one tenth of the Jews will survive the last half of the Tribulation. Among the Gentiles, the death rates will be much higher. (Isaiah 28:17)

God is about to humble the whole world, **<u>including Israel</u>**! God will keep his promises to Israel and a remnant will be saved and possess the land forever, <u>all of it</u>!

"The hail shall sweep away the refuge of lies, and the waters shall overflow the hiding place". (Isaiah 28:17)

God is going to judge Israel in the middle of the tribulation. That is when the Antichrist will betray Israel. Their conspiracy will not save them, **<u>but God will</u>** as soon as they change their attitude about him.

For the first 3 ½ years of the Tribulation, those professing to be Christians will be singled out as primary targets. Gradually the persecuted list will include other non-Muslim minorities, but not the Jews or the Catholics.

The Jews and Catholics each have their own agreement with the Antichrist which offers a false sense of safety for 7 years. In the middle of the 7 years, the Antichrist will betray Israel, the Vatican and everyone else.

"And your covenant with death shall be disannulled, and your agreement with hell shall not stand; when the overflowing scourge shall pass through, then ye shall be trodden down by it". (Isaiah 28:18)

TERROR!

"For thus saith the LORD; We have heard a voice of trembling, of fear, and not of peace". (Jeremiah 30:3)

The last 3 ½ years of the tribulation will be the absolute worst time of all human history. By the Middle of the 7 years, The Muslims will have total control and terror will max out! The Antichrist will betray the Jews, In the middle of the "week", the 70th Week of (Daniel 9:27)

"Ask ye now, and see whether a man doth travail with child? wherefore do I see every man with his hands on his loins, as a woman in travail, and all faces are turned into paleness"? (Jeremiah 30:6)

This is, I believe, a description of the aftermath of a Muslim takeover of Jerusalem in the middle of the Tribulation. Muslims have a history of torturing and mutilating their victims in the name of Allah. They either kill or enslave their captives.

"I will go and return to my place, till they acknowledge their offence, and seek my face: ***in their affliction they will seek me early****".* (Hosea 5:15)

"Therefore hath ***the curse*** *devoured the earth, and they that dwell therein are desolate: therefore the inhabitants of the earth are burned,* ***and few men left****".* (Isaiah 24:6)

There will be few men left because Muslims kill the men and take the women as slaves. After Jesus Christ returns, there will be one surviving male for every seven females. Male slaves are made into eunuchs.

"Howl ye; for the day of the LORD is at hand; it shall come as a destruction from the Almighty" (Isaiah 13:6)

Salvation Is In The Name Of Yeshua

Yeshua literally means "Yahweh [the Lord] is Salvation".

"And it shall come to pass, that whosoever shall call on the name of the LORD shall be delivered: for in mount Zion and in Jerusalem shall be deliverance, as the LORD hath said, and in the remnant whom the LORD shall call". (Joel 2:32)

The English translation of Yeshua is Jesus.

41.2 The United States Is The "Little Horn"

The Bible warns us about a time called the seventieth week of Daniel. Some call it "the great tribulation". This is when an Islamic Antichrist will deceive, conquer and control the world for **seven years.**

The Koran and the Hadiths speak of a twelfth Imam. He will take over the West first with false promises of "***peace and safety***". (See ETBH Ch. 35-38) Eventually, he will take over the whole world to a degree. Some of the world will be under his total command and some of it not so much. Russia, China, Persia and their allies will maintain some of their autonomy but they will all have to worship the Antichrist and receive his mark or be killed.

The Oslo accords signed by **Yitzhak Rabin**, the fifth Prime Minister of Israel, is between Israel and the Palestine Liberation Organization with Yasser Arafat signing for the P.L.O. and Bill Clinton as the sponsor and witness. On September 13, 1993 in front of the White House; they conducted the signing ceremony; shook hands and said… "**peace and security**". That is the same as saying "***peace and safety***" (1Thessalonians 5:3)

The Antichrist will promise to enforce or "***confirm***" the Oslo Accords or "*the covenant with* "***many***" for seven years or "*for one week*".

The Oslo accords was named 2600 years ago in (Daniel 9:27)as … "*the covenant with* ***many***"*:* The Hebrew word translated as "***many***" in this 2600 year old verse is the Hebrew word "rab". In Hebrew, you can put a masculine or feminine suffix on a word and turn it into a person's name. In this case, Rabin is the masculine name form for the Hebrew word "rab" which means many in Hebrew.

In November 1995, Rabin was assassinated because of the peace process which fizzled shortly thereafter. The Oslo Accords, to this day, has never been enforced or **confirmed**.

"And he shall ***confirm*** *the covenant with* ***many*** *for one week: and* ***in the midst of the week*** *he shall cause the sacrifice and the oblation to cease, and for the overspreading of abominations he shall make it desolate, even until the consummation, and that determined shall be poured upon the desolate".* (Daniel 9:27)

The Antichrist will be the one who will promise to enforce the Rabin Treaty for **seven years**. In the **middle** of the **seven years** of Tribulation

(70th week of Daniel) the Islamic Antichrist will betray Israel and take it over militarily. He will betray Israel in the middle of the 7 **year treaty/ guarantee.**

Remember, when they signed the Oslo Accords, they shook hands and said "**peace and security**". That is the same as saying ***"peace and safety"***

(1Thessalonians 5:3) *"For **when** they shall say, **Peace and safety; then sudden destruction** cometh upon them, as travail upon a woman with child; and they shall not escape".*

I expect for Temple construction to begin with the **Jewish New Year** and to only require three days to complete.

The ***"sudden destruction"*** spoke of in (1Thessalonians 5:3) may be the result of a coordinated global Jihad response to the third Temple being built. I believe that it will. Nine days after the Rapture and nine days after the Temple construction began, they will be saying "peace and security" on Yom Kippur.

(2Thessalonians 2:3) *"Let no man deceive you by any means: for that day shall not come, except there come a **falling away first**, and that man of sin **be revealed**, the son of perdition";*

On Yom Kippur the antichrist will be revealed to the Muslim world as the twelfth Imam, their great Muslim Messiah. To the non-Muslim world, he will promise peace.

*"And I saw, and behold a **white horse**: and he that sat on him had a bow; and **a crown was given unto him**: and he went forth conquering, and to conquer".* (Revelation 6:2)

It seems, by all indications, that the United States is the Nation or "crown" referred to in (Revelation 6:2)...***"a crown was given unto him"***. (See ETBH Ch. 37-38) Dr. Hank Lindstrom believed that the Antichrist would come out of the United States of America and so do I.

As a summation of earlier discussions in this book on this subject, the ***"white horse"*** is a reference to Islam. Mohamed's horse that he is said to have rode to heaven is named "Lightening". That is Buraq in Arabic and it is pronounced the same as the illegal alien that used to occupy the White House.

The ***"crown"*** is referring to the control of the United States being ***"given"*** over to the Antichrist. The rider is holding an empty bow. He is showing that it has no arrows to indicate that although he is armed, he promises not to use it.

With promises of peace, the Antichrist gains control of the U.N. and the military of the United States and executes his own plan just after sundown on Yom Kippur, nine days after the rapture.

One of these years in the near future, nine days after the Rapture, on Yom Kippur, just after sundown is when I believe the ***"sudden destruction"*** that is spoken of in (1Thessalonians 5:3) will begin. (See ETBH Ch. 41)

I suspect that the Antichrist may be involved in the dividing up of Europe into the ten regions spoken of in Daniel 7. Nevertheless, plans to divide up the planet into units of ten have been evolving since before the French Revolution.

*"After this I saw in the night visions, and behold a **fourth beast**, dreadful and terrible, and strong exceedingly; and it had great **iron** teeth: it devoured and brake in pieces, and stamped the residue with the feet of it: and it was diverse from all the beasts that were before it; and it had **ten horns**".* (Daniel 7:7)

*"I considered the horns, and, behold, <u>there came up among them</u> **<u>another</u>** little horn, before whom there were <u>three of the first horns plucked up by the roots:</u> and, behold, in this horn were eyes like the eyes of <u>man,</u> and a mouth <u>speaking great things</u>".* (Daniel 7: 8)

Through flatteries, deception and betrayal, the Antichrist will take over the West, starting with the U.S. The Bible says that ***… "a crown was given unto him"*** (Revelation 6:2).

The ten horns of (Daniel 7:7) are the ten regions of the Roman Empire that became known as Europe. Presently, Europe is not divided into the 10 regions of the Western Roman Empire The boundaries have changed many times over the centuries. Sir Isaac Newton believed that old Roman boundaries would be restored that divided the Roman Empire into ten regions in the "last days".

Out of those original ten came "***another*** *little horn", or an eleventh nation.* That nation is described as being of European decent; a younger nation but much more powerful than all of them combined; a global superpower.

That eleventh nation can only be the United States. We are a young superpower nation that came out of Europe. (See ETBH Ch. 35-38) The Antichrist will be in full control of our military. He will use it to take over three of those European countries, probably the "G3". When that happens, the other 7 nations will capitulate without a fight.

The "G3" countries are France, England and Germany. We already have bases in all three. Thousands of years ago the Bible accurately describes the fact that there would be three powerful and influential countries that are vulnerable to such an attack from within at this time. All of those countries have large Muslim populations that are ready for Jihad. The stage is set.

After taking over Europe and staying in Rome for a while, the Antichrist will move into the Mideast. He goes back to Iraq. He will take over the land that used to be called the Assyrian Empire. Because of that conquest, the Antichrist will be called "*the Assyrian*"...

"***O Assyrian***, *the rod of mine anger, and the staff in their hand is mine indignation*" (Isaiah 10:5)

"*I will send him against an hypocritical nation, and against the people of my wrath will I give him a charge*" (Isaiah 10:6)

God is angry because Israel has been ignoring him. Israel is trusting in their secret deals to save themselves from the coming Islamic terror. God will allow the Islamic Antichrist to bring terror to an unbelieving world and to an unbelieving Israel during this period of time called… "The Seven Years of Tribulation" or "The Seventieth Week of Daniel". "Week" means seven.

After the Antichrist takes over the U.S., he will take over three of the ten European Nations and the other seven will give up without a fight. He will stay in Rome for a while and then the Antichrist will take Iraq, Syria, Lebanon, Saudi Arabia and Egypt.

*"Therefore hath the **curse devoured the earth**, and they that dwell therein are desolate: therefore the inhabitants of the earth are burned, and **few men left**"*. (Isaiah 24:6)

The word ***"curse"*** is translated from the Hebrew word "***alah***" (Strongs H423) which has one less "L" than the Arabic form, but is pronounced the same.

Satan is the curse of (Daniel 9:11) and is identified here by the name of "alah". In the Quran, chapter nine, verse one hundred and eleven is a verse often quoted by jihadi terrorists. Quran 9:111 explains that Muslims must be willing to kill and be killed in the fight to subjugate non-Muslims because they have been purchased by Allah. Muslim obedience in the fight against non-Muslims is required in exchange for entering paradise.

There are hundreds of passages that require Muslims to fight against non-Muslims to kill or subjugate them. Chapter nine, Verse one hundred and twenty three …"Oh believers! Fight the disbelievers around you and let them find firmness in you."

God has identified Islam with the phrase… ***"the curse devoured the earth".*** That means that Islam will cover the entire world during the last half of the Tribulation and that includes Russia and China! It will affect men 7 times more than women! Seven times more men will be killed than women! Islam kills the men and takes the women as "possessions of their right hand" (slaves) according to the Koran.

After the tribulation ends, estimates of human survival into the Millennium, after believers are separated from unbelievers, is from 20 to 100 million worldwide. Out of eight survivors, only one will be male. (See ETBH Ch. 6.17)

(Zechariah 5: 3) "*Then said he unto me, This is the* ***curse*** *that goeth forth* ***over the face of the whole earth****: for every one that stealeth shall be cut off as on this side according to it; and every one that sweareth shall be cut off as on that side according to it".*

The Hebrew word for "***curse***" is "alah" in verse 3 above. This passage is talking about **Islam taking over the entire planet** during the last half of the Tribulation. That is three and a half years or 42 months of Sharia law to the max.

Beheading is more identifying evidence that Islam is the "*curse*" spoken of in Daniel 9:11.

"And I saw thrones, and they sat upon them, and judgment was given unto them: and I saw the souls of them that were ***beheaded for the witness of Jesus****, and for the word of God, and which had not worshipped the beast, neither his image, neither had received his mark upon their foreheads, or in their hands; and they lived and reigned with Christ a thousand years".* (Revelation 20:4)

Muslims are waiting and watching for **Al-Imam Al-Mahdi** which is the Muslim Messiah. Muslims believe that he will establish the 12th caliphate and lead them to rule the world. There are prophecies written in the hadiths about Al-Imam Al-Mahdi.

An Islamic website at http://www.irshad.org/islam/prophecy/mahdi.htm has a collection of hadith regarding Mahdi…

Hadhrat Abdullah bin Mas'ood reports from the prophet who said…

"The world will not come to pass until a man from among my family, whose name will be my name, rules over the Arabs."

(Tirmidhi Sahih, Vol. 9, P. 74; Abu Dawud, Sahih, Vol. 5, P. 207; also narrated by Ali b. Abi Talib, Abu Sa'id, Umm Salma, Abu Hurayra)

Ali b. Abi Talib has related a tradition from the Prophet who informed him: "The promised Mahdi will be among my family. God will make the provisions for his emergence within **a single night**." (Ibn Majah, Sahih, Vol. 2, P. 519)

Abu Sa'id al-Khudari[(RA)] narrated that the Prophet said:

"Our Mahdi will have a broad forehead and a pointed (prominent) nose. He will fill the earth with justice as it is filled with injustice and tyranny. **He will rule for seven years**."

Hadhrat Umme Salmah[(RA)] narrates that Rasulullah[(SAW)] said… "**After the death of a Ruler there will be some dispute between the people**. At that time a citizen of Madina will flee (from Madina) and go to Makkah. While in Makkah, certain people will approach him between Hajrul Aswad and Maqaame Ibraheem, and forcefully pledge their allegiance to him".

"Thereafter a huge army will proceed from Syria to attack him but when they will be at Baida, which is between Makkah and Madina, they will be swallowed into the ground".

"On seeing this, the Abdaals of Shaam as well as large numbers of people from Iraq will come to him and pledge their allegiance to him. Then a person from the Quraish, whose uncle will be from the Bani Kalb tribe will send an army to attack him, only to be overpowered, by the will of Allah. This (defeated) army will be that of the Bani Kalb".

(**My comment on this hadith regarding Mahdi**) "God will make the provisions for his emergence within **a single night**." Now that sounds like the "***sudden destruction***" spoken of in (1Thessalonians 5:3) "**The Mahdi will rule for seven years**". The antichrist of the Bible will also rule for seven years. This is talking about the same person from two opposite perspectives. (Abu Dawud, Sahih, Vol. 2, p. 208; Fusul al-muhimma, p. 275)

(My comment on the hadith regarding Mahdi) **Once the 12th Imam of the 12th caliphate is <u>revealed</u> as such**, he will have a standing army consisting of all Muslims, everywhere in the world that believes he is the 12th Imam. Muslims are looking for the 12th Imam and eagerly awaiting to hear his orders to kill "infidels" which they have all sworn to obey with religious fervor.

(2 Thessalonians 2:3) *"Let no man deceive you by any means: for that day shall not come, except there come a* ***falling away first****, and that* ***man of sin be <u>revealed</u>****, the son of perdition";*

The "***falling away***" spoken of here is the Rapture of the Church **before** all of this killing and terror begins. The part where it says that *"the man of sin be* ***revealed****"*, I believe that the False Prophet reveals the antichrist to the Muslim world as the 12th Imam of the 12th Caliphate, nine days after the Rapture, on Yom Kippur.

Yom Kippur is when they will say "Peace and Safety". When Yom Kippur ends at sundown, the 70th week of Daniel begins. Just hours after the sunsets is when Jihad begins with darkness and power outages.

That is a lot of instant power over Muslim violence! The Antichrist has his armies already placed in every country including the United States, Russia, Europe and China.

According to Scripture, the Antichrist will have a *"crown"* given to him.

Notice that in the Quran and the hadiths, the 12th Imam is related to Mohamad; he rules for 7 years and there is the death of a ruler involved in his gaining power in "**<u>a single night</u>**". That sounds like the beginning of Jihad in the United States to me. I am afraid that someone may try to hurt our President and Vice-president.

If something happened to Trump and Pence, then Nancy Pelosi will be in charge! I could see her and the Democrats giving the country over to Barry if there were no Christians opposing them. The Rapture of the Church could play a role in removing the Christian opposition.

"The promised Mahdi will be among my family. God will make the provisions for his emergence within __a single night__." (Ibn Majah, Sahih, Vol. 2, P. 519)

Notice that the Mahdi will make his move in "**__a single night__**". That sounds like "***sudden destruction***" to me, after sundown ending Yom Kippur.

Furthermore, there is a short war that involves Syria that is miraculously won. That sounds like the war described in the Bible where Egypt leads most Arab nations against Israel and loses.

I believe that this may happen after the 10 days of Awe, at the end of Yom Kippur, after sundown. On the other hand, Dr. Lindstrom believed that the war with Egypt, Syria and most of the Arab Nations will be closer to the middle of the Tribulation. We will just have to wait and see.

41.3 Sudden Destruction!

If the Jews rebuilt their Temple for the third time, the Muslims would launch a surprise attack on the next opportunity for surprise. A coordinated Muslim attack throughout the West could create enough anarchy to request martial law and U.N. involvement. That is **an emergency that meets all of the requirements.**

Obama had a martial law plan on his White House website titled "The Board of Governors". The Board of Governors divides the United States into 10 regions. By dividing the country into 10 regions you **nullify the constitution** which is an agreement between 50 states not 10 regions.

Each region is ruled by a governor that answers to Obama. This could also be the template for dividing up Europe into 10 regions. Once this new system establishes unity between the U.S. and the 10 Kingdoms of Europe, other countries will either want to join for benefits or be forced to join to avoid consequences.

The French Revolution also divided things by 10 and chopped off the heads of dissenters. The French Revolution was the work of Free Masons and their greatest influence seems to have come from Erasmus Darwin, Charles Darwin's grandfather. Erasmus worked behind the scenes in both the American and French revolutions and is responsible for the theory of Evolution made popular by his grandson. The writings by Erasmus Darwin about evolution influenced the French Revolution. Erasmus did most of his writing during the French Revolution, and died from exhaustion about a year after the Revolution ended.

Obama's "Board of Governors" divides the U.S. into 10 regions. It may be the template for Martial Law for any country that he can destabilize internally with help from the left; the world media; Hollywood; the Globalists; the U.N. ; the Vatican and more…

Obama would have the support of the Pope and Catholics; he would have the support of the far left; the main stream News Media; the Globalists; the secret societies like Masons and the Muslims, just to name a few. These groups already control our country in more ways than can be readily noticed.

If Obama became head of the U.N. at the time when all of this anarchy breaks loose, he could gain control of the United States, while it is under martial law, Democrat control and without a constitution. He would just go with "The Board of Governors". He would secretly make the U.S. the twelfth caliphate. The Muslims are looking for the 12th Imam to establish the next caliphate and lead them in their fight to subdue all non-Muslims. The Bible describes Islam taking over the world during the Tribulation. (Isaiah 24:6)

Only the caliphate can conduct offensive Jihad. The 12th Imam will reduce the 3 options offered to "the people of the Book" down to only 2 options, convert to Islam or lose your head.

The caliphate will move from the U.S. to Rome and then to Iraq. The Antichrist/caliphate will conquer the land that used to make up the

Assyrian Empire. He will be called "***the Assyrian***" because of conquest and he will take over Iraq.

Egypt will start a war against Israel sometime during the first half of the Tribulation. This Egyptian/Arab war against Israel will happen **after the Rapture**, which marks the beginning of "*the Day of the Lord*", and **before the middle of the 7 year treaty/guarantee.**

There is a span of about 3 ½ years of possible dates for this short war to take place. There is debate about when this war occurs. I personally believe that this war may occur as soon as the end of Yom Kippur after sundown on the year of the Rapture. That would be at the beginning of the Tribulation. I don't know, I am speculating. Dr. Lindstrom believed that it would be closer to the middle of the Tribulation and he was speculating also.

If **the next Egypt/Israeli war** occurs at that time it would literally fulfill the "*sudden destruction*" prophecy. (1Thessalonians 5:3) If this Egyptian led war is actually a part of a global Muslim Jihad launched against the West, then the timing would require martial law in the United States and Europe. If Obama controls the Muslims, Catholics, the leftists, the **Secret Societies** and News Media he could easily get control of the U.N. and then the U.S. if something happens to Trump and Pence.

The antichrist will take over 3 European countries very quickly. Probably the "G3" nations which are France, England and Germany. Once again, it is interesting how there is a division of the European Union called "G3" that matches the number that the Bible says will be taken over first. It is easy to see how those 3 nations could be taken because our military is already based in those countries. Keep in mind, Daniel was written about 2600 years ago.

The seven countries that make up the remainder of the Western territory of the Ancient Roman Empire will surrender to the Antichrist without a fight. It appears today that they are mostly bankrupt and have no reason to resist. Although the European Union is not currently divided into 10 "Kingdoms" yet, I believe they will be soon because the Bible is never wrong.

The war against Israel will be led by Egypt. Most of the Arab Nations, will follow Egypt against Israel sometime during the first 3 ½ years of the 7 year Treaty/guarantee. I believe that the planning for that attack may already be underway. The Bible tells us the outcome.

Egypt will lead most of the Arab Nations against Israel. The Antichrist will stop them with amazing speed and power. It will be a quick war and all will be amazed. This is when the Antichrist takes Saudi Arabia, Syria, Iraq and Lebanon. This war may be what brings the Antichrist into the Middle East and to Iraq.

"He shall enter also into the glorious land, and many countries shall be overthrown: ***but these shall <u>escape</u> out of his hand,*** *even* ***Edom****, and* ***Moab****, and the chief of the children of* ***Ammon****"*. (Daniel 11:41)

Jordan will escape the Antichrist! Egypt will not escape.

"He shall stretch forth his hand also upon the countries: and ***the land of Egypt shall not escape****"*. (Daniel 11:42)

"But he shall have power over the treasures of gold and of silver, and over all the precious things of Egypt: and the Libyans and the Ethiopians shall be at his steps". (Daniel 11:43)

When the Antichrist takes Egypt, he prepares to attack Libya and Ethiopia next …

…*"But tidings out of the* ***east*** *and out of the* ***north*** *shall trouble him: therefore he shall go forth with great fury to destroy, and utterly to make away many."* (Daniel 11:44)

The Antichrist stops his advance because he fears Russia, China and he shall… *"honor the* ***god of forces****: and a god whom his fathers knew not"* (Daniel 11:38) Allah is a god of force and the antichrist will demand worship or death. God says that the antichrist will be cruel!

It appears like Putin Diplomacy works for a while, however in the middle of the 70th week of Daniel, the antichrist is indwelt by satan himself. I expect for things to happen pretty fast after that. I would expect the attempt to kill all Christians and Jews to be sudden. I would also expect for the antichrist to replace any non-Muslim world leaders with Muslims.

The method for replacing world leaders is to bribe, blackmail, coerce and kill whoever is required. I believe that world leaders who are not under the control of the antichrist are surrounded by persons who secretly are. After standing up to the antichrist and making him back down, the leader of Russia will be in the sights of satan himself.

It sounds like Putin may be willing to protect Christians. I pray that he will accept Christian refugees who will be trying to escape from the antichrist in the West. I also pray that Putin and Trump and their families will all get saved whether sooner or later, just not too late!

The whole world will be tested during the seventieth week of Daniel. But it looks like the people of the United States will get it first. The people of the world will look on with horror as the Islamic terror spreads anarchy, starvation, war and disease throughout the West.

About 1.95 Billion will die worldwide during those first 3 ½ years of the 7 year Treaty guarantee. I believe that a disproportionate amount of that 1.95 Billion deaths will be in the United States and Europe. Will Russia, Jordan, China, Israel and other countries accept Christian/American refugees? I am praying that they will and other countries as well.

People around the world will see the horrors faced by the American people during the Tribulation and be glad they are not here. Those in America may not be able to get out because the Bible says..."*and they shall not escape*". (1Thessalonians 5:3)

For those who will be left behind, I think that there will only be about 9 days after the Rapture that escape from the West is likely. After Yom Kippur, sundown the 70th Week of Daniel begins with *"sudden destruction"*. I believe that it will be very difficult to escape America after Yom Kippur following the Rapture. The borders may be closed and our border wall will be keeping people in.

The Antichrist will move from Rome to Babylon which is modern day Iraq. That is where it all started with the Tower of Babel in the heart of Iraq.

The U.S. is going back to Iraq. That is where Daniel's Vision occurred. He saw an image made of 5 different materials standing in Iraq. The feet were made of part iron and part clay. The United States of America is the feet of iron and clay. In other words, there is going to be division in the kingdom. There will be the conflict between Islam and the non-Muslims. Globalists will stirrup race wars and anarchy. Perhaps worldwide, but definitely in the U.S, Europe and Canada. The Antichrist will be given a crown. Dr. Lindstrom taught that the U.S. would be that crown.

41.4 The Antichrist Will Be Cruel.

The things that God calls evil, the Antichrist and satan call good. The things that God calls good, the Antichrist, satan and their minions call evil, they are opposites! God calls satan *"the god of this world"* and *"the father of lies"*. He is the enemy of his own Creator. Think of it as disorder attacking Order.

Our Creator Yeshua is the God and Creator of all of the Universe in human flesh. He is the God of Abraham, Isaac and Jacob. He talked to Moses through a burning bush. He claims to be God, and Lucifer was his Creation. Lucifer acquired the ability to refashion himself and others against his Creator and sin entered the world. It was Disorder fighting against Order and it still is today.

Christians and Jews from the U.S. and Europe will face persecution unlike anything ever seen before. There will never be another time like it again. (Matthew 24:21) God says that the ***Assyrian*** will be ***cruel!***

"And the Egyptians will I give over into the hand of a ***cruel lord****; and a* ***fierce king*** *shall rule over them, saith the Lord, the LORD of hosts".* (Isaiah 19:4)

When God calls someone ***"cruel"*** and ***"fierce"*** then you know that he is going to really be bad!

The Antichrist will be Cruel! Egypt is mentioned by name and will suffer greatly under the Antichrist.

*"And the Egyptians will I give over into the hand of a **cruel** lord; and a fierce king shall rule over them, saith the Lord, the LORD of hosts".* (Isaiah 19:4)

When God calls someone cruel, you can be sure that he will be very cruel!

41.5 Trump

There is no doubt in my mind that God is using Trump, whether he is saved or not. Just because he is known as a womanizer, doesn't mean that he did not get saved as a child. I pray that he is.

The Bible speaks of a restrainer being taken out of the way at the rapture. (2Thessalonians 2:7) Trump certainly appears to be a restrainer of sorts. I don't believe that this passage is specifically speaking about Trump. I believe that it is talking about the Holy Spirit working through believers. I don't know if Trump is saved or not, but it looks like Trump is taking on all of the evil in the world by himself.

I really don't know what to think about Trump, but he and his supporters appear to be standing in the way of Obama, Pope Frances, the New World Order and their climate change/population reduction agenda . Obama and Pope Frances Promoted the U.N. Sustainable Growth Agenda 2030 from the very beginning. This is the one that replaced Agenda 21. The U.N. is pushing the global warming hoax along with other hoaxes to justify a totalitarian world government to reduce the world's population.

God called King Nebuchadnezzar "*my servant*" while he was still an unbelieving pagan. Nebuchadnezzar got saved years later through Daniel's testimony, prayers and God's intervention. The point being, if Trump and/or Pence are not saved at the Rapture, then all of the Christians who have been watching their backs and praying for them will be gone. The Bible says that evil will take over during the Tribulation. You won't be able to trust anyone.

Trump represents Western Civilization's last chance at survival and it doesn't appear like our leaders are giving him the support that he is entitled to. The loyalty of those around our President is questionable at best.

Trump is opposing two powerful men in particular who have their minions working all over the government. Trump is dismantling Obama Care and reversing some of Obama's legacy. Part of that legacy is the global warming hoax. Pope Frances is in on the hoax and Trump is telling the Pope that he is wrong. Trump has publically disagreed with the Pope on climate change, calling it a hoax, on several occasions.

It appears to me that Trump is holding back both the antichrist and the false prophet. I don't know if he is saved or not, but it sure looks like God is using Trump and his supporters to slow down this evil that has taken over our beloved country.

I know that it looks like Trump is going to keep America great, but the Bible says otherwise…

(2Timmothy 3: 13) *"But evil men and seducers shall wax worse and worse, deceiving, and being deceived"*

(Mica : 7:3) "*That they may do evil with both hands earnestly, the prince asketh, and the judge asketh for a reward; and the great man, he uttereth his mischievous desire: so they wrap it up*".

Consider what would happen if the Rapture were to happen, let's say for example, this year (2020). <u>If</u> Trump and Pence are both saved, they will both be raptured and the speaker of the House becomes President. If either one or both of them is not saved, and is left behind at the Rapture, they will no longer have believers watching their back. The radical left and Muslims will be free to do as they wish and they will be cruel. I doubt that either

Trump or Pence will have much time to figure out what happened before the Islamic left gets to them.

This is a good time for our leaders to talk to God and get saved and make sure that your families are also saved. Satan is the father of lies. If you try to use the "Art Of The Deal" on satan you will fail. Satan and his followers will destroy everyone who deals with them.

The Quran speaks of a leader being killed and civil war breaking out in the end times. Some of the Hadiths comment on that…

Hadhrat Umme Salmah(RA) narrates that Rasulullah(SAW) said… "**After the <u>death of a Ruler</u> there will be some <u>dispute between the people</u>**".

This either sounds like Trump and Pence get raptured out along with the rest of us or a coup by assassination. A leader is killed so that someone else can take over. It also sounds like the people are going to fight about who is responsible for the leader's death or disappearance and who will replace him. This sounds like a coup, anarchy and revolution.

The Democrats, ANTIFA, the radical left, the controlled media; Hollywood pedophiles; Obama appointed bureaucrats; the Vatican and other religious leaders; radical groups and Muslims want Trump out so much that they are willing to lie and break the law to frame and impeach the President. I am afraid that these evil people, the globalists, will do <u>anything</u> to stop Trump from restoring our republic.

If Trump and Pence are trusting in what Jesus did for us all, on the Cross of Calvary 2,000 years ago, they will be raptured with the rest of the born again believers and miss the horror of the 70th week of Daniel, If either one is trusting in their works, then they are not saved and will be left behind to face the Antichrist.

If they die without trusting in Jesus <u>alone</u>, then they will end up in hell. All of those who are left behind must change their minds about who Jesus is and what he did for them on the Cross of Calvary. You must believe God's record about his Son to be born again. There are plenty of false records that will not save but there is only one record that is truly from God and it is found in the Bible.(John 3:16)

There is trouble coming! These ANTIFA groups that travel around the world stirring up trouble in country after country. They are well funded and organized. The Globalist plan is to create a utopia for themselves and exterminate anyone they don't need. Jews and Christians are specifically targeted by the Quran.

The Catholic Church will be exempt from persecution at first. In the middle of the Tribulation that will change. Persecution will extend to all non-Muslims who refuse to worship the antichrist and receive his mark, that includes ANTIFA and all of the radical left that help the Muslims come to power.

Judeo/Christian Refugees will be looking for a new place to live to escape the persecution, bloodshed and terror.

Countries who welcome Christian and Jewish refugees into their country will be making a good decision. Christians and Jews believe in good work ethics, God and family. God will reward those who protect Christians and Jews. Perhaps the countries who offer refuge to Christian and Jews may not have to endure all of God's judgements.

41.6 Jordan Will Be Safe From The Antichrist

King Abdullah II will have to decide whether or not to allow the Jews to escape from the antichrist into Jordan. This happens in the middle of the Tribulation, he decides to accept them. Perhaps he will also accept other believers as well.

The tiny country of Jordan contains the only real estate mentioned by name in the Bible that God says will escape the Antichrist! God will protect that tiny nation because it will be a refuge for the remnant of Israel in the middle of the Tribulation.

Jordan could also be a refuge for Christians trying to escape the Antichrist. People will be trying to escape the United States and Europe to get away from the Antichrist during the Tribulation. Jordan sounds like a nice place to go because ***"Edom, Moab, and Ammon"*** **will escape the Antichrist!**

Real-estate in Jordan will be precious during the Tribulation!

"He shall enter also into the glorious land, and many countries shall be overthrown: but ***these shall escape out of his hand, even Edom, and Moab, and the chief of the children of Ammon****".* (Daniel 11:41)

Wow! No other country was mentioned by name to escape the antichrist! The real estate value is going to take a big jump in Jordan. Unfortunately, Jordan is a Muslim country and that may be a challenge for Westerners.

The rock city called Petra is also in the mountains of Jordan and that is where the Jews are told to escape to in the middle of the Tribulation. God will be protecting the tiny country of Jordan during the Tribulation!

"***When*** *ye therefore shall see the* ***abomination of desolation****, spoken of by Daniel the prophet, stand in the holy place, (whoso readeth, let him understand :)"*(Matthew 24:15)

"*let them which be in Judaea flee into the mountains:*" (Matthew 24:16)

This is talking about the mountains of Jordan where the rock city of Petra is located. Jordan is where Edom, Moab and Ammon are located and they will escape the Antichrist.

One third of Israel will follow these directions and be protected by God. Two thirds will not. They will be killed by the Antichrist in cruel fashion to spread terror throughout the world. This is said to be a time of refining Israel and the world for the Kingdom.

"Let him which is on the housetop not come down to take any thing out of his house: (Matthew 24:17)

"Neither let him which is in the field return back to take his clothes" (Matthew 24:18)

This sounds to me like those who appear to be field workers will escape. But anyone carrying something will be stopped. That includes, anyone carrying children. Children should be evacuated <u>before</u> things get this bad.

Believers should pray for King Abdullah II, he needs Jesus just like every other human being. We will all be judged by God. Every human being needs God's righteousness which He freely gives to all those who believe the record that God gave of his Son.

(1John 5:11) *"And this is the record, that God hath given to us eternal life, and this life is in his Son"*

41.7 Russian President Vladimir Putin;

President Putin is in charge of the 2nd most powerful Military power on the Planet. Russia has made a lot of weapons' that no one wants to be on the receiving end of, but **where will Russians spend eternity?** Statements made and actions taken by President Putin indicate that he may be sympathetic with Christian values.

Many Russians call themselves atheist, agnostic or Russian Orthodox. Russians like everyone else needs to ask themselves, "were we Created for a purpose or are we the result of meaningless accidents that all just happened to come together at the right moments"?

The Bible has accurately told the future by naming Russia's capital cities and a lot more, so consider what God has to say about Russia's future.

During those first three and a half years of the Tribulation, Christians will be looking for a place of refuge. I believe that God will bless any Nation, leader or person that offers refuge to God's people who are fleeing the Antichrist during the Tribulation. Russia will oppose the Antichrist during the first three and a half years of the Tribulation.

In the middle of the Tribulation, the Antichrist will be indwelt by satan. From that time on things really get bad! Any world leaders that have managed to hang on will be overthrown from within. At this point, if a refuge has not already been established in north eastern Siberia, there will be no place for Russians to escape to.

Russia Is First Mentioned In The Bible. Long before Russia existed, Ezekiel talks about Russia using the ancient Hebrew name that is still used today. The location of the capital cities were also **described. Long before they existed or anyone ever said the word "Moscow".** Just remember that God said it first. The Bible says that Russia is in the Uttermost North part of the earth relative to Jerusalem.

Thousands of years **before** they existed the Bible gives the names of Russia's top two cities, Meshech and Tubal (Moscow and Tobolsk). The city of Moscow is named and described as being directly north of Jerusalem **before the city even existed**! Moscow is directly north of Jerusalem. The Modern Hebrew word for Russia used today is the same one used in Ezekiel 38 over 2600 years ago.

More than a century ago, **The Old Scofield reference Bible copyrighted in 1909** has this note on page 883, footnote 1(Ezekiel 38:2)...

..."That the primary reference is to the Northern (European) powers, headed up by Russia, all agree. The whole passage should be read in connection with (Zechariah 12:1-4; 14:1-9); (Matthew 24:14-30); (Revelation 14:14-20; 19:17-21). "Gog" is the prince, "Magog", his land. The reference to Meshech and Tubal (Moscow and Tobolsk) is a clear mark of identification. Russia and the Northern powers have been the latest persecutors of dispersed Israel, and it is congruous both with divine justice and with the covenants (Genesis 15:8), note; (Deuteronomy 30:3) note that destruction should fall at the climax of the last mad attempt to exterminate the remnant of Israel in Jerusalem. The whole prophecy belongs to the yet future "*day of Jehovah*" (Isaiah 2:10-22); (Revelation 19:11-21), and to the battle of Armageddon (Revelation 16:14; 19:19) note, but includes also the final revolt of the nations at the close of the kingdom-age (Revelation 20:7-9)".

The Old Scofield reference Bible copyrighted in 1909

Today we can look at maps to understand the location of Russia. Twenty Six Hundred years ago, when Ezekiel 38 was written, maps did not exist. People usually did not travel far from home. But the Bible talks about Russia and describes it as being the "*Uttermost*" north of Jerusalem.

If you were to draw a straight line connecting Jerusalem with the North Pole, that line would go through Moscow. What are the chances of that happening? God also mentions Russia's future. As I have demonstrated in previous chapters, Bible Prophecy is detailed, precise and has a 100% accuracy record! Only God can do that! Somebody needs to start preparing a place in Siberia.

41.8 Armageddon Will Destroy Five Sixths Of Russia!

The Bible says that the Antichrist will take over Syria. I know that is bad news after all of the money and hard work that Russia has put into that place, but the Bible is never wrong. **Damascus will be destroyed <u>completely</u>**. That is where the attacks against Israel came from.

"The burden of Damascus. Behold, Damascus is taken away from being a city, and it shall be a ruinous heap" (Isaiah 17:1)

The problem is that Assad let Hamas use his country to stage rocket attacks on Israel. God warns…

…"And I will bless them that bless thee, and curse him that curseth thee: and in thee shall all families of the earth be blessed". (Genesis 12:3)

After the Egyptian war against Israel; after 3 ½ years of the antichrist increasing his powerbase; in the middle of the Tribulation; there will be a war in heaven and satan will be cast down to the earth. This is when satan indwells the antichrist; breaks the treaty; invades Israel; and takes over and desecrates the Temple. This occurs in the middle of the "*seventieth week of Daniel*" or middle of the 7 year treaty guarantee. (Daniel 9:27) This is also when everyone will have to choose to either obey the antichrist and take his mark or run for their lives and hide.

Putin has spoken up in favor of Christian's and family values in Russia. I am guessing that President Putin won't go along with the antichrist. The Bible says that this will be a time of deception and betrayal. I believe that the antichrist will secretly use bribery, blackmail and coercion to influence the politics in other countries including Russia.

If Putin stands in the way of the N.W.O. he will be their main target. Lots of money will be used to corrupt people close to him. In the middle of the Tribulation, I believe that Putin may lose power by coup or assassination attempt and be replaced by a Muslim.

If Putin is aware of God's warnings and prepares a place for Christians to hide till the last 3 ½ years of the Tribulation are finished, he may also have a place for himself to escape to. The Bible indicates that Siberia is the perfect place to hide until the end of the Tribulation and Armageddon because Siberia will survive.

"*And he shall plant the tabernacles of his palace between the seas in the glorious holy mountain; yet he shall come to his end, and none shall help him.*" (Daniel 11:45)

The first part of this passage is talking about the middle of the Tribulation when the abomination of desolation, takes place. That is when the antichrist sets up his image in the third Temple.(Daniel 9:27) (Mark 13:14) The second half is talking about what is going to happen to the antichrist three and a half years later when Jesus returns to defeat him at the battle of Armageddon.

Armageddon is the final battle of this age. It happens at the end of the 7 years of Tribulation also known as the 70th week of Daniel. It lasts for one hour and destroys one third of all mankind. Once again, that number will be disproportionately distributed around the world. The United States gets it first, but Russia gets it last. Five sixths of Russia will be destroyed at Armageddon.

Until now the battle of Armageddon did not seem possible. No one believed that anyone could produce a two hundred million man army; No one believed that the Euphrates River would ever dry up; no one believed that 1/3rd of humanity could be killed in a one hour battle, but they believe it now. China has maintained the ability to field and supply a 200,000,000 man army since the 1960's. That is the same number that the Bible uses. (Revelation 9:14-16) The Euphrates River has 5 dams. Any of these could stop the river. These dams are often under the control of terrorist groups trying to extort money from local governments.

It takes about 30 minutes to send an I.C.B.M. halfway around the world. A Nuclear war could easily destroy 1/3rd of humanity in an hour.

Thousands of years ago, the Bible names Russia's allies Persia, Ethiopia, Libya, Turkey (Ezekiel 38:5-6) and China (Revelation 9:14-16). They will attack the antichrist in Jerusalem. It sound like Russia will gain F35 stealth technology through Turkey. Maybe that is why Russia will use a massive air assault.

*"Thou shalt **ascend** and come like a storm, **thou shalt be like a cloud to cover the land**, thou, and all thy bands, and many people with thee".* (Ezekiel 38:9)

*"And thou shalt **come up** against my people of Israel**, as a cloud to cover the land**; it shall be in the latter days, and I will bring thee against my land, that the heathen may know me, when I shall be sanctified in thee, O Gog, before their eyes"* (Ezekiel 38:16)

This attack is not just because the antichrist is evil, it is also because the leader of Russia has plans for personal gain. He is gaining support from other countries because nobody likes the antichrist and they all want him gone but the Russian leader wants more than just that. The Russian leader wants all the booty for himself. He wants to plunder Jerusalem and that makes God angry!

*"Son of man, **set thy face against Gog**, the land of Magog, the chief prince of Meshech and Tubal, and **prophesy against him**,* (Ezekiel 38:2)

*"And say, Thus saith the Lord GOD; Behold, **I am against thee**, O Gog, the chief prince of Meshech and Tubal"* (Ezekiel 38:3)

This passage indicates that God is angry with the leader of Russia at the Battle of Armageddon. He and his land will be severely judged by God! (Ezekiel 38:18-19)

*"Therefore, thou son of man, prophesy against Gog, and say, Thus saith the Lord GOD; Behold, **I am against thee, O Gog**, the chief prince of Meshech and Tubal":* (Ezekiel 39:1)

I don't believe this is Putin. I believe it is his Muslim replacement. Hopefully Putin will get saved and realize what is going to happen in time to escape.

*"And I will turn thee back, and leave but **the sixth part of thee**, and will cause thee to come up from the north parts, and will bring thee upon the mountains of Israel":* (Ezekiel 39:2)

"And I will send a fire on Magog, and among them that dwell carelessly in the isles: and they shall know that I am the LORD".(Ezekiel 39:6)

Only one sixth of Russia will survive! But God tells us which part will survive. God has a designated safety area for Russian believers and maybe others who believe as well. If a Russian believes the Bible then he will want to be close to North Eastern Siberia as the end of the 7th year approaches. Remember God always does what he says!

*"**I the LORD have spoken it, and I will do it**"*. (Ezekiel 36:36)

41.9 Siberia Will Survive Armageddon

The defeated Russian Army will temporarily escape to northeastern Siberia…

"But I will remove far off from you the northern army, and will drive him into a land barren and desolate, with his face toward the east sea, and his hinder part toward the utmost sea, and his stink shall come up, and his ill savour shall come up, because he hath done great things". (Joel 2:20)

Joel was written about 2800 years ago when maps were non-existent. How could Joel have knowledge about the geography of North Eastern Russia 2800 years ago unless it came from God? The North Eastern part of Russia is described well enough to identify this area as Siberia. This is where the Russian army will go to escape God's armies for the time being. Siberia will be the only place left for them to go. The rest of Russia will be destroyed. The area described is about **one sixth of Russia**.

What will happen to all of those armies? There will be millions and millions of dead and dying one hour after the battle begins. That is when Jesus Christ, His Armies and the New Jerusalem are seen by the whole world, coming down to orbit the earth.

The New Jerusalem will be a cube measuring 1500 miles on every side making it slightly larger than the moon; it will produce its own light and there will be no darkness on earth as it orbits for 1,000 years.

The Vatican has its own telescope known by the acronym of… "L.U.C.I.F.E.R. They will probably spot the New Jerusalem coming and call it an alien invasion. Jesus and his armies will be attacked by the world's armies. NASA's Asteroid Redirect Program will probably be used for that along with Trump's new Space Force.

Down through the centuries critics tried to find fault with the Bible. They would ask how could men attack Jesus coming in the air? Well now there is Space Force and NASA's Asteroid Redirect Program.

The first thing that Christ does is stop the nuclear and conventional war against the antichrist who will be ruling from the Temple in Jerusalem. You may think that standing up against a perverted, satanic psychopath is a good thing. I certainly do. But the leader of Russia "*has an evil thought*", according to the Bible. That "*evil thought*" indicates to me that he is attacking for other reasons and he probably is an unbeliever!

Jesus and his armies break up the fight just before all life on earth is extinguished. All of the nations of the earth attack Jesus Christ and his armies. The result is the destruction of the armies that were gathered against Israel. The nations will be judged by how they treated Israel, among other things. **The mass graves will fill valleys.** (Ezekiel 39:1-29)

"And it shall come to pass in that day, that I will give unto Gog a place there of graves in Israel, the valley of the passengers on the east of the sea: and it shall stop the noses of the passengers: and there shall they bury Gog and all his multitude: and they shall call it The valley of Hamongog". (Ezekiel 39:11)

"And seven months shall the house of Israel be burying of them, that they may cleanse the land" (Ezekiel 39:12).

There is a lot of death in that hour, but it looks like the north east part of Siberia will escape. That sounds like a good place to develop.

Will Russia Accept And Protect Christian Refugees? Russia might be a safe place for Christians during the first 3 and a half years of the Tribulation. After the first three and a half years of the Tribulation, Siberia will be the only safe place in Russia, because Islam will eventually take over the world during the last half (3 ½ years) of the Tribulation, including Russia. Finally, only the north eastern portion of Siberia will survive Armageddon.

41.10 Xi Jinping, President Of The People's Republic Of China

In all of history no one ever believed that China could produce a 200,000,000 million man army, but it can now. Since the sixties, the

Chinese have maintained the ability to field a two hundred million man army. The same number given in (Revelation 9:14-16 & 16:12)

China will send a 200,000,000 man army across a dried up Euphrates River. (Revelation 9:14-16 & 16:12)

The Future Of Your 200,000,000 Man Army is God's judgement. Those who have put their trust in Jesus Christ as their personal Savior will be saved. Those who rejected Jesus will be lost.

Jesus is the Savior of the world, who was crucified, died, buried and three days later rose from the dead as a payment for all of the sins of all of humanity. The moment that you believe, you are saved forever whether or not you tell anyone.

China should consider accepting Christian refugees from wherever they may come from.

42

"Howl Ye; For The Day Of The Lord Is At Hand!"

42.1 Men Will Seek God Because Of Fear.

People generally don't think about death all the time, but when they become very ill or have a near death experience, that changes. God knows this. We all know that men have an appointment with death. Sometimes there is a warning like prolonged illness and sometimes it is the result of a sudden accident, without warning. Throughout our lives we know that our turn will come someday.

God warns us in many ways. God knew from the beginning that someday, He would have to clean up planet earth. God provided prophets who wrote down things long ago. Things that men could not know at the time that it was written, things like the future, science and technology.

God has warned us about sin, righteousness and God's judgement. Time is about to run out. Jesus will return seven years after the Rapture to stop Armageddon; to save Israel and set up His Kingdom.

During those seven years men will be in fear of dying and forced to make a decision about who Jesus is and what He did for them. This will be a seven year battle between God and Satan over the hearts, minds and souls of men.

The Church will be Raptured out of this world, nine days before the Tribulation begins. During those nine days satan will use false prophets with lying wonders to deceive the world. The Antichrist and False Prophet will use fake miracles as though they were from God.

God will counter satan's minions with two Old Testament witnesses who will perform real miracles from God; and one hundred and forty four thousand witnesses from Israel to preach the Gospel throughout the whole world; and twenty eight judgements from God! After all of that, some will believe, but many will rebel and die without hope.

The 70th Week of Daniel will be last seven years of the dispensation of law! The Tribulation will be seven years of genocide and terror, climaxing at Armageddon. Nothing in all of human history can be compared to it. All of the goats must be separated from the sheep before the Messianic Kingdom can begin. God will use 144,000 young Jewish men to warn the world. The constant threat of death from starvation, disease and the Antichrist will encourage men to listen to them.

*"So persecute them with thy tempest, and **make them <u>afraid</u>** with thy storm".* (**Psalm 83:15**)

*"Fill their faces with shame; **<u>that they may seek thy name,</u> O LORD**".* (**Psalms 83:16**)

I expect for the Day of the Lord to begin with widespread power outages. It just seems to fit with darkness and Jihad. I believe the antichrist may control the power grid during the first 3 ½ years but after that God will get more involved with the increasing Judgements. One thing is for certain, it will be dark from several causes and it will be very dark.

The power will probably be out and clouds of smoke and whatever else that is in the air, will obscure any light from the sun, moon and stars.

"***Woe** unto you that desire the day of the LORD! to what end is it for you? **the day of the LORD is <u>darkness,</u> and not light***". (Amos 5:18)

"As if a man did flee from a lion, and a bear met him; or went into the house, and leaned his hand on the wall, and a serpent bit him."(Amos 5:19)

This is a time of testing and trials. There are many who call themselves "Christians" but are actually referring to their accepted, Biblical based value system of good works and "moral" living. That is not the same as believing in God's Only Begotten Son as God and Savior. Everyone must understand and believe the Gospel of Grace to be born again. If not, you will be left behind at the Rapture.

I have encountered many who believe in a post tribulation Rapture. People of that persuasion usually have other doctrinal errors as well. The most dangerous error is believing in a false gospel of human merit.

(Galatians 1:6) warns us about... ***"another gospel"***. (Galatians 1:7)warns us about those who... "***would pervert the gospel***".

(Galatians 1:8) *"But though we, or an angel from heaven, preach **any other gospel** unto you than that which we have preached unto you, let him be **accursed**"*

Verse nine repeats what was just said in verse eight emphasizing the importance of not preaching "***any other gospel***". Anyone who does, is accursed. This means that **there is only <u>one</u> Gospel**, any variations are "*accursed*" by God.

(Galatians 1:11)" *But I certify you, brethren, that the **gospel** which was preached of me **is not after man***".

In other words, no one has the right to change the Gospel because it came from God, not from man. If a preacher changes the Gospel, then he has perverted it and it will not save!

(1Corinthians 1: 17) *"For Christ sent me not to baptize, but to preach the gospel: **<u>not</u> with wisdom of words, lest the cross of Christ should be <u>made of none effect</u>.**"*

God is warning us that using men's words and wisdom in the place of God's can make God's Gospel message misunderstood and "***<u>made of none</u>***

***effect.*"**. In other words, a false gospel will not save and there is a lot of them out there and they all lead to hell.

(2Corinthians 11:3) *"But I fear, lest by any means, as the serpent beguiled Eve through his subtilty, so your minds should be corrupted from the **simplicity** that is in Christ"*

The Gospel is simple. God loves us and paid for our sins on the cross and gives us eternal life as a free gift when we believe that Jesus paid for our sins in full and gives us Eternal Life. Eternal Life can't be earned and can't be lost.

(2Corinthians 11:4) *"For if he that cometh preacheth **another Jesus**, whom we have not preached, or if ye receive **another spirit**, which ye have not received, or **another gospel**, which ye have not accepted, ye might well bear with him.*

The Jesus of the Bible is God in flesh who paid for all the sins of humanity on the Cross of Calvary, once for all and forever. The Jesus of the Bible gives Eternal Life Freely to all those who believe and He will never take it back!

The Gospel is simple. Just believe in who Jesus is and what he did for us and you have Eternal Life the moment you do and forever after. It is a Gift, it is Free!

The Spirit of God seals every believer during this Age the moment we believe. This is when the believer becomes born again. This Spiritual birth is invisible, it is silent, it is peaceful and it is Eternal.

(Matthew 7:13) *"Enter ye in at the strait gate: for wide is the gate, and broad is the way, that leadeth to destruction, and many there be which go in thereat:"*

(Matthew 7:14) *"Because strait is the gate, and narrow is the way, which leadeth unto life, and few there be that find it"*

There are many false preachers who are changing the gospel. There are many false gospels that cannot save because they add human merit or question the fact that Jesus is God!

(Matthew 7:15) *"Beware of false prophets, which come to you in sheep's clothing, but inwardly they are ravening wolves".*

False prophets put on a good outward appearance but they are lost and misleading the lost. These people are living in a pseudo-Christian religious fantasy and deceiving others along the way.

(Matthew 7:22) *"Many will say to me in that day, Lord, Lord, have we not prophesied in thy name? and in thy name have cast out devils? and in thy name done many wonderful works?"*

(Matthew 7:23) *"And then will I profess unto them, **I never knew you**: depart from me, ye that work iniquity".*

These people clearly were trusting in their own good works which they did in the name of Jesus Christ. Jesus calls their good works "*iniquity*". They never believed God's Gospel of Free Grace. They were preaching a different gospel message of works because it helps with fund raising. God only saves those who are trusting to be saved, not those who are working to be saved.

(Romans 4:4) *"**Now to him that worketh** is the reward **not** reckoned of **grace**, but of **debt**".*

"Debt" means no Grace. Those who stand before God in their own righteousness will fall short of the Righteousness of God. God's standard is absolute perfection. We need the righteousness of God because… "*the wages of sin is death*" (Romans 6:23) This is talking about the second death which is the final hell.

The Gospel of "*Grace*" is the message that Eternal Life is the Gift of God, not of works and not of yourself. Eternal Life cannot be of works and Grace at the same time. When Jesus was hanging on the cross he said, "***it is finished***". If you believe that, then you have Grace and Eternal Life.

If you think that what Jesus did on the cross was not good enough; if you think that you have done or will do something additional to ensure that you are saved; then you have not believed the Gospel of Grace, you are mixing your "*dead works*" with God's perfect work of Grace.

(Romans 11: 6) *"And if by grace, then is it no more of works: otherwise grace is no more grace. But if it be of works, then is it no more grace: otherwise work is no more work".*

"*no more grace*" means no salvation because dead works will not save. Only one thing can save and that is Jesus Christ alone through just faith alone, not of yourself or anyone or anything else.

The Pharisees had a false religious system of laws, rules and works that they pretended to keep to get into heaven. They did not keep the rules themselves, but they pretended to. The Pharisees had a merit based false system for earning their way to heaven. But it was all a lie. It was a "heavenly way" to hell. It was the opposite of what it pretended to be.

(Matthew 16:12)*"Then understood they how that he bade them not beware of the leaven of bread, but of the* ***doctrine*** *of the Pharisees and of the Sadducees".*

(Luke 12: 1) *"Beware ye of the leaven of the Pharisees, which is* ***hypocrisy****".*

The "*leaven of the Pharisees*" is hypocrisy or pretending to be better than you are. Thinking that you are good enough to enter Heaven based on your efforts will take you to hell. That is what the Pharisees were preaching works for salvation.

(Galatians 5: 9) *"A little leaven leaveneth the whole lump"*

A person may call themselves a "Christian", do good works and even sing in the choir, but if one thinks that their behavior has anything to do with their salvation then they are trusting in their works and are not saved and are going to go through the Tribulation if they don't change their mind first. If they die believing that their works helps them to become or stay saved, then Jesus will say… *"I never knew you: depart from me, ye that work iniquity"* (Matthew 7:23)… and they will be cast into the Lake of Fire.

The Bible says that the works of those who are working to be saved are "***dead works***".

"Therefore leaving the principles of the doctrine of Christ, let us go on unto perfection; not laying again the foundation of repentance from ***dead works****, and of faith toward God,"* (Hebrews 6:1)

"How much more shall the blood of Christ, who through the eternal Spirit offered himself without spot to God, purge your conscience from ***dead works*** *to serve the living God?"* (Hebrews9:14)

The Bible says that "*many*" will be deceived into trusting in their own works and not in Christ alone! They are preparing for what they think the Tribulation will be like, but once again, where you find doctrinal error, it usually is accompanied by other error.

Some of these people think that they are ready. But they have been misled. Many are storing up gold, silver, guns, ammo and supplies to barter with. They just don't understand what is about to happen. God understands and tells us if we are listening.

"Your gold and silver is cankered; and the rust of them shall be a witness against you, and shall eat your flesh as it were fire. Ye have heaped treasure together for the ***last days****"*. (James 5:3)

"Neither their silver nor their gold shall be able to deliver them in the day of the LORD'S wrath; but the whole land shall be devoured by the fire of his jealousy: for he shall make even a speedy riddance of all them that dwell in the land". (Zephaniah 1:18)

I have talked to preppers who believe that they will do just fine because they think that they are ready. They envision a scenario similar to Hollywood's rendition of the Tribulation. By the middle of the Tribulation gold and silver will be worthless and dangerous to possess. When the dollar collapses they believe that their gold and silver will buy what they need, but the Bible says…

"***They shall cast their silver in the streets**, and their gold shall be removed: their **silver and their gold shall not be able to deliver them in the day of the wrath of the LORD**: they shall not satisfy their souls, neither fill their bowels: because it is the stumblingblock of their iniquity*". (Ezekiel 7:19)

I have talked to many people who are "getting ready" for what they think is ahead. They act like it is going to be a big hurricane party. If they don't know what the Bible says, they cannot imagine what is about to happen. The 7 years of Tribulation is unlike anything that the World has ever seen nor will ever see again.

After the 70th Week of Daniel and Armageddon are past and the Angels gather up all of the unbelievers; about 99% of the world's population will be destroyed. 25% will be destroyed during the first 3 ½ years and it continues to escalate and climax at Armageddon. In one hour, one third of humanity will be destroyed at Armageddon. After that the Angels gather all of the unbelievers who are left at the end of the Tribulation and cast them into the Lake of Fire.

During that seven years many will be terrorized and lied to and treated with cruelty till they are killed by the Antichrist. The Antichrist will win until Jesus Christ returns to defeat him at the end of the 7 years.

*"Woe unto you that desire the **day of the LORD**! to what end is it for you? the **day of the LORD** is **darkness**, and not light."* (Amos 5:18)

"As if a man did flee from a lion, and a bear met him; or went into the house, and leaned his hand on the wall, and a serpent bit him". (Amos 5:19)

*"Shall not the **day of the LORD** be **darkness**, and not light? even very dark, and no brightness in it?"* (Amos 5:20)

I believe there will be widespread power outages for extended periods of time during the tribulation.

*"Behold, the **day of the LORD** cometh, **cruel** both with wrath and fierce anger, to lay the land desolate: and he shall destroy the sinners thereof out of it."* (Isaiah 13:9)

Cruelty toward non-Muslims is a sign of religious piety among Muslims. When Muslims attack non-Muslims in Europe, they all swarm their helpless victim as a show of Religious piety and fervor. They want to be seen hitting and kicking the non-Muslim so that no one will doubt their loyalties. Later the same could happen to them if someone decides to get revenge for an unrelated issue by false accusations.

https://www.youtube.com/watch?v=xat2T31waGM

https://www.youtube.com/watch?v=AxYb-Yr-OoY

https://www.youtube.com/watch?v=St2h8HsCRgM

42.2 Islam Is Coming!

The first thing to think about once you realize that the Rapture is past and you are left behind is that you must understand the **simple** Gospel! You must trust in Christ **alone** for your salvation! Do not trust in your past, present or future behavior! Salvation is the **Gift** of God, it is free, just faith in Jesus **alone**, nothing else! Everyone gets saved like the thief on the cross who believed in Jesus. He was nailed to the cross, he knew that he had nothing to give. He trusted Jesus and Jesus said to him…

…"*To day shalt thou be with me in paradise*". (Luke 23: 43)

We have nothing to offer to God in addition to what he has already done for us on the cross for our salvation! To try to offer something of yourself in addition to what God has already done for you is the same as making yourself equal to God. That is sin!

You need to humble yourself, and approach God like the thief nailed to the cross. You must realize that you have nothing to offer for your salvation.

You need Grace! You need a gift! The whole purpose of the Tribulation is to humble people so that in their helpless condition they will listen to God like the thief nailed to the cross!

Salvation is a gift, it is free! "Christian service" has nothing to do with your salvation! Christian service is a privilege and God pays a wage to his children for their service. God offers rewards for believers who faithfully serve and try to obey God which is their reasonable service. But that can only happen <u>after</u> someone has trusted in Jesus Christ <u>alone</u>!

If you consider yourself a Christian and are trusting in Christ <u>**PLUS**</u> anything else such as your behavior, then you are not saved regardless of how active you are in your church. God calls your works "*filthy rags*". (Isaiah 64:6) and "*dead works*". (Hebrews 6:1 and 9:14)

You must accept God's Grace by believing God's Record about His Son. That Record says …

(John 3:16) *"For God so loved the world that He gave His only begotten Son, that whosoever* ***believeth*** *in Him shall not perish but* ***have*** *<u>everlasting</u> life".*

The next thing to think about is escaping and hiding, not trying to stand and fight. My grandfather used to say, "A good run is better than a bad stand!" When trying to escape the Antichrist there are several things to consider. The Bible says that the Antichrist will "***prevail against the saints***"

"*I beheld, and the same horn made war with the saints, and* ***prevailed against them***" ;(Daniel 7:21)

If God says that **the antichrist will prevail**, it makes sense to be thinking about escape and not attacking! Why would anyone go to a fight that they know that they will lose? After the Rapture the survival priorities are: **first** get saved!; **second**, get away from the antichrists control; **third**, get to a safe location; **fourth**, find one of the 144,000 young Jewish male missionaries for up to date information; **fifth**, secure food, water and shelter; **sixth**, locate other believers; **seventh**, establish a chronology of future events.

*"And **he shall speak great words against the most High**, and shall **wear out the saints** of the most High, and think to **change times and laws**: and they shall be given into his hand until a time and times and the dividing of time*". (Daniel 7:25)

When I read this passage that talks about the little horn speaking great words against the most High, I think about Obama's speech where he made fun of the Bible.

https://www.youtube.com/watch?v=KTyYbThVR-o

The part of this passage that talks about the little horn thinking about changing times and laws, I believe that it is talking about **sharia**, **Islamic law** and the Islamic calendar.

The term "***wear out the saints"*** is a reference to **Dhimmitude.** Dhimmitude is the institutionalized, systematic, psychological slow torture, terror and wearing down of the unbelievers of Islam. Jews and Christians are specifically mentioned in the Koran and in the Hadiths that require either their conversion, subjugation or deaths.

Dhimmitude is what Shari Law mandates from those who do not pray to Allah. It is a state of continual harassment, humiliation, terror and extortion against non-Muslims to systematically wear them down into submission. This passage (Daniel 7:25)also says that the "little horn" will prevail against the Saints for three and a half years.

The worst place on earth to be during the tribulation is in the Nation of the Antichrist, which is the U.S.! After the Rapture, I expect the False Prophet to begin performing fake miracles signs and wonders that will climax on Yom Kippur.

For centuries the Jews have observed the **"Ten Days of Awe"**. The Ten Days of Awe begins with the first day of Rosh Hashanah and continues through to the end of Yom Kippur.

That is the time that I believe the False Prophet and the Antichrist will suddenly begin a series of satanic, miracles, signs and wonders for the purpose of deceiving the world into following the Antichrist.

I believe that Yom Kippur will probably be… *"When they shall say, Peace and safety"* (1Thessalonians 5:3). That is when the Antichrist is "*revealed*" and promises to enforce the Rabin Peace treaty also known as the Oslo Accords. They may also announce "***Peace and Security***" on the United Nations Day of International Peace on September 21. I believe that it most likely will be on Yom Kippur.

However, at sundown the 70th week of Daniel will begin…"*Then sudden destruction cometh upon them, as travail upon a woman with child; and* ***they shall not escape***"! Those who are left behind, need to believe the gospel and escape the U.S. and Europe **before** sundown on Yom Kippur of that year because the Bible says ."***They shall not escape***". (1Thessalonians 5:3)

Those who are left behind and are beginning to realize what is going on should try to get as far away from the U.S. as fast as possible before sundown Yom Kippur! Consider the possibility of the borders being closed, airplanes grounded and no way to escape!

Those left behind may only have about nine days left to get out of Europe and the nation of the Antichrist. The **"Ten Days of Awe",** will be the time to escape from the United States and Europe. This is the time that the Antichrist and the False prophet will be doing miracles, signs and wonders to deceive the world into following the Antichrist.

Also consider the communist doctrine of cultural purification that led to Genocide in the former Soviet Union, Cambodia and other places taken over by communism. **Americans who believe in traditional, Christian values will be exterminated first!**

Jordan is the only country mentioned in the Bible as escaping the Antichrist. If you can't make it to Jordan, try to get south of the equator, in the mountains or in the jungles of the Amazon. **Siberia will survive Armageddon and may also escape the Antichrist**. Israel will escape the Antichrist for the first 3 ½ years of the tribulation and then the Antichrist will take over Israel also. The believing Jews in Israel will follow the Biblical escape plan and 1/3rd will be safe in the rock city of Petra during the last 3 ½ years of the Tribulation.

Nuclear fallout may remain in the northern hemisphere.

Dr. Lindstrom once indicated that he didn't believe that more than a hundred million or so would survive the Tribulation to live into the Millennium. He said that he did not have a specific passage that he could demonstrate that with but when he considered all of the descriptions of all of the judgments he just did not think there would be many survivors left. He estimated that the world's population was about 100,000,000 at the time of Christ.

Dr. Lindstrom went through the trouble of calculating how many bodies would be required to produce a river of blood 200 miles long and as deep as a horses bridle as described in (Revelation 14:17-20). He first calculated the volume of blood in the river of the valley of Jehoshaphat and then divided that volume by 5 liters per person to give the total number of people that would be slain by Christ at his return. He said it would take billions of people to produce that much blood.

I think that it is important to mention that there is a river of water formed with the new valley of Jehoshaphat. I believe that this river of blood is partly water.

"Then another angel came out of the temple which is in heaven, he also having a sharp sickle. And another angel came out from the altar, who had power over fire, and he cried with a loud cry to him who had the sharp sickle, saying, "Thrust in your sharp sickle and gather the clusters of the vine of the earth, for her grapes are fully ripe." So the angel thrust his sickle into the earth and gathered the vine of the earth, and threw it into ***the great winepress***

of the wrath of God. *And the winepress was trampled outside the city, and blood came out of the winepress, up to the horses' bridles, for one thousand six hundred furlongs".* (Revelation 14:17-20) (1600 furlongs = 160 miles)

When Christ returns, his feet will touch on the Mount of Olives and it will crack open and create a valley in the middle (Zechariah 14:3-9). A river will begin to flow from under the Temple in Jerusalem and go in two directions. One side will flow into the Dead Sea and the other will flow through the newly created valley through the Mount of Olives. I believe that some of this water will mix with the blood to help wash it down.

The idea of only 100 million people surviving the Tribulation disturbed me greatly. I spent a considerable amount of time thinking about the 28 judgments; Armageddon and this river of blood and I could see that the numbers did not look very good. I prayed about it and eventually found the answer. But first let's briefly consider the 28 Judgments.

42.3 Death Tolls During The Tribulation.

The first four judgments which are known as the 4 seals cover the first 3 ½ years. The fifth seal is in the middle, the sixth seal is associated with Armageddon and the 7th seal is associated with the second return of Christ.

As a result of those **first four judgments, 25% of the world's population is extinguished** by the Antichrist during the first 3 ½ years! By todays population of 7.8 billion, that would come out to 1.95 billion. That number translates to a little more than 1.5 million per day for 3 ½ years straight.

The middle of the Tribulation is when Satan enters into the Antichrist, and this is when the killing and terror really gets much worse! There are no numbers for death tolls during this time but it will be much greater than the first half. In the middle of the Tribulation is when the Antichrist betrays Israel and the Vatican. From this point on, Jews and Catholics will also be hunted and killed.

Between the middle of the Tribulation and Armageddon there are16 judgments. **One third** of the world's population will be killed at the battle of Armageddon. This accounts for 4 judgments. Between the Battle of Armageddon and the feast of Tabernacles there will be four more judgments which accounts for the river of blood.

Those who are left behind at the Rapture and become believers, God will provide 144,000 Jewish missionaries to the world. They will preach the Gospel and provide information about the timing and nature of these judgments before they happen. This will have survival benefits for believers. These 144,000 Jewish missionaries demonstrate that God has turned back to Israel to be his representatives to the world. From this point on, if someone wants to learn about God, they will want to find one of these Jews.

42.4 The Jewish Survival Rate Is The Key

*"And I will bring the **third part** through the fire and refine them with silver... they shall call on my name, and I will hear them; I will say it is my people, and they shall say the Lord is my God"*. (Zechariah. 13:9)

The slaughter of Jews will begin in the middle of the Tribulation. Two thirds of the Jews living in Israel will be slaughtered. Only one third of the Jews will survive! That group of predominately believing Jews will escape to Petra and survive. God will feed and protect them for the last 3 ½ years.

*"And concerning the **tithe of the herd**, or of the flock, even of **whatsoever passeth under the rod**, the **tenth** shall be holy unto the LORD"* (Leviticus 27:32)

Worldwide there will be nine tenths of Israel killed by the Antichrist. The one tenth that survives are all believers.

*"And **I will cause you to pass under the rod**, and I will bring you into the bond of the covenant:"* (Ezekiel 20:37)

This is talking about the regathering of Israel's remnant from out of the countries of the world. Only one tenth of the Jews will survive worldwide. I am going to consider that 9 tenths of all Jews will be killed during those seven years. I believe that the two thirds killed in Israel indicate an increased survival rate to the believing Jews living in Israel who obey the command to escape to Petra when the Antichrist enters the Temple. But I will consider that number to be included in the one tenth of believing Jewish survivors worldwide.

Therefore when you consider that, we can calculate the number of Jewish believers that survive the Tribulation by using existing population numbers divided by 10. That will roughly give us the total number of surviving Jewish believers worldwide. This includes both men and women.

To calculate the number of believing Gentile Survivors, consider the following verse…

"Thus saith the LORD of hosts; In those days it shall come to pass, that ***ten men*** *shall take hold out of all languages of the nations, even shall take hold of the skirt of him that is a Jew, saying, We will go with you: for we have heard that God is with you"*. (Zechariah 8:23)

This passage indicates that for every believing Jewish male survivor, there will be ten believing Gentile survivors. Furthermore…

…*"And in that day* ***seven women*** *shall take hold of* ***one man****, saying, We will eat our own bread, and wear our own apparel: only let us be called by thy name, to take away our reproach"*. (Isaiah 4:1)

In addition to having the ratio of Jewish men to Gentile men, we now have a way to calculate the ratio of males to females. From this passage, we can see that polygamy will be required for at least the first Generation of survivors because men will be scarce.

42.5 Consider The Number Of Surviving Jewish Believers.

From the information found in the scriptures we can estimate how many believing Jewish men; how many believing Jewish woman; how many believing Gentile men and how many believing Gentile women will survive the Tribulation based on the current Jewish population.

I am estimating that there is currently about 20,000,000 Jews living throughout the world. There are plenty of people who would disagree with that current population estimate, but that is the number that I am using. Some estimates are as low as 13.5 million and others are as high as 16 million. Fifteen to twenty years ago the total worldwide Jewish population was around 20 million. I don't believe that the numbers could have declined that much. I don't accept any of the current numbers for reasons that are too involved to mention here. If you want to use a different number, to start with, the math is still the same. I am going to assume a current Jewish world population of 20,000,000 to calculate my estimate because I am rounding up to what I consider to be significant figures. My calculations are only estimates to demonstrate the magnitude of killing during this time.

If you consider the passage that says that only one tenth of Jews will survive(Zechariah 8:23), that leaves us with only 2,000,000 Jewish men and women combined. Now consider the wording about 7 women clinging to one man (Isaiah 4:1). That means that out of every 8 surviving Jews, there will be one man and 7 women.

1/8 x 2,000,000 gives us **the total number of surviving Jewish Males which equals 250,000**. The other seven eights equals **1,750,000 Jewish women survivors.**

The total number of Jewish <u>males</u> who survive the Tribulation is <u>250,000</u>. That means that the chances for survival for Jewish males, worldwide, is .025 or 1 out of 40.

The total number of Jewish <u>females</u> who survive the Tribulation is <u>1,750,000</u>. That means that the chances for survival for Jewish females, worldwide, is .175 or 7 out of 40.

42.6 The Believing Gentile Survival Rate

Now consider the passage that says that 10 Gentile men will cling to one Jew. That means that out of 11 surviving men there will be one Jew and ten Gentiles.

"*Thus saith the LORD of hosts; In those days it shall come to pass, that* ***ten men*** *shall take hold out of all languages of the nations, even shall take hold of the skirt of him that is a Jew, saying, We will go with you: for we have heard that God is with you*". (Zechariah 8:23)

Now we can calculate the number of surviving Gentile male believers from the number of surviving Jewish male believers.

If you multiply ten times the number of believing Jewish male survivors that will give you the number of believing Gentile male survivors. Therefore we multiply 250,000 times 10 and that gives you **2,500,000 believing Gentile male survivors.**

With a current world population at about 7.8 billion, I will assume for the sake of estimating that about half of that number is males. Therefore we currently have a world population of about 3.9 billion male Gentiles. So we divide 2,500,000 by 3,900,000,000 and that gives us a survival rate of .00064 or one out of 1,560 believing Gentile males will survive.

The number of believing Gentile female survivors is calculated by multiplying 7 time the number of believing Gentile male survivors. So 7 times 2,500,000 equals **17,500,000 believing, female Gentile survivors.**

When you divide that number(17,500,000) by the estimated world population of 3,900,000,000 you get a survival rate of .0045 or one in 223.

These numbers represent Jewish and Gentile **believers only. All unbelievers** will be gathered up by the Angels and taken to the valley of Jehoshaphat to be killed by Jesus Christ himself!(*Joel*3:12)

42.7 The Antichrist Will Hunt All Jewish And Gentile Believers.

Here are the estimated populations of Jews and Gentiles, also divided by Gender. **Jewish Men**= 250,000; **Jewish women** = 1,750,000 survivors; **Gentile men** = 2,500,000 survivors; **Gentile women** = 17,500,000. According to these estimates, the total population of believers entering into the Millennial Kingdom of Jesus = approximately 22,000,000.

This is an approximation based on current population numbers (2020) which may be accurate give or take a few million or so because of disagreements about how to determine who is qualified to be called a Jew. But I think that Dr. Hank Lindstrom had the right idea when he said "not more than a hundred million", based on the estimated world population numbers during the time of Christ. My estimates are much lower!

Furthermore, we can consider that one eighth of the survivors are men and seven eights are women. Therefore we have 2,750,000 men left and 19,250,000 women. (Jews and Gentiles combined)

The **chances of survival** of Believing, **Jewish <u>men</u> is roughly about <u>one in forty</u>.**

The **chances of survival** for believing, **Gentile <u>men</u>** is **one out of <u>1,560</u>**. Gentiles are going to be hit the worst.

"*Therefore hath the **curse** devoured the earth, and they that dwell therein are desolate: therefore the inhabitants of the earth are burned, and **few men left***". (Isaiah 24: 6)

The word "curse" as used in (Isaiah 24:6) is translated from the Hebrew word "***Alah***" which has one less l than the Arabic word but is pronounced the same and has the same meaning except the Hebrew word has negative connotations and the Arabic word is positive in their culture.

One explanation for the disproportionate number of men killed may be because of the way Muslims treat conquered men. The men are either killed or castrated and turned into slaves to be abused for their pleasure before finally killing them later. This may be the explanation for the descriptive language used to describe the takeover of Jerusalem during the middle of the Tribulation.

"*Ask ye now, and see whether a man doth travail with child? wherefore do I see every man with his hands on his loins, as a woman in travail, and all faces are turned into paleness?*" (Jeremiah 30:6)

However, women are kept as sex slaves and abused. The Bible talks about the cruel religion of the antichrist. The Bible says that they regard the god of force. Now who do you think that is? The most popular videos among Muslims involve the torture, mutilation and killing of those who don't accept their ideology. The Bible lets us know. If you want to know what the Tribulation will be like, take a look at these youtube videos about what Muslims are doing to women! … Sorry, they took down the videos. There is a lot of censuring going on.

This is a time of evil; a time of deception; a time of betrayal; a time of cruelty and a time of terror! By the time Armageddon comes around there won't be many people left. Those who do live through the Tribulation will be mostly unbelievers. The few believers who are able escape death will have seen more horror than Americans and Europeans can imagine!

When you consider that the Bible says that all life on Earth would be destroyed if God did not stop Armageddon, then you can understand these extreme numbers. Furthermore, consider also what the population control plans are for the United Nations and who is making those plans!

The club of Rome and Lucis trust formulated Agenda 21 and 2030 for the U.N. Lucis Trust was formerly Lucifer Publishing Company. As a member of the Club of Rome, Ted Turner made the statement that the world did not need more that about 500 million people. The rest were useless eaters. He later changed that number down to 250 to 300 million. It seems that whoever can come up with the most extreme numbers and extreme lies gets special consideration or rewards from some secret group. The bottom line is that we have been betrayed by our leaders, our representatives and our experts. The upcoming Genocide will be no accident. This has been planned for a very long time.

Our leaders will disarm us just like African civilians were disarmed before Muslim immigrants were introduced. The well-educated, Christianized and civilized African citizens were first disarmed and then slaughtered by well-armed Muslims. The Muslims have taken over and destroyed those once modern and prosperous cities and turned them into ruins. The same thing happened in Lebanon and Bosnia. It is beginning to happen in Europe and Australia which also have been disarmed. The United States is about to end up worse than Europe but that fact is undiscernible without a good knowledge of Bible Prophecy. Most people who see the danger, think that it is still years or decades away, but I don't think so!

42.8 God's Warning To Those Who Persecute Believers.

God says that there are degrees of punishment in Hell. (Matthew 11:20–24)

Those who persecute God's people will receive the maximum punishment in Hell!

"For ye, brethren, became followers of the churches of God which in Judaea are in Christ Jesus: for ye also have suffered like things of your own countrymen, even as they have of the Jews:" (1Thessalonians 2:14)

"Who both killed the Lord Jesus, and their own prophets, and have persecuted us; and they please not God, and are contrary to all men:" (1Thessalonians 2:15)

"Forbidding us to speak to the Gentiles that they might be saved, to fill up their sins alway: for the ***wrath is come upon them to the uttermost****"*. (1Thessalonians 2:16)

Once again, those who persecute Christians and fight against God's message of salvation by grace through just faith in the finished work of Jesus Christ on the Cross of Calvary, will face the maximum punishment in Hell for all of eternity!

The degree of punishment in Hell is also directly proportionate to the amount of truth that was rejected by the unbeliever…

(Matthew 11:22) *"But I say unto you, It shall be more tolerable for Tyre and Sidon at the day of judgment, than for you"*.

(Matthew 11:23) *"And thou, Capernaum, which art exalted unto heaven, shalt be brought down to hell: for* ***if the mighty works, which have been done in thee, had been done in Sodom, it would have remained until this day****"*

(Matthew 11:24) *"But I say unto you, That it shall be more tolerable for the land of Sodom in the day of judgment, than for thee"*.

There are degrees of punishment in hell. The more truth that is rejected the greater the punishment.

43

28 Judgments Gets Man's Attention!

43.1 28 Judgments Spread Out Over 7 Years

Speculating about our Lord's return for the church is all a part of "Watching". Satan doesn't want us to watch, speculate or talk about the Rapture of the Church. Satan will direct his minions to scorn and attack those who openly speculate about his return or anything else in God's word. This is the greatest witnessing tool of the Last Days. Satan's minions will oppose those who use it.

"So teach us to number our days, that we may apply our hearts to wisdom" (Psalms 90:12).

The 28 judgements are spread out over 7 years. They consist of Seven Seals; Seven Trumpets; Seven Vial Judgements; and Seven Thunder Judgements.

The Seven Seals seem to be associated with the Antichrist and his crimes against God and man that span the 7 years of Tribulation.

The white horse is the first of the 7 seals and the first of the four horseman! The First Seal represents the beginning of the 70th week of Daniel. The White Horseman is the Antichrist being revealed on Yom Kippur of that year. He will be revealed and promoted by the False Prophet. According to the literal interpretation of the Bible, the False Prophet comes out of the Roman Catholic Church.

Dr. Lindstrom has repeatedly said that the False Prophet will definitely come out of the Roman Catholic Church. Furthermore, he believed that he would probably be the Pope himself. There are too many unusual things about Pope Frances the First to be just a coincidence.

I believe that the four horsemen seals, are spread out over the first 3 ½ years of the Tribulation. I believe that the first four seals or horsemen represent the things the Antichrist does to achieve power. He is the rider and he rises to power over the first three and a half years of the Tribulation.

The order, color and names of the Four Horseman are...

First Seal, The **White Horse**; "*he went forth conquering, and to conquer*". (Revelation6:2)

Second seal is the **Red Horse.** The Red Horse represents War!

The **Third Seal** is the **Black horse**; food rationing.

Fourth Seal or 4th **Pale Horse** represents mass beheadings! The color "pale" is actually a pale green, like the color of the flag of Saudi Arabia.

As a result of the first four Horseman, **25% of mankind is killed.**

The fifth seal is in the **middle of the Tribulation**. This is when Satan indwells the Antichrist and he takes over Jerusalem and the Temple is desecrated.

The **sixth seal** is associated with **Armageddon**

The Great White Throne Judgement,

The **Seventh Seal** is associated with the **beginning of the Kingdom**. Before the Kingdom, the Antichrist and False Prophet are cast alive into the Lake of Fire.

43.2 First Seal, The Antichrist Revealed On The White Horse,

*"And I saw when the Lamb opened one of the seals, and I heard, as it were the noise of **thunder**, one of the four beasts saying, Come and see"*. (Revelation 6:1)

*"And I saw, and behold a **white horse**: and he that sat on him had a bow; and a crown was given unto him: and he went forth conquering, and to conquer"*. (Revelation 6:2)

An Islamic connection to the Antichrist and the world system that he controls should be kept in mind while trying to understand what is happening here and where this is leading.

The Koran says that Mohamad rode a white horse to heaven. The horses name was "Burak" which means "lightening" and is pronounced the same as "Barack".

The next Caliph of the soon to be created Global Islamic Caliphate is believed to be **Muhammad al-Mahdi** or the Twelfth **Imam**. The **Mahdi** will lead Muslims in their war against the "nonbelievers" of Islam. Muslims all over the world are waiting to hear his command for Jihad. They are praying for war against the Infidels but that can't happen until someone is recognized as the twelfth Caliph. Until that time, "good Muslims" must not wage war unless it is "defensive". That is called Defensive Jihad.

There are three stages of Jihad. The first stage is the secret or peaceful stage. Some call it "Stealth Jihad". During this early stage the message of peace and tolerance is preached. This first stage lasts until the numbers and strength of the Muslims and their allies is great enough to begin asserting their "religious rights".

The second stage is the one that I believe we are in now. That is where the power and influence of Muslims and their allies is great enough to begin using violence, terror, intimidation and force to get concessions made that position themselves to take over by suprise.

According to the Bible, a satanic leader will take over the world during the seven years of Tribulation that is also known as the 70th week of Daniel. The Biblical description of the Antichrist seems to match the descriptions of the Mahdi found in the Koran.

There are many Muslims who are expecting the Mahdi to rule the world for **seven years**. The Mahdi will force everyone to obey **Sharia law** by chopping off heads, opposite hands and feet, stoning's and slavery. Wife beatings, forced Genital mutilation and more horror than most Westerners can imagine are incorporated into Sharia law including slavery!

For most Shia Muslims, the **Mahdi** was born but disappeared and will remain hidden from humanity until he reappears to bring **Sharia law** to the world. This doctrine known as the Occultation is a central doctrine in Iran. For Twelver Shia, this "hidden **Imam**" is Muhammad al-**Mahdi**, the **Twelfth Imam**. According to their religion, he will conquer the world for Islam in the last days.

It is interesting how the Biblical description of the future seems to be similar to the Koran's except the Bible calls this guy the "Antichrist" and the seven years of his rule will destroy most of humanity. If Jesus Christ himself did not stop him, the whole world would be destroyed!

The use of ***"Alah"*** in connection to the 70th week of Danial is another clear indication of an Islamic connection to God's judgement of the world (Isaiah 24:6). It is no coincidence that ***"Alah"*** is found again in (Daniel 9:11). It is also no coincidence that (Revelation 9:11) contains the word ***"Apollyon"***. There seems to be a connection here. "*Apollyon*" roughly translates into "the destroyer".

The Day of the LORD begins at sundown, at the sound of the last trump of Rosh Hashanah on the year of the Rapture. I believe that this is when construction on the Third Temple begins.

"Woe unto you that desire the day of the LORD! to what end is it for you? the day of the LORD is ***darkness, and not light.****"* (Amos 5:18)

*"Shall not the day of the LORD be **darkness, and not light? even very dark, and no brightness in it**?* (Amos 5:20)

I would not be surprised if the power grid goes down at about the same time of night as the Las Vegas Massacre of September 29, 2017 occurred. The time was 10:05pm.

The **"The Ten Days of Awe"** (Yamim Noraim) or "Days of Repentance" begins on the first day of Rosh Hashanah and will end on Yom Kippur. I believe that will be the day that the man of sin will be revealed by the False Prophet, the last day of "The Ten Days of Awe." Sundown will mark the end of Yom Kippur and the beginning of the Seventieth Week Of Daniel. That is when I expect them to say "*Peace and safety*"(Security), but just after sundown comes "*sudden destruction*".

The "*man of sin*" will be revealed to the world the day he assumes the responsibility of enforcing the treaty of Yitzhak Rabin for 7 years. I believe that this is the same treaty referred to as "***the covenant with many***" in (Daniel 9:27).

Rabin's name in Hebrew means "many". It is the same Hebrew word.

This will happen on the 6th Feast, the Feast of Yom Kippur or (Atonement). I believe that is the day they will say "*peace and security*". (1Thessalonians 5:3) Or it may be on the United Nations International Day of Peace which occurs every September 21.

*"For when they shall say, Peace and safety; then **sudden destruction** cometh upon them, as travail upon a woman with child; and they shall not escape"*. (1Thessalonians 5:3)

Yitzhak Rabin, Bill Clinton, and Yasser Arafat Negotiated and finally signed the Oslo Accords on September 13, 1993 (Yom Kippur) this is "*the covenant with many*" spoken of in (Daniel 9: 27). At the televised signing ceremony back in 9/13/1993, they shook hands and said ***"Peace and Security"***

"And he shall confirm the covenant with ***many*** *for one week"* (Daniel 9:27)

One week is seven years.

The Antichrist will "*confirm the covenant with many*" on Yom Kippur with the help of the Roman Catholic False Prophet.

This event which will occur on Yom Kippur, marks the beginning of the 70th week of Daniel, also known as the Great Tribulation. The 70th week of Daniel begins after sundown, the day the Antichrist is revealed and I believe that is the day they will say "*Peace and Safety*" as described in (1 Thessalonians 5:3)…

"*For when they shall say,* ***Peace and safety****; then* ***sudden destruction*** *cometh upon them, as travail upon a woman with child; and they shall not escape*".

I believe this "***sudden destruction***" begins after sundown at the end of Yom Kippur. The Day of the Lord' begins with terror and darkness! (1Thessalonians 5:3)

"*Howl ye; for the day of the LORD is at hand; it shall come as a destruction from the Almighty*". (Isaiah 13:6()

So who do you think the Antichrist is being revealed to? It is not the believers because we will all be gone at the Rapture 9 days earlier. Some believe that they know who the Antichrist is. The Antichrist will be revealed to the unbelievers that are left behind. In particular the Antichrist will be revealed to the followers of Islam as the twelfth caliphate. I believe that is where the "*sudden destruction*" comes from. The Muslims can't go to war without a Caliphate, only defensive Jihad is permitted. That is where Islam's twelfth Imam comes in. The Antichrist is the coming global twelfth caliphate and he will be revealed to the Muslims as such.

The white horse represents the Antichrist and his Promises of Peace. The empty bow demonstrates that the Antichrist has military power, but he is promising not to use it, but he does! He gains power by false promises of Peace, flattery and betrayal.

The Antichrist will also be revealed to the followers of the Roman Catholic Church as per the agreement made between the Woman and the Beast of (Revelation 17). Catholics will be temporarily exempt from persecution while Muslims abuse non-Muslims. The Catholics will have an agreement with the Antichrist that he will honor for about 3 ½ years. In the middle of the Tribulation, he betrays the Catholics, Israel and the Jews scattered abroad.

The Antichrist will also be revealed to the followers of Contemporary Pseudo Christianity, and Pseudo Contemporary Judaism, the followers of New Age Philosophy and all of the false religions of the world and any others who have been blinded by the god of this world. They are all conditioned to believe his lies, and accept him to be their leader. Only those who call upon the name of the Lord will be Saved and they can do that at any time!

43.3 Second Seal: The Red Horse, War

(1Thesalonians 5:3) speaks of **"sudden destruction"!**

The Koran instructs the initiated in a warfare of deception, betrayal and sudden, violent surprise attacks. It teaches them to lie in wait and pretend to be one of them. Terrorist sleeper cells appear to be "moderate Muslims".

The Koran tells a story of Mohamed. When Mohamed discovered that his enemy was too strong he would make oaths of Peace and join them. Pretend to be one of them until he was strong enough to win by betrayal.

A story about how Madinah, was captured uses all of the key elements of warfare taught in the Koran. First there is Deception. Next there is Trust gained through the Deception. After that Mohamed would use that trust to make gains toward the enemy and finally betray those who were befriended. Muslims are told to always be ready to use those gains against the enemy when the time of Jihad comes!

Jihad is at the core of the Koran and the core of Islam. The term Jihad literally means "struggle". The term can be used to describe a spiritual struggle of an individual or the struggle against unbelievers until they have all been subdued.

Islam is the only "religion" in the world that mandates violence against unbelievers! It is also the only religion in the world to identify another religion or Nationality to specifically mandate violence against it. Islam mandates violence against Christians and Jews specifically calling them "People of the Book" and any other unbelievers. Islam is also the only "religion" that specifically denies Biblical doctrines and condemns those who believe them.

There are three kinds of Jihad.

The **first kind of Jihad** involves migrating in the way of Allah. This first stage involves lying about Islam's intent and concealing it from the "enemy" (unbeliever) and preaching peace and tolerance. Mohamed said that war is deceit and Allah is the best of schemers.

The **second kind of Jihad** is called **Defensive Jihad**. Defensive Jihad is when the Muslim strength is strong enough to do sporadic attacks to weaken the enemy and gain converts, but not strong enough to overpower the unbelievers.

The **third kind of Jihad** is when the Muslims have gained the trust of their enemies and are in a position and are strong enough to win a sudden, surprise attack from within. The Koran has rules about concealing their intent to their victims. Anyone who tells the truth about their intent for migrating will be called a "slanderer" and killed. To the Muslim, slander is unwelcomed truth and the offender can be killed. **Obama said…"the future must not belong to those who slander Islam."** Telling the truth about Islam can get you killed!

Islam teaches that Muslims should Generally be truthful to other Muslims, unless one is lying to "smooth over differences."

Taqiyya is lying to unbelievers to gain their trust so that they can be conquered in the name of Allah and establish **Sharia law.**

Mohamed made a peace treaty with the occupants of Madinah, he pretended to accept their religion. He even prayed with the residences, until the time that their numbers increased and their victims guard was down. During prayer time, he attacked their former non-Muslim "friends" and killed or enslaved them.

"*And when he had opened the second seal, I heard the second beast say, Come and see*". (Revelation 6:3)

"*And there went out **another horse that was red**: and power was given to him that sat thereon to take peace from the earth, and that they should **kill one another**: and **there was given unto him a great sword***". (Revelation 6:4)

I believe that that sword represents an Islamic Jihad. They like to use the sword for decapitating their victims. The sword is the center of attention when it comes to Jihad, it is the icon of religious war in the name of Allah. The sword represents piety concerning their belief system. Notice that it says in verse 4, ***"that they should kill one another"***. In the Greek this is talking about sudden, violent, cruel, worldwide Anarchy that involves a sword. This is not one army attacking another. This is everybody attacking each other. It sounds like the work of a community organizer to me. An Islamic Jihad! That is the "chaos" that will proceed the N.W.O.

Our southern border is an open highway for terrorists wanting to enter our country. They have been coming in for quite some time now. I suppose that they are bringing in automatic weapons, R.P.G's and all of the bomb building technology and materials that they can "sneak" by paid-off officials.

All they have to do is attack our grid! We could be facing a potential catastrophe in an instant! With a little more help from the perps the situation will turn into as big of a catastrophe as is ordered. The terrorists will be well trained, organized, equipped, vaccinated, motivated and cruel.

The populations of the western countries have already been betrayed by our leaders by bringing in terrorists and secretly aiding the enemy to establish a N.W.O. on the ruins of the old order.

If that isn't enough terror, consider that we are in the middle of the biggest arms race in the history of the world! All sides are trying to come up with the best weapons for killing. Nobody knows what our military can do right now or what they will do in the near future! I can see Biblical events at the door.

We are seeing the preparation for The Battle of Armageddon! It comes at the end of the seven years of Tribulation; it will be nuclear; it will be worldwide; it will destroy one third of mankind; it will last for only one hour and it will be stopped by Jesus Christ himself.

"Proclaim ye this among the Gentiles; Prepare war, wake up the mighty men, let all the men of war draw near; let them come up" (Joel 3:9)

"*Beat your plowshares into swords, and your pruninghooks into spears: let the weak say, I am strong*". (Joel 3:10)

"*Assemble yourselves, and come, all ye heathen, and gather yourselves together round about: thither cause thy mighty ones to come down, O LORD*". (Joel 3:11)

Right now we are in the middle of the biggest arms race in history! Wars and rumors of war will continue to accelerate until Armageddon.

43.4 Third Seal The Black Horse, Famine!

So a little recap here. First the Antichrist is revealed, Second the Jihad begins at sundown, third, Sudden anarchy and war; and now **famine**!

Well naturally you would expect a disruption of every public service you can think of. No electrify, **no water**, no A.C. in summer, no heat in the winter, no refrigeration, **no food, no clean water,** no medicine or hygiene and no security.

"And when he had opened the ***third seal****, I heard the third beast say, Come and see. And I beheld, and lo a black horse; and he that sat on him had a pair of balances in his hand"*. (Revelation 6:5)

"And I heard a voice in the midst of the four beasts say, A measure of wheat for a penny, and three measures of barley for a penny; and see thou hurt not the oil and the wine". (Revelation 6:6)

The "Board of Governors" Martial Law plan specifies that the President will have control over food, water, communication, transportation, population and medical resources.

According to the Bible, it will take a full day's wages to buy your daily food. There will be no savings. The elite will continue to enjoy luxuries.

When it talks about not hurting the oil and wine, I believe that the elite will have all of the comforts. But I also think that this passage may be speaking of more than that. The term oil can also refer to petroleum. Petroleum is where all of those comforts come from. Oil is the center of world attention right now.

This **Third Seal will be rationing** for everyone except the elite. If terrorists were to shut down our power grid and create anarchy at the same time, there could be a food shortage. People could starve to death and according to the Bible, they will! There is not enough food stored. Cannibalism is described during the tribulation. The government will control all food, all commerce, all news, all communications and all of everything else.

A lack of food and other essentials will be incentive for survivors to believe the lies of the False Prophet and Antichrist. People will want to just jump on board the N.W.O. fast train to avoid starvation or worse, but that will require the mark of the Beast!

43.5 Fourth Seal The Pale Horse Death

*"And when he had opened the **fourth seal**, I heard the voice of the **fourth beast** say, Come and see."* (Revelation 6:7)

*"And I looked, and behold a **pale horse**: and his name that sat on him was Death, and Hell followed with him. And power was given unto them over the fourth part of the earth, **to kill with sword**, and **with hunger**, and with death, and with the beasts of the earth"*. (Revelation 6:8)

This fourth Seal represents the Fourth Horseman which is announced by the Fourth Beast (vs. 7) and is given power over the fourth part of the earth. OK that is four times four. What is the significance of that? Numbers are significant in God's word. But what could it mean?

This is talking about a couple of things. This horse is pale green and Mohamed's flag was also pale green just like the flag of Saudi Arabia. This seal consists of plagues of a biological nature, I believe. Only the terrorists will be vaccinated, everyone else will be sick, hungry and thirsty. Nobody is going to want to fight or even be able to.

I don't think that it will take the Antichrist very much time to disarm the public. Twenty five percent will die during those first 3 ½ years. That is called the "*beginning of sorrows*".

"*All these are the beginning of sorrows*". (Matthew 24:8)

The "pale" horse is pale green. In the Greek, the first definition for **Strong's G5515 – chlōros** ... is green. The second definition is pale yellow/green.

Mohamed's flag is said to be pale green and Islam controls about one fourth of the Earth's population. These facts have led some to consider this to be identifying Islam as the source of all this death.

The Four Horseman represent 4 Seal judgements that actually represents activities of the Antichrist during those first 3 ½ years. These first four judgements cover the first half of the Tribulation and result in **the deaths of 25% of the world's population.**

43.6 Fifth Seal, Satan enters into the Antichrist, mid-tribulation

Satan enters into the Antichrist in the middle of the Tribulation. He breaks his treaties, he conquers the Mideast and finally Jerusalem! This is when God's two Prophets are killed.

The Antichrist moves into the Third Temple, he claims to be God and he requires all to worship him and receive his mark. Those who disagree will be hunted, abused and killed! This is the Abomination of Desolations spoken of by Daniel the Prophet. This happens in the middle of the tribulation. The description of the aftermath follows…

"*And when he had opened the **fifth seal**, I saw under the altar the souls of them that were slain for the word of God, and for the testimony which they held*": (Revelation 6:9)

"*And they cried with a loud voice, saying, How long, O Lord, holy and true, dost thou not judge and avenge our blood on them that dwell on the earth?*" (Revelation 6:10)

"*And white robes were given unto every one of them; and it was said unto them, that they should rest yet for a little season, until their **fellowservants** also and their **brethren**, that should be killed as they were, should be fulfilled*". (Revelation 6:11)

God's two Prophets and the remainder of the 144,000 young, unmarried, virgin Jewish missionaries of God will be killed by the **Middle of the Tribulation**. The bodies of the two Prophets of God will be left in the street for all to see. People from all walks of life and from all over the world will see them on satellite T.V. and internet.

The technology to do the things described in the Bible did not exist at the time this was written, but it does now! Nevertheless, even though God spoke of these things thousands of years ago and we see it all coming to pass, people still don't seek God. Many have followed Satan's lies and are comfortable with that.

God raises the two Witnesses from the dead on live T.V. worldwide! Everyone will see it. Yet many still will not believe.

(Revelation 12) Satan loses another fight in heaven. He is cast down to the earth and indwells the Antichrist. The two Saints are also killed, all of this happens in the middle of the seven years.

43.7 God's Prophets Are Killed

"*And when he had opened* ***the fifth seal****, I saw under the altar the souls of them that were slain for the word of God, and for the testimony which they held*": (Revelation6:9)

"*And they cried with a loud voice, saying, How long, O Lord, holy and true, dost thou not judge and avenge our blood on them that dwell on the earth?*"(Revelation 6:10)

It looks like there is a lot of terror, a lot of starvation, a lot of dehydration, and a General lack of all of the nice things that we have become accustomed to.

I believe that Terrorists will create the Anarchy. They will knock out our power grid and we will have a catastrophe. The N.W.O. gang wants to do away with the U.S. constitution. All they need is a good excuse for Martial Law. They have cultivated groups, gangs and terrorists abroad such as I.S.I.S. or Boko Haram for that purpose and they have already done it here as well. These Muslim fighters are coming across our southern boarders with the intensions of war. They are just waiting for the 12th Caliphate to be identified and to say "Jihad!"

In the middle of the Tribulation, there is a war in heaven. I believe that this war is either in the first or second heaven. It definitely is not in the third Heaven. Furthermore, this war may or may not occur in our dimensions.

I believe that we may soon be required to specify dimensional coordinates to describe events in the future because of an increased awareness of things not previously understood.

This is when Satan enters into the Antichrist. This also corresponds with the **5th seal** and the Antichrist breaking his treaty with Israel and killing God's two prophets.

"And their dead bodies shall lie in the street of the great city, which spiritually is called Sodom and Egypt, where also our Lord was crucified". (Revelation 11:8)

I think that it is important to mention that this is talking about Jerusalem. It is comparing Jerusalem to Egypt because of all of the false religions and to Sodom because of Sodomy. God also compares the United States to Babylon. (See chapters 41-44)

According to Ezekiel, Israel would return to the Promised Land in unbelief and they have. Most Jews living in Israel are either atheists or agnostics. The "Star of David" that Israel proudly displays on their flag is actually not the star of David. King David never had such a star but his son Solomon did and it was called Solomon's seal. Solomon's seal was the star of Molech.

"*But ye have borne the tabernacle of your Moloch and Chiun your images, the star of your god, which ye made to yourselves*". (Amos 5:26)

"*Yea, ye took up the tabernacle of Moloch, and the star of your god Remphan, figures which ye made to worship them: and I will carry you away beyond Babylon*". (Acts 7:43)

God is a jealous God, he will deal with Israel during the Tribulation along with everyone else. God has a lesson for everyone. The easiest lesson is to simply believe the Gospel!

"And they of the people and kindreds and tongues and nations shall see their dead bodies three days and an half, and shall not suffer their dead bodies to be put in graves". (Revelation 11:9)

The technology required for everyone on earth to see what is happening in the streets of Jerusalem did not exist until recently.

"And they that dwell upon the earth shall rejoice over them, and make merry, and shall send gifts one to another; because these two prophets tormented them that dwelt on the earth". (Revelation 11:10)

"And after three days and an half the Spirit of life from God entered into them, and they stood upon their feet; and great fear fell upon them which saw them". (Revelation 11:11)

"And they heard a great voice from heaven saying unto them, Come up hither. And they ascended up to heaven in a cloud; and their enemies beheld them". (Revelation 11:12)

It sounds like live T.V. coverage to me. You would think that this would cause people to believe, but it doesn't. People know that they can't believe what they see on T.V.

43.8 Sixth Seal Is Armageddon

It is important to know that the Sixth **Seal**; Sixth **Trumpet** and the Sixth **Bowel** Judgement all seem to be associated with **Armageddon!** Therefore, I believe that the sixth thunder judgement may also be associated with Armageddon.

*"And I beheld when he had opened the **sixth seal**, and, lo, there was a great earthquake; and the sun became black as sackcloth of hair, and the moon became as blood;"* (Revelation 6:12)

"And the stars of heaven fell unto the earth, even as a fig tree casteth her untimely figs, when she is shaken of a mighty wind." (Revelation 6:13)

"*And the heaven departed as a scroll when it is rolled together; and every mountain and island were moved out of their places*" (Revelation 6:14)

OK that was a description of Armageddon! "*The heaven departed as a scroll when it is rolled together*" sounds like a mushroom cloud from a nuclear explosion. A nuclear explosion would explain how mountains and islands could be "*moved*".

"*And the kings of the earth, and the great men, and the rich men, and the chief captains, and the mighty men, and every bondman, and every free man, hid themselves in the dens and in the rocks of the mountains;*" (Revelation 6:15)

"*And said to the mountains and rocks, Fall on us, and hide us from the face of him that sitteth on the throne, and from the wrath of the Lamb*": (Revelation 6:16)

"*For the great day of his wrath is come; and who shall be able to stand?*" (Revelation 6:17)

That was the second return! Unbelievers will be hiding in their bunkers and the Angels of the Lord will be dragging them out and taking them off to be judged.

43.9 Seventh Seal Is The Second Return

It is important to know that the 7th Seal, the 7th Trumpet, the 7th bowel judgement and the 7th thunder judgement all seem to be associated with the 2nd return of Christ.

If you know the order, you are one step closer to understanding their significance.

43.10 Meanwhile, Things Are Happening In Heaven.

Even though the following scene occurs in Heaven after John is shown the Seven Seals, I believe this occurs in the **middle of the Tribulation**. John has just been shown events that span the entire seven years. These 7 **events are called the Seven Seal Judgements**. Now God is going to show John events in Heaven that occur in preparation for the Seven Trumpet Judgements which I believe will begin in the middle and span the last 3 ½ years of the Tribulation.

Remember! **All of the "6" judgements are associated with Armageddon**;

...all of the "7" Judgements are associated with the 2nd return of Christ.

The **First Seal Judgement** is **The Antichrist Revealed (White Horse) It will mark the beginning of the 70th week of Daniel and the Antichrist being revealed to the Muslim and New Age followers!**

He will promise **"Peace and Security"**, but he will do the opposite! He will bring "***Sudden Destruction***" in the form of Terror, Starvation and Death! He will be a Secret Muslim. He will cultivate men's trust for the purpose of betrayal!

"*And after these things I saw four angels standing on the four corners of the earth, holding the four winds of the earth, that* ***the wind should not blow on the earth****, nor on the sea, nor on any tree*". (Revelation 7:1)

Here is another reference to unusual weather on the Earth during the 70th week of Daniel. According to this, I expect to see a day or a week, or some set number of significance, to represent this amount of time that the earth remains without wind. No air movement means air will stagnate! Sail boats will be stuck out at sea with limited supplies. Many could be discomforted in some way or at least be inconvenienced. Some could die! If there is suddenly no wind, it is going to get people's attention.

Man has been trying to weaponize the weather since the 60's. H.A.R.P. may be able to pump energy into the Jet Stream and make it go faster or further. Man may increase the Jet Stream's energy with microwaves, but only God can stop all the wind on earth! God is getting men's attention. God is trying to communicate with the world. God is warning mankind that he is on his way! Every man needs to be ready because our Creator is coming! God will provide 144,000 Jewish Missionaries to fill in the details about these judgements to win and protect new believers.

"*And I saw another angel **ascending from the east**, having **the seal of the living God**: and he cried with a loud voice to the **four angels, to whom it was given to hurt the earth and the sea***", (Revelation 7:2)

"*Saying, Hurt not the earth, neither the sea, nor the trees, **till we have sealed the servants of our God in their foreheads***". (Revelation 7:3)

The one hundred and forty four thousand Jewish missionaries will be the first to get this seal for believers. I don't know what it is but I believe these fellas are going to distribute this seal to all believers. That is quite a job. I believe this seal will protect those who possess it from at least one judgement, probably more.

"*And I heard the number of them which were sealed: and there were sealed **an hundred and forty and four thousand** of all the tribes of the children of Israel*" (Revelation 7:4)

"*Of the tribe of Juda were sealed twelve thousand. Of the tribe of Reuben were sealed twelve thousand. Of the tribe of Gad were sealed twelve thousand*". (Revelation 7:5)

Twelve thousand from each of the 12 tribes makes up the One Hundred and forty four thousand Jewish missionaries. These fellas are the first to receive the seal of God in their foreheads. They are not only going to carry the Gospel, but they may be in charge of distributing this "seal of God" if it is a physical thing.

*"After this I beheld, and, lo, a great multitude, which no man could number, of all nations, and kindreds, and people, and tongues, stood before the throne, and before the Lamb, **clothed with white robes**, and palms in their hands;"* (Revelation 7:9)

"And cried with a loud voice, saying, Salvation to our God which sitteth upon the throne, and unto the Lamb". (Revelation 7:10)

*"And all the angels stood round about the throne, and about the elders and the **four beasts**, and fell before the throne on their faces, and worshipped God",* (Revelation 7:11)

"*Saying, Amen: Blessing, and glory, and wisdom, and thanksgiving, and honor, and power, and might, be unto our God for ever and ever. Amen*". (Revelation 7:12)

*"And one of the elders answered, saying unto me, **What are these which are arrayed in <u>white robes</u>?** and whence came they*"? (Revelation 7:13)

*"And I said unto him, Sir, thou knowest. And he said to me, **These are they which came out of <u>great tribulation</u>**, and **have washed their robes, and made them <u>white</u> in the blood of the Lamb**".* (Revelation 7:14)

"*Therefore are they before the throne of God, and serve him day and night in his temple: and he that sitteth on the throne shall dwell among them*". (Revelation 7:15)

"*They shall hunger no more, neither thirst any more; neither shall the sun light on them, nor any heat*". (Revelation 7:16)

"*For the Lamb which is in the midst of the throne shall feed them, and shall lead them unto living fountains of waters: and God shall wipe away all tears from their eyes*". (Revelation 7:17)

We will eat, drink and work in Heaven. God is promising to protect them and care for them forever. We will work for God forever. Eventually, He will also wipe away bad memories.

43.11 Trumpets are "Attention-getters".

Trumpets are devices that are used to get people's attention. The nature of these judgements qualifies them as attention getters. The whole world is going to know that something is going on. This will be a time of "**Great Tribulation"** and terror! People will want to know what is going on. The terrifying events described in the Bible during this time will cause men to ask questions. Some will turn to God others will believe the mainstream media and their satanic lies. Men will wonder..."is this of God, or is it just bad luck or aliens or something?"

"*And I saw the seven angels which stood before God; and to them were given **seven trumpets***". (Revelation 8:2)

"*And another angel came and stood at the altar, having a golden censer; and there was given unto him much incense, that he should offer it with the prayers of all saints upon the golden altar which was before the throne*". (Revelation 8:3)

These prayers are represented visually as smoke coming out of a censor. I believe that we will be able to see and hear those prayers in heaven.

"*And the smoke of the incense, which came with the prayers of the saints, ascended up before God out of the angel's hand*". (Revelation 8:4)

"*And the angel took the censer, and filled it with fire of the altar, and cast it into the earth: and there were voices, and thunderings, and lightnings, and an **earthquake***" (Revelation 8:5)

"*And the seven angels which had the **seven trumpets** prepared themselves to sound*". (Revelation 8:6)

Trumpet Judgment # 1

"***The first angel sounded**, and there followed hail and fire mingled with blood, and they were cast upon the earth: and the **third part of trees was burnt up, and all green grass was burnt up***" (Revelation 8:7)

The use of the word "blood" here does not necessarily mean literal blood. It could mean that but the same word is also used to describe things that are red by their nature such as red wine or a red moon. It could be that we are going to be hit by meteorites or a comet that breaks up in the atmosphere and turns the color of the waters red. This seemingly natural disaster from Space will cause more water to turn red.

Elijah stopped the rain for 3 ½ years when he prophesied in Israel during his life on earth. There are references that indicate that he will do that again during the first 3 ½ years of the Tribulation.

Remember that Elijah and Moses will be in Jerusalem during the first 3 ½ years of the Tribulation. When the Antichrist invades Jerusalem in the Middle of the Tribulation he will kill them both. The whole world will see it live on TV. A celebration will begin in all of the countries controlled by the Antichrist all over the globe. **They will exchange gifts** as part of this celebration.

I wonder if there is a Christmas connection here? Christmas is not a Christian doctrine. December 25, has traditionally been the day pagans celebrate the birth or rebirth of Nimrod the Sun god which dates the practice all the way back to Babylon.

Over the centuries the pagan religion spread around the world, the name has changed to accommodate the local languages, but the story and practices have remained pretty much the same.

Nimrod's wife and mother is Semiramis who eventually became known as the goddess Astarte which became Asherah/Ashtoreth which became Isis/ Ishtar which became Easter. Isis is known as the goddess of liberties and her statue is in New York harbor.

*"And they of the people and **kindreds and tongues and <u>nations</u> shall <u>see</u>** their dead bodies **three days and an half**, and shall not suffer their dead bodies to be put in graves"* (Revelation 11:9)

"And they that dwell upon the earth shall rejoice over them, and make merry, and shall send gifts one to another; because ***these two prophets tormented them that dwelt on the earth****"*. (Revelation 11:10)

The bodies of The Two Witnesses are left in the street for all to **see**. Until recently it was not possible for "*all nations, kindreds and tongues*" to **see** what is going on in Jerusalem. This sounds like a global event with everyone watching CNN worldwide. That was not possible 50 years ago. For centuries Bible critics have said that these kinds of things were not possible. They said that it was all nonsense. Today we realize that it is possible to do that. The critiques are silenced on that one point, but just remember that **God said it first.**

"And after three days and an half the Spirit of life from God entered into them, and they stood upon their feet; and great fear fell upon them which saw them". (Revelation 11:11)

The earth will be dried up because of a lack of rain. The Nile River has never dried up but the Bible said that it would dry up during the Tribulation. (Isaiah 19:5-10) It has been drying up for the last 50 years. The drought conditions create tinderbox conditions for forest fires. The meteors start the fires. Not only the rivers turn red, but the dust from these meteors may linger in the air.

I believe that the first Trumpet occurs in the middle of the Tribulation and is associated with the "***Abomination of Desolations***" by either immediately preceding or probably following it. This judgement seems to be Gods response to what the Antichrist does to Christians and Jews in the middle of the Tribulation.

Trumpet Judgements 2 – 5, I believe, are spread out over the last 3 ½ years.

The Second Trumpet Judgment

This judgement may be limited to the Mediterranean Sea. So either a third part of the Mediterranean Sea or a third part of something else will be affected. One third of all aquatic life will die and cause quite a stink!

Water will be contaminated, there will be less food available from the Sea and seaports will be affected.

"*And the* ***second angel sounded****, and as it were a great mountain burning with fire was cast into the sea: and the* ***third part of the sea became blood***"; (Revelation 8:8)

"*And the* ***third part of the creatures which were in the sea****, and had life, died; and the third part of the ships were destroyed*". (Revelation 8:9)

This is a good time to mention that N.A.S.A. warned years ago that our Solar System is moving through a part of our galaxy that is densely populated with debris of all sizes. The craters found on earth and dated, are said to correspond with the cycle of our Solar System traveling in and out of the regions of our Galaxy that are densely populated. So here we go again.

Third Trumpet Judgment

"*And the third angel sounded, and there fell a great star from heaven, burning as it were a lamp, and it fell upon the third part of the rivers, and upon the fountains of waters*"; (Revelation 8:10)

"*And the name of the star is called* ***Wormwood****: and the third part of the waters became wormwood; and many men died of the waters, because they were made bitter*". (Revelation 8:11)

Now we are talking about a full sized asteroid! This thing is as big as a mountain and it is nasty too! It's called "Wormwood" for a reason! It poisons the water! It pollutes our environment and it is called "Wormwood"!

Here is a quote from a publication titled "Wormwood Forest: A Natural History of Chernobyl, by Mary Mycio". The publication date was Sept. 9, 2005. – The following excerpt is from the chapter entitled "Biblical Botany": In this chapter Mary Mycio says…

"In the wake of the Chernobyl explosion, few people in the officially atheist Soviet Union had Ukrainian-language Bibles. But some of those who did noted that the word "wormwood" in the Wormwood star of the book of Revelation was translated as "polyn" and was a very close botanical cousin to Chernobyl. Suddenly, the biblical prophecy seemed to acquire new meaning: wormwood was radiation, and it's mention in the Bible predated the nuclear apocalypse that would end the world. The story spread like wildfire through the Soviet rumor mill and as far as Washington, D.C., where President Ronald Reagan was said to have believed it, too." (A direct quote from Wormwood Forest: A Natural History of Chernobyl, by Mary Mycio

I don't have a strong opinion about the connection between the word "Chernobyl" and the word used for "wormwood" in that passage. A connection does seem plausible.

Fourth Trumpet Judgment

This may be the result of that big asteroid hitting the earth. It is believed that a much larger asteroid hit the earth about 65,000,000 years ago. That asteroid kicked up so much dust and debris that it cut off the suns light and the earth suddenly froze! This asteroid is not that big, but it kicks up some debris which obscures 1/3 of the sky.

"*And the fourth angel sounded, and the third part of the sun was smitten, and the third part of the moon, and the third part of the stars; so as the third part of them was darkened, and the day shone not for a third part of it, and the night likewise*". (Revelation 8:12)

N.A.S.A. has an Asteroid Redirect Mission which is moving ahead aggressively. It is interesting that N.A.S.A. has become so interested in destroying incoming targets from space. The Bible says that the armies of the world will stop fighting each other at Armageddon to attack Christ returning in the sky with his armies.

"Behold, he cometh with clouds; and ***every eye shall see him****, and they also which pierced him: and all kindreds of the earth shall wail because of him. Even so, Amen* (Revelation 1:7)

The technology to attack targets coming in from outer space did not exist at the time this was written but it is being tested now.

The Next 3 Trumpet Judgments Are Called The 3 Woes

"And I beheld, and heard an angel flying through the midst of heaven, saying with a loud voice, ***Woe, woe, woe****, to the inhabiters of the earth by reason of the other voices of the trumpet of the three angels, which are yet to sound!"* (Revelation 8:13)

Fifth Trumpet Judgment (First Woe)

"And **the fifth angel sounded**, and I saw a star fall from heaven unto the earth: and to him was given the key of **the bottomless pit**". (Revelation 9:1)

The Fifth Trumpet Judgement involves the bottomless pit. Now is a good time to review (chapters 18.9 & 18. 10 & 47.7)

Jesus has the key to the bottomless pit. Either this is Christ himself or it is someone who Christ gave the key to. Now we are seeing direct contact with the underworld! I think that it is important to understand who has the keys. The keys, I think, implies that the place is locked up. The person with the key controls access. In other words, there is no passage or access from either side. Jesus controls all of that with the key.

The bottomless pit, I believe, is the Southern part of a hollow shaft that connects the surface with the earth's hollow center. The Bible describes the hollow center somewhat in the story of Lazarus and the Rich man. (Luke 16:20-31) I believe that a hollow shaft, from the surface at the South Pole, travels through the hollow center and connects it with the surface on the other side at the North Pole. The North Pole entrance is obscured by water. The South Pole entrance is obscured by snow and mountains.

Traditionally, people believed that the earth was hollow. Scorners often cook up Pseudo-science to contradict the Bible. But the Bible is always proven to be right. This issue of whether the earth is full of hot, liquid lava or just hollow like the Bible says will be no exception.

"*And he opened the bottomless pit; and there arose a smoke out of the pit, as the smoke of a great furnace; and the sun and the air were darkened by reason of the smoke of the pit*". (Revelation 9:2)

"*And there came out of the smoke locusts upon the earth: and unto them was given power, as the scorpions of the earth have power*" (Revelation 9:3)

For years I have known about these special locusts that sting repeatedly for five months. They will be very painful stings but will not be lethal. For five months they will sting those who do not have the seal of God in their foreheads. I think that you get the seal instructions from the one hundred and forty four thousand Jewish missionaries sent out by God's two witnesses.

If you get left behind because you never understood and believed the Gospel, this is God's way of getting your attention. If you don't find out about the seal of God before this judgement begins, you will be bitten! That is good encouragement.

"*And it was commanded them that they should not hurt the grass of the earth, neither any green thing, neither any tree; but* ***only those men which have not the seal of God in their foreheads****.* (Revelation 9:4)

"*And to them it was given that they should not kill them, but that they should be tormented five months: and their torment was as the torment of a scorpion, when he striketh a man*". (Revelation 9:5)

"*And in those days shall men seek death, and shall not find it; and shall desire to die, and death shall flee from them*". (Revelation 9:6)

I think this passage is saying that people will really be in so much pain and torment that they will wish they were dead! But these are not lethal stings. If you shoot them, they will die like anyone else but not from the stings.

"*And the shapes of the locusts were like unto horses prepared unto battle; and on their heads were as it were crowns like gold, and their faces were as the faces of men*". (Revelation 9:7)

"*And they had hair as the hair of women, and their teeth were as the teeth of lions*". (Revelation 9:8)

"*And they had breastplates, as it were breastplates of iron; and the sound of their wings was as the sound of chariots of many horses running to battle.* (Revelation 9:9)

It sounds like these things are going to be hard to kill. This doesn't sound like anything that any of us have seen. I have considered the possibility that they could be man-made and mechanical in nature because we have the technology to do that. The opposition to that notion that I get for that suggestion is that the creatures come out of the bottomless pit and God controls that.

There has been a lot of occult interest about the South Pole starting with the Nazis. First Hitler, then Admiral Byrd and now Obama. Why did Obama go to the southern tip of Argentina, the closest point to Antarctica and give his speech facing due south?

These "locusts" may be man-made or made by God. If they are man-made, God's people will take control. If they are God made, then they certainly are doing his will. Attacking unbelievers is a good thing. Maybe they will be persuaded to believe on Jesus.

"*And they had tails like unto scorpions, and there were stings in their tails: and their power was to* ***hurt men five months***". (Revelation 9:10)

Five months is a long time if you are being tormented by these things. That might cause one to think about that "Seal of God" and wonder about how to get one for yourself!

"And they had a king over them, which is the angel of the bottomless pit, whose name in the Hebrew tongue is ***Abaddon****, but in the Greek tongue hath his name* ***Apollyon****".* (Revelation 9:11)

It is interesting to note that Daniel chapter 9 verse 11 has a reference to "**Alah**" and Revelation chapter 9 verse 11 has a reference to **Apollyon**. These appear to be parallel verses. A statue of the Hindu god Lord Shiva, the "god of Destruction" or called "The Destroyer" is prominently displayed outside of the C.E.R.N. Hadron Partial Collider. Of further interest should be the fact that the town where C.E.R.N. is located is called-Saint Genis Pouilly, which In Roman times was called Apolliacum, the town and a temple there were both dedicated to Apollyon. Wow! There is a connection here, but what does it mean?

"One ***woe*** *is past; and, behold, there come two woes more hereafter".* (Revelation 9:12)

Sixth Trumpet Judgment (Second Woe)

This judgement is associated with Armageddon. It is predictive in nature. It is saying that the Euphrates River will be dried to prepare the way for a two hundred million man army from China.

"And the sixth angel sounded, and I heard a voice from the four horns of the golden altar which is before God", (Revelation 9:13)

"Saying to the sixth angel which had the trumpet, Loose the four angels which are bound in the great river Euphrates". (Revelation9:14)

"And the four angels were loosed, which were prepared for an hour, and a day, and a month, and a year, ***for to slay the third part of men****".* (Revelation 9:15)

"*And the number of the army of the horsemen were **two hundred thousand thousand**: and I heard the number of them.*" (Revelation 9:16)

China has boasted a 200,000,000 man and woman army since the 1960's. Until this century, many things mentioned in the Bible were ridiculed as impossible.

"*And thus I saw the horses in the vision, and them that sat on them, having breastplates of fire, and of jacinth, and brimstone: and the heads of the horses were as the heads of lions; and out of their mouths issued fire and smoke and brimstone*" (Revelation 9:17)

"By these three was the third part of men killed, by the fire, and by the smoke, and by the brimstone, *which issued out of their mouths*". (Revelation 9:18)

Could this be lasers being used by the Antichrist military?

"*For their power is in their mouth, and in their tails: for their tails were like unto serpents, and had heads, and with them they do hurt*". (Revelation 9:19)

That sounds like maybe a missile with fins and a warhead.

"*And the rest of the men which were not killed by these plagues yet repented not of the works of their hands, that they should not worship devils, and idols of gold, and silver, and brass, and stone, and of wood: which neither can see, nor hear, nor walk*": (Revelation 9:20)

"*Neither repented they of their murders, nor of their sorceries, nor of their fornication, nor of their thefts*". (Revelation 9:21)

OK, they asked for it! They did not want God. This is their bed. They made it, now they can have the N.W.O. and all of its privileges for those who worship the Antichrist and take his mark. But if one doesn't want to go along with that, they will be hunted, tormented, humiliated and killed.

Islam is the opposite of Biblical Christianity. According to Islam, to believe the Christian Doctrine of Salvation by faith is to be guilty of the most abominable sin of all! Satan doesn't want people to find the way to salvation.

Jesus is the way! He came to the Earth in Human flesh; to teach and have fellowship with his creation; and to pay for the sins of the world.

Satan doesn't want anyone to believe that eternal salvation is free; it is a gift from God; it is by Just Faith; not of works; not of yourself; you get it when you believe, not later; you can never lose it; it is eternal and it is a done deal!

To believe the gospel is to be born again forever! It can't be undone! Satan doesn't like for people to understand that because once they understand that eternal life is a free gift that God freely gives to all those who put their trust in his payment for sin, they become saved!

God has offered eternal life free to anyone who will put their trust in him. Those who have not put their trust in Christ by this time are in serious trouble! After Jesus stops Armageddon, his Angels will gather all of the unbelievers together. Every unbeliever, every single one throughout the world will be gathered in the Valley of Jehoshaphat where Christ will slay them all and judge them to determine their degree of punishment in Hell!

"Assemble yourselves, and come, all ye heathen, and gather yourselves together round about: thither cause thy mighty ones to come down, O LORD". (Joel 3:11)

"*The* ***second woe is past****; and, behold,* ***the third woe cometh*** *quickly*". (Revelation 11:14)

Seventh Trumpet Judgment. (The Third Woe)

"And the seventh angel sounded; and there were great voices in heaven, saying, The kingdoms of this world are become the kingdoms of our Lord, and of his Christ; and he shall reign for ever and ever". (Revelation 11:15)

This is where Christ takes his kingdom at the end of the 7 years of tribulation. For unbelievers this will be the ultimate "*Woe*"!

43.12 Seven Thunder Judgments

John is about to write about the Thunder Judgements when Christ stops him. The Thunder Judgements are not described. They remain a mystery, a surprise for Satan.

I believe that the first six of these thunder judgements may occur during the 2nd half of the tribulation, before Armageddon and the seventh may occur before the Feast of Tabernacles.

"*And I saw another mighty angel come down from heaven, clothed with a cloud: and a rainbow was upon his head, and his face was as it were the sun, and his feet as pillars of fire*": (Revelation 10:1)

"*And he had in his hand a little book open: and he set his right foot upon the sea, and his left foot on the earth*", (Revelation 10:2)

"*And cried with a loud voice, as when a lion roareth: and when he had cried, **seven thunders** uttered their voices*" (Revelation 10:3)

"*And when the **seven thunders** had uttered their voices, I was about to write: and I heard a voice from heaven saying unto me, **Seal up those things** which **the seven thunders uttered,** and **write them not***". (Revelation 10:4)

It looks to me like this may be Jesus himself! It looks like he is preparing to attack Satan's armies with these 7 Thunder Judgements!

"*And the angel which I saw stand upon the sea and upon the earth lifted up his hand to heaven*", (Revelation 10:5)

"*And sware by him that liveth for ever and ever, who created heaven, and the things that therein are, and the earth, and the things that therein are, and the sea, and the things which are therein, that there should be time no longer:*" (Revelation 10:6)

If this isn't Jesus, then whoever it is, he wants to get busy right now! I believe that the **7 thunder Judgements** are like 7 God sized Wild Cards! No one knows what they are or when they will be. It would be helpful to know when and where not to be to avoid these judgements. I believe that is where the 144,000,000 young Jewish Missionaries from all twelve tribes of Israel, plays an important role. Knowing where to go and when to get there is also important!

"*But in the days of the voice of the* ***seventh angel****, when he shall begin to sound, the mystery of God should be finished, as he hath declared to his servants the prophets*" (Revelation 10:7)

By the time the 7th angel sounds, people should know that there is a God in Heaven. They should know because this is what all of the Prophets of God were talking about! There should be no doubt, but apparently many still don't humble themselves to God and believe the Gospel. There will be plenty of nonbelievers when Christ comes back. They will be the majority. When God says that they will be cruel, then you can believe that it is going to be very cruel!

"*And the voice which I heard from heaven spake unto me again, and said, Go and take the little book which is open in the hand of the angel which standeth upon the sea and upon the earth*". (Revelation 10:8)

"*And I went unto the angel, and said unto him, Give me the little book. And he said unto me, Take it, and eat it up; and it shall make thy belly bitter, but it shall be in thy mouth sweet as honey*" (Revelation 10:9)

"*And I took the little book out of the angel's hand, and ate it up; and it was in my mouth sweet as honey: and as soon as I had eaten it, my belly was bitter*". (Revelation 10:10)

"*And he said unto me, Thou must prophesy again before many peoples, and nations, and tongues, and kings*". (Revelation 10:11)

Some teach that this "little book" is the Book of Daniel.

43.13 Seven Vial Judgments

"*And I saw another sign in heaven, great and marvellous,* ***seven angels having the seven last plagues****; for in them is filled up the wrath of God*". (Revelation 15:1)

"*And I saw as it were a sea of glass mingled with fire: and them that had gotten the victory over the beast, and over his image, and over his mark, and over the number of his name, stand on the sea of glass, having the harps of God*". (Revelation 15:2)

"*And they sing the song of Moses the servant of God, and the song of the Lamb, saying, Great and marvellous are thy works, Lord God Almighty; just and true are thy ways, thou King of saints*". (Revelation 15:3)

"*Who shall not fear thee, O Lord, and glorify thy name? for thou only art holy: for all nations shall come and worship before thee; for thy judgments are made manifest*". (Revelation 15:4)

"*And after that I looked, and, behold, the temple of the tabernacle of the testimony in heaven was opened*": (Revelation 15:5)

"*And the* ***seven angels*** *came out of the temple, having the* ***seven plagues****, clothed in pure and white linen, and having their breasts girded with golden girdles*". (Revelation 15:6)

"*And one of the four beasts gave unto the seven angels* ***seven golden vials full of the wrath of God****, who liveth for ever and ever.* (Revelation 15:7)

"*And the temple was filled with smoke from the glory of God, and from his power; and no man was able to enter into the temple, till the* ***seven plagues of the seven angels*** *were fulfilled*". (Revelation 15:8)

"*And I heard a great voice out of the temple saying to the seven angels, Go your ways, and pour out the vials of the wrath of God upon the earth*". (Revelation 16:1)

God is still trying to get people's attention, but they are not listening to God. They are listening to Satan and they are being deceived!

The First Vial

"*And the first went, and poured out his vial upon the earth; and there fell **a noisome and grievous sore** upon the men which had the mark of the beast, and upon them which worshipped his image*". (Revelation 16:2)

This mark is an express image under the skin. So it is identification, stored records and probably much more. The chip probably has a lithium battery. Electronic warfare or pulses from an unstable Sun could damage electronics in the chip or battery and cause the battery to leak. A leaking lithium battery under your skin **could cause a grievous sore**!

The Second Vial.

"***And the second angel poured out his vial upon the sea;** and it became as **the blood of a dead man:** and every living soul died in the sea*". (Revelation 16:3)

Don't drink the water! I can't help noticing that there is going to be a water problem. The problem is that the water will be turning red and stinking horribly! The smell is compared to a dead man's blood. That sounds bad! The fish and sea life will start dying. But some of the people who drink the water will also die!

The Third Vial

"*And the third angel poured out his vial upon the rivers and fountains of waters; and they became blood*". (Revelation 16:4)

"And I heard the angel of the waters say, Thou art righteous, O Lord, which art, and wast, and shalt be, because thou hast judged thus". (Revelation 16:5)

"For they have shed the blood of saints and prophets, and thou hast given them blood to drink; for they are worthy". (Revelation 16:6)

"And I heard another out of the altar say, Even so, Lord God Almighty, true and righteous are thy judgments". (Revelation 16:7)

The Bible has written plenty about this time that we are living in. Some of the details come from God's Prophets who wrote it down as much as 3500 years ago. Nevertheless, today we are seeing things mentioned in the Bible coming to pass. Some of these things no one expected to see before the Rapture unless it was just before. For example, many rivers and lakes are already turning red in an unprecedented fashion.

THE FOURTH VIAL

"And ***the fourth angel poured out his vial upon the sun****; and power was given unto him to scorch men with fire"*. (Revelation 16:8)

"And men were scorched with great heat, and blasphemed the name of God, which hath power over these plagues: and they repented not to give him glory". (Revelation 16:9)

Now this is an unstable Sun being described. A solar pulse is the sign of an old and unstable star. According to the Bible, a little more than a thousand years later the earth will be consumed by fire. That sounds like our sun going through the "red giant stage" of an old star. In such a case, the sun would change from its yellow color to red and expand out and burn up the first three or four planets.

THE FIFTH VIAL

God is really trying to let these guys know who the boss is, but they just aren't getting it!

"And the fifth angel poured out his vial upon the seat of the beast; and his kingdom was full of darkness; and they gnawed their tongues for pain", (Revelation 16:10)

"And blasphemed the God of heaven because of their pains and their sores, and repented not of their deeds". (Revelation 16:11)

The Sixth Vial

The River Euphrates has never dried up in all of human history. So this is another prophecy that people had a hard time believing. Yet now we have five dams along the river that could be used to shut off the water supply. I just recently heard about a terrorist group who back in March had taken over 6 out of 8 large dams on the Tigris and Euphrates Rivers and is threatening to shut off the water. They are continually attacking a seventh dam.

"And the sixth angel poured out his vial upon the great river Euphrates; and the water thereof was dried up, that the way of the kings of the east might be prepared". (Revelation 16:12)

Remember that the "Kings of the East" is China. The Bible gave the exact number of the army at a time when none of this seemed possible.

"And I saw three unclean spirits like frogs come out of the mouth of the dragon, and out of the mouth of the beast, and out of the mouth of the false prophet". (Revelation 16:13)

"For they are the spirits of devils, working miracles, which go forth unto the kings of the earth and of the whole world, to gather them to the battle of that great day of God Almighty". (Revelation 16:14)

"Behold, I come as a thief. Blessed is he that watcheth, and keepeth his garments, lest he walk naked, and they see his shame". (Revelation 16:15)

"And he gathered them together into a place called in the Hebrew tongue Armageddon". (Revelation 16:16)

This battle will last only one hour and will wipe out 1/3rd of all men left alive at that time.

The Seventh Vial

"*And* ***the seventh angel poured out his vial into the air****; and there came a great voice out of the temple of heaven, from the throne, saying,* ***It is done***". (Revelation 16:17)

"*And there were voices, and thunders, and lightnings; and there was a great* ***earthquake****, such as was not since men were upon the earth, so mighty an earthquake, and so great*". (Revelation 16:18)

"*And the great city was divided into three parts, and the cities of the nations fell: and great Babylon came in remembrance before God, to give unto her the cup of the wine of the fierceness of his wrath*". (Revelation 16:19)

"*And every island fled away, and the mountains were not found*". (Revelation 16:20)

"*And there fell upon men a great hail out of heaven, every stone about the weight of a talent: and men blasphemed God because of the plague of the hail; for the plague thereof was exceeding great*". (Revelation 16:21)

This is the final plague. It is over. Now it is time for the judgement!

44

Armageddon, The Harvest!

44.1 "Beat Your Plowshares Into Swords"

There is currently 7.6 billion people said to be living on planet earth. After the battle of Armageddon, God's Angels will separate the survivors into two groups.

First the unbelievers will be taken by the Angels to Jesus for their destruction. After that there will only be about 20-30 million survivors left to live on into the Millennium.

(Megiddo. Gr. for Armageddon Hebrew)

Before the battle of Armageddon starts, there will be an arms race unlike anything ever seen before in human history! We are seeing that arms race now!

"*Beat your plowshares into swords, and your pruninghooks into spears: let the weak say, I am strong*" (Joel 3:10)

The weapons technology has already advanced far beyond anything that the public can imagine. If the enemy doesn't know about your capabilities, he can't devise a defense against it. The Biblical descriptions of Armageddon have seemed too fantastic to be true for some people. But now, technology has made it seem that almost anything is possible.

44.2 Arms Race In Space

"Proclaim ye this among the Gentiles; Prepare war, wake up the mighty men, let all the men of war draw near; let them come up": (Joel 3:9)

The Tribulation will start out with an arms race already in full gallop! There will be wars, rumors of wars and an arms race like never before! The technology is currently far beyond anything that we can imagine. By the time you find out something new that they can do, there is already something else, much better coming up right behind the new one. It is now a steady stream of innovations that can be weaponized. Everyone is trying to outperform their opponent in military technology.

I believe that the Western Empire of the Antichrist will have surprisingly superior technology. Some of that technology will be utilized early in the Tribulation, but there will be undisclosed surprises saved for the final battle. Everyone will have their own nasty surprise. The Antichrist will have control of all of the technology and all the military might of the United States.

44.3 The Other Nations

The 70th Week of Daniel will begin with war, just after sundown on Yom Kippur some year in the future. This is when the Islamic Antichrist takes over the West, starting with the U.S.. Toward the middle of the Tribulation the Antichrist will take Egypt and prepare to take Ethiopia and Libya, but will be stopped by threats from Russia and China.

The Antichrist is Diabolical, evil, powerful and cruel… and he will be in control of the U.S. military during the Tribulation. The United States by that time will probably look like a war zone from coast to coast but its military will be advanced, intact and the Antichrist will be in control.

About 2600 years ago, Long before Russia existed, the Bible talks about the Russia that we know of today. You can't deny this prophecy because of the detail and the length of time involved. Back in 1909 the Scofield

Reference Bible was copyrighted. Written over one hundred years ago ... See chapter 46 for Scofield Reference Bible, footnote I, from page 883" in the Scofield Reference Bible.

That note was written <u>before</u> the Bolshevik Revolution and before the technology existed to do the things mentioned in Ezekiel and other places in the Bible. You can't deny that this is talking about Russia because it gives Russia's name and the names of its two major cities long before they existed. The Bible gives the direction of Russia as north, relative to Jerusalem. If you held a string on the North Pole with one finger and held the other end on Jerusalem with the other finger, the string would cross right through Moscow because it is directly North of Jerusalem. The Bible is an amazing book that you can trust what it says, and stake your life on it.

According to what Ezekiel wrote 2600 years ago, the **Russian alliance** will be joined by Persia, Ethiopia and Libya. The Bible says that Russia will stop the Antichrist/Little Horn (West) from taking over Ethiopia and Libya, however the Antichrist/Little Horn (West) will have already taken over **Syria**, Lebanon, Saudi Arabia, Egypt and Israel. Currently Syria is Russia's Ally, but the Bible says that the U.S. will take it over after it takes over Europe. Russia will lose Syria, but it may gain other Allies.

"And say, Thus saith the Lord GOD; Behold, I am against thee, O Gog, the chief prince of Meshech and Tubal": (Ezekiel 38:3)

"And ***I will turn thee back, and put hooks into thy jaws, and I will bring thee forth****, and all thine army, horses and horsemen, all of them clothed with all sorts of armour, even a great company with bucklers and shields, all of them handling swords"*: (Ezekiel 38:4)

I think that the "hook" mentioned here is the oil. The word for horses means "swift in flight" and is also translated as "swallow" as in the bird.

"Persia, Ethiopia, and Libya with them; all of them with shield and helmet:" (Ezekiel 38:5)

The ***"Kings of the East"*** which can only be **China** with her **200,000,000 man army,** China's allies and a group that is called the **Kings of the South**, which are probably **Persia** and their allies will join in this fight.

The Bible says that a house divided cannot stand. Satan's "house" is divided. After the middle of the Tribulation, satan will be in control of the world government and will require worship. I don't believe that President Putin or President Xi Jinping or Trump will go along with any of that so I expect for some key leaders to be replaced by coup, assassination or by other means before the Battle of Armageddon occurs.

That preemptive attack occurs one hour before Christ stops it. God intervenes because Israel's remnant is at stake. In one hour 1/3rd of those remaining alive will die and 5/6ths of Russia will be included in that number. Siberia sounds like the one sixth that will survive. That makes Siberia a good place to escape to if you can deal with the weather.

*"But I will remove far off from you the **northern army**, and will drive him into a land barren and desolate, with his face toward the **east sea**, and his hinder part toward the **utmost sea**, and his stink shall come up, and his ill savour shall come up, because he hath done great things*". (Joel 2:20)

That passage is describing Siberia's location 2600 years ago. There were no maps for the prophet to look at. So how did he know? He sure did not go up there and take a look. He knew because God told him.

"*Thou shalt **ascend** and come like a storm, thou shalt be **like a cloud to cover the land**, thou, and all thy bands, and **many people** with thee*" (Ezekiel 38:9).

This army from the North is heading south to Israel, yet the verbiage says that they will "***<u>ascend</u>*** *and come like a storm, thou shalt be like* ***a <u>cloud</u> to cover the land***"

This is a clear reference to a very large air born attack. Why else would the language say that these guys are going to "***<u>ascend</u>***" when they are traveling South? The next words are "***Thou shalt be like a <u>cloud</u> to cover the land***".

Well that is literally what they are going to do. They will ascend into the air for an air born attack. This sort of thing was unimaginable at the time that this was written. Down through the years critiques of the Bible have used passages like this to ridicule the Bible, but no one is laughing at this possibility anymore. Just remember that God said it first!

"*Therefore, thou son of man, prophesy against Gog, and say, Thus saith the Lord GOD; Behold, I am against thee, O Gog, the chief prince of Meshech and Tubal*": (Ezekiel 39:1)

"*And I will turn thee back, and* ***leave but the sixth part of thee****, and will cause thee to come up from the north parts, and will bring thee upon the mountains of Israel:*" (Ezekiel 39:2)

This is talking about the Russian Army that will be defeated by Christ's returning army. They all run to Siberia because that is all that is left of Russia! If a Russian Believer gets sent to Siberia during the last months of the Tribulation he will probably survive. 5/6ths of Russia will be destroyed! Siberia is going to be all of Russia that will survive Armageddon! Siberian real-estate is a good investment!

44.4 All Nations Will Be Gathered Against Israel.

"*Then shall they deliver you up to be afflicted, and shall kill you: and* ***ye shall be hated of all nations*** *for my name's sake*". (Matthew 24:9)

"*For I will gather all nations against Jerusalem to battle; and the city shall be taken, and the houses rifled, and the women ravished; and half of the city shall go forth into captivity, and the residue of the people shall not be cut off from the city*". (Zechariah 14:2)

"*Then shall the LORD go forth, and fight against those nations, as when he fought in the day of battle*". (Zechariah 14:3)

THERE ARE 4 DIFFERENT WORLD ALLIANCES INVOLVED

The **Kings of the North** (Russia), who may lead the resistance will be allied with **Libya**, **Ethiopia** and **Turkey**, just to mention a few. (Ezekiel 38:5-6)

The **Kings of the South**, Persia (Iran) and their allies…

The **Kings of the East** (China) and their allies

The **Antichrist as dictator of the Western Nations** and part of the Middle East. He will be possessed by Satan who will be ruling from the Temple in Jerusalem when he is attacked.

The Western Nations which were formerly the Ancient Roman Empire is modern day Europe. The Eastern Nations are the descendants of the Assyrian Empire which is Babylon or modern day Iraq, Saudi Arabia, Egypt, Syria, Lebanon...

44.5 A 200 MILLION MAN ARMY CROSSES A DRIED UP EUPHRATES RIVER.

The Bible gives the exact number of soldiers in the Chinese army that crosses over the dried up Euphrates River. The number of men in that army is 200,000,000. And **that is the exact number that the Chinese Army can produce today**.

"And the number of the army of the horsemen were two hundred thousand thousand: and I heard the number of them". (Revelation 9:16)

A little reminder here about when this was written… It was written 2000 years ago when there wasn't that many people on the earth. Only God can tell the future like that. I don't hear anything even close to that coming out of the Koran. The Bible is full of Prophecy that has already been accurate about the past. There is plenty more prophecy coming in the very near future that will also be fulfilled with the same precision.

"And the sixth angel poured out his vial upon ***the great river Euphrates****; and* ***the water thereof was dried up****, that the way of the kings of the east might be prepared"*. (Revelation 16:12)

Recently, hydroelectric dams have been built on the Euphrates river. These dams control the water and electric power. The dams are being taken over by rebels. Last I heard, rebels control 6 out of 8 large hydroelectric dams on the Tigris and Euphrates Rivers. The rebels are threatening to shut off the water.

Before those dams were built, such an event was not possible! People in the past could not imagine how this could be. But now it is and it is the timing of all of these different events that should be getting people's attention.

44.6 Important Things To Know About Armageddon

The Tribulation will end with the 6th Seal Judgment; and the 6th Trumpet Judgement; and the 6th Vial Judgement and perhaps the 6th Thunder Judgement and the Battle of Armageddon. Somewhere in those Judgements which will be most intensified as we approach Armageddon, will be a preemptive attack Led by Russia and Her allies against the Antichrist.

Now remember that The Antichrist has taken over Jerusalem and has set up his image in the third temple and is requiring everyone to worship his image.

The Russians say "enough is enough" and decides to do something about this nonsense. Russia gathers up support and allies for the job. Now at this point it would seem that Russia would be a world hero. Up to this point they may be. However God sees more than we do. God sees the motivations of the Russian leadership and those motives are not good.

Remember that the whole world is under economic control by the Antichrist and they are required to worship him. I expect for existing leaders to be replaced by politicians who are willing to do that sort of thing. For example

I would expect Trudeau, Clintons and their crowd to fit in just fine with the Antichrist, but Putin, Xi and Netanyahu probably will not.

The Russian leadership will not try to save the world from the Antichrist because they just don't believe in that sort of thing. The Russian leaders wants control of the oil and other resources of Israel. They want to get rid of the competition and get a bigger piece of the pie for themselves. The problem is that to accomplish their goal, they are going to have to strike at God's land and threaten the people of God's promise and that makes God angry!

"*And thou shalt come from thy place out of the north parts, thou, and many people with thee, all of them riding upon horses, a great company, and a mighty army*" (Ezekiel 38:15)

This passage has been supposed by many to refer to the Middle of the Tribulation but that is an error. In verse 20 the presence of the Lord is mentioned and that doesn't happen till Armageddon, at the **end** of the Tribulation.

Russia strikes Megiddo Air Base, which is one of Israel's principle airbases located in the valley of Jezreel also called Megiddo and Armageddon. Armageddon will be the greatest tragedy in all of human history!

44.7 The Approximate Time When Armageddon Begins.

It is true that you cannot know the day or the hour, but what about the week, the month and the year?

"*But of that day and hour knoweth no man, no, not the angels of heaven, but my Father only*". (Matthew 24:36)

People often quote the passage to say that you can't know the day and hour of the Rapture, but this is not talking about the Rapture. This passage is talking about the second return of Christ, which is seven years after the

Rapture. Christ retakes earth from satan at end the battle of Armageddon, seven years after the Rapture!

Just like we have already discussed concerning the Rapture which will happen 7 years earlier, strict interpretation is essential. It is important for those who call on Jesus during the Tribulation to know about Armageddon and **when it will happen!**

The battle of Armageddon will be fought **between the Feast of Atonement and the Feast of Tabernacles**. That is thirteen days out of the year that it could happen. You may not know the day and the hour but I believe that it will occur on one of those 13 days.

It must happen between those two Feast days because the Tribulation begins and ends on Yom Kippur and the millennium begins on the feast of Tabernacles.

Seven Hebrew years (360 day years) later, or 2520 days later, the Tribulation or if you prefer, the ***"70th week"* of Daniel will end**. If you want to count from the rapture, just add 9 to 2520 for 2529 as a start counting point from the last trump of Rosh Hashanah. Remember, it could happen on any day between the last two feasts of Israel.

Tribulation Saints will have information to help them survive, such as take cover by the Feast of Atonement and be prepared to stay underground until notified that it is safe. The only source of good information is the Bible and the 144,000 Jewish missionaries sent out by God during the first half of the Tribulation. These folks will have been dead since the 5th seal and not available in person at this critical time. However, the words of the 144,000 are from God so their words will still be around. The Tribulation Saints will find out what they said and write it down!

44.8 Armageddon; 1/3rd Of Mankind Killed In An Hour!

The battle of Armageddon will last one hour and wipe out a third of all mankind. (Revelation 9:14-16)

"And the sixth angel sounded, and I heard a voice from the four horns of the golden altar which is before God" (Revelation 9:13)

"Saying to the sixth angel which had the trumpet, Loose the four angels which are bound in the great river Euphrates". (Revelation 9:14)

"And the four angels were loosed, which were prepared for ***an hour****, and a day, and a month, and a year, for* ***to slay the third part of men****".* (Revelation 9:15)

"And the number of the army of the horsemen were ***two hundred thousand thousand****: and I heard the number of them".* (Revelation 9:16)

"And the sixth angel poured out his vial upon the great river Euphrates; and the water thereof was dried up, that the way of the ***kings of the east*** *might be prepared".* (Revelation 16:12)

According to this passage written about 2,000 years ago, the Euphrates River will dry up so a 200,000,000 man army could cross over. At the time that this was written, there were not that many people on the earth and no one believed that the great river Euphrates would ever dry up.

Make no mistake about it, **Armageddon will be Nuclear!** Radiation sickness is described in the Bible! Furthermore, the type of weapons and the type of warfare used at Armageddon has never been seen before. The world will display all of the technology they are amassing to destroy one another. Everybody will be trying to surprise their enemies with new and innovative ways to kill each other.

"And except those days should be shortened, there should no flesh be saved: but for the elect's sake those days shall be shortened". (Matthew 24:22)

The **Battle will last for one hour and destroy 5/6th of Russia**. Siberia is described as the only part of Russia to survive. The northeastern part of Siberia will be a place of refuge to those who survived Armageddon.

"And except those days should be shortened, there should no flesh be saved: but for the elect's sake those days shall be shortened". (Matthew 24:22)

.The technology to do those kinds of things were inconceivable to anyone living more than 300 years ago or even 100 years ago. Yet the Bible talks about it with accuracy. The Bible has an established record of accuracy. The ability to tell the future with 100% accuracy is God's confirmation that the Prophet is His.

The Bible fits together perfectly when you follow the literal interpretation. All false doctrine stems from leaving the good tree of strict, literal interpretations that leads to eternal life! Just Faith! Those who don't believe one hour before Armageddon begins, probably won't.

During the battle Christ Appears in Heaven. Some of the unbelievers will know that they are in serious trouble. Many, I believe will believe the lies of the Antichrist and consider this to be an Alien invasion and attack Jesus and His armies!

44.9 NASA Is Testing Its Asteroid Redirect Program.

N.A.S.A. is in the testing stage of its "Redirect" program. As we have known for some time now, we are entering a densely populated region of the Galaxy. The cycle of the earth passing through this region is reflected by the fossil record and by Geological evidence of past collisions.

N.A.S.A. has decided not to go through this region of space unarmed! Although the system is not fully implemented and upgrades are expected, it has proved to be functional.

The idea is to first, locate and track large objects that may collide with the Earth. The challenge is to gather this information in time to do something about it! That technology is improving rapidly. Next challenge is to get a large enough nuclear explosive device to the right spot in space at the right time.

N.A.S.A. is developing robotic, artificial intelligence along with other new technologies, to perform the tasks required.

Not only are nuclear warheads going to be employed but a towing method is being developed as well. A whole new field of science will develop around asteroids. All of these technologies can have military applications.

Space warfare has arrived and will be fully functional just in time to be used against Christ returning to earth.

After the Tribulation ends on Yom Kippur and before the Feast of Tabernacles, Russia will lead the attack against the Antichrist in Jerusalem.

44.10 The Great Harvest!

There is currently 7.6 billion people said to be living on planet earth. After the battle of Armageddon the survivors will be separated by God's Angels. First the unbelievers will be taken by the Angels *to the valley of Jehoshaphat* for their destruction. After that there will only be about 20 million left to live on into the Millennium.

There is said to be more people alive on the earth right now than all of history combined.

The Bible describes the time that we are living in as the time of Harvest…

"*Let the heathen be wakened, and come up to the valley of Jehoshaphat: for there will I sit to judge all the heathen round about*". (Joel 3:12)

"*Put ye in the sickle, for **the harvest is ripe:** come, get you down; for the press is full, the fats overflow; for their wickedness is great*". (Joel 3:13)

God has sent prophets. He has provided witnesses which gave their lives to spread the Gospel.

"*Another parable put he forth unto them, saying, The kingdom of heaven is likened unto a man which sowed good seed in his field*":(Matthew13:24)

"*But while men slept, his enemy came and sowed tares among the wheat, and went his way*". (Matthew13:25)

"*But when the blade was sprung up, and brought forth fruit, then appeared the tares also.*" (Matthew13:26)

"*So the servants of the householder came and said unto him, Sir, didst not thou sow good seed in thy field? from whence then hath it tares?*" (Matthew13:27)

"*He said unto them, An enemy hath done this. The servants said unto him, Wilt thou then that we go and gather them up?*" (Matthew13:28)

"***But he said, Nay; lest while ye gather up the tares, ye root up also the wheat with them***". (Matthew13:29)

"*Let both grow together until the harvest: and in the time of harvest I will say to the reapers,* ***Gather ye together <u>first</u> the tares****, and bind them in bundles to burn them: but gather the wheat into my barn*". (Matthew13:30)

"*Then Jesus sent the multitude away, and went into the house: and his disciples came unto him, saying, Declare unto us the parable of the tares of the field*"(Matthew13:36)

"*He answered and said unto them, He that soweth the good seed is the Son of man;*" (Matthew13:37)

"*The field is the world; the good seed are the children of the kingdom; but the tares are the children of the wicked one*;" (Matthew13:38)

"*The enemy that sowed them is the devil; the harvest is the end of the world; and the reapers are the angels* "(Matthew13:39)

"*As therefore the tares are gathered and burned in the fire; so shall it be in the end of this world*" (Matthew13:40)

"*The Son of man shall send forth his angels, and they shall gather out of his kingdom all things that offend, and them which do iniquity;*" (Matthew13:41)

"*And shall cast them into a furnace of fire: there shall be wailing and gnashing of teeth*". (Matthew13:42)

"*Then shall the righteous shine forth as the sun in the kingdom of their Father. Who hath ears to hear, let him hear*". (Matthew13:43)

45

The Lords Return!

45.1 The Vatican Owns A Telescope Called L.U.C.I.F.E.R.

The Vatican owns and operates an infrared, thermal imaging telescope located in Mt. Graham in south eastern Arizona. The name given to the Vatican's telescope was L.U.C.I.F.E.R. which is an acronym that stands for … **L**arge Binocular Telescope Near-infrared Spectroscopic **U**tility with **C**amera and **I**ntegral **F**ield unit for **E**xtragalactic **R**esearch

When the New Jerusalem comes from Heaven in the north. It will be visible to the naked eye. But the Antichrist and the False Prophet will probably see it first. That is because the false prophet has his own giant infrared telescope and they will be busy looking for it. They know where to look for it, so as soon as it appears they will be the first to "discover" it.

Being the first to discover something comes with a lot of talking rights. Everyone is going to want to hear all about it. Because the so-called "infallible" False Prophet Pope is "infallible" in all spiritual matters, he is going to authoritatively declare this to be a spiritual emergency, the demon possessed False Prophet will give another lie.

The false prophet will lead the way to misinformation and deception. People are being conditioned to believe in Alien abductions so it is plausible that the False Prophet will be the one who falsely identifies the New Jerusalem and God's armies as an alien invasion.

I found a website that did some research about this. It turns out that the Vatican is looking for extraterrestrials to visit the earth and they want to be the first to see and greet them. There are a lot of websites that deal with that sort of thing.

45.2 The New Jerusalem Is Seen As It Approaches

There will be no Aliens until Jesus comes back, at least not real ones. Jesus comes back with Cherubim, Seraphim, Angels and all of their vehicles or whatever they will bring with them. That should be interesting to see! It will be strange and scary for unbelievers!

Before Armageddon, all aliens will be fake! The real ones come with Jesus and they will gather all unbelievers into the valley of Jehoshaphat to be slayed by Jesus himself! The believers will be rescued.

The world will be misinformed by the False Prophet and the Antichrist so when they see the New Jerusalem coming out of heaven in the north, they will probably think that it is an alien spacecraft and attack it!

The New Jerusalem will be a cube measuring 1500 miles on each side. Everyone will see that coming from the North! People may be conditioned to think of it as something like the "Borg" from Star Trek because it is so large and cube shaped. The similarity in shape and size to the villains in a popular sci-fi story with a cult like following makes me suspicious. Anyway, the New Jerusalem is a little bigger than our moon and I believe that it will be visible long before it gets here. I believe it will orbit the earth.

Do you think that this is just another coincidence that the Vatican has already positioned themselves with a telescope to make that call? Or do you think that the Vatican is a place that is full of demon possessed perverts who secretly worship Satan?

Heaven is Northward and I believe that is the direction the New Jerusalem, Christ's armies and Christ himself will all come from". See (Leviticus 1:11); (Psalms 26:7; 48:1-2); (Isaiah 14:12-14); (Psalms 75:5); (Job 26:7)

"And the city lieth foursquare, and the length is as large as the breadth: and he measured the city with the reed, twelve thousand furlongs. The length and the breadth and the height of it are equal". (Revelation 21:16)

This thing is big! It contains 3.375 billion cubic miles of space! It has twelve round gates. That is two round openings for each side. (Revelation 21:21)

They're going to see this thing coming from a long distance away because it is big and it produces its own light (Revelation 21:23)! To the unbelievers who believed all the lies, this will be terrifying because it will mean their destruction. To the believers it means rescue!

45.3 "Every Eye Shall See Him"

*"Behold, he cometh with clouds; and **every eye shall see him**, and they also which pierced him: and all kindreds of the earth shall wail because of him. Even so, Amen"*. (Revelation 1:7)

All of the people in the world will have live T.V. coverage. They will see this on the Antichrist T.V. News network. They will see Jesus and his armies coming to get all of the **unbelievers first!**

There will be no mistake when Christ returns to stop Armageddon because every eye will see him! The preterists believe that these prophecies concerning Christs return are already past and fulfilled, but nobody saw it or recorded it. CNN, ABC, NBC and all of the others will be there. People will watch this on T.V. all over the world. This was not possible with the technology available when John wrote Revelation. God spoke about advancements in technology first!

*"But thou, O Daniel, shut up the words, and seal the book, even to the time of the end: **many shall run to and fro**, and **knowledge shall be increased**"*. (Daniel 12:4)

45.4 The 2nd Return Of Christ!

After the **sixth** "seal"; and **sixth** "trumpet"; and **sixth** bowl and **sixth** "thunder"; judgements are all done, then begins the 4 "#7's" Judgement's which will deal with the return of Jesus Christ to the Earth, and taking that first step onto the Mount of Olives. As he does, he will command the mountain to be removed as described in (Zachariah 14:4) the Bible. And it will do as he commands, just like it was described in the Bible! (Matthew 17:20)

"And his feet shall stand in that day upon the mount of Olives, which is before Jerusalem on the east, and the mount of Olives shall cleave in the midst thereof toward the east and toward the west, and there shall be a very great valley; and half of the mountain shall remove toward the north, and half of it toward the south" (Zachariah 14:4)

That valley will be called the Valley of Jehoshaphat.

Today it is commonly known that there is a rift in the Mount of Olives that runs east and west just like the Bible says.

"*Woe unto them that call evil good, and good evil; that put darkness for light, and light for darkness; that put bitter for sweet, and sweet for bitter!"* (Isaiah 5:20)

Marxism and Islam are in direct opposition to the Bible and what God has said. What the Bible calls good is called evil by both Islam and Marxism. What the Bible calls evil is called good by Islam and Marxism.

Jesus's Angels said that He would return in like manner. That means that He will come down slowly as the World watches and contemplates what this means. To those who are believers, this means rescue, but they will have to wait their turn. **The unbelievers will be taken First**!

To those who are unbelievers, this will mean terror! As a lost person begins to realize the Truth, I imagine that they will be in horror as they realize that the big picture is not what they thought.

They will be judged as unbelievers because they rejected truth! They did not want the truth so they got a lie. This is where lies ultimately lead…

"*And the third angel followed them, saying with a loud voice, If any man worship the beast and his image, and receive his mark in his forehead, or in his hand*", (Revelation 14:9)

"*The same shall drink of the wine of the wrath of God, which is poured out without mixture into the cup of his indignation; and he shall be tormented with fire and brimstone in the presence of the holy angels, and in the presence of the Lamb:*"!! (Revelation 14:10)

"*And the smoke of their torment ascendeth up for ever and ever: and* ***they have no rest day nor night, who worship the beast and his image****, and whosoever receiveth the mark of his name*". (Revelation 14:11)

After a warning like that, I would think that people would begin to recall things that they chose to ignore. Perhaps warnings from Christian friends who tried to warn about this. To them, this will be a horrifying sight! They will all want to hide but they can't!

…"*and his feet shall stand in that day upon the mount of Olives, which is before Jerusalem on the east, and the mount of Olives shall cleave in the midst thereof toward the east and toward the west, and there shall be a very great valley; and half of the mountain shall remove toward the north, and half of it toward the south*". (Zechariah 14:4)

Not too long ago seismologists discovered such a rift. The rift in the Mount of Olives already exists just like the Bible describes it. Men did not have the technology to discover this until recently, but the Bible talks about it 2600 years ago. **Just remember… God said it first!**

45.5 The 2nd Return; Unbelievers Are Taken And Believers Left

The valley mentioned in (Zechariah 14:4) will be called the Valley of Jehoshaphat. This is where all of the unbelievers are taken by the angels to be killed by Jesus Christ himself. Those fellas who are looking for 70 virgins are in for a big disappointment! Everyone who has rejected truth because they were willingly deceived will be brought to the Valley of Jehoshaphat by God's Angels!

The Bible says the unbelievers will be gathered up by the Angels **first**, and brought to this new valley running East and West through the Mount of Olives.

"*So shall it be at the end of the world: the angels shall come forth, and sever the wicked from among the just*", (Matthew 13:49)

45.6 Jesus Will Separate The Wheat From The Tares

<u>The First order of business</u> when Jesus Christ Returns will be to stop the Battle of Armageddon and **<u>gather all unbelievers</u>** into one place. This is where the wheat and tares are separated! **The tares are gathered <u>first</u>.**

"In the time of harvest I will say to the reapers, ***Gather ye together <u>first</u> the <u>tares</u>****, and bind them in bundles to burn them: but gather the wheat into my barn."* (Matthew 13:30)

I think that it is important to recognize **<u>who</u> is taken <u>first</u>** and who is left. **The tares are taken first and they represent the unbelievers!**

"Then shall two be in the field; the one shall be taken, and the other left. (Matthew 24:40)

After Christ returns the first order of business will be to remove all unbelievers. Their unbelief makes them unrighteous, spiritually dead in their own sins and iniquity; condemned, corrupt and an offence in the eyes of God…

"*The Son of man shall send forth his angels, and they shall gather out of his kingdom* ***all things that offend****, and* ***them which do iniquity***"; (Mathew 13:41)

This passage is talking about God's Angels gathering **<u>all unbelievers</u>** into one place on the earth. That spot is called the **"Valley of Jehoshaphat"**, but **<u>is doesn't exist yet</u>**.

They will be gathered by God's Angels and taken to the newly formed Valley of Jehoshaphat where they will be killed by Jesus Christ himself, and then cast into the Lake of Fire!

"*And shall cast them into a furnace of fire: there shall be wailing and gnashing of teeth.*" (Matthew 13:42)

The Antichrist and the False Prophet will be cast in first and they are the only ones said to be cast in alive!

Just like the valley of Jehoshaphat, God talks about places before they even existed. Some places were written about thousands of years before they even had a name! Well that is how it is with Russia, Moscow Tobolsk and other places as well. The lake of fire is just as real.

The Jesus of the Bible says that he is God and he is the only way. Jesus is God the Creator and Savior who paid for our sin! He says that if you believe that and trust in Him alone for your salvation, he gives you, as a free gift, Eternal Life!

"For by grace are ye saved by faith and that not of yourselves, it is the gift of God, not of works, lest any man should boast". (Ephesians 2:8, 9)

You don't even have to tell anyone. Just tell God. This is between you and Him. He is waiting and listening. But if you do share this wonderful news with others, God pays in wages and eternal rewards and pays lavishly!

If you do decide to tell others, God will reward you. If you don't tell others, you will still be saved but you will lose rewards.

"*Watch therefore: for ye know not what hour your Lord doth come*". (Mathew 24:42)

"*But know this, that if the goodman of the house had known in what watch the thief would come, he would have watched, and would not have suffered his house to be broken up*". (Mathew 24:43)

"*Therefore be ye also ready: for **in such an hour as ye think not** the Son of man cometh"*. (Mathew 24:44)

"*Who then is a **faithful** and **wise** servant, whom his lord hath made ruler over his household, to give them meat in due season?"* (Mathew 24:45)

"Blessed is that servant, whom his lord when he cometh shall find so doing". (Mathew 24:46)

"*Verily I say unto you, that **he shall make him ruler over <u>all</u> his goods***". (Mathew 24:47)

"*But and if that **evil** servant shall say in his heart, My lord delayeth his coming*"; (Mathew 24:48)

"*And shall begin to smite his fellowservants, and to eat and drink with the drunken*" (Mathew 24:49)

"*The lord of that servant shall come in a **day** when he looketh not for him, and in an **hour** that he is not aware of,* (Mathew 24:50)

"*And shall cut him asunder, and appoint him his portion with the hypocrites: **there shall be weeping and gnashing of teeth***". (Mathew 24:51)

The issue about the Gift of God is the most important issue for any individual or nation to consider! Young or old, anyone who can distinguish his left hand from his right is qualified for an **<u>Appointment with God</u>!** There are many false gospels to lead you to Hell but there is only one true Gospel that will take you to Heaven. **The true Gospel is Simple; it is free; and it will set you free!** No "religious organization will have authority over you unless you join a true body of believers to worship and fellowship with in accordance with the scripture.

"But I fear, lest by any means, as the serpent beguiled Eve through his subtilty, so your minds should be corrupted from the ***simplicity*** *that is in Christ".* (2Corinthians 11:3)

Newton once said..."Truth is ever to be found in the **simplicity**, and not the multiplicity and confusion of thing",

"It is the perfection of God's works that they are all done with the **greatest simplicity. He is the God of order and not of confusion**. And therefore as they would understand the frame of the world must endeavor to reduce their knowledge to all possible simplicity, so must it be in seeking to understand these visions". **(Newton)**

***The Gift of God is Eternal Life*! It is Free! It is that simple! Just faith!**

"For the wages of sin is death but ***the Gift of God is Eternal Life*** *through Jesus Christ our LORD"* (Romans 6:23)

Just Faith, it is that simple! It is free! Eternal Life cannot be earned and cannot be lost. It is free, it is Eternal and you get it the instant that you put your trust in Jesus as your Savior!

"Now to him that worketh is the reward not reckoned of grace, but of debt" (Romans4:4)

Eternal Life can only be received as a Gift, God does not save those who are trying to save themselves. He only saves those who are Trusting to be saved. It is that simple!

"...believe on the Lord Jesus Christ and thou shalt be saved..." (Acts 16:31)

Knowing God and receiving the Free Gift of Eternal Life is the most important thing in your life. After you are saved, the most important thing is helping others to find the truth.

The Bible says that there will be **a famine in the land in The Last Days;** God is not talking about a famine for food**; He is talking about a famine for Truth!** I believe that more than 90% of the people who call themselves Christians are really lost. The Bible calls them "False Brethren". So don't think that you have been presented the whole truth about the Bible. God reveals the future, He did in the past and he still is today.

God gives Eternal Life as a free gift to those who put their trust in Jesus! Your works or behavior have nothing to do with it! It Is <u>**FREE**</u>!!

You can't earn it and you can't lose it because it is <u>**FREE**</u>**! Eternal Life is The <u>GIFT OF GOD</u>!**

Eternal Life can't be earned and it can't be lost. **It must be received as a gift from God without you claiming any credit.** You can receive it the moment that you believe on Jesus and ask God to save you. Privately humble yourself before God, not men and trust on Jesus Christ as your God and your Savior in human flesh. Believe that Jesus Died and Rose again on the third day, just like the Bible says. If you do, then you are born again.

You can be sure that the Bible is true and what it says will happen, will happen. Rejecting Jesus Christ as your Savior will send you to Hell and that is the only thing that will.

The main thing that you need to know is that everyone needs Jesus, even smart, tough leaders. I am serious! I'm not talking about "Religion", I am talking about the real deal! I mean the literal interpretation of the Bible. That is what Sir Isaac Newton believed he was very wise in that regard. I pray that you will be too!

Sir Isaac Newton (1642 – 1747) once said… "About the times of the End, a body of men will be raised up who will turn their attention to the prophecies, and insist upon their literal interpretation, in the midst of much clamor and opposition."

Newton believed in the literal interpretation of the Bible and so do I.

"He who thinks half-heartedly will not believe in God; but he who really thinks has to believe in God." **(Newton)**

God says…

"Thus saith the LORD, Let not the wise man glory in his wisdom, neither let the mighty man glory in his might, let not the rich man glory in his riches:"(Jeremiah 9:23)

"But let him that glorieth glory in this, that he understandeth and knoweth me, that I am the LORD which exercise lovingkindness, judgment, and righteousness, in the earth: for in these things I delight, saith the LORD" (Jeremiah 9:24)

45.7 Jesus Will Destroy All Unbelievers At His Return.

"Behold, the day of the LORD cometh, cruel both with wrath and fierce anger, to lay the land desolate: and he shall destroy the sinners thereof out of it". (Isaiah 13:9)

The 2nd Return of Christ happens at the end of the Tribulation. It marks the end of the 70th week of Daniel. A "week" is 7 years. At the Second Return, Angels will be present to assist with the gathering of all the unbelievers that survived the Tribulation and Armageddon.

The Parable of the Wheat and Tares depicts the tares as unbelievers and has them gathered first and destroyed. (Matthew 13: 30)

At the second return the Lord will have all the surviving unbelievers gathered together and killed. n

"And another angel came out from the altar, which had power over fire; and cried with a loud cry to him that had the sharp sickle, saying, Thrust in thy sharp sickle, and gather the clusters of the vine of the earth; for her grapes are fully ripe". (Matthew 14:18)

"And the angel thrust in his sickle into the earth, and gathered the vine of the earth, and cast it into the great winepress of the wrath of God". (Matthew 14:19)

"And the winepress was trodden without the city, and blood came out of the winepress, even unto the horse bridles, by the space of a thousand and six hundred furlongs". (Matthew 14:20)

This is what happens to those who are taken at the second return.

At the second return the unbelievers will be taken away just like in the days of Noah and destroyed.

"But as the days of Noe were, so shall also the coming of the Son of man be". (Matthew 24: 37)

"For as in the days that were before the flood they were eating and drinking, marrying and giving in marriage, until the day that Noe entered into the ark", (Matthew 24: 38)

"And knew not until the flood came, and took them all away*; so shall also the coming of the Son of man be". (Matthew 24: 39)*

"Then shall two be in the field; the one shall be taken, and the other left". (Matthew 24: 40)

"Two women shall be grinding at the mill; the one shall be taken, and the other left". (Matthew 24: 41)

Notice that it is the unbelievers that were taken in Noe's flood, not the believers. This is **not** the Rapture. This is the second Return of Christ. Jesus will be coming in the clouds seven years after the rapture to divide up the survivors. That is when the Angels will separate the believers from the unbelievers!

Unbelievers are taken away by Angels to be killed by Jesus and believers are left to repopulate the earth. Only believers will be left to live on into the Millennium!

46

The Great White Throne Judgment Of The Lost!

46.1 The First Two To Be Cast Into The Lake Of Fire

The Antichrist and the False Prophet are the first two humans to be cast into the Lake of Fire. They are the only ones said to be cast in alive!

There are degrees of punishment in Hell. Those that oppose God and his people will get the max! Those who persecute God's people will receive the maximum punishment in Hell!

"For ye, brethren, became followers of the churches of God which in Judaea are in Christ Jesus: for ye also have suffered like things of your own countrymen, even as they have of the Jews:" (1Thessalonians 2:14)

"Who both killed the Lord Jesus, and their own prophets, and have persecuted us; and they please not God, and are contrary to all men:" (1Thessalonians 2:15)

"Forbidding us to speak to the Gentiles that they might be saved, to fill up their sins alway: for the ***wrath is come upon them to the uttermost****".* (1Thessalonians 2:16)

"Therefore hath ***the curse devoured the earth****, and they that dwell therein are desolate: therefore the inhabitants of the earth are burned, and* ***few men left****".* (Isaiah 24:6)

It is interesting to note that the word "Curse" here is the same Hebrew word used in (Daniel 9:11) and also translated as "Curse". They both come from the Hebrew word Alah. Strong's H423 - 'alah אָלָה

After the Great White Throne Judgment, Dr. Lindstrom once said that he did not think that more than about one hundred million people would be left after all the unbelievers are rounded up by the Angels and taken to the valley of Jehoshaphat. That was the personal opinion of an electrical engineer that dedicated his life to studying, understanding and rightly dividing God's word.

Dr. M. Hank Lindstrom's video's, audios and published papers are available on his website; www.biblelineministries.org

I recently figured out how to calculate the survival rate of the Tribulation. (About 20 million **survivors**, see chapter 47)

For follow up questions or new information or contact about the contents of this book … go to

www.endtimesbiblehandbook.org

46.2 The Millennium begins on the Feast of Tabernacles...

"And he shall judge among the nations, and shall rebuke many people: and they shall beat their swords into plowshares, and their spears into pruninghooks: nation shall not lift up sword against nation, neither shall they learn war any more".(Isaiah 2:4)

After the Thousand years is past, The "New Jerusalem will leave earth orbit and go else where. The New Jerusalem is a cube measuring 1500 miles on every side, with twelve round gate holes. The believers will

46.3 Eternity Future

God is going to start over with a new heaven and a new earth…l

(Isaiah 65:17) *"For, behold, I create new heavens and a new earth: and the former shall not be remembered, nor come into mind..*

(Isaiah 66:22) *"For as the new heavens and the new earth, which I will make, shall remain before me, saith the LORD, so shall your seed and your name remain".*

(2Peter 3:13) *"Nevertheless we, according to his promise, look for new heavens and a new earth, wherein dwelleth righteousness".*

(Revelation 21:1) *"And I saw a new heaven and a new earth: for the first heaven and the first earth were passed away; and there was no more sea".*

(1Thessalonians 4:17) *"and so shall we ever be with the Lord".*

Index Of Scriptures (Chapters 1-46)

1

2

A

C

D

E

G

H

J

L

M

N

P

R

T

Z

Recommended Resources

Trusted Sources Of Information.

This list will be updated …

www.endtimeshandbook.org

The Old King James Bible

Old Scofield Reference Bible (KJV)

Dr. M. H. Lindstrom. www.biblelineministries.org

The Handbook of Personal Evangelism by Dr. A. Ray Stanford

The Rapture Question by Walvoord

Books by John F Walvoord; http://walvoord.com/

Hal Lindsey; http://www.hallindsey.com/

Thomas Ice; "The End Times Controversy" http://www.pre-trib.org/about/dr-ice

David Hunt; Particularly "The Woman Rides the Beast" plus his whole collection. https://www.thebereancall.org/

J. Dwight Pentecost; "Things to Come";

H.A. Ironside; His whole collection

J. Vernon McGee http://www.ttb.org/

Les Feldick … lesfeldick.org

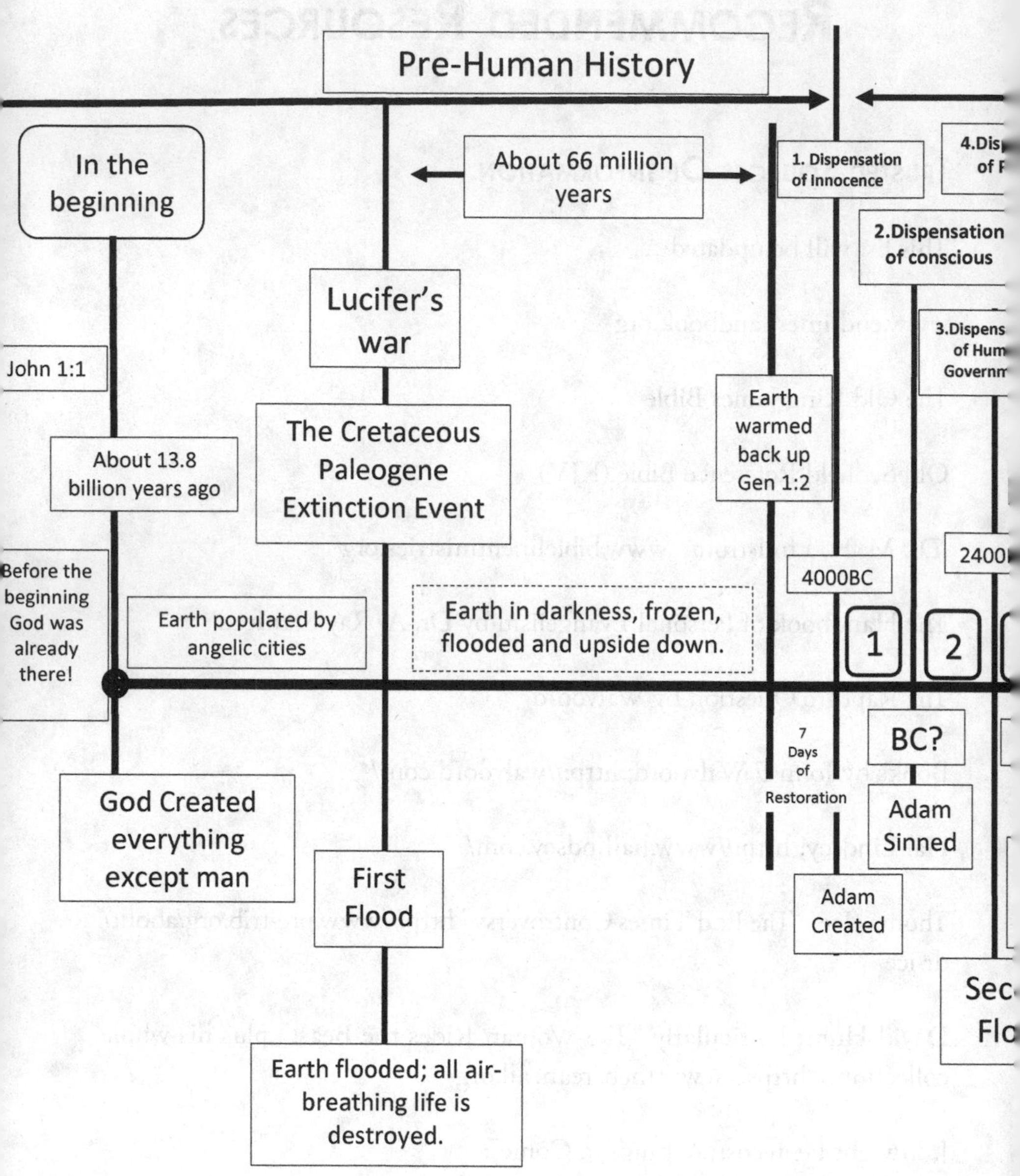

Pre-Human History
In the beginning
About 66 million years
1. Dispensation of Innocence
2.Dispensation of conscious
Lucifer's war
John 1:1
About 13.8 billion years ago
The Cretaceous Paleogene Extinction Event
Earth warmed back up Gen 1:2
Before the beginning God was already there!
Earth populated by angelic cities
Earth in darkness, frozen, flooded and upside down.
4000BC
1
2
7 Days of Restoration
BC?
Adam Sinned
God Created everything except man
First Flood
Adam Created
Earth flooded; all air-breathing life is destroyed.

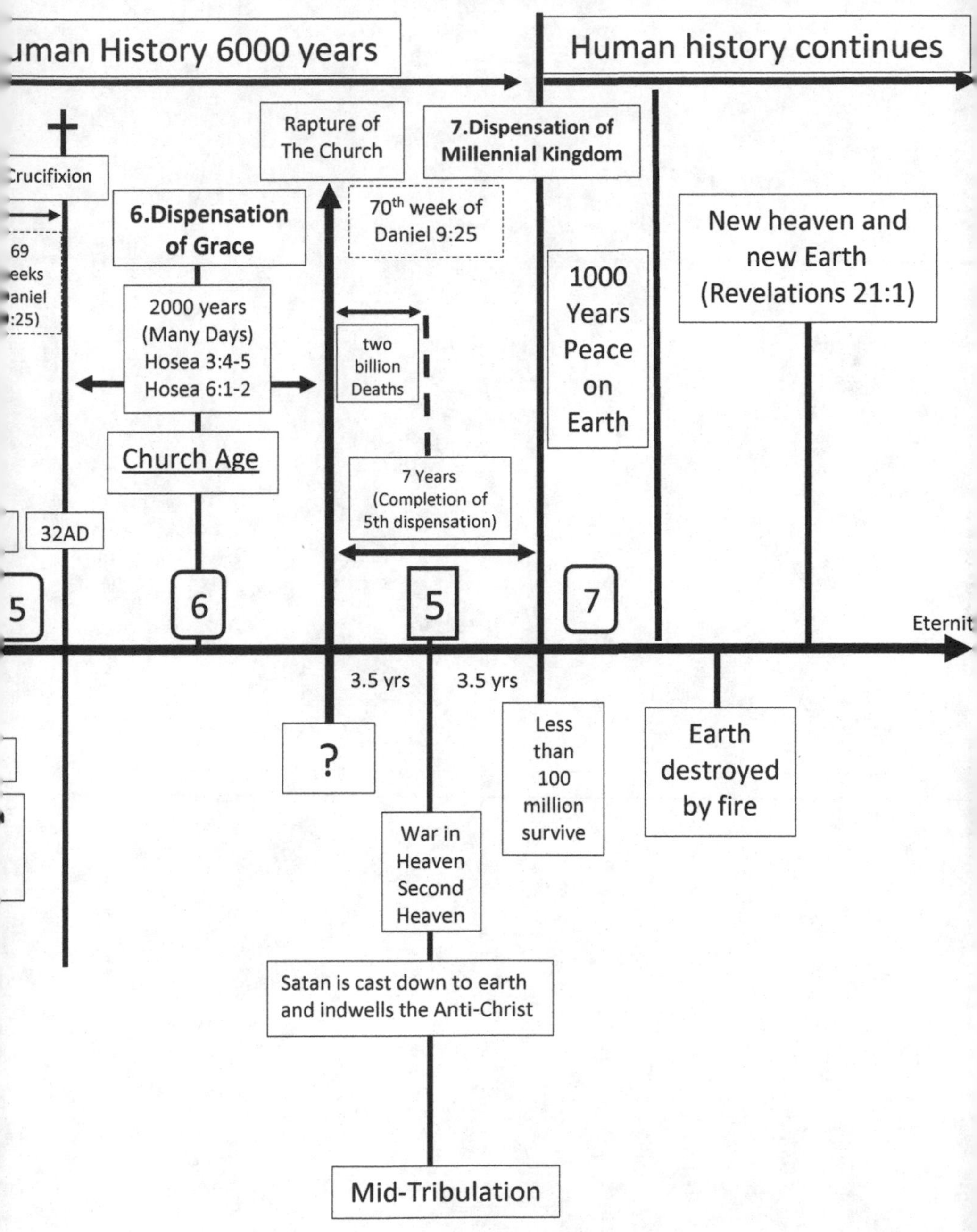

uman History 6000 years
Human history continues
Rapture of The Church
7.Dispensation of Millennial Kingdom
rucifixion
6.Dispensation of Grace
70th week of Daniel 9:25
69 eeks aniel :25)
New heaven and new Earth (Revelations 21:1)
1000 Years Peace on Earth
2000 years (Many Days) Hosea 3:4-5 Hosea 6:1-2
two billion Deaths
Church Age
7 Years (Completion of 5th dispensation)
32AD
5
6
5
7
Eternit
3.5 yrs
3.5 yrs
?
Less than 100 million survive
Earth destroyed by fire
War in Heaven Second Heaven
Satan is cast down to earth and indwells the Anti-Christ
Mid-Tribulation

About the Author

(continued from front)

Chad graduated from Plant High School in 1971 and attended Florida Bible College (FBC) for the following year; Later earned an AA degree from Hillsborough Community College (HCC) in a Pre-Engineering curriculum. After one semester in the college of engineering at U.S.F., Chad decided to open a Health Food Store and study nutrition.

In 1984 Republic Health Foods Inc. was established. Chad spent the next 13 years studying Life Extension Nutrition and the Bible.

Republic Health Foods was a ministry for spreading God's word through the Health Food Store. In 1997 the business was sold with the intent of reopening in a new location but that did not materialize. Chad's Tree Service began and has supported the Trowell family until recently.

www.endtimesbiblehandbook.org

Chad Is available for...

Group Meetings, Bible Studies, Sunday Schools...

And

The Creation and Bible Prophecy Conference

For a custom fit program for your group or organization...

Contact us at...

www.endtimesbiblehandbook.org

Printed in the United States
By Bookmasters